Print Reading for Construction

Residential and Commercial

Write-in Text with 140 Large Prints

Walter C. Brown
Professor Emeritus
Division of Technology
Arizona State University

Daniel P. Dorfmueller
d.p. dorfmueller co., inc.
Technical Author and Consultant
Lebanon, Ohio

Publisher
The Goodheart-Willcox Company, Inc.
Tinley Park, IL
www.g-w.com

Manufactured in the United States of America.
Library of Congress Catalog Card Number 2017029718
ISBN 978-1-63126-922-6
5 6 7 8 9 – 19 – 22 21 20 19

The Goodheart-Willcox Company, Inc. Brand Disclaimer: Brand names, company names, and illustrations for products and services included in this text are provided for educational purposes only and do not represent or imply endorsement or recommendation by the author or the publisher.

The Goodheart-Willcox Company, Inc. Safety Notice: The reader is expressly advised to carefully read, understand, and apply all safety precautions and warnings described in this book or that might also be indicated in undertaking the activities and exercises described herein to minimize risk of personal injury or injury to others. Common sense and good judgment should also be exercised and applied to help avoid all potential hazards. The reader should always refer to the appropriate manufacturer's technical information, directions, and recommendations; then proceed with care to follow specific equipment operating instructions. The reader should understand these notices and cautions are not exhaustive.

The publisher makes no warranty or representation whatsoever, either expressed or implied, including but not limited to equipment, procedures, and applications described or referred to herein, their quality, performance, merchantability, or fitness for a particular purpose. The publisher assumes no responsibility for any changes, errors, or omissions in this book. The publisher specifically disclaims any liability whatsoever, including any direct, indirect, incidental, consequential, special, or exemplary damages resulting, in whole or in part, from the reader's use or reliance upon the information, instructions, procedures, warnings, cautions, applications, or other matter contained in this book. The publisher assumes no responsibility for the activities of the reader.

The Goodheart-Willcox Company, Inc. Internet Disclaimer: The Internet resources and listings in this Goodheart-Willcox Publisher product are provided solely as a convenience to you. These resources and listings were reviewed at the time of publication to provide you with accurate, safe, and appropriate information. Goodheart-Willcox Publisher has no control over the referenced websites and, due to the dynamic nature of the Internet, is not responsible or liable for the content, products, or performance of links to other websites or resources. Goodheart-Willcox Publisher makes no representation, either expressed or implied, regarding the content of these websites, and such references do not constitute an endorsement or recommendation of the information or content presented. It is your responsibility to take all protective measures to guard against inappropriate content, viruses, or other destructive elements.

Cover images (clockwise from top left): Margrit Hirsch/Shutterstock.com; yuttana Contributor Studio/Shutterstock.com; Richard Thornton/Shutterstock.com; Flashon Studio/Shutterstock.com; Alhim/Shutterstock.com

Back cover images: Lev Kropotov/Shutterstock.com (top); Taras Kolomiyets/Shutterstock.com

Library of Congress Cataloging-in-Publication Data

Names: Brown, Walter C., author. | Dorfmueller, Daniel P., author.

Title: Print reading for construction : residential and commercial : write-in text with 140 large prints / by Walter C. Brown, Daniel P. Dorfmueller.

Description: Seventh edition. | Tinley Park, Illinois : The Goodheart-Willcox Company, Inc., [2019] | Includes index.

Identifiers: LCCN 2017029718 | ISBN 9781631269226

Subjects: LCSH: Building--Details--Drawings. | Blueprints.

Classification: LCC TH431 .B76 2019 | DDC 692/.1--dc23 LC record available at https://lccn.loc.gov/2017029718

Preface

Print Reading for Construction is a textbook for those who wish to learn basic print reading and increase their knowledge of construction drawings. The term *print reading,* as used in this textbook, refers to interpreting and visualizing construction drawings.

Print Reading for Construction is a combination text and workbook, or "write-in text." The text tells and shows how, and the workbook provides space for meaningful print reading, sketching, and estimating activities. Actual construction prints used with the text can be found in the **Large Prints** folder. They provide realistic job experience. The text is equally applicable for students studying construction, estimating, or construction management.

Print Reading for Construction is organized into sections based on a progression of topics from simple to complex. A list of *Technical Terms* opens each unit. *Learning Objectives* are presented to provide an overview of the content and define the most important skills you will learn in the unit. Each unit is followed by *Test Your Knowledge* questions. These brief questions will help you evaluate how well you understand the topics presented in the unit.

Most units include two or more *Activities* that provide you with practical print reading experience. Many of these will make use of the residential and commercial prints contained in the **Large Prints** folder. The write-in text format with perforations allows you to remove pages from the book, complete assignments, and turn them in directly to the instructor.

Using the Large Prints

The **Large Prints** folder contains seven sets of construction prints for the building projects referenced in this text. There are 140 total prints in the folder package. Included are sets for four residential building projects and three commercial building projects.

Print reading activities corresponding to each project are presented in this text. Three of the projects are referenced in the print reading activities in Units 8–17:

- The Sullivan Residence
- The Marseille Residence
- Delhi Flower and Garden Centers Greenhouse and Sales Building

The print reading activities for these projects appear at the end of each unit and are intended to be completed after studying the corresponding unit.

There are four building projects referenced in the *Advanced Print Reading Projects* section of this text. These include two residential projects and two commercial projects:

- Advanced Project A—Residence Mercedes Pointe
- Advanced Project B—The North House Residence
- Advanced Project C—Office and Warehouse
- Advanced Project D—Cincinnati Manor Building

The print reading activities for these projects are more extensive in scope and are intended for additional print reading practice.

Organization of Large Prints

Prints in the **Large Prints** folder are arranged so that each building project is printed on one side of the sheets. This organization helps in locating information across multiple sheets. The projects are grouped in the **Large Prints** folder as follows:

Group A—Left Pocket:

Front of sheet:

- The Sullivan Residence (5 prints)
- The Marseille Residence (9 prints)
- Advanced Project A—Residence Mercedes Pointe (8 prints)
- Advanced Project B—The North House Residence (13 prints)

Back of sheet:

- Delhi Flower and Garden Centers Greenhouse and Sales Building (30 prints)
- Advanced Project C—Office and Warehouse (5 prints)

Group B—Right Pocket:

Front of sheet:

- Advanced Project C—Office and Warehouse (35 prints)

Back of sheet:

- Advanced Project D—Cincinnati Manor Building (35 prints)

Indexes for Large Prints

The following tables list the prints in the **Large Prints** folder for the projects referenced in Units 8–17. Indexes for the prints associated with the advanced print reading activities appear later in the text where the projects are referenced.

The Sullivan Residence Print Index (5 Prints)

Print Label	Sheet Number	Sheet Title
SUL-1	1	Exterior Elevations, Details, and Notes
SUL-2	2	Foundation Plan
SUL-3	3	First Floor Plan
SUL-4	4	Truss Profiles, Roof Plan, and Details
SUL-5	5	Exterior Elevations and Wall Section

The Marseille Residence Print Index (9 Prints)

Print Label	Sheet Number	Sheet Title
MAR-1	1	Exterior Elevations, Details, and Notes
MAR-2	2	Foundation Plan
MAR-3	MP2	Foundation Mechanical/Plumbing Plan
MAR-4	3	First Floor Plan
MAR-5	MP3	First Floor Mechanical/Plumbing Plan
MAR-6	4	Second Floor Plan
MAR-7	MP4	Second Floor Mechanical/Plumbing Plan
MAR-8	5	Truss Profiles, Roof Plan, and Details
MAR-9	6	Exterior Elevations and Sections

Delhi Flower and Garden Centers Greenhouse and Sales Building Print Index (30 Prints)

Print Label	Sheet Number	Sheet Title	Print Label	Sheet Number	Sheet Title
DEL-1	1	Title Sheet	DEL-16	S1.1	Foundation Plan
DEL-2	4	Existing Conditions/Demolition Plan	DEL-17	S1.2	Foundation Plan
DEL-3	5	Grading and Erosion Control Plan	DEL-18	S2.1	Roof Framing Plan
DEL-4	A0.1	Supplemental Specifications	DEL-19	S2.2	Framing Plans and Truss Elevations
DEL-5	A1.1	First Floor Plan	DEL-20	S3.1	Foundation Details
DEL-6	A1.2	First Floor Plan, Details	DEL-21	S4.1	Framing Details
DEL-7	A1.3	Mezzanine Plan	DEL-22	S5.1	Structural Notes
DEL-8	A2.1	Roof Plan	DEL-23	P1.1	Plumbing Plans
DEL-9	A3.1	Reflected Ceiling Plan	DEL-24	P2.1	Sanitary Isometric, Fixture Schedule, Plumbing Specs
DEL-10	A4.1	Exterior Elevations	DEL-25	M1.1	First Floor Mechanical Plan
DEL-11	A4.2	Exterior Elevations	DEL-26	M1.2	Mezzanine Mechanical Plan and Schedules
DEL-12	A5.1	Interior Elevations and Schedules	DEL-27	E1.1	First Floor Lighting Plan
DEL-13	A6.1	Enlarged Plans and Interior Elevations	DEL-28	E1.2	Lighting Plans and Schedules
DEL-14	A9.1	Interior Wall Sections	DEL-29	E2.1	First Floor Power Plan
DEL-15	A9.2	Exterior Wall Sections	DEL-30	E4.1	Riser Diagram, Panel Schedules, and Notes

G-W Integrated Learning Solution

Together, We Build Careers

At Goodheart-Willcox, we take our mission seriously. Since 1921, G-W has been serving the career and technical education (CTE) community. Our employee-owners are driven to deliver exceptional learning solutions to CTE students to help prepare them for careers. Our authors and subject matter experts have years of experience in the classroom and industry. We combine their wisdom with our expertise to create content and tools to help students achieve success. Our products start with theory and applied content based upon a strong foundation of accepted standards and curriculum. To that base, we add features and tools designed to help promote effective and efficient teaching and learning. G-W recognizes the crucial role instructors play in preparing students for careers. We support educators' efforts by providing time-saving tools that help them plan, present, and assess with traditional and digital activities and assets. We provide an entire program of learning in a variety of print, digital, and online formats, including economic bundles, allowing educators to select the right mix for their classroom.

Student-Focused Curated Content

Goodheart-Willcox believes that student-focused content should be built from standards and accepted curriculum coverage. **Print Reading for Construction** also uses a building block approach with attention devoted to a logical teaching progression that helps students build upon their learning. We call on industry experts and teachers from across the country to review and comment on our content, presentation, and pedagogy. Finally, in our refinement of curated content, our editors are immersed in content checking, securing figures that convey key information, and revising language and pedagogy.

About the Authors

Daniel P. Dorfmueller is an instructor and consultant for the construction industry based in Lebanon, Ohio. He provides training in print reading for construction and print reading for concrete courses. Dan has been involved in the construction industry for more than 40 years and has been teaching students how to read construction prints for more than 20 years. He worked for Baker Concrete Construction in the greater Cincinnati area for 17 years.

Dan is a Fellow member of the American Concrete Institute and has served as a speaker at the World of Concrete. He has written several articles and served as an adjunct instructor at Northern Kentucky University. Dan holds a bachelor's degree in architecture from the University of Cincinnati.

During his career, Dr. Walter C. Brown was a leading authority in the fields of drafting and print reading. He served as a consultant to industry on design and drafting standards and procedures and held a variety of professional offices of state and national associations.

Reviewers

The authors and publisher wish to thank the following industry and teaching professionals for their valuable input into the development of **Print Reading for Construction**.

Michael M. Byrnes
Construction Instructor
Construction Careers Academy
San Antonio, TX

Mike DeMattei
Construction Management Instructor
John A. Logan College
Carterville, IL

Jodie Eiland
Carpentry and Masonry Trades Instructor
Gordon Cooper Technology Center
Shawnee, OK

G. Scott Fleming
Program Coordinator
Fleming College
Peterborough, ON

Robert P. Gresko
Instructor, Building Construction Technology
Pennsylvania College of Technology
Williamsport, PA

Timothy Mehring
Technology and Engineering Instructor
Germantown High School
Germantown, WI

Mary Jill Rydall
Professor, Carpentry
Canadore College
North Bay, ON

Ted Saunders
Building Construction Instructor
Canadore College
North Bay, ON

Robert A. Wozniak
Associate Professor/Architecture, Building Science + Sustainable Design
Pennsylvania College of Technology
Williamsport, PA

Acknowledgments

The authors wish to express appreciation to the following individuals, architectural firms, and construction companies that provided assistance and supplied prints and mechanical manuals on all aspects of construction.

Alvin & Company, Inc.
American Welding Society
Flo Anderson
APA-The Engineered Wood Association
Armco Steel Corporation
Autodesk, Inc.
Baker Concrete Construction, Inc.
Batson & Associates Architects
Charles O. Biggs, AIA
Brick Institute of America
CertainTeed Corporation
CESO Engineers & Architects
Christensen, Cassidy, Billington and Candelaria, Inc., Architects
CITY Properties Group, LLC
Concrete Reinforcing Steel Institute
CR architecture + design
Cypress Specialty Steel Company
Dale/Incor
Environmental Design Consultants: Kral, Zepf, Frietag & Associates
Evans International Homes
Garlinghouse Plan Service
Gosnell Development Corporation
Dr. Charles W. Graham, Associate Professor, Texas A&M University
Robert Ehmet Hayes and Associates, Architects
Helgeson and Biggs, Architects, Inc.
Herndon Engineering Services, Inc.
Hewlett-Packard Development Company
iPlanTables
Jack Klasey
KZF Design, Inc.
LiveRoof, LLC
Marathon Steel Company
Marvin Windows and Doors
MeadowBurke
Steve Olewinski
Patterned Concrete of Cincinnati
The Ridge Tool Co.
Rookwood Building Group, LLC
John J. Ross, AIA Architect
RTKL Associates, Inc.
SCAFCO Corporation
Schweizer Associates Architect, Inc.
Don Singer Architect
Smith & Neubek & Associates
Studio 4, LLC
Vanderbuilt Homes, Inc.
Whitacre Engineering
Wire Reinforcement Institute

The authors wish to express special appreciation to the companies that provided construction prints for the projects used in this textbook:

Cincinnati Manor Building
Neyer Architects, Inc.
THP Limited, Inc.
IBI Group, Cincinnati

Delhi Flower and Garden Centers Greenhouse and Sales Building
Delhi Flower and Garden Center
Arch/Image 2 Architects
Pinnacle Engineering Services, Inc.
PE-Services
Abercrombie & Associates, Inc.

Goodheart-Willcox Office and Warehouse Facility
Charles E. Smith, Areté 3 Ltd.

The Marseille Residence
Studer Residential Designs, Inc.
Schadler Plumbing
Doc Rusk Heating & Cooling

The North House Residence
Norris & Dierkers Architects/Planners, Inc.

Residence Mercedes Pointe
Eagle Custom Homes
McGill Smith Punshon, Inc.
Superior Designs, LLC

The Sullivan Residence
Studer Residential Designs, Inc.
Schadler Plumbing
Doc Rusk Heating & Cooling

Dan Dorfmueller (best known as "dorf") thanks his wife, Deb, and his children and grandchildren for supporting the many hours spent on the revision of this book.

Features of the Textbook

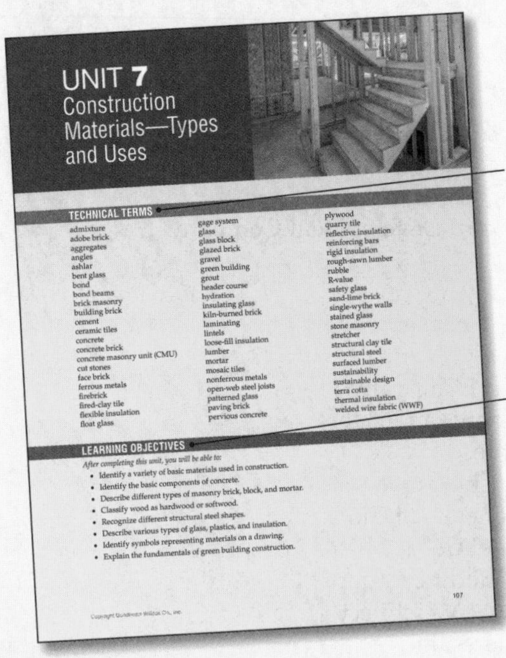

UNIT 7
Construction Materials—Types and Uses

TECHNICAL TERMS

admixture
adobe brick
aggregates
angles
ashlar
bent glass
bond
bond beams
brick masonry
building brick
cement
ceramic tiles
concrete
concrete brick
concrete masonry unit (CMU)
cut stones
face brick
ferrous metals
firebrick
fired-clay tile
flexible insulation
float glass
gage system
glass
glass block
glazed brick
gravel
green building
grout
header course
hydration
insulating glass
kiln-burned brick
laminating
lintels
loose-fill insulation
lumber
mortar
mosaic tiles
nonferrous metals
open-web steel joists
patterned glass
paving brick
pervious concrete
plywood
quarry tile
reflective insulation
reinforcing bars
rigid insulation
rough-sawn lumber
rubble
R-value
safety glass
sand-lime brick
single-wythe walls
stained glass
stone masonry
stretcher
structural clay tile
structural steel
surfaced lumber
sustainability
sustainable design
terra cotta
thermal insulation
welded wire fabric (WWF)

LEARNING OBJECTIVES

After completing this unit, you will be able to:

- Identify a variety of basic materials used in construction.
- Identify the basic components of concrete.
- Describe different types of masonry brick, block, and mortar.
- Classify wood as hardwood or softwood.
- Recognize different structural steel shapes.
- Describe various types of glass, plastics, and insulation.
- Identify symbols representing materials on a drawing.
- Explain the fundamentals of green building construction.

107

Technical Terms list the key terms to be learned in the unit. Review this list after completing the unit to be sure you know the definition of each term.

Learning Objectives clearly identify the knowledge and skills to be obtained upon completion of the unit. After completing a unit, review these to ensure that you have mastered the new material.

Careers in Construction features provide information about career opportunities in the construction trades.

Illustrations clearly and simply communicate the specific topic. Photographic images have been updated in this edition to show the latest technology and products in the construction industry.

Green Building features highlight key items related to green building technology, sustainability, energy efficiency, and environmental issues.

Notes explain or expand on important aspects of a topic.

Test Your Knowledge Questions reinforce the content covered in the unit.

Activities allow you to apply knowledge and solve problems.

Print Reading Activities provide opportunities to apply print reading skills using prints in the **Large Prints** folder that accompanies the text.

Advanced Projects are provided at the end of the text for additional practice in reading prints and completing exercises.

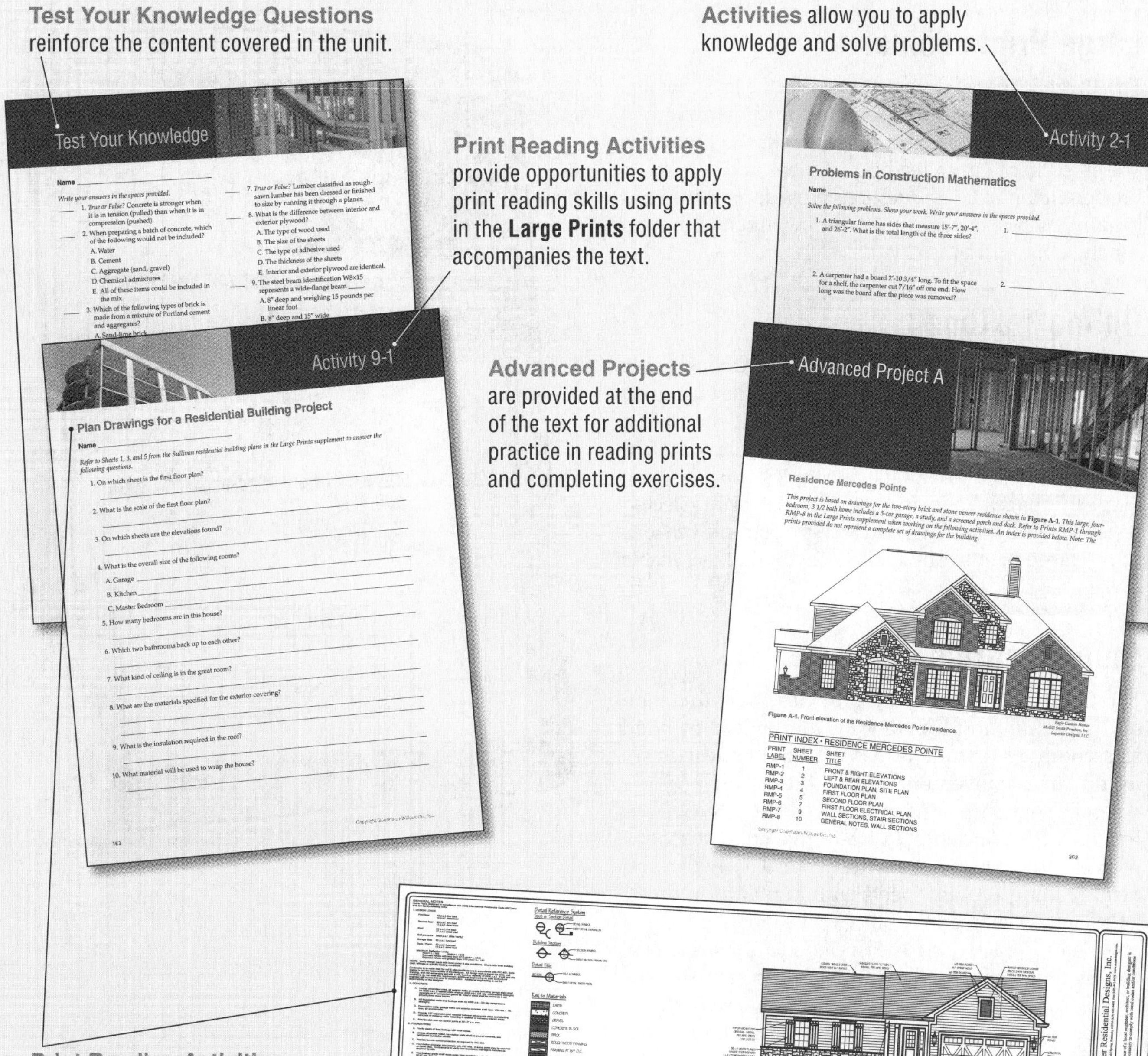

Print Reading Activities reference prints in the **Large Prints** folder. Use the corresponding prints to locate information and answer the questions.

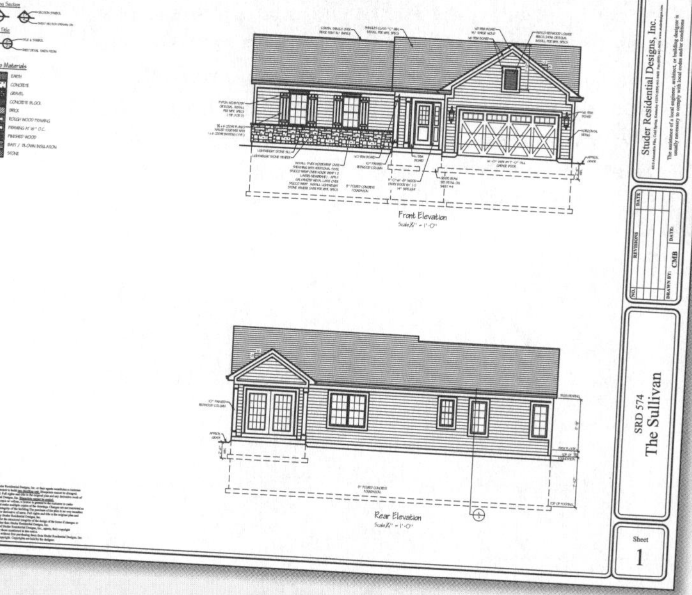

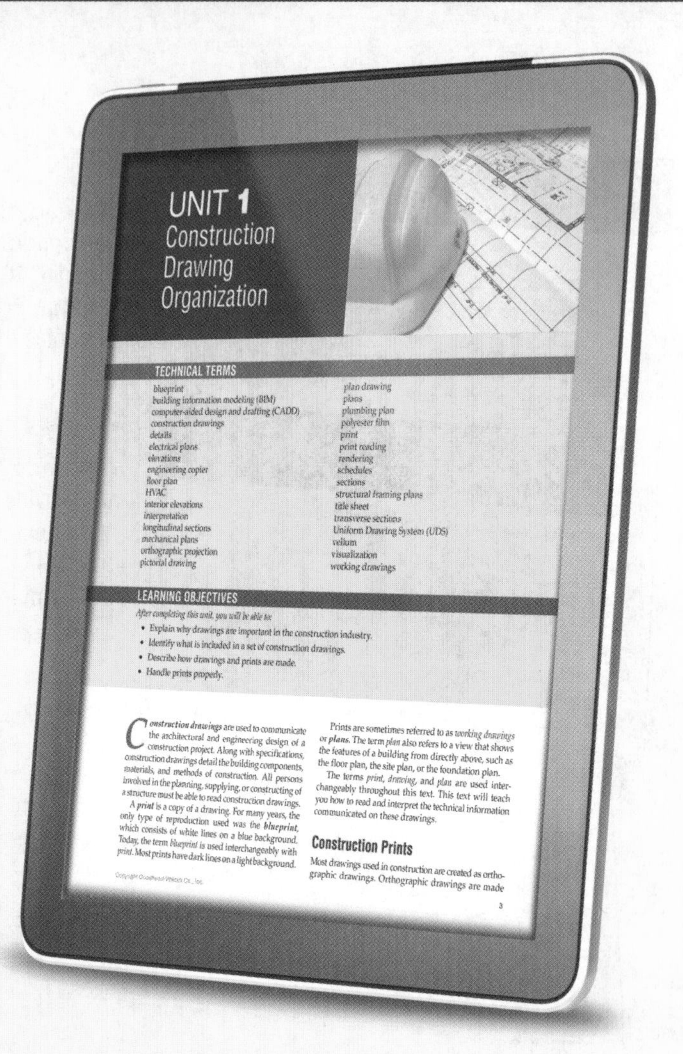

UNIT 1
Construction Drawing Organization

TECHNICAL TERMS

blueprint
building information modeling (BIM)
computer-aided design and drafting (CADD)
construction drawings
details
electrical plans
elevations
engineering copier
floor plan
HVAC
interior elevations
interpretation
longitudinal sections
mechanical plans
orthographic projection
pictorial drawing
plan drawing
plans
plumbing plan
polyester film
print
print reading
rendering
schedules
sections
structural framing plans
title sheet
transverse sections
Uniform Drawing System (UDS)
vellum
visualization
working drawings

LEARNING OBJECTIVES

After completing this unit, you will be able to:

- Explain why drawings are important in the construction industry.
- Identify what is included in a set of construction drawings.
- Describe how drawings and prints are made.
- Handle prints properly.

Construction drawings are used to communicate the architectural and engineering design of a construction project. Along with specifications, construction drawings detail the building components, materials, and methods of construction. All persons involved in the planning, supplying, or constructing of a structure must be able to read construction drawings.

A ***print*** is a copy of a drawing. For many years, the only type of reproduction used was the ***blueprint***, which consists of white lines on a blue background. Today, the term *blueprint* is used interchangeably with *print*. Most prints have dark lines on a light background.

Prints are sometimes referred to as *working drawings* or ***plans***. The term *plan* also refers to a view that shows the features of a building from directly above, such as the floor plan, the site plan, or the foundation plan.

The terms *print*, *drawing*, and *plan* are used interchangeably throughout this text. This text will teach you how to read and interpret the technical information communicated on these drawings.

Construction Prints

Most drawings used in construction are created as orthographic drawings. Orthographic drawings are made

3

Large Prints Folder

The **Large Prints** folder supplied with the text contains foldout prints from residential and commercial construction. The prints are used with the questions presented in the print reading activities and Advanced Projects in the text. The questions provide opportunities to gain practical experience in reading and interpreting construction prints.

Online Textbook

This online version of the printed textbook gives students access anytime, anywhere whether using an iPad, netbook, PC, or Mac computer. Using the Online Textbook, students can easily navigate from a linked table of contents, search specific topics, quickly jump to specific pages, zoom in to enlarge text, print selected pages for offline reading, and access electronic versions of the large prints. The Online Textbook is available at www.g-wonlinetextbooks.com.

Online Learning Suite

The Online Learning Suite provides the foundation of instruction and learning for digital and blended classrooms. All student instructional materials are found on a convenient online bookshelf and are accessible at home, at school, or on the go. The Online Learning Suite includes an interactive online textbook with electronic large prints, Test Your Knowledge and print reading activity sheets with digital form fields, e-flash cards, drag-and-drop activities, and a variety of other learning activities. Also included are instructional videos to engage students for learning success. The videos can be used by students to preview concepts for upcoming classroom lectures or as an on-demand review for challenging topics. The Online Learning Suite effectively brings digital learning to students and is easy for instructors to use.

Instructor Materials

ExamView® Assessment Suite

Quickly and easily prepare, print, and administer tests with the ExamView® Assessment Suite. With hundreds of questions in the test bank corresponding to each unit, you can choose which questions to include in each test, create multiple versions of a single test, and automatically generate answer keys. Existing questions may be modified and new questions may be added.

Instructor's Presentations for PowerPoint®

These presentations help visually reinforce key concepts. The presentations are designed to allow for customization to meet daily teaching needs. They include objectives and images from the textbook.

Instructor's Resource CD

One resource provides instructors with time-saving preparation tools such as answer keys, lesson plans, and other teaching aids. Electronic versions of the prints supplied in the **Large Prints** folder are included.

Online Instructor Resources

Online Instructor Resources are time-saving teaching materials organized in a convenient, easy-to-use online bookshelf. Lesson plans, answer keys, presentations for PowerPoint®, ExamView® Assessment Suite software with test questions, and other teaching aids are available on demand, 24/7. Accessible from home or school, Online Instructor Resources provide convenient access for instructors with busy schedules.

Brief Contents

Contents

SECTION 1
INTRODUCTION TO PRINT READING

SECTION 2
PRINT READING BASICS

SECTION 3
SPECIFICATIONS AND MATERIALS

SECTION 4
READING PRINTS

SECTION 6
ADVANCED PRINT READING PROJECTS

SECTION 7
REFERENCE SECTION

Features

Green Building

Careers in Construction

SECTION 1
Introduction to Print Reading

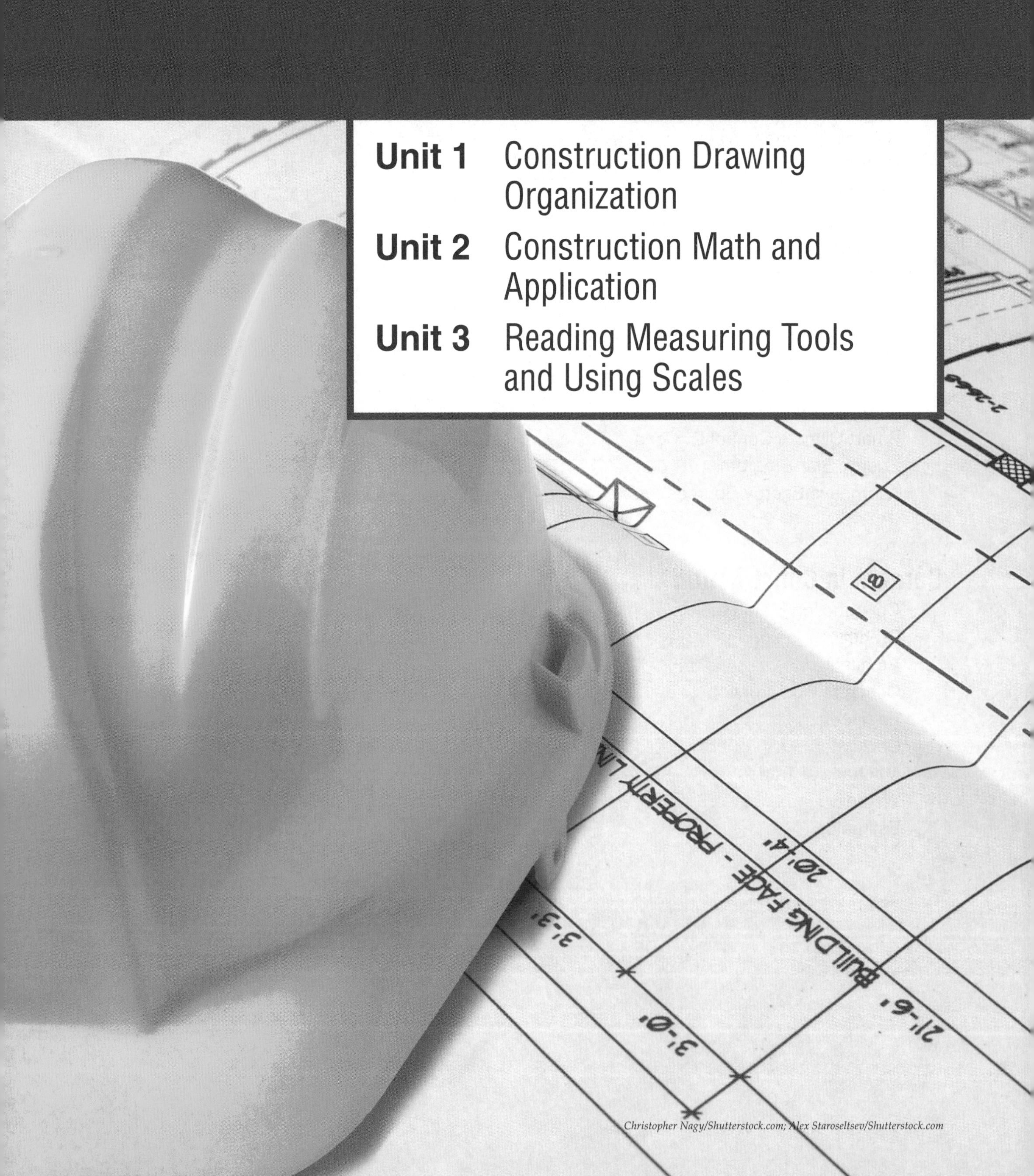

Christopher Nagy/Shutterstock.com; Alex Staroseltsev/Shutterstock.com

UNIT 1 Construction Drawing Organization

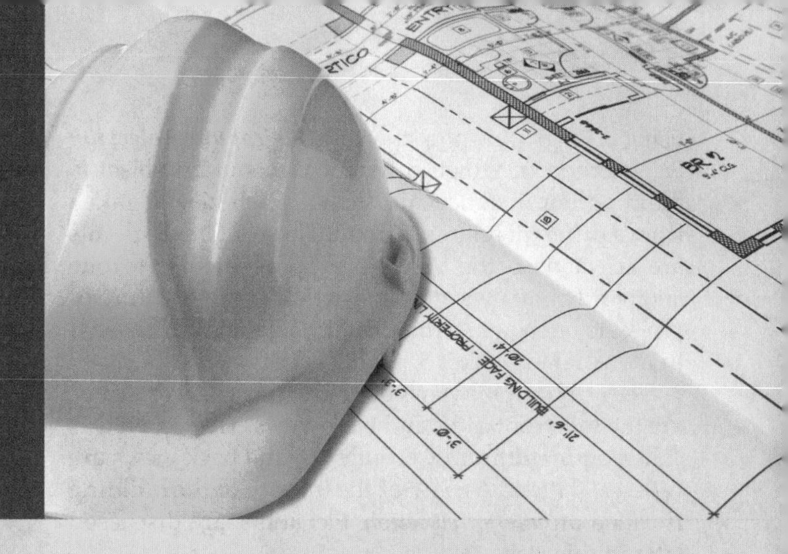

TECHNICAL TERMS

blueprint
building information modeling (BIM)
computer-aided design and drafting (CADD)
construction drawings
details
electrical plans
elevations
engineering copier
floor plan
HVAC
interior elevations
interpretation
longitudinal sections
mechanical plans
orthographic projection
pictorial drawing
plan drawing
plans
plumbing plan
polyester film
print
print reading
rendering
schedules
sections
structural framing plans
title sheet
transverse sections
Uniform Drawing System (UDS)
vellum
visualization
working drawings

LEARNING OBJECTIVES

After completing this unit, you will be able to:

- Explain why drawings are important in the construction industry.
- Identify what is included in a set of construction drawings.
- Describe how drawings and prints are made.
- Handle prints properly.

Construction drawings are used to communicate the architectural and engineering design of a construction project. Along with specifications, construction drawings detail the building components, materials, and methods of construction. All persons involved in the planning, supplying, or constructing of a structure must be able to read construction drawings.

A ***print*** is a copy of a drawing. For many years, the only type of reproduction used was the ***blueprint***, which consists of white lines on a blue background. Today, the term *blueprint* is used interchangeably with *print*. Most prints have dark lines on a light background.

Prints are sometimes referred to as *working drawings* or ***plans***. The term *plan* also refers to a view that shows the features of a building from directly above, such as the floor plan, the site plan, or the foundation plan.

The terms *print*, *drawing*, and *plan* are used interchangeably throughout this text. This text will teach you how to read and interpret the technical information communicated on these drawings.

Construction Prints

Most drawings used in construction are created as orthographic drawings. Orthographic drawings are made using orthographic projection. ***Orthographic projection*** is a method in which different views of an object (a building, for instance) are shown. Each view is taken from a different reference point. These reference points are selected so the views are perpendicular to one another. This allows all of the details of the structure or project to be shown. Orthographic projection is discussed in more detail in Unit 5.

The orthographic views used in construction drawings are the top, front, side, and back views. The top view is called a ***plan drawing***. Front, side, and back views are called *elevations*. A view of the interior of the building is called an *interior elevation*. Elevations are discussed later in this unit.

Other views are used in addition to plan views and elevations to clarify construction of a building. A view that shows the interior construction of a building feature, such as a wall, is called a *section*. An enlarged view of a section or a plan view is called a *detail*. Sections and details provide information that cannot be clearly shown on other drawings and are discussed later in this unit.

Another type of drawing used in building projects is a pictorial drawing. A ***pictorial drawing*** is used to help the viewer visualize the structure or the project in its entirety. This type of drawing is enhanced with trees, shrubs, shading, and other materials to make it appear more realistic. A ***rendering*** is a pictorial drawing that shows what a structure will look like when the project is finished, **Figure 1-1**. Pictorial drawings typically do not show any construction details and are primarily used for presentation purposes.

Identifying Information in a Set of Prints

Small construction projects usually include all necessary information on a single plan drawing, an elevation, and a few details. Larger construction projects that are more complicated require many plans, elevations, sections, and details, collectively called ***working drawings***. Working drawings are divided into sections according to the types of construction being performed. A letter classification identifies the drawings in each section. The following letter classifications are specified in the Uniform Drawing System (UDS), discussed later in this unit. The letter classifications in **bold** type are the ones most commonly used in building construction.

- **G—General.** Project phasing, contractor staging areas, schedules, fencing, photographs, code summary, symbol legends, and site maps.
- H—Hazardous Materials. Handling, removal, and storage of hazardous materials.
- V—Survey/Mapping. Surveyed and digitized points and features.
- B—Geotechnical. User defined.
- **C—Civil.** Structure removal, site clearing, excavation, site grading, roads, waterways, sanitary and storm sewer, pavers, plot plans, and details. (See Figure 1-3.)
- L—Landscape. Landscaping, planting, site hardscapes, and irrigation.
- **S—Structural.** Concrete, steel and wood structure, and details. (See Figure 1-4.)
- **A—Architectural.** Floor plans, elevations, finishes, building sections, schedules, and details. (See Figures 1-5, 1-6, 1-7, 1-8, 1-9, and 1-10.)
- I—Interiors. Interior demolition, furnishings, graphics, and interior design.
- Q—Equipment. Equipment installed inside and outside of the building, such as athletic, bank, dry cleaning, kitchen, medical, and playground equipment.
- F—Fire Protection. Fire alarm and suppression systems.
- **P—Plumbing.** Waste and water supply systems.

Autodesk, Inc.

Figure 1-1. A rendering shows how the finished structure will appear.

- D—Process. Process piping systems, equipment, and instrumentation.
- **M—Mechanical.** Heating, ventilation, and cooling systems.
- **E—Electrical.** Power and lighting systems.
- W—Distributed Energy. Distributed energy systems and structures, such as electrical substations.
- T—Telecommunications. Audio and visual systems, security systems, and network cabling.
- R—Resource. Existing drawings, such as architectural, structural, and real estate drawings.
- X—Other Disciplines. User defined.
- **Z—Contractor/Shop Drawings.** Drawings made by subcontractors. Shop drawings are used by the tradeworker to install the work and are trade and supplier specific. For instance, when fabricating reinforcing steel, the supplier will make detailed diagrams for each portion of the footing, walls, and columns to show how the reinforcing steel is to be installed. These are different from the structural engineer's design drawings. This additional detail will help the tradeworker install the reinforcing steel correctly.
- O—Operations. User defined.

Note that there is some crossover between letter identifications. For example, all drawing content that is part of the (C) Civil, (L) Landscape, and (G) Geotechnical drawings could be on (C) Civil drawings. All drawing content that is part of the (P) Plumbing and (D) Process drawings could be on (P) Plumbing drawings. This depends on the complexity of the building and how much the information needs to be separated out from other trades.

Additional Drawing Identification Methods

Although working drawings are typically identified using the letter classifications previously discussed, some architects and engineers employ a numbered system without the use of letters. For example, each drawing in a project may be identified simply as Sheet *n*, where *n* represents the number assigned to the drawing.

When letter classifications are used, it is common to employ a sheet numbering system that uses a series of numbers to identify the drawing type and sheet number. These numbers follow the letter classification. The first number following the letter classification identifies the drawing type. This number is a single digit:

- 0—General
- 1—Plans
- 2—Elevations
- 3—Sections
- 4—Large-scale views
- 5—Details
- 6—Schedules and diagrams
- 7—User defined
- 8—User defined
- 9—3D views

The next number identifies the sheet number. For example, using this system, sheet C1.01 indicates a plan view drawing (1) in the civil drawings section (C) assigned sheet number 01. Sheet C1.02 indicates a plan view drawing (1) in the civil drawings section (C) assigned sheet number 02, and so on. Sheet A2.02 indicates an elevation (2) in the architectural drawings section (A) assigned sheet number 02.

Whether or not drawings in a set are identified with a letter classification and numbered system, individual drawings in a project are typically identified on a ***title sheet***. See **Figure 1-2**. Generally, the title sheet provides a detailed list of drawings and other information about the project. In the example shown, a detailed list identifies each drawing by sheet number and name. This type of list may also identify the most recent issue date and revision information. This is helpful when identifying changes to drawings for change orders or determining the scope of work related to an estimate, proposal, and contract.

A title sheet also typically lists names of the professionals involved in the project, such as the architect, landscape architect, structural engineer, mechanical engineer, electrical engineer, and owner of the project. Other items included on title sheets include abbreviation lists, material and symbol legends, a rendering of the completed building, and a site map locating the project.

Typical Prints

Prints are usually arranged in the approximate order of construction. A set of prints consists of a title sheet and general (G-1, G-2, etc.), civil engineering (C-1, C-2, etc.), structural engineering (S-1, S-2, etc.), architectural (A-1, A-2, etc.), electrical (E-1, E-2, etc.), mechanical (M-1, M-2, etc.), and plumbing (P-1, P-2, etc.) prints.

General (**G**) prints identify overall information about a project. They show items such as project phasing, contractor staging areas for construction materials and equipment, contractor parking, and project fencing.

Civil engineering (**C**) prints include site plans and show items such as utilities, easements, grading, landscaping, and site details. A typical site plan for a commercial building is shown in **Figure 1-3**. The site plan can also include grade contour lines, walks, and driveways. Property lines, building setbacks, and utility locations are also shown.

Structural (**S**) prints include foundation plans and above-grade framing plans. Structural prints show items such as structural concrete and structural steel and the building support system. See **Figure 1-4**. Structural prints include sections and details to show construction requirements.

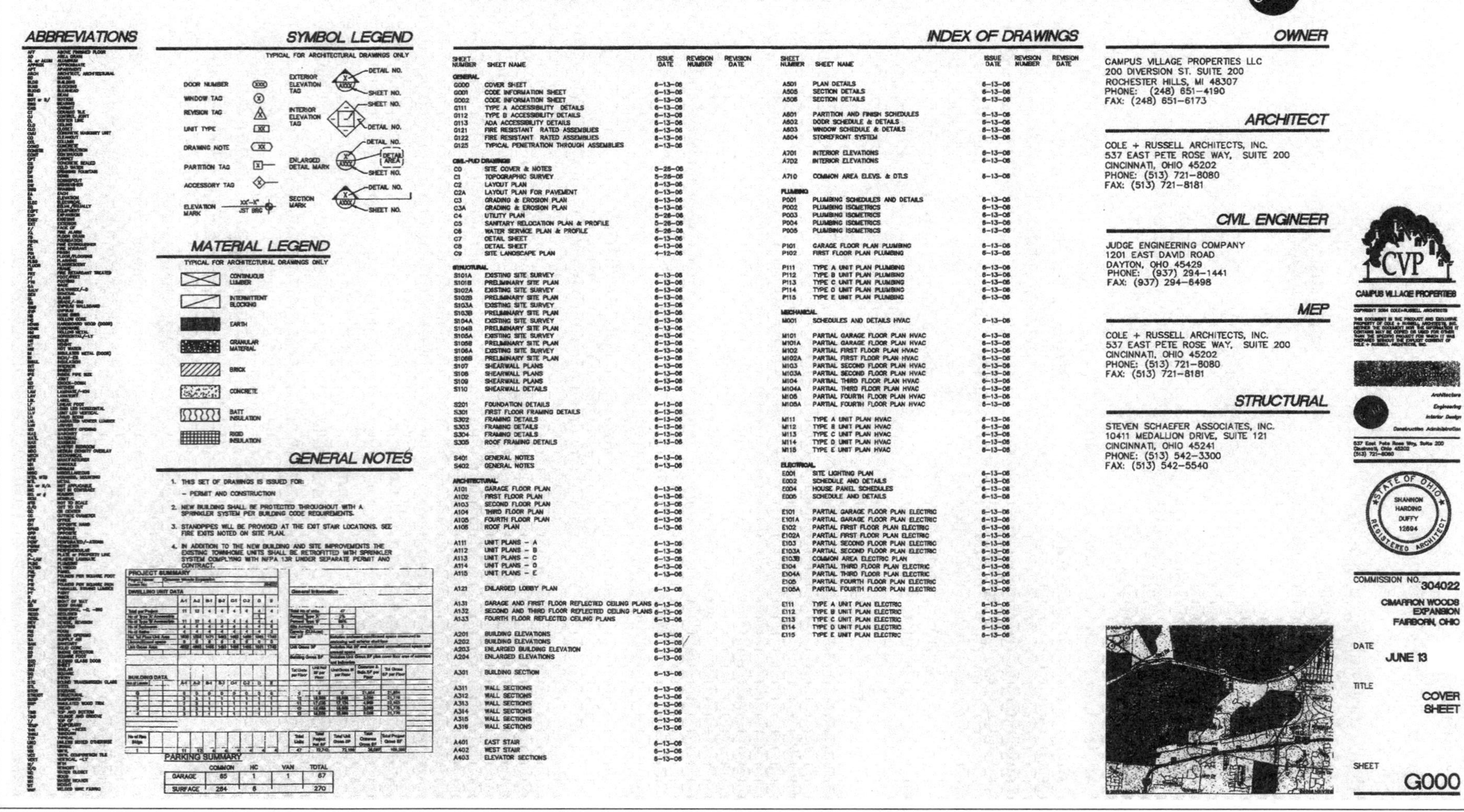

CR architecture + design

Figure 1-2. A title sheet for a set of drawings for a commercial building project.

For simple residential buildings, the foundation and basement plans are usually included on the same drawing. This plan is used to show the foundation walls, footings, piers, and fireplaces.

Architectural (**A**) prints include floor plans, elevations, building sections, wall sections and detail sections, door and window schedules, and room finish schedules. In residential construction, the architectural prints usually make up the majority of working drawings.

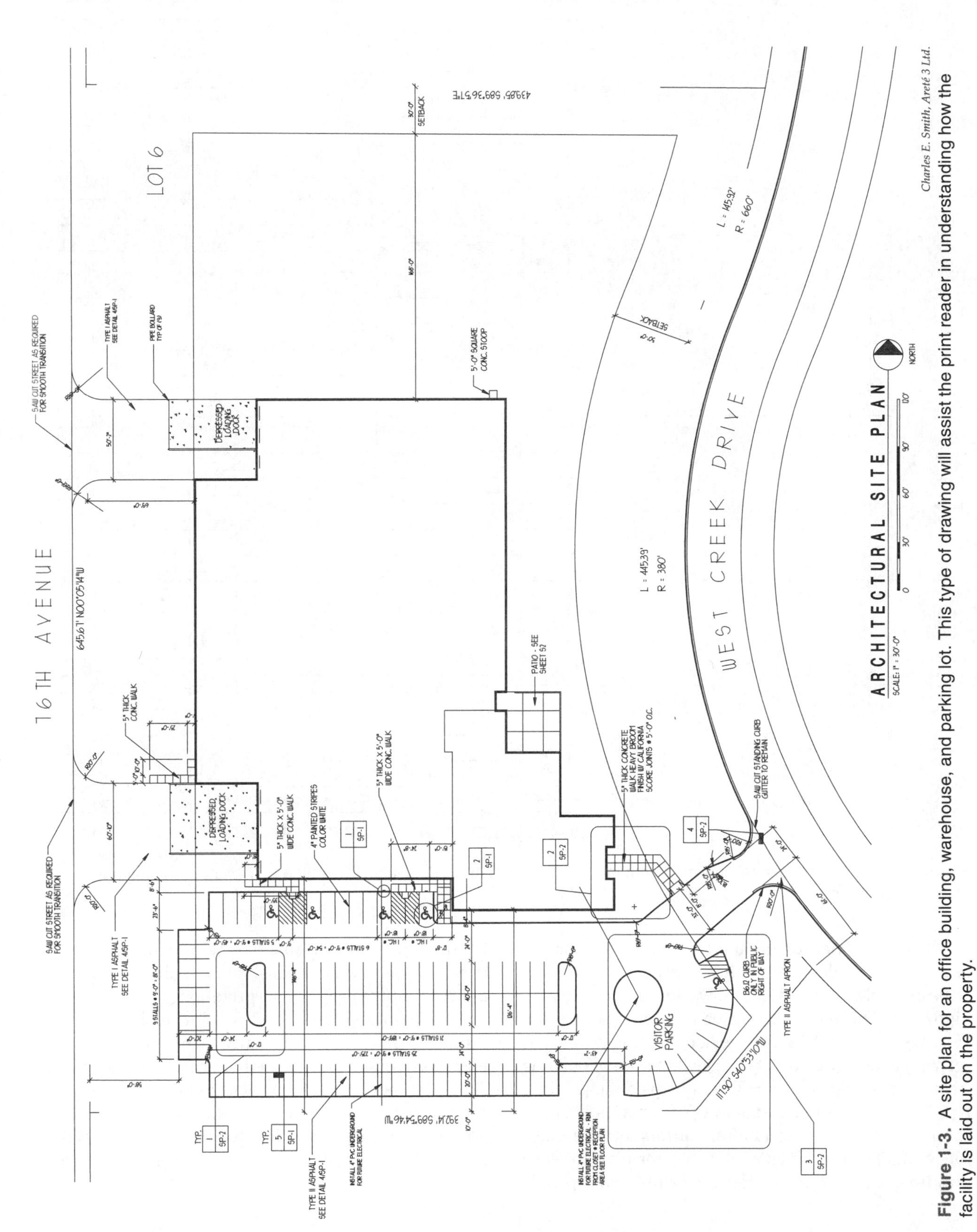

Charles E. Smith, Areté 3 Ltd.

Figure 1-3. A site plan for an office building, warehouse, and parking lot. This type of drawing will assist the print reader in understanding how the facility is laid out on the property.

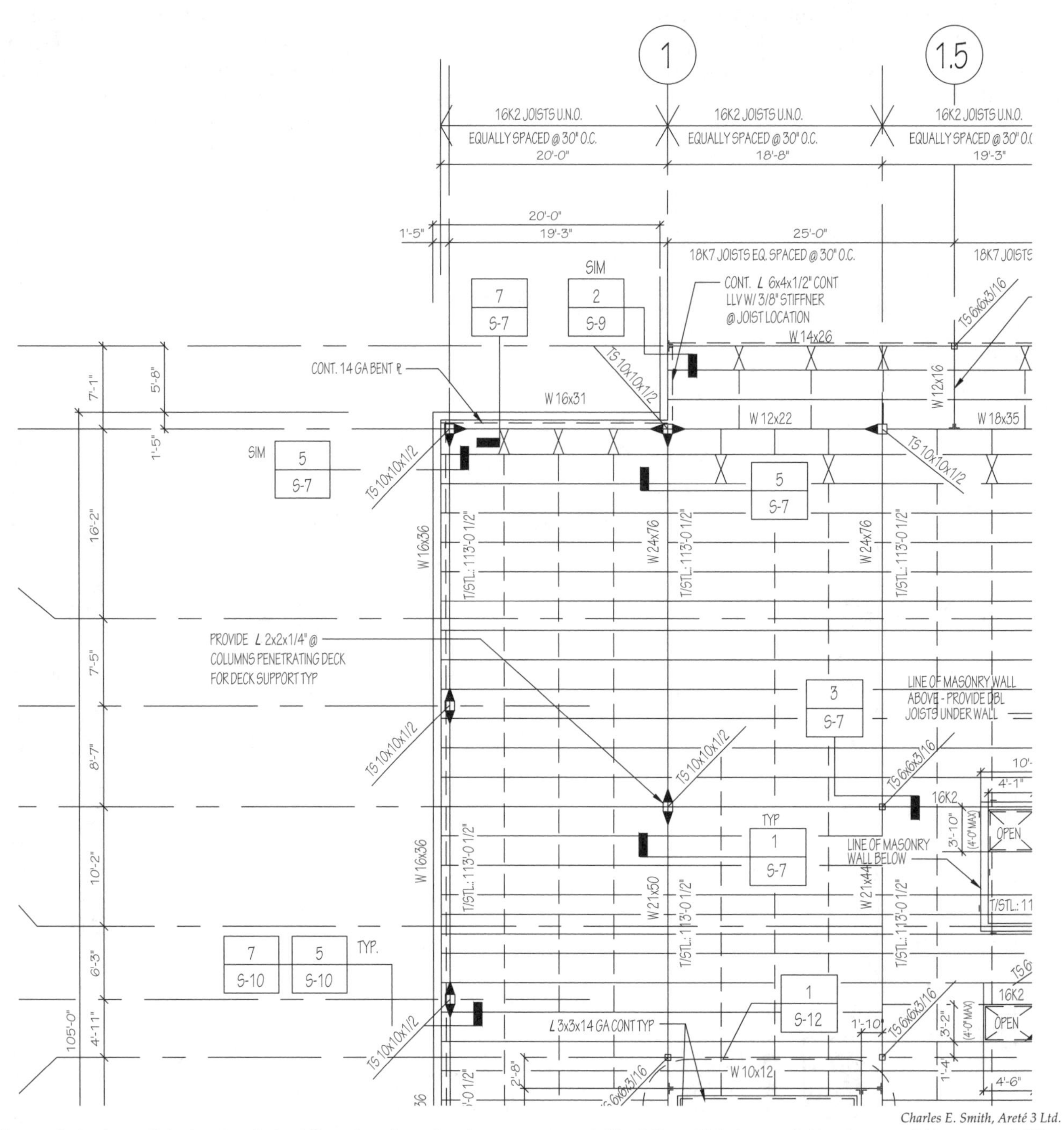

Charles E. Smith, Areté 3 Ltd.

Figure 1-4. A partial structural steel floor framing plan for a commercial building. This type of drawing would be part of (S) Structural Drawings.

The ***floor plan*** provides a great deal of information and acts as a reference for the location of additional enlarged plans, sections, elevations, and details, **Figure 1-5**. The floor plan is actually a horizontal section view taken 42″ to 48″ above the floor looking down. The section plane may be offset (change levels) if the building involves a split-level floor. The floor plan shows floor finishes, walls, doors, stairways, fireplaces, built-in cabinets, and some mechanical equipment. Drawings for multistory buildings include a floor plan for each building level. When reviewing prints, most people begin with the floor plans. The floor plans provide the overview needed to establish the visualization that will assist in future project interpretation.

Elevations depict the exterior features of the building, **Figure 1-6**. Usually, a minimum of four elevation drawings is needed to show the design of all sides of the structure. More elevation views are required for unusual designs, such as internal courtyards, or angular buildings. ***Interior elevations*** of the building are used when additional interior wall surface detail is needed.

Sections are views showing the building as if it were cut apart, **Figure 1-7**. They show walls, stairs, and other details not clearly shown in other drawings. Sections are usually drawn in larger scale than the elevations and plan drawings. Sections taken through the narrow width of an entire building are known as ***transverse sections***. Those through the long dimension are known as ***longitudinal sections***.

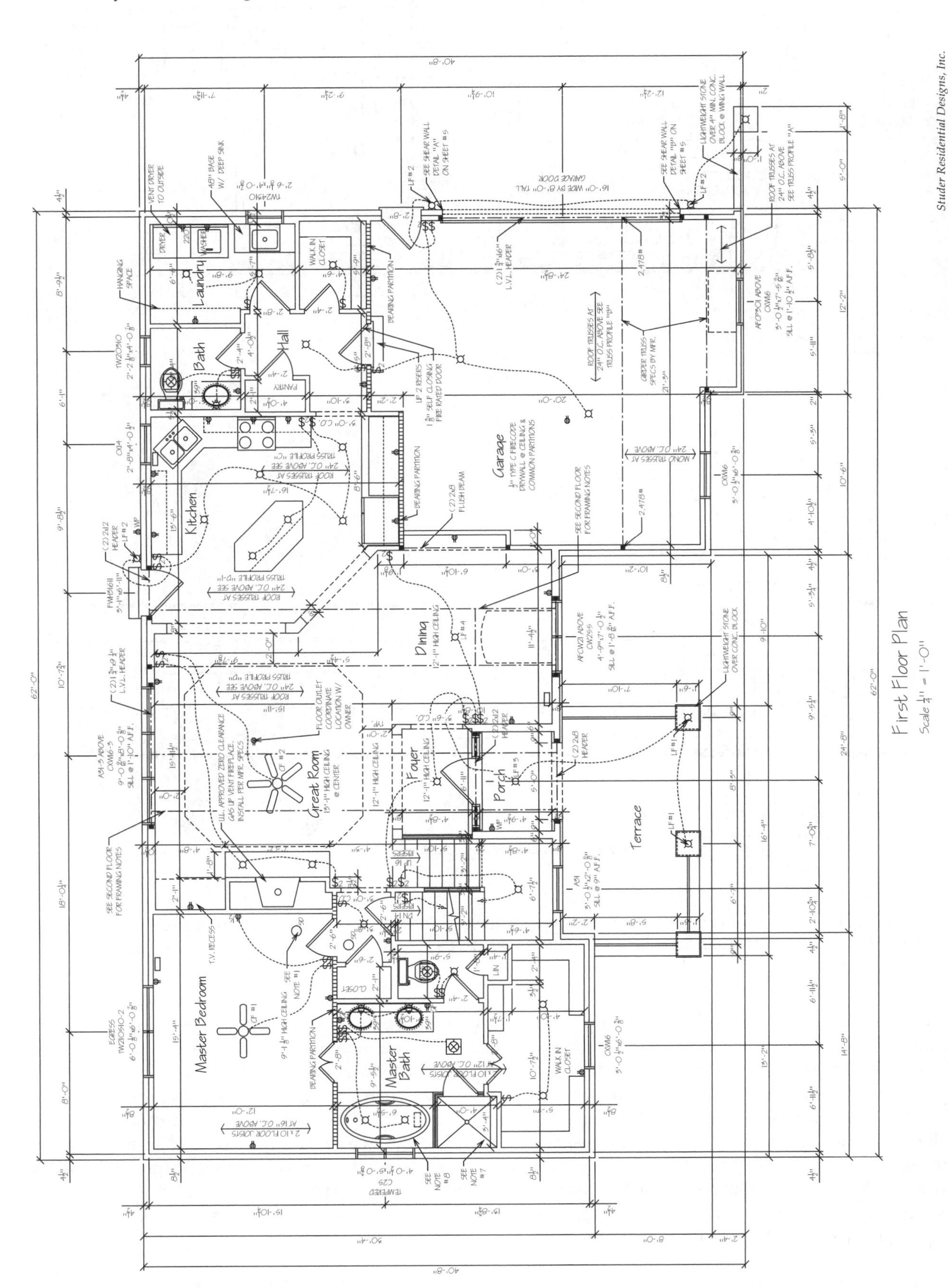

Figure 1-5. The floor plan is one of the most common and informative drawings used in construction. Print readers should start with the floor plans when trying to understand how a building is constructed. On a residential floor plan, the electrical plan can also be shown.

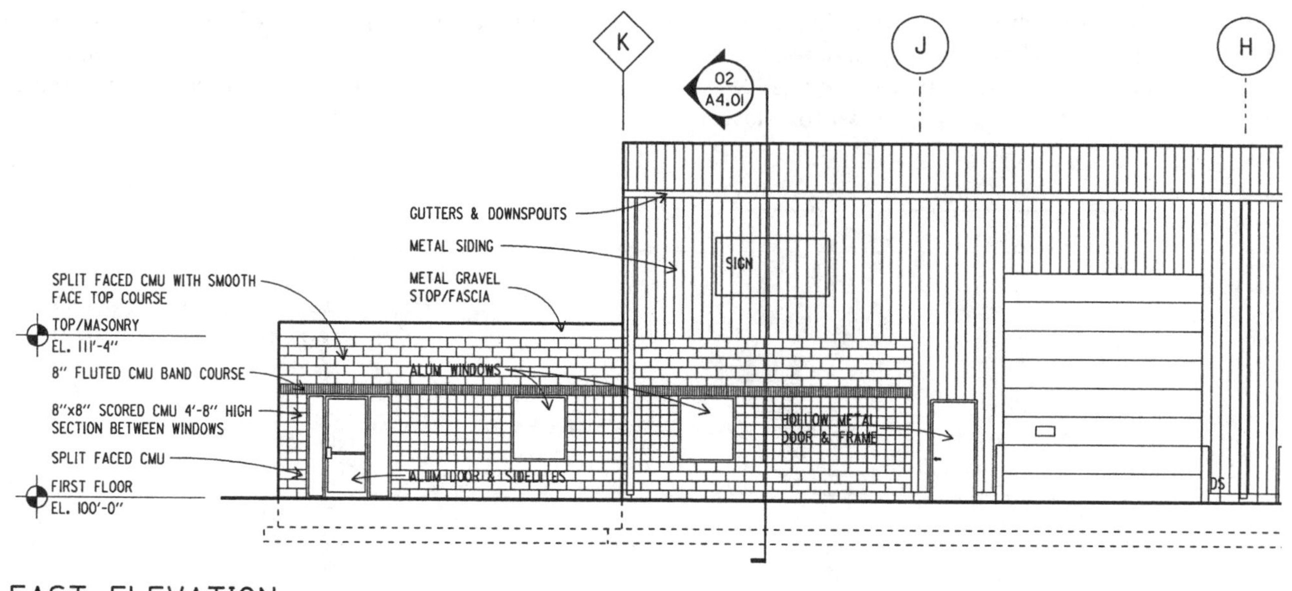

A

KZF Design, Inc.

B

Charles E. Smith, Areté 3 Ltd.

Figure 1-6. Elevation drawings. Elevations are part of (A) Architectural Drawings. A—This elevation shows the exterior facade of a commercial building. Note the different materials that are defined on this drawing. B—An exterior elevation drawing for an office building.

Details are required for complex building components and unusual construction, such as an arch, a cornice, a structural steel connection, or a retaining wall depicting how the architecture connects to the structure. Details are drawn to a larger scale, such as 1″ = 1′-0″, to clearly describe the building components and features. See **Figure 1-8**.

Schedules are lists of materials needed in the construction process. A schedule normally lists the item, an identification mark, size, number required, and any other useful information. Each item in the schedule is referenced on the plan and elevation drawings. Different types of schedules include door schedules, **Figure 1-9**, window schedules, and lighting fixture schedules. Schedules are also used for other purposes, such as showing the materials required in each room of the building. A schedule used for this purpose is called a room finish schedule, **Figure 1-10**. Almost all commercial buildings will have a room finish schedule.

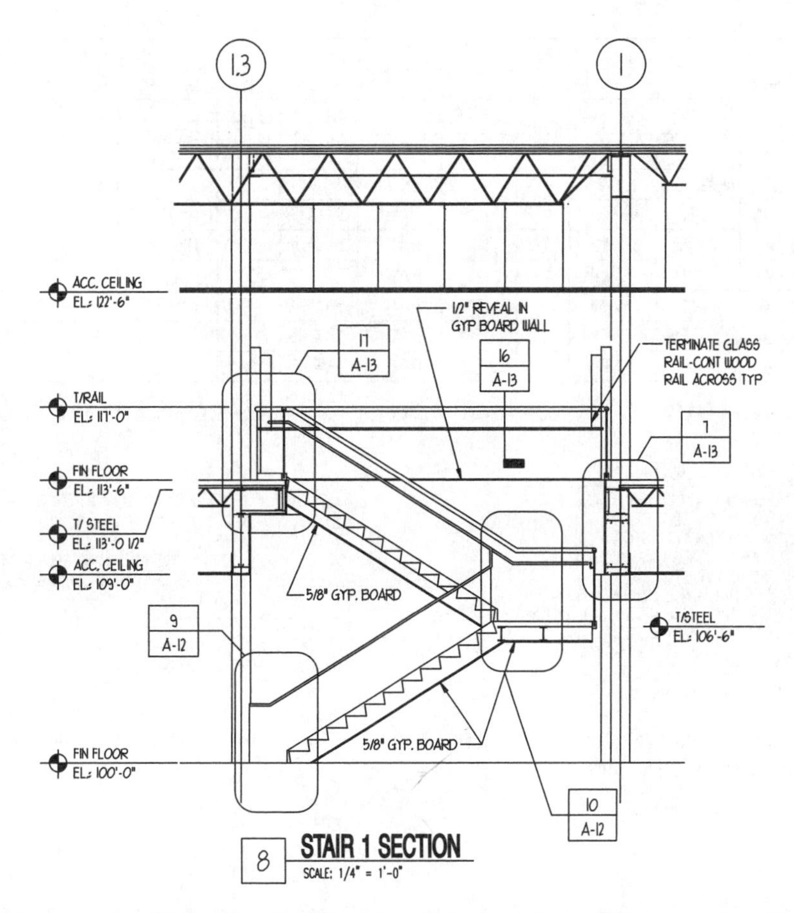

Charles E. Smith, Areté 3 Ltd.

Figure 1-7. Sections indicate areas within a structure to better define how the building goes together. This section illustrates the stair construction. A section of this kind would be found in (A) Architectural Drawings.

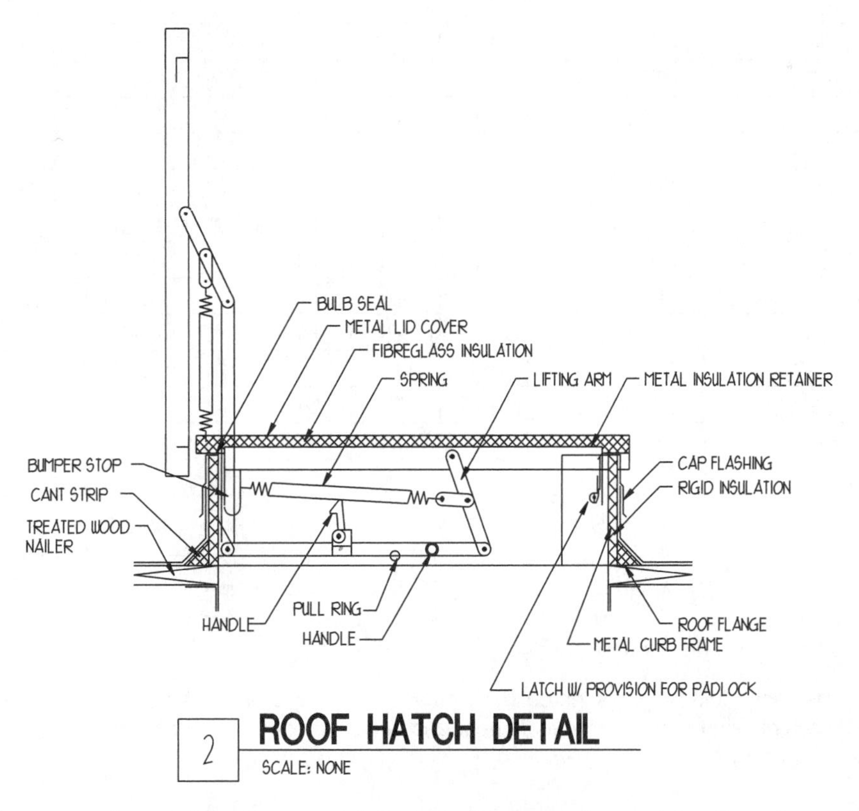

Charles E. Smith, Areté 3 Ltd.

Figure 1-8. As sections are enlarged, they become details indicating more close-up detailing of a particular portion of a building. This kind of detail would be found in (A) Architectural Drawings.

SHELL BUILDING DOOR SCHEDULE												
DOOR No.	DOOR					FRAME				FIRE RATING	HARDWARE	REMARKS
	TYPE	SIZE	THICK	MATERIAL	FINISH	TYPE	MATERIAL	FINISH	DETAIL			
100A	A	6'-0"x7'-0"	1 3/4"	AL/GLASS	ENAM	-	AL	ENAM	-		R	CURTAIN WALL - SEE SHEET A-6
121A	B	3'-0"x7'-0"	1 3/4"	HM	PAINT	A	HM	PAINT	3,4/A-16		B,H,M,N,P,T	-
122A	B	3'-0"x7'-0"	1 3/4"	HM	PAINT	B	HM	PAINT	3,4/A-16	C	B,H,M,N,P,T	1 HOUR
122B	B	3'-0"x7'-0"	1 3/4"	HM	PAINT	A	HM	PAINT	-	C	B,C,H,T	1 HOUR
127A	B	3'-0"x7'-0"	1 3/4"	HM	PAINT	A	HM	PAINT	3,4/A-16		B,H,M,N,P,T	-
127B	C	10'-0"x12'-0"	1 3/4"	STEEL	PREF.	-	STEEL	PAINT	1,2/A-16		-	HI-LIFT O.H. DOOR
127C	C	9'-0"x10'-0"	1 3/4"	STEEL	PREF.	-	STEEL	PAINT	1,2/A-16		-	HI-LIFT O.H. DOOR

H.M.= HOLLOW METAL, MTL.= METAL, PREF.= PREFINISHED, WD.= WOOD, AL.= ALUMINUM, ST/V.= STAIN & VARNISH, GL.= GLASS, STL.=STEEL, B. ENAM.= BAKED ENAMEL FINISH

HARDWARE SCHEDULE

A LCN SERIES 1070 CLOSER (STANDARD)
B LCN SERIES 1071 CLOSER (H.C.)
C GLYNN JOHNSON WALL STOP #60W
E VON DUPRIN SERIES 99 PANIC BAR
F 2 PAIR BUTT HINGE McKINNEY #T2714
G 1 1/2 PAIR BUTT HINGE McKINNEY #T2714
H 1 1/2 PAIR BUTT HINGE HEAVY DUTY McKINNEY #T4A3786
K HAGGAR #30S 3 1/2"x15" PUSH PLATE
L HAGGER #30S PUSH PL w/ #3G PULL
M REESE #815A POLYPRENE WEATHER STRIP
N REESE #S424A THRESHOLD
O HARDWARE PROVIDED BY MANUFACTURER
P KEYED LOCK SET
Q REESE #377C SWEEP
R MANUFACTURER TO PROVIDE PUSH BAR, PULL HANDLE H.C. CLOSER, THRESHOLD, WEATHER STRIPPING, SWEEP, THUMB TURN LOCK SET
S SCHLAGE S SERIES LATCH (MEDIUM DUTY)
T SCHLAGE L SERIES LATCH (HEAVY DUTY)
SUGGESTED HARDWARE OR EQUAL

Charles E. Smith, Areté 3 Ltd.

Figure 1-9. This schedule contains the details and detail references for all the doors in the building. This kind of schedule would be found in (A) Architectural Drawings.

ROOM FINISH SCHEDULE										
ROOM No.	ROOM NAME	FLR.	BASE	WALLS				CL'G	CL'G HT.	REMARKS
				N	E	S	W			
100	VESTIBULE	QT	VC	PT	GL	PT	GL	ACT	9'-0"	SURF MTD PEDIMAT
101	WAITING	CPT	VC	PT	GL	PT	-	ACT	9'-0"	-
102	RECEPTION	CPT	V	-	-	PT	PT/WD	ACT	9'-0"	-
103	OFFICE	CPT	V	PT	PT	PT	PT	ACT	9'-0"	-
104	VP SALES	CPT	V	PT	PT	PT	PT	ACT	9'-0"	-
105	CLASSROOM	CPT	V	PT	PT	PT	PT	ACT	9'-0"	-
106	CONFERENCE	CPT	V	PT	PT	PT	PT	ACT	9'-0"	-
107	OFFICE	CPT	V	PT	PT	PT	PT	ACT	9'-0"	-
108	OFFICE	CPT	V	PT	PT	PT	PT	ACT	9'-0"	-
109	CONT. OFFICE	CPT	V	PT	PT	PT	PT	ACT	9'-0"	-
110	ADMIN. VP	CPT	V	PT	PT	PT	PT	ACT	9'-0"	-
111	OPEN OFFICE	CPT	V	PT	PT	PT	PT	ACT	9'-0"	-
112	COMPUTER	CPT	V	PT	PT	PT	PT	ACT	9'-0"	-
113	CONFERENCE	CPT	V	PT	PT	PT	PT	ACT	9'-0"	-

QT = QUARRY TILE
CT = CERAMIC TILE
CPT = CARPET
VCT = VINYL COMPOSITE TILE
GL = GLASS
V = 4" VINYL BASE
PT = PAINT
OPEN = EXPOSED CONSTRUCTION
ACT = 2x2 ACOUSTICAL CEILING TILE
VT = VINYL TILE
BL = CONCRETE BLOCK
CS = SEALED CONCRETE
DW = DRYWALL
VC = VINYL COVE BASE

Charles E. Smith, Areté 3 Ltd.

Figure 1-10. A room finish schedule provides the builder with an easy way to understand the different materials used within each room in the building. This kind of schedule would be found in (A) Architectural Drawings.

CAREERS IN CONSTRUCTION

Construction Manager

A construction manager is an experienced professional who manages and coordinates entire construction projects. Usually, this responsibility involves management of a construction management team. The construction management team oversees construction and scheduling, changes during construction, and coordination between disciplines and trades. The team also manages the budget for the construction company and sometimes the owner's budget.

A successful construction manager not only manages the project, but also manages many people, including subcontractors and material suppliers. The size and complexity of the project determine the number of individuals required for successful project execution. In a typical project, other managers working under the supervision of the construction manager include a project engineer, a project estimator, a superintendent, and a foreman. The construction manager is involved in all phases of construction from approval of the initial design concepts to completion of the building.

Traditionally, construction managers came up through the building trades by becoming a skilled worker in a specific occupation, such as a carpenter, plumber, electrician, or surveyor. Today, however, many colleges offer construction management degree programs. In many firms, candidates are required to have a bachelor's degree in construction management as well as experience in the construction industry.

bikeriderlondon/Shutterstock.com

Construction managers are responsible for managing entire building projects and coordinating the work of others on the construction management team.

Schedules are usually included as part of a set of working drawings. Door schedules frequently are included on the plan drawings. Window schedules generally appear on the elevation drawings. On larger projects, schedules are listed on their own sheet.

Structural framing plans may be included in a set of plans for the framing of floors, the roof, and various wall sections. Structural framing plans are part of the Structural (**S**) prints.

The ***plumbing plan*** shows the hot and cold water system layout, the sewage disposal system, and plumbing fixture locations. For typical residential projects, the entire plumbing plan is typically shown on one drawing, and plumbing fixtures are often shown on the architectural floor plan. For more complex structures, separate plans for each system are used.

The ***mechanical plans*** include heating, ventilating, and air-conditioning (***HVAC***) plans. These plans show the location of mechanical equipment for heating and cooling systems in the building.

The ***electrical plans*** show electrical wiring, fixtures, and devices. The electrical plans include the lighting plan, reflected ceiling plan, and panel schedules. On larger projects, riser diagrams and load calculations may also be included. For smaller jobs and residential projects, the electrical plan may appear on the architectural floor plan.

Uniform Drawing System (UDS)

The ***Uniform Drawing System (UDS)*** is a standardization of drawing guidelines developed by the Construction Specifications Institute (CSI). The UDS consists of eight interrelated modules. These modules contain standards, guidelines, and various tools for organizing and presenting drawing information used in planning, designing, and constructing facilities.

Module 01—Drawing Set Organization

Establishes set content and order, sheet identification, and file naming for the set of construction drawings. The sheet identification guidelines provide a standard format based on letter classifications and a numbered system, as discussed earlier in this unit.

Module 02—Sheet Organization

Provides formats for sheets, including the sheet title block and production data areas, along with their content.

Module 03—Schedules
Standardizes schedule formats for consistency of content and terminology used for construction drawings.

Module 04—Drafting Conventions
Standardizes conventions used in drawings, such as drawing orientation, layout, symbols, material indications, line types, dimensioning standards, drawing scales, diagramming, notations, and cross-referencing systems.

Module 05—Terms and Abbreviations
Provides consistent spelling and terminology, standardizes abbreviations, and establishes common usage.

Module 06—Symbols
Addresses common symbols, classifications, graphic representation, and organization as used in creating, understanding, and fulfilling the intent of documents.

Module 07—Notations
Standardizes notation classification, use of notes, notation format, notation components, and notation terminology. Also addresses notation location and the linking of notations to specifications.

Module 08—Code Conventions
Identifies types of general regulatory information that should appear on drawings, locates code-related information in a set of drawings, and provides standard graphic conventions.

The UDS is a standardized construction document accepted throughout the building industry. It is one of the major publications included in the US National CAD Standard.

Making Prints

Most drawings today are created using ***computer-aided design and drafting (CADD)*** systems. CADD software enables the architect or engineer to develop a design relatively quickly, and in a manner that allows for changes to be made more easily. Drawings created using CADD are generated on a plotter or other printing device. See **Figure 1-11**.

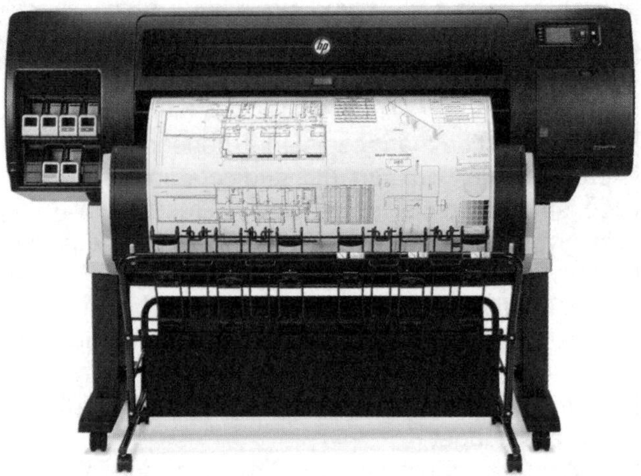

Hewlett-Packard Development Company

Figure 1-11. High-quality prints can be made from CADD drawings using a printing device such as this large-format inkjet printer.

Drawings made in CADD are created at full size. When the drawing is plotted, the appropriate scale is specified to fit the drawing on the selected sheet size. Using a drawing scale is discussed in Unit 3. Typical sheet sizes used for construction drawings are shown in **Figure 1-12**.

Prior to the emergence of CADD, drawings were created manually by hand, directly on paper. Drawings created this way were made on translucent paper (***vellum*** or ***polyester film***). Original drawings were kept in file storage and prints were reproduced from the originals using a large-format ***engineering copier*** (a photocopier designed to handle larger sizes of paper).

Some offices continue to keep original paper drawings on file and use an engineering copier to make prints when needed. Drawings can be reduced or enlarged when photocopied. For example, a large D-size (24″ × 36″) drawing can be reduced to a smaller print for easier handling.

It is important to note that due to the nature of copying machines and printing devices, prints may not always be exact duplicates of the original drawing. Slight enlargement or reduction may occur. Therefore, you should never scale a dimension from a print unless you have verified that the drawing has, in fact, printed to scale. This can be verified by scaling several dimensions on the print to guarantee its accuracy.

Electronic Documentation Processes

Developments in computer technology are changing the way in which prints are accessed and used in the field. Because most drawings today are created as electronic files using design software, they can be readily exchanged between design teams and contractors. In some cases, electronic files are used in place of paper prints at the job site. Files can be stored on a local

Drawing Sheet Sizes	
Size Designation	**Size in Inches**
A	9 × 12
B	12 × 18
C	18 × 24
D	24 × 36
E	36 × 48
E1	30 × 42
E2	26 × 38
E3	27 × 39

Goodheart-Willcox Publisher

Figure 1-12. Typical drawing sheet sizes.

network or an offsite server accessed using the Internet. Many drawings today are kept in "cloud storage" on a cloud service network so that all parties involved in design and construction can retrieve the most current information online. Drawings can be opened and viewed on personal computer monitors, laptops, and handheld devices such as smartphones and tablets. In some projects, the design and construction teams may use large-format display tables to view prints. See **Figure 1-13**. Computer-generated prints are typically saved to a format that does not require the design software to open files. A common format is the Adobe® portable document format (PDF).

Newer technologies are also impacting processes used in design, construction, and building management. In traditional CADD processes, two-dimensional (2D) drawings are created to represent the different views of a building. A newer practice is to design the building as a three-dimensional (3D) model using parametric modeling software. Instead of creating a set of drawings using traditional 2D CADD software, the designer uses the 3D model to generate the drawing documentation automatically. This approach is used in building information modeling. ***Building information modeling (BIM)*** is a process in which a 3D model provides a virtual representation of a building and is used in design, construction, and operation of the building. BIM offers a number of advantages to architectural designers and building contractors. Creating a 3D building model allows the designer and owner to evaluate realistic representations of a building before it is built. In addition, all information and data related to the project, including the drawing documentation, is linked to the 3D building model. When a change is made to the model, all of the drawing documentation and building data updates to reflect the change. This helps reduce design errors and simplifies change management processes traditionally used to update prints. BIM also improves coordination between designers and contractors and helps prevent conflicts in construction. Throughout the project, all parties involved in design, construction, and building management have access to the building data. This improves efficiency and helps reduce costs.

In commercial construction, BIM offers significant advantages to building owners. Once construction is completed, a BIM model can be used for facility management tasks such as scheduling maintenance, troubleshooting, and estimating energy costs. Instead of referring to paper prints, maintenance managers use the building model and facility management system

GREEN BUILDING

Distributing Prints

In traditional practice, construction drawings prepared using CADD software are transferred to paper in order to present plans to a client. To conserve paper and minimize printer use, think carefully before making paper prints for distribution. Construction drawings can be shared with other design teams and clients via e-mail or a cloud-based file sharing service.

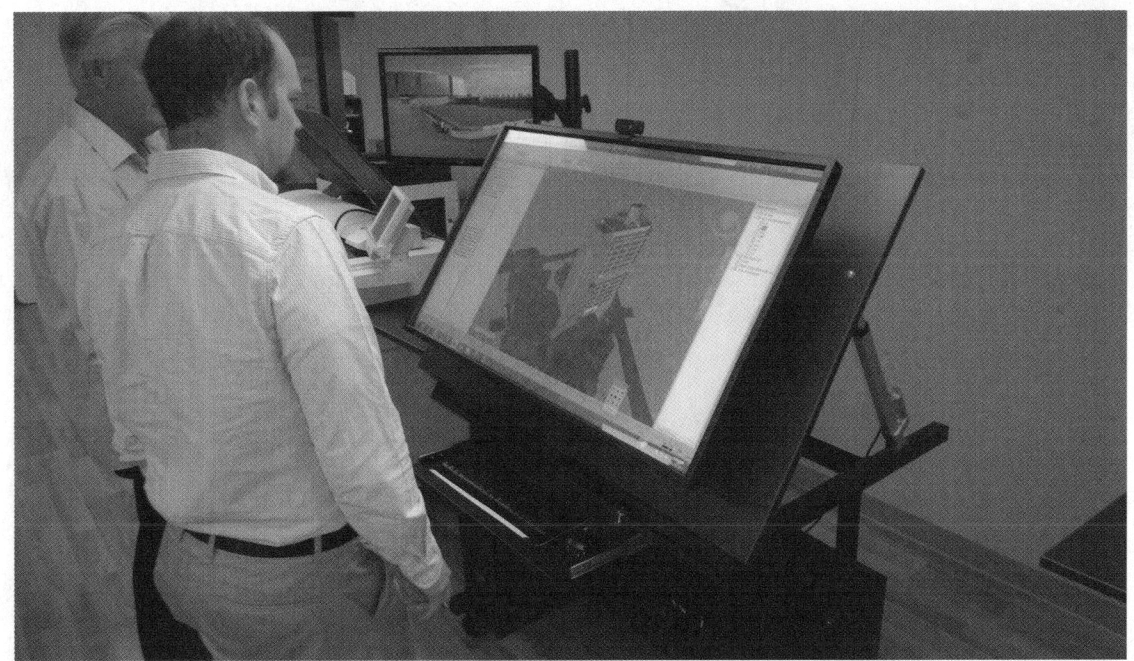

Photo courtesy of iPlanTables

Figure 1-13. This wide-format display table has a touchscreen monitor and is equipped with a second monitor for additional display.

to retrieve information. This is usually more efficient because the system can be searched for the necessary data. In addition, the information is usually more accurate because the building model is kept up-to-date by engineers to reflect changes, such as new equipment or building renovations.

Reading Prints

Print reading is the gathering of information from prints or a set of drawings. It involves two principal elements: visualization and interpretation. Reading prints develops these abilities. The more experience you have, the better your skill level.

Visualization is the ability to create a mental image of a building or project from a set of working drawings. A study of print reading principles and learning to sketch will help you visualize construction drawings and details.

Interpretation is the ability to understand lines, symbols, dimensions, notes, and other information on the working drawings. Each of these areas will be discussed in this textbook.

Handling Prints

Prints and related specification sheets are as important as the tools used to build a building. With proper care, prints can be kept usable for a long period of time.

There are several guidelines for handling prints:

- Never write on a print unless you are an authorized professional to do so, such as an estimator or a project engineer noting a revision to the drawing.
- Keep prints clean. Soiled prints are difficult to read and contribute to errors.
- Do not eat or drink near prints.
- Fold or roll prints carefully. Rolling is best.
- Do not lay sharp tools or pointed objects on prints.
- Keep prints out of direct sunlight except when using them. Prints will fade and deteriorate if left in the sun.
- When prints are not in use, store them in a clean, dry place.

Test Your Knowledge

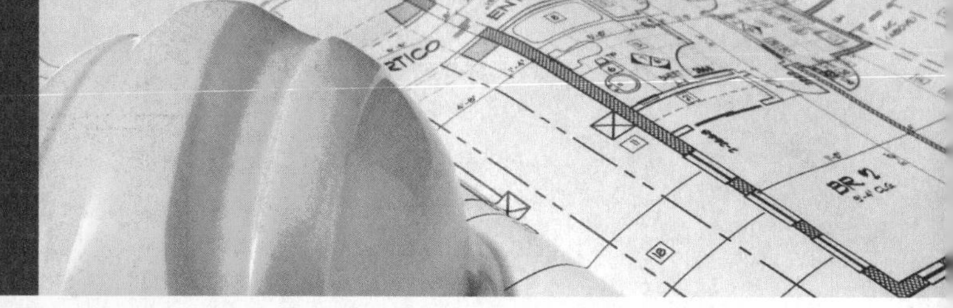

Name ______________________________

Write your answers in the spaces provided.

____ 1. Which is *not* a common type of drawing?

A. Mechanical plan

B. Plumbing plan

C. Painting plan

D. Foundation plan

E. They are all common drawings.

____ 2. Under what conditions can the electrical plan be included on the floor plan?

A. When the building is a smaller job or a residential project

B. When print paper is scarce

C. When the architect is also an electrician

D. When a single company is doing all the construction

E. There are no conditions when the electrical plan can be included on the floor plan.

____ 3. Which drawing would show the location of a building sprinkler system?

A. Electrical plan

B. Foundation plan

C. Site plan

D. Plumbing plan

E. A building sprinkler system would not be shown on any of these drawings.

____ 4. Which occupation does *not* need to know how to read construction prints?

A. Welder

B. Construction estimator

C. Owner of a lighting store

D. Concrete supplier

E. All of these occupations require print reading.

____ 5. A list of materials included on a drawing is called ____.

A. notes

B. an elevation

C. a section

D. a schedule

E. None of the above.

____ 6. *True or False?* A rendering would show all the details needed to build an office building.

____ 7. *True or False?* The locations of utilities are generally shown on the site plan.

____ 8. *True or False?* Most construction drawings are created using orthographic projection.

____ 9. *True or False?* Prints will not be damaged if they are left out in direct sunlight for an extended period.

____ 10. *True or False?* Working drawing is another term for print.

____ 11. In a set of working drawings, under which drawing classification would the steel framing be found?

A. Civil drawings

B. Architectural drawings

C. Plumbing drawings

D. Structural drawings

E. Mechanical drawings

____ 12. Which type of schedule would identify the required floor material?

A. Door schedule

B. Room finish schedule

C. Column schedule

D. Wall schedule

E. Floor schedule

____ 13. *True or False?* When first opening a set of drawings, you should review the floor plans to visualize what the building is going to look like when it is finished.

____ 14. *True or False?* Drawings made in CADD are created at reduced size and plotted at the appropriate scale.

____ 15. When handling prints, when is it acceptable to write on them?

A. When estimating the project cost

B. When noting the name of a supplier

C. When there is no other paper available

D. It is always acceptable to write on prints as long as it is with a pencil.

Tom Grundy/Shutterstock.com

Constructing a set of stairs requires the ability to carefully read prints and perform any needed calculations.

UNIT 2
Construction Math and Application

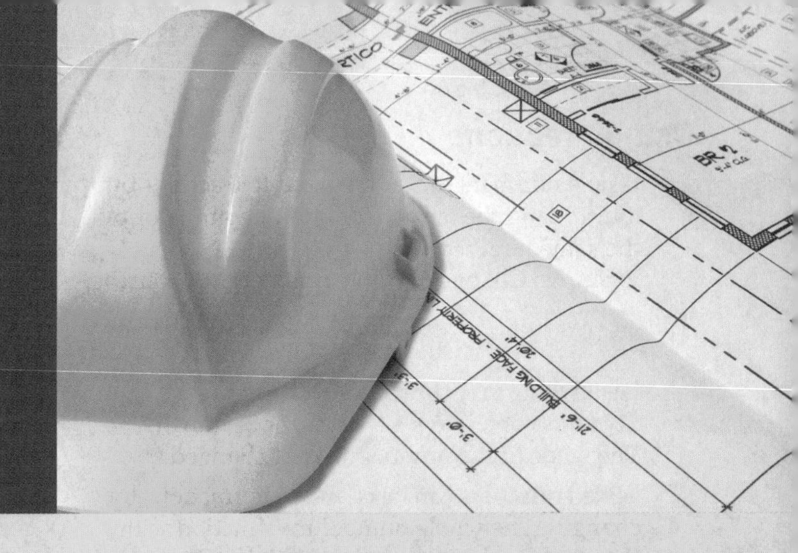

TECHNICAL TERMS

arc
area
circumference
decimal fraction
denominator
diameter
improper fraction
mixed number
numerator
pi
proper fraction
radius

LEARNING OBJECTIVES

After completing this unit, you will be able to:

- Convert between improper fractions and mixed numbers.
- Add, subtract, multiply, and divide decimal fractions.
- Calculate dimensions.
- Calculate areas and volumes of objects.
- Relate math to construction problems.

Construction workers and estimators often need to make calculations when working with prints. This unit deals with construction-oriented calculations, including fractions and decimals in both customary and metric units.

Fractions

Fractions are written with one number over the other, such as $\frac{9}{16}$. The number on the bottom (*16*) is called the ***denominator***. It indicates the number of equal parts into which a unit is divided. The number on top (*9*) is called the ***numerator***. It indicates the number of equal parts taken, **Figure 2-1**. In the fraction shown (*9/16*), nine of the sixteen parts are taken.

A ***proper fraction*** is one whose numerator is less than its denominator, such as 7/16 or 3/4. An ***improper fraction*** is one whose numerator is greater than its denominator, such as 5/4 or 19/16. A ***mixed number*** is a number that consists of a whole number and a proper fraction, such as 2 3/4 or 5 1/8. Proper fractions represent numbers between zero and one. Improper fractions and mixed numbers represent numbers larger than one.

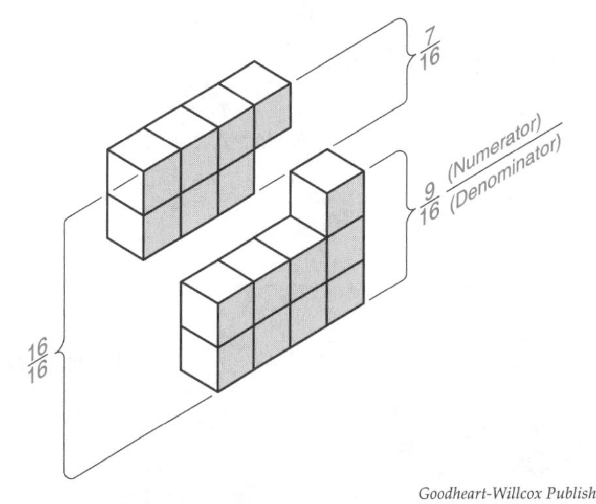

Goodheart-Willcox Publisher

Figure 2-1. The fraction 9/16 represents 9 pieces of a whole divided into 16 equal pieces.

Using Fractions

- Whole numbers can be changed to fractions by multiplying the numerator and denominator by the same number.
  Example: Change 6 (whole number) into fourths.
  $$\frac{6}{1} \times \frac{4}{4} = \frac{24}{4}$$
  Each whole unit contains 4 fourths.
  Six units contain 6 × 4 fourths, or 24 fourths.
  The value of the number has not changed.
- Mixed numbers can be changed to fractions by changing the whole number to a fraction with the same denominator as the fractional part and adding the two fractions.

  Example: Convert 3 5/8 to an improper fraction.
  $$3\frac{5}{8} = \left(\frac{3}{1} \times \frac{8}{8}\right) + \frac{5}{8} = \frac{24}{8} + \frac{5}{8} = \frac{29}{8}$$
  Three units contain 3 × 8 eighths, or 24 eighths. Adding the 5/8 part of the mixed number to 24/8 gives us 29/8.
- Improper fractions can be reduced to a whole or mixed number by dividing the numerator by the denominator:
  $$\frac{17}{4} = 17 \div 4 = 4\frac{1}{4}$$
- Fractions can be reduced to the lowest form by dividing the numerator and denominator by the same number:
  $$\frac{6}{8} = \frac{6 \div 2}{8 \div 2} = \frac{3}{4}$$
  The value of a fraction does not change if the numerator and denominator are divided by the same number, since this is the same as dividing by 1.
- Fractions can be changed to higher terms by multiplying the numerator and denominator by the same number:
  $$\frac{5}{8} = \frac{5 \times 2}{8 \times 2} = \frac{10}{16}$$
  The value of a fraction is not changed by multiplying the numerator and denominator by the same number.

ADDING FRACTIONS

To add fractions, the denominators must all be the same.

Example: $\frac{5}{16} + \frac{3}{8} + \frac{11}{32} = ?$

The *least common denominator (LCD)* into which these denominators can be divided is 32. Change all fractions to have 32 in the denominator:

$$\frac{5}{16} \times \frac{2}{2} = \frac{10}{32}$$

$$\frac{3}{8} \times \frac{4}{4} = \frac{12}{32}$$

Now that the fractions have the same denominator, add their numerators. The common denominator is used in the sum:

$$\frac{10}{32} + \frac{12}{32} + \frac{11}{32} = \frac{33}{32}$$

Convert to a mixed number:

$$\frac{33}{32} = 1\frac{1}{32}$$

PRACTICE PROBLEMS

Adding Fractions

Add the following fractions. Reduce answers to lowest form.

1. $\frac{3}{4} + \frac{1}{8} + \frac{1}{2} =$
2. $\frac{7}{8} + \frac{3}{16} =$
3. $\frac{5}{12} + \frac{3}{8} + \frac{3}{4} =$
4. $\frac{3}{10} + \frac{9}{10} + \frac{1}{20} =$
5. $\frac{7}{16} + \frac{3}{32} + \frac{1}{4} =$
6. $1\frac{3}{4} + \frac{7}{8} + 1\frac{1}{16} =$
7. $\frac{5}{32} + \frac{7}{64} + \frac{7}{8} =$
8. $1\frac{3}{8} + \frac{3}{32} + \frac{7}{16} =$
9. $3\frac{1}{16} + \frac{9}{16} + \frac{1}{2} =$
10. $5\frac{1}{5} + \frac{3}{10} + 8\frac{1}{2} =$
11. $4\frac{5}{8} + 20\frac{7}{32} =$
12. $\frac{3}{8} + \frac{7}{64} + \frac{9}{16} =$
13. $12\frac{7}{8} + 25\frac{3}{8} =$
14. $\frac{21}{32} + \frac{9}{64} + \frac{1}{4} =$
15. $\frac{3}{8} + 1\frac{1}{2} + \frac{7}{16} + \frac{7}{8} =$

SUBTRACTING FRACTIONS

To subtract fractions, the denominators must all be the same.

Example: $\frac{3}{4} - \frac{5}{16} = ?$

The least common denominator into which these denominators can be divided is 16.

Change 3/4 so that its denominator is 16:

$$\frac{3}{4} \times \frac{4}{4} = \frac{12}{16}$$

Subtract the numerators and retain the common denominator:

$$\frac{12}{16} - \frac{5}{16} = \frac{7}{16}$$

PRACTICE PROBLEMS

Subtracting Fractions

Subtract the following fractions. Reduce answers to lowest form.

1. $\frac{3}{8} - \frac{1}{4} =$
2. $\frac{3}{4} - \frac{5}{16} =$
3. $1\frac{7}{8} - \frac{13}{16} =$
4. $3\frac{1}{2} - \frac{9}{16} =$ (*borrow* $\frac{16}{16}$ *from* 3)
5. $10\frac{3}{8} - 7\frac{3}{32} =$
6. $5 - 2\frac{3}{8} =$
7. $12\frac{1}{16} - 8\frac{1}{2} =$
8. $4\frac{1}{4} - 3\frac{1}{16} =$
9. $20\frac{7}{8} - 11\frac{3}{64} =$
10. $15\frac{5}{8} - 5\frac{1}{2} =$

MULTIPLYING FRACTIONS

Fractions can be multiplied as follows:

1. Change all mixed numbers to improper fractions.
2. Multiply all numerators to get the numerator of the answer.
3. Multiply all denominators to get the denominator of the answer.
4. Reduce the fraction to lowest terms.

Example:

$$\frac{1}{2} \times 3\frac{1}{8} \times 4 = ?$$

$$\frac{1}{2} \times \frac{25}{8} \times \frac{4}{1} = \frac{100}{16}$$

$$\frac{100}{16} = 6\frac{4}{16} = 6\frac{1}{4}$$

PRACTICE PROBLEMS

Multiplying Fractions

Multiply the following fractions. Reduce answers to lowest terms.

1. $\frac{3}{4} \times \frac{1}{2} =$
2. $2\frac{5}{8} \times \frac{1}{4} =$
3. $\frac{7}{8} \times 5 =$
4. $6\frac{3}{4} \times \frac{1}{3} =$
5. $12\frac{1}{2} \times \frac{1}{2} =$
6. $4\frac{3}{4} \times \frac{1}{2} \times \frac{1}{8} =$
7. $16 \times \frac{3}{4} =$
8. $9\frac{5}{8} \times \frac{1}{2} =$
9. $10 \times \frac{4}{5} =$
10. $\frac{14}{3} \times 6 =$

DIVIDING FRACTIONS

Fractions can be divided as follows:

1. Change all mixed numbers to improper fractions.
2. Invert (turn upside-down) the divisor and multiply.

Example:

$5\frac{1}{4} \div 1\frac{1}{2} = ?$

$\frac{21}{4} \div \frac{3}{2} =$

$\frac{21}{4} \times \frac{2}{3} = \frac{42}{12}$

$\frac{42}{12} = 3\frac{6}{12} = 3\frac{1}{2}$

PRACTICE PROBLEMS

Dividing Fractions

Divide the following fractions. Reduce answers to lowest terms.

1. $2\frac{3}{4} \div 6 =$
2. $12 \div \frac{3}{4} =$
3. $16\frac{1}{8} \div 2 =$
4. $8\frac{2}{3} \div \frac{1}{3} =$
5. $16\frac{1}{4} \div 20 =$
6. $\frac{7}{8} \div \frac{7}{16} =$
7. $15 \div 1\frac{1}{4} =$
8. $21 \div 3\frac{1}{8} =$
9. $5\frac{1}{4} \div \frac{3}{8} =$
10. $3\frac{5}{8} \div 2 =$

Decimal Fractions

The denominator in a ***decimal fraction*** is 10 or a multiple of 10 (100, 1000, etc.). When writing decimal fractions, omit the denominator and place a decimal point in the numerator.

3/10 is written 0.3 (three tenths)

87/100 is written 0.87 (eighty-seven hundredths)

375/1000 is written 0.375 (three hundred seventy-five thousandths)

4375/10000 is written 0.4375 (four thousand three hundred seventy-five ten thousandths)

Whole numbers are written to the left of the decimal point and fractional parts are to the right:

5253/1000 is written 5.253 (five and two hundred fifty-three thousandths)

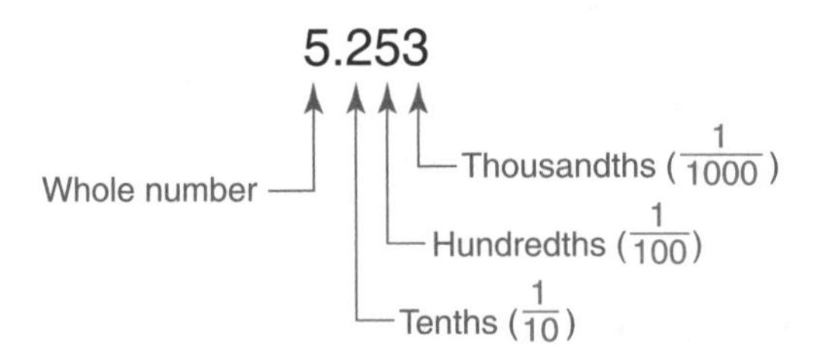

ADDING AND SUBTRACTING DECIMALS

Decimals are added and subtracted in the same manner as whole numbers. With decimals, however, the decimal points must be aligned vertically.

Example:

Add:

$$\begin{array}{r} 7.3125 \\ 1.25 \\ 0.625 \\ +\ 3.375 \\ \hline 12.5625 \end{array}$$

Subtract:

$$\begin{array}{r} 8.625 \\ -\ 2.25 \\ \hline 6.375 \end{array}$$

The decimal point in the answer is directly below the decimal points in the problem.

PRACTICE PROBLEMS

Adding and Subtracting Decimals

Solve the following problems:

Add:

1. 4.5625 + 0.875 + 2.75 + 5.8137 =
2. 1.9375 + 3.25 + 0.375 =
3. 7.0625 + 0.125 + 8.0 =
4. 11.342 + 16.17 + 0.4207 =
5. 0.832 + 0.4375 + 0.27 =

Subtract:

6. 27.9375 – 16.937 =
7. 3.306 – 1.875 =
8. 4.0 – 0.0625 =
9. 10 – 0.75 =
10. 2.25 – 1.125 =

MULTIPLYING DECIMALS

Decimals are multiplied in the same manner as whole numbers. The decimal points are disregarded until the multiplication is completed. To find the position of the decimal point in the answer, count the total number of decimal places to the right of the decimal point in the numbers being multiplied; then set off this number of decimal places in the answer, starting at the right.

Example:

$$\begin{array}{r} 6.25 \\ \times\ 1.5 \\ \hline 9.375 \end{array}$$

× 1.5 (3 decimal places in the two numbers)
9.375 (3 decimal places)

PRACTICE PROBLEMS

Multiplying Decimals

Solve the following problems:

1. $\begin{array}{r} 4.825 \\ \times\ 1.75 \\ \hline \end{array}$
2. $\begin{array}{r} 12.05 \\ \times\ 4.124 \\ \hline \end{array}$
3. $\begin{array}{r} 167 \\ \times\ 0.25 \\ \hline \end{array}$
4. $\begin{array}{r} 0.838 \\ \times\ 5.9 \\ \hline \end{array}$
5. $\begin{array}{r} 65.96 \\ \times\ 0.37 \\ \hline \end{array}$
6. $\begin{array}{r} 0.375 \\ \times\ 6 \\ \hline \end{array}$
7. $\begin{array}{r} 4.95 \\ \times\ 1.35 \\ \hline \end{array}$
8. $\begin{array}{r} 3.75 \\ \times\ 100 \\ \hline \end{array}$
9. $\begin{array}{r} 93.18 \\ \times\ 0.07 \\ \hline \end{array}$
10. $\begin{array}{r} 5639.25 \\ \times\ 10 \\ \hline \end{array}$

DIVIDING DECIMALS

Dividing decimals is identical to dividing whole numbers, except that the decimal point must be properly placed in the quotient (answer).

Example: 26 ÷ 6.5 = 4.0

To place the decimal point in the quotient, count the number of places to the right of the decimal point in the divisor. Add this number of places to the right of the decimal point in the dividend and place the decimal point directly above in the quotient.

Example: 36.5032 ÷ 4.12 = ?

```
                    Quotient
Divisor →      8.86
         4.12 )36.50'32  ← Dividend
      ①      − 32 96 ② ③
                 3543
               − 3296
                 2472
               − 2472
                    0
```

1. Move decimal to right end of the divisor.
2. Move decimal in the dividend the same number of places it was moved in the divisor.
3. The decimal in the quotient is directly above the newly located decimal in the dividend.

PRACTICE PROBLEMS

Dividing Decimals

Solve the following problems:

1. 9.45 ÷ 2.7 =
2. 7.9392 ÷ 0.96 =
3. 654.5 ÷ 35 =
4. 172.8 ÷ 2.4 =
5. 1386.0 ÷ 1.65 =
6. 25924.64 ÷ 31.6 =
7. 331.266 ÷ 80.6 =
8. 821.7 ÷ 83 =
9. 4401.25 ÷ 503 =
10. 2585.52 ÷ 26.6 =

Calculating Dimensions

Working drawings in construction projects are typically dimensioned in feet and inches. It is often necessary to add and subtract measurements in feet and inches to calculate overall distances and determine "missing" dimensions in dimension strings. Dimensions are made up of whole numbers and fractions. They are added and subtracted in the same way as adding and subtracting whole numbers and fractions.

When adding dimensions, inch values and feet values are added separately, beginning with the inch values first.

Example: Add 12′-3″ to 9′-6″.

```
  12′-3″
+  9′-6″
  21′-9″
```

When adding inch values, convert inches to feet as needed. Every 12 inches is equal to 1 foot.

Example: Add 12′-3″ to 9′-10″.

```
  12′-3″
+ 9′-10″
  21′-13″ (Take 12″ from 13″ and convert to 1′.
           Add 1′ to 21′)
  22′-1″ (The leftover amount in inches is 1″)
```

Subtracting dimensions works in a similar manner. Calculate the value in inches first. When necessary, borrow from the foot value to subtract an inch value.

Example: Subtract 9′-1″ from 12′-3″.

```
  12′-3″
−  9′-1″
   3′-2″
```

Example: Subtract 9′-6″ from 12′-3″.

```
  12′-3″
−  9′-6″
   = ?
```

Borrow 12″ (1′) from the foot value and add to the inch value.

```
 11′-15″
-  9′-6″
 2′-9″
```

Fractions in dimensions are calculated in the same manner as other fractions. The main item to remember is that you must keep the denominators of fractions the same when adding and subtracting.

Example: Add 3′-4 1/4″ to 7′-9 1/2″.

Feet	Inches	Fraction
3	4	1/4
7	9	2/4
10	13	3/4
or 11′-1 3/4″		

Addition and subtraction operations are commonly performed to determine "missing" dimensions in dimension strings. When adding, convert inches to feet as needed. When subtracting, borrow from the foot value to subtract an inch value as needed.

Example: Find the missing dimension in the string of dimensions below.

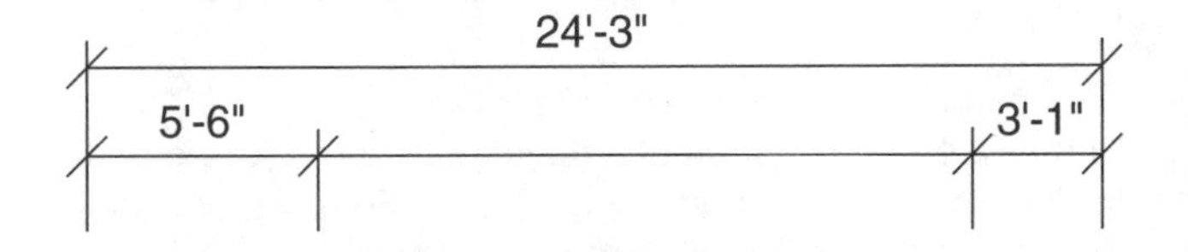

Add:

5′-6″ + 3′-1″ = 8′-7″

Subtract:

Overall dimension = 24′-3″

```
 24′-3″         23′-15″
- 8′-7″   or   - 8′-7″
                15′-8″
```

Converting Dimensions to Decimals

In some cases, it may be easier to work with decimal values when calculating dimensions. Dimensions in feet and inches can be converted into decimals by converting inches into decimal feet. This is done by dividing the inch value in the dimension by 12. The resulting decimal is added to the foot value.

Example: Convert 12′-3″ to decimal feet.

3 ÷ 12 = .25 = .25′

.25′ + 12′ = 12.25′

The inch values must first be converted to decimal feet before adding or subtracting decimal feet dimensions. Decimal feet dimensions are added or subtracted in the same manner as whole numbers and decimal fractions. The conversion table in **Figure 2-2** lists common inch values, decimal inch equivalent values, and decimal foot equivalent values. As when adding dimensions in whole inches and feet, values in decimal inches and feet are added separately.

Inch to Decimal Values				
Fractional Inch Value	**Decimal Inch Value**	**Decimal Foot Value**	**Inch Value**	**Decimal Foot Value**
1/16″	.0625″	.0052′	1″	.0833′
1/8″	.125″	.0104′	2″	.1667′
3/16″	.1875″	.0156′	3″	.2500′
1/4″	.25″	.0208′	4″	.3333′
5/16″	.3125″	.0260′	5″	.4167′
3/8″	.375″	.0313′	6″	.5000′
7/16″	.4375″	.0365′	7″	.5833′
1/2″	.5″	.0417′	8″	.6667′
9/16″	.5625″	.0469′	9″	.7500′
5/8″	.625″	.0521′	10″	.8333′
11/16″	.6875″	.0573′	11″	.9167′
3/4″	.75″	.0625′	12″	1.000′
13/16″	.8125″	.0677′		
7/8″	.875″	.0729′		
15/16″	.9375″	.0781′		

Goodheart-Willcox Publisher

Figure 2-2. Common inch values and equivalents in decimal inches and decimal feet. Decimal foot values are added separately from decimal inch values.

See the Reference Section in this textbook for a listing of common fractions, decimal fraction equivalents, and metric equivalents.

Area Measurement

Often, it is necessary to know the amount of paving, floor space, window opening, or wall space in a particular part of the project. This measurement is known as ***area***, and it is given in square units (square feet, square yards, or square meters, for example).

Square and Rectangular Areas

To compute the area of a rectangle or square, multiply the length of one side times the length of an adjacent side (length × width). The lengths must have the same units of measure. The resulting area calculation is in square units. For example, if you multiply two lengths in feet together, the area will be in square feet. If the lengths are given in inches, the area will be in square inches. If the lengths are given in feet and inches, the lengths must be converted to decimal feet and then multiplied so the answer will be in square feet.

Example: 12′-3″ × 10′-6″ would convert to 12.25′ × 10.5′.

Example: Determine the area of the room shown in **Figure 2-3A**.

$$\begin{array}{r} 12\text{ ft} \\ \times\ 10\text{ ft} \\ \hline 120\text{ ft}^2 \end{array}$$

The area of a wall is computed in the same way, except that the area of all openings (doors and windows) must be subtracted from the total.

Example: Determine the area of the wall surface shown in **Figure 2-3B**.

Total Wall Area:

$$\begin{array}{r} 20\text{ ft} \\ \times\ 8\text{ ft} \\ \hline 160\text{ ft}^2 \end{array}$$

Window Area:

$$\begin{array}{r} 5\text{ ft} \\ \times\ 4\text{ ft} \\ \hline 20\text{ ft}^2 \end{array}$$

(Total Wall Area) – (Window Area)

$$\begin{array}{r} 160\text{ ft}^2 \\ -\ 20\text{ ft}^2 \\ \hline 140\text{ ft}^2 \end{array} \text{ of wall surface}$$

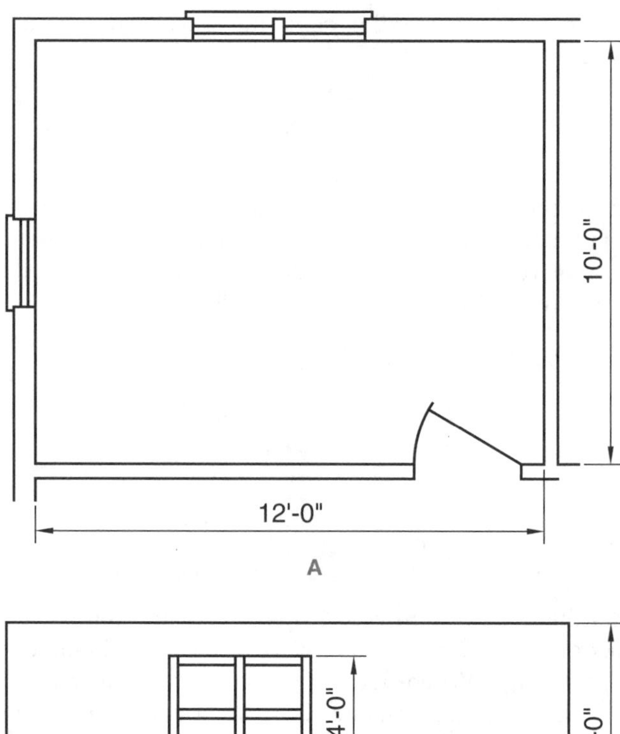

Figure 2-3. Finding area of a rectangular area. A—Formula for finding area of a floor: A = L × W. B—Formula for finding area of a wall: A = L × W – area of openings.

Triangular Areas

A triangular area can be computed by multiplying the height times the base and then dividing by two. **Figure 2-4** illustrates the formula.

Example: Compute the area of the end of the gable roof shown in **Figure 2-4**.

$$5\text{ ft} \times 24\text{ ft} = 120\text{ ft}^2$$

$$120\text{ ft}^2 \div 2 = 60\text{ ft}^2$$

Circles and Circular Areas

The characteristics of circles are shown in **Figure 2-5**. The ***circumference*** is the distance around the circle. The ***diameter*** is the length of a line running between two points on the circle and passing through the center. The ***radius*** is one-half the length of the diameter.

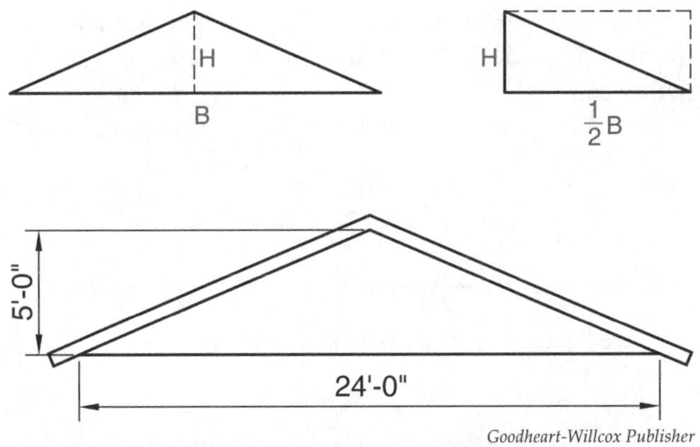

Figure 2-4. Formula for finding area of a triangular area: A = (B × H) ÷ 2.

When determining the circumference, area, or volume of a circular object, the number π (pi) is used in the formula. ***Pi*** is the ratio of circumference to diameter, and is equal to 3.1416.

Circumference of a circle = π × d

Example: Determine the circumference of a circle that has a 6′-6″ diameter.

$$3.1416 \times 6.5' = 20.42'$$

A portion of a circle is called an ***arc***. Lengths along an arc can be calculated by determining the circumference and multiplying the circumference by the degree percentage of the circle.

Arc length = 3.1416 × d; divide by 360/number of degrees in arc

Example: Determine the length of an arc that has a radius of 4′-3″ and has an arc of 90°.

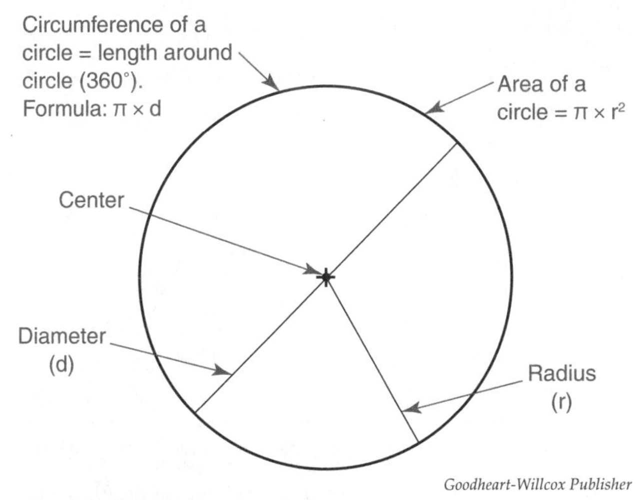

Figure 2-5. To determine area and circumference of a circle (or portions of a circle), you must be familiar with the properties of the circle, such as radius and diameter.

First, determine the circumference of a complete circle:

$$3.1416 \times (4.25' \times 2) = 26.70'$$

Next, determine the portion of the arc:

$$360°/90° = 4 \text{ or } 25\%$$

Finally, divide the circumference by the portion of a circle:

$$26.7'/4 = 6.68' \text{ or } 6'\text{-}8''$$

The area of a circle can be found by multiplying π times the radius squared.

Area of a circle = π × r²

Example: Determine the area in square feet of the circular patio shown in **Figure 2-6**.

Patio diameter = 30′, radius = 15′

$$3.1416 \times 15^2 = \text{Area}$$

$$3.1416 \times 225 \text{ ft}^2 = \text{Area}$$

$$\begin{array}{r} 3.1416 \\ \times \quad 225 \text{ ft}^2 \\ \hline 706.86 \text{ ft}^2 \text{ patio} \end{array}$$

A portion of an area of a circle can be figured the same way by computing the area of the circle and then multiplying it by the portion of the complete circle.

Volume Measurement

Volume is a cubic measure. It is found by multiplying area by depth. The following example will compute the volume of ready-mix concrete required for a 4″ slab for the patio in **Figure 2-6**. When calculating volume, you must be certain that all of the numbers being multiplied together have the same units. Since the area

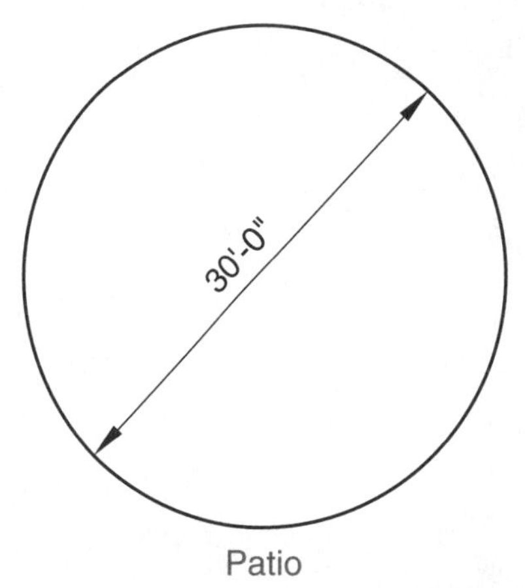

Figure 2-6. Formula for finding area of a circle: A = π × r².

of the patio is known to be 706.86 ft^2, the 4″ depth will be changed to 0.333′.

$$\begin{array}{r} 706.86\ \text{ft}^2\ \text{(area of patio)} \\ \times\ 0.333\ \text{ft} \\ \hline 235.38\ \text{ft}^3\ \text{in patio slab} \end{array}$$

Since concrete is sold by the cubic yard, it is necessary to change the cubic feet to cubic yards. There are 27 cubic feet in a cubic yard, so divide the number of cubic feet by 27:

$$\begin{array}{r} 8.71 = 8.71\ \text{cubic yards (cy)} \\ 27\overline{)235.38} \\ -\ 216 \\ \hline 193 \\ -\ 189 \\ \hline 48 \\ -\ 27 \\ \hline 21 \end{array}$$

To allow for enough concrete for the patio slab, the computed volume is rounded up to the next largest cubic yard. This is done to account for waste and helps allow for enough material. In this example, the calculation is rounded up to 9 cy.

Area and volume calculations for metric measurements are made in the same manner using the appropriate units.

Example:

Patio diameter = 10 meters

Thickness = 10 centimeters

Determine area of patio:

$$\begin{array}{r} 3.1416 \\ \times\ 25\ \text{m}^2\ \text{(radius squared)} \\ \hline 78.54\ \text{m}^2\ \text{(area of patio)} \end{array}$$

Convert thickness to meters:

10 cm = 0.1 m

Multiply area by thickness:

$$\begin{array}{r} 78.54\ \text{m}^2 \\ \times\ 0.1\ \text{m} \\ \hline 7.854\ \text{m}^3\ \text{(volume of patio)} \end{array}$$

Note that fewer calculations are required with metric units.

Test Your Knowledge

Name ________________________________

Write your answers in the spaces provided. Reduce all answers to the lowest terms.

____ 1. Add: $\frac{3}{8} + \frac{3}{8}$

A. $\frac{6}{64}$

B. 0

C. 1

D. $\frac{3}{4}$

E. None of the above.

____ 2. Subtract: $\frac{13}{16} - \frac{3}{8}$

A. $\frac{7}{16}$

B. $\frac{10}{8}$

C. $1\frac{2}{3}$

D. $\frac{19}{16}$

E. None of the above.

____ 3. Multiply: $\frac{2}{3} \times \frac{9}{16}$

A. $\frac{1}{3}$

B. $\frac{11}{48}$

C. $\frac{11}{19}$

D. $\frac{3}{8}$

E. None of the above.

____ 4. Divide: $\frac{4}{9} \div 3$

A. $1\frac{1}{3}$

B. $\frac{4}{27}$

C. $\frac{7}{27}$

D. $\frac{1}{9}$

E. None of the above.

____ 5. Add: $3\frac{3}{4} + 2\frac{1}{3}$

A. $5\frac{4}{7}$

B. $5\frac{3}{12}$

C. $6\frac{1}{12}$

D. $5\frac{1}{4}$

E. None of the above.

____ 6. Add: 23.98 + 1.123 + 4.003

A. 29.403

B. 7.524

C. 75.24

D. 29.106

E. None of the above.

____ 7. Subtract: 23.837 – 4.77

A. 23.360

B. 19.067

C. 19.063

D. 233.60

E. None of the above.

____ 8. Multiply: 1.23 × 2

A. 2.23

B. 24.6

C. 246

D. 2.46

E. None of the above.

____ 9. Divide: 12.64 ÷ 4

A. 316

B. 50.56

C. 16.64

D. 3.16

E. None of the above.

____ 10. Divide: 88.56 ÷ 2.4

A. 36.9

B. 212.544

C. 90.76

D. 3.69

E. None of the above.

Activity 2-1

Problems in Construction Mathematics

Name ______________________________

Solve the following problems. Show your work. Write your answers in the spaces provided.

1. A triangular frame has sides that measure 15′-7″, 20′-4″, and 26′-2″. What is the total length of the three sides?

   1. ______________________________

2. A carpenter had a board 2′-10 3/4″ long. To fit the space for a shelf, the carpenter cut 7/16″ off one end. How long was the board after the piece was removed?

   2. ______________________________

3. Fifteen strips, 1 1/4″ wide, are to be ripped from a sheet of plywood. If 1/8″ is lost with each cut, how much of the plywood sheet is used in making the 15 strips? (Assume 15 cuts are necessary.)

   3. ______________________________

4. An interior wall of a building is made up of metal studs with 5/8″ gypsum board on each side. See the drawing below. If the actual width of a stud is 3 5/8″, what is the total thickness of the wall?

   4. ______________________________

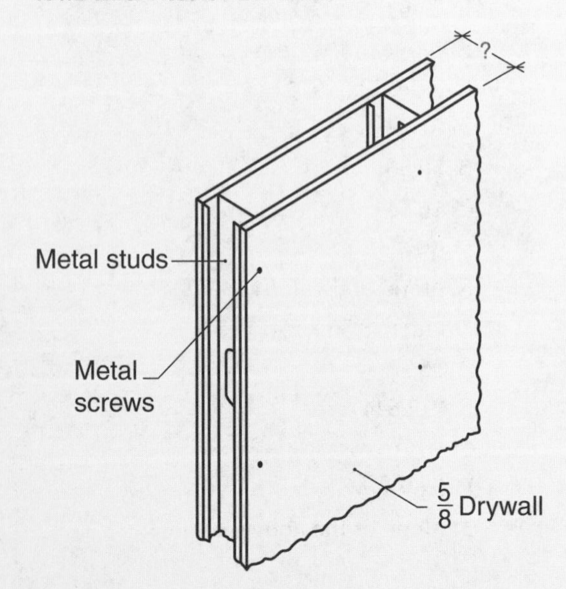

5. A carpenter worked 10 weeks on a particular job, 5 1/2 days per week and 7 3/4 hours per day. How many hours did the carpenter work on the job?

5. ______________________

6. What is the area of a rectangular floor that is 7′-3″ long and 4′-2″ wide?

6. ______________________

7. There are 15 risers in a set of stairs running from the basement to the first floor. Each riser is 7 1/4″ high. What is the distance between floors?

7. ______________________

8. The distance between two floors is 9′-0 1/2″. If 14 risers are to be used in a set of stairs, what is the height of each riser?

8. ______________________

9. How many 1 × 2 shelf cleats 8″ long can be cut from a 1 × 2 board 16′ long? How much is left? (Disregard waste in saw cut.)

9. ______________________

10. A contractor removed 357 cubic yards of earth from a building site. If the contractor's trucks can haul 17 cubic yards per load, how many truckloads of earth were moved?

10. ______________________

11. Figure the amount of concrete (in cubic yards) required to pour a floor slab of the following dimensions: 18′ × 24′ × 4″.

11. 

12. How many cubic yards of concrete are needed to pour a slab 60′ × 80′ × 5″ thick?

12. 

13. How many gallons of paint are required to paint one side of a block wall of the following dimensions: 6′ × 172′? The paint being used will cover 200 ft^2 per gallon.

13. 

14. Compute the gallons of sealer required to seal a floor 120′ × 250′. The sealer being used will cover 175 ft^2 per gallon.

14. 

15. How much concrete is required to pour the slab shown below?

15. 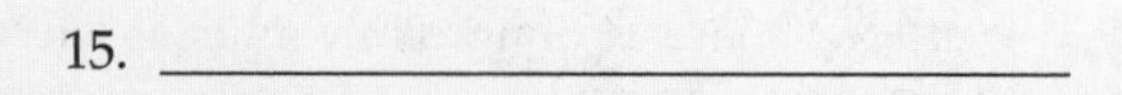

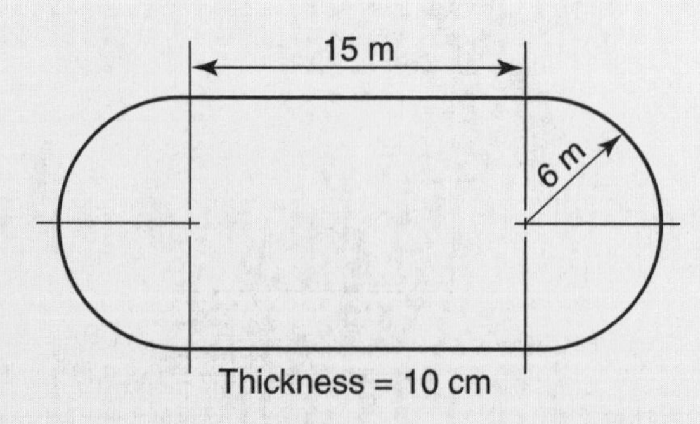

16. In a room 12′-4″ by 13′-9″ with 8′-0″ high ceilings, there is one door 3′ × 7′ and two windows 4′ square. Answer the following questions. Assume there is no waste in materials, and round off your answer to the next largest unit. Show both exact and rounded figures.

A. How much wall surface in square feet is there to paint?

B. What is the floor area?

C. How many gallons of paint will be required to cover the walls with 2 coats of paint if each gallon covers 200 square feet of surface?

D. How many lineal feet of cove base floor trim will be required?

16. A. ____________

B. ____________

C. ____________

D. ____________

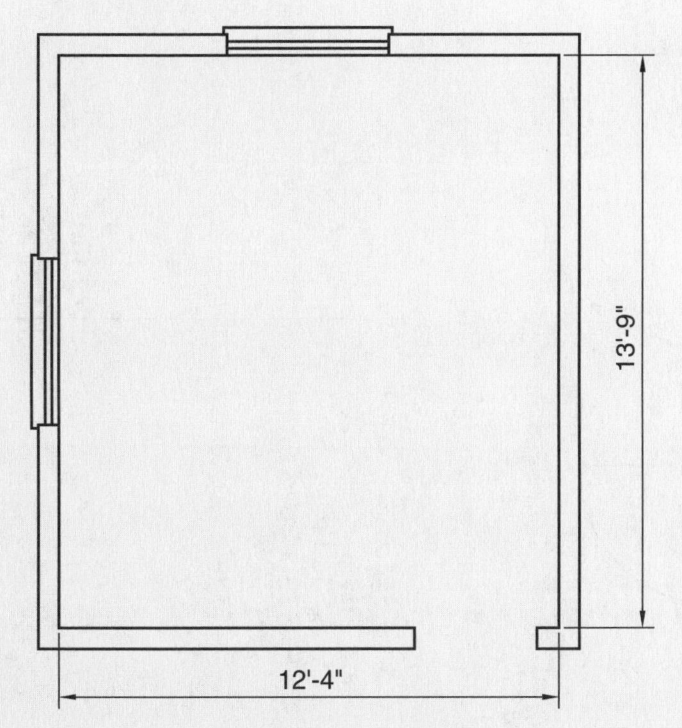

17. If a concrete column is 20″ in diameter and 18′-3″ in height, how many cubic yards of concrete would be needed to fill the form? Round off to the next largest cubic yard.

17. 

18. What is the circumference of the following shape?

18. ____________

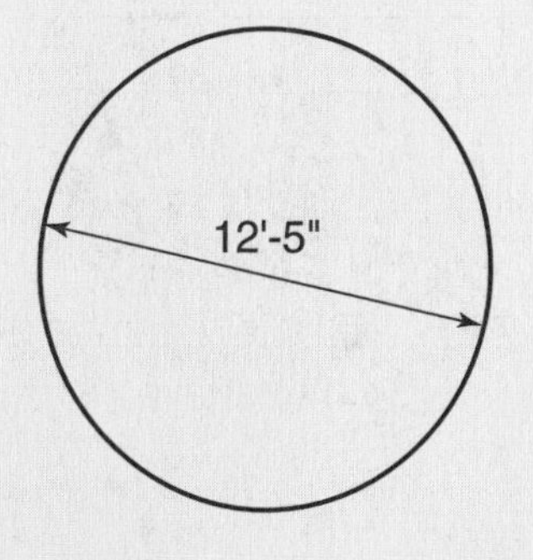

19. What is the total lineal distance, in feet, along the curb shown? Round off to the next largest foot.

19. 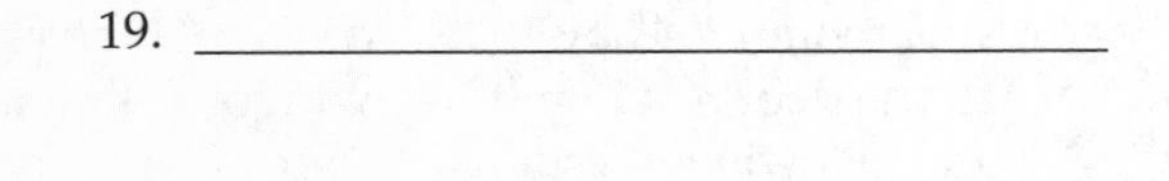

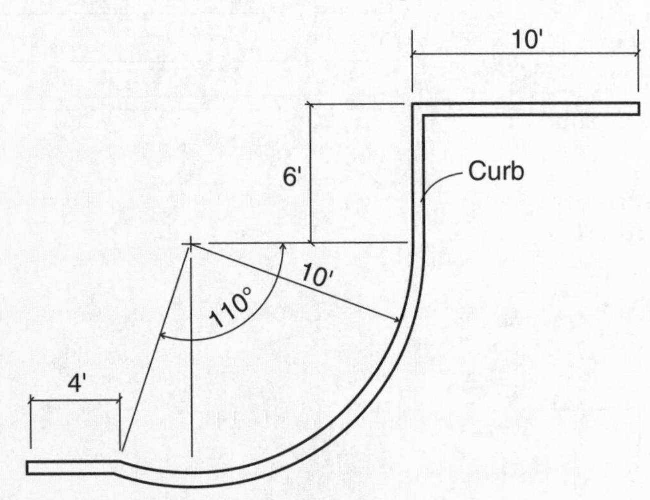

20. What is the area in square feet of the sidewalk shown below? Round off to the next largest foot.

20. 

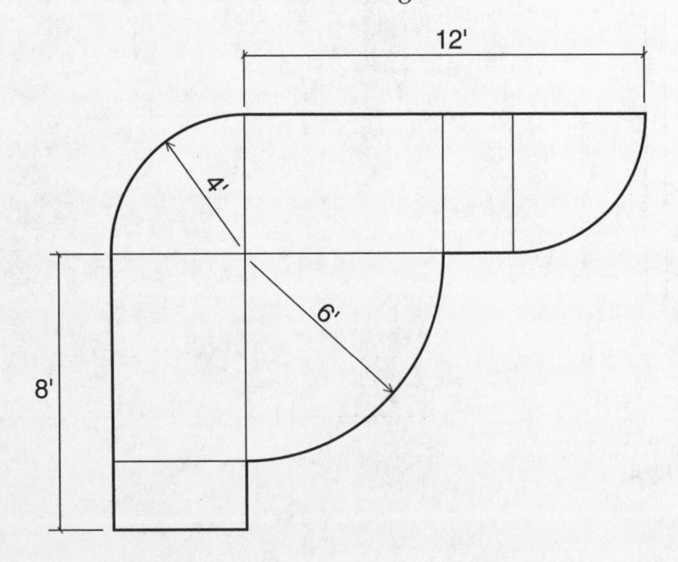

21. Find the missing dimension in the dimension strings below.

21. A. 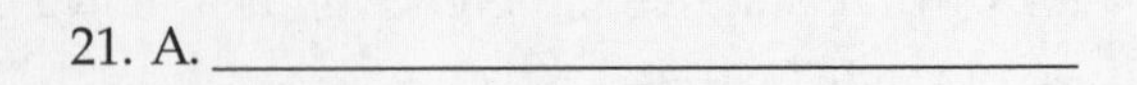

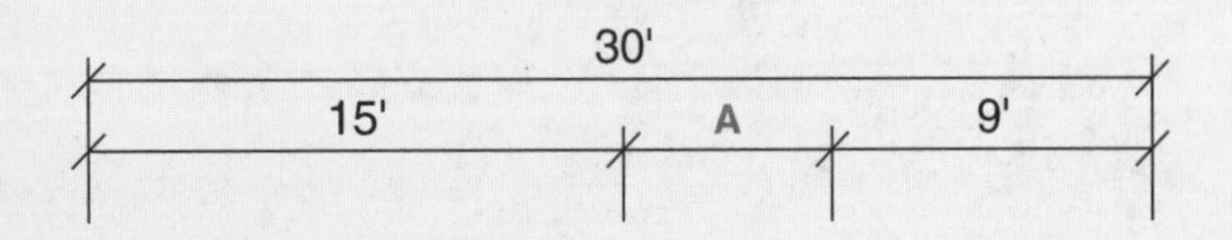

22. Find the missing dimension in the dimension strings below.

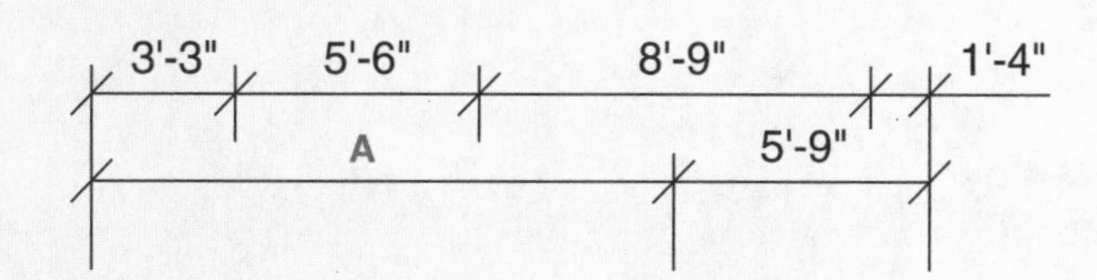

22. A. ____________________

23. Find the missing dimensions in the dimension strings below.

75'-8"
A
56'-3"
17'-8"
B
30'-10"
C

23. A. ____________________
B. ____________________
C. ____________________

24. Find the missing dimensions in the dimension strings below.

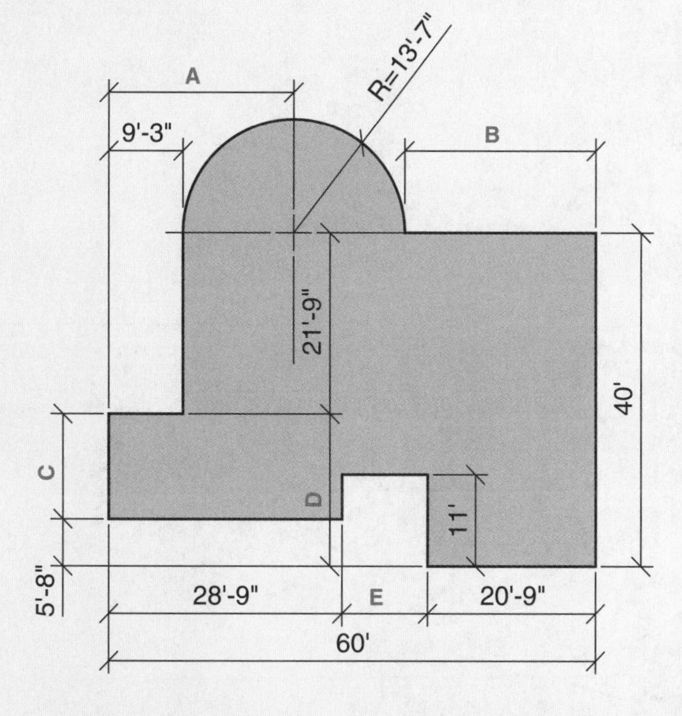

24. A. ____________________
B. ____________________
C. ____________________
D. ____________________
E. ____________________

25. Find the area of the partial circle.

25. ______________________

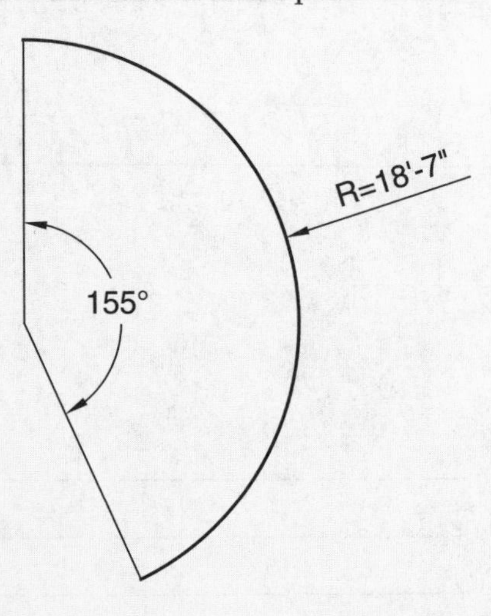

26. Find the area of the trapezoid.

26. ______________________

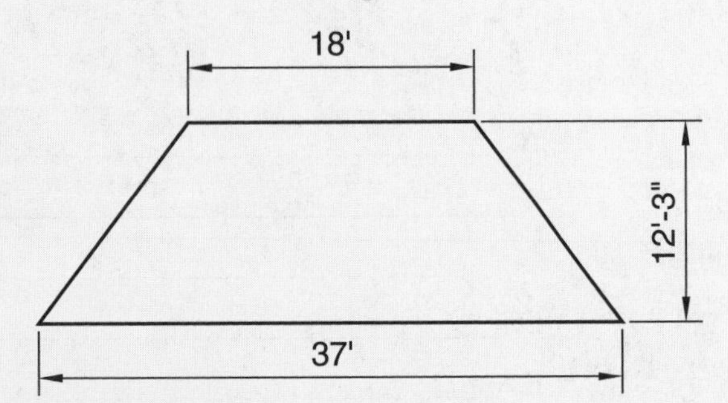

27. Find the volume of the triangle.

27. ______________________

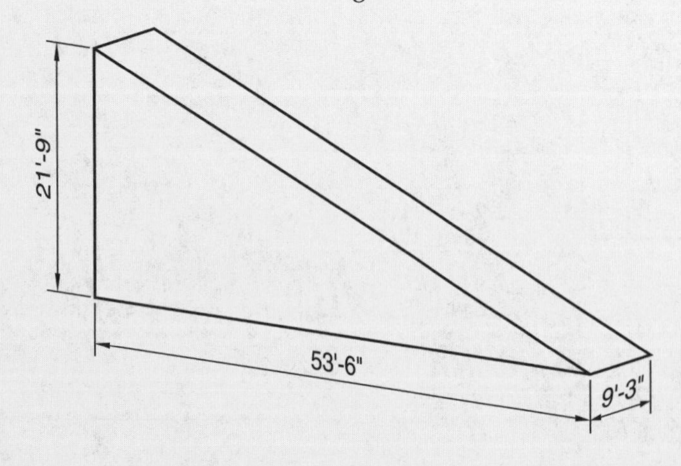

UNIT 3
Reading Measuring Tools and Using Scales

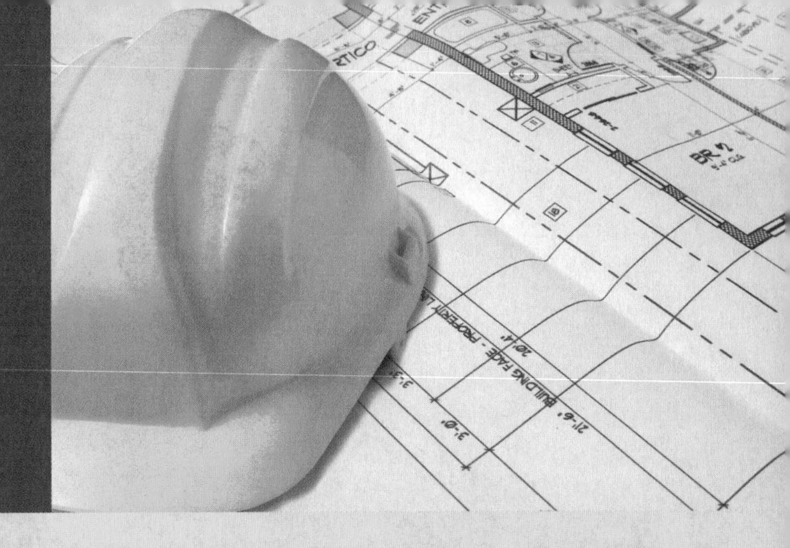

TECHNICAL TERMS

architect's scale
centimeter (cm)
engineer's scale
fractional rule
full-divided scale
meter (m)
metric rule
metric scale
millimeter (mm)
open-divided scale
scale

LEARNING OBJECTIVES

After completing this unit, you will be able to:

- Read both customary (English) and metric rules and tapes.
- Convert between customary and metric units.
- Identify the drawing scale of a print.
- Read dimensions on a print.
- Make measurements using an architect's scale.
- Make measurements using an engineer's scale.

Many different types of measuring tools are used in the construction industry. These include framing squares, bench rules, steel rules, and tapes, **Figure 3-1**. This unit is intended to help you review the methods of reading these measuring tools.

In the customary (also called "English") measurement system, the distances are divided into feet, inches, and fractional parts of an inch. The rule used with this system is called the ***fractional rule***. In the metric system, the divisions are in meters, centimeters, and millimeters. The rule used with this system is called the ***metric rule***.

This unit also introduces how to use scales to read dimensions on prints. You will learn how to identify the drawing scale used on a print. You will also learn how to read and lay out measurements using different types of scales.

Fractional Rule

Measurements in residential construction and most commercial construction seldom must be more accurate than one-sixteenth of an inch. A fractional rule that is divided into 16ths will work for most building construction. On the rule shown in **Figure 3-2**, the inch is divided into 16 parts. Thus, each small division is 1/16th of an inch. Notice that each of the marks shown in **Figure 3-2** has an indicator ("tick mark") of a different length. To read the fractional rule, follow these steps:

1. Study the major divisions of the inch. These represent 1/4″ (4/16), 1/2″ (8/16), and 3/4″ (12/16).
2. Note that there are four small divisions within each 1/4″ division. Each of these divisions represents 1/16″. Two of these divisions equal 1/8″ (same as 2/16).

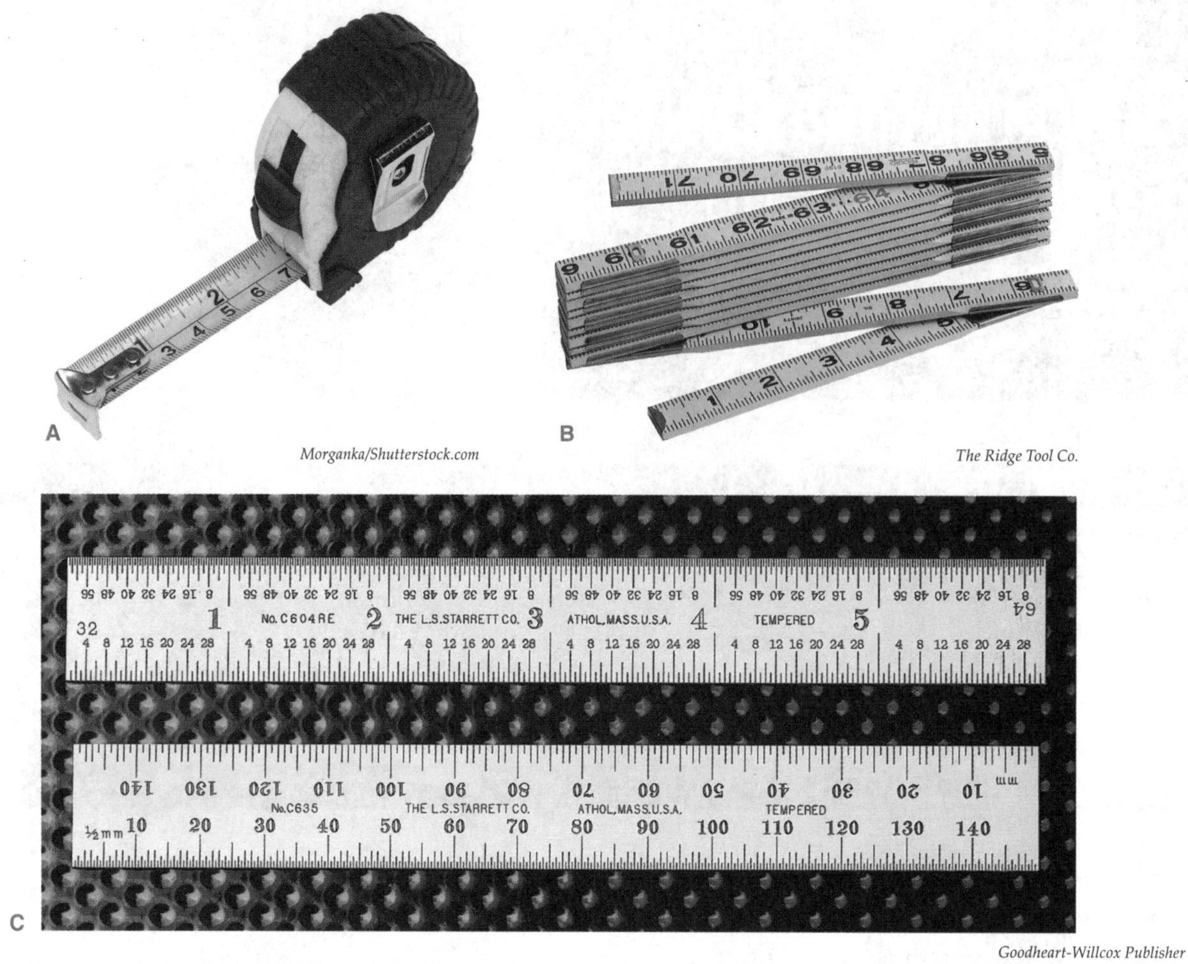

Morganka/Shutterstock.com

The Ridge Tool Co.

Goodheart-Willcox Publisher

Figure 3-1. Some measuring tools used in construction. A—Measuring tape. B—Folding rule. C—Steel rules. The 6″ fractional rule shown at the top is divided into 1/64″ divisions along the top and 1/32″ divisions along the bottom. The metric rule shown at the bottom is divided into millimeter divisions along the top and half-millimeter divisions along the bottom.

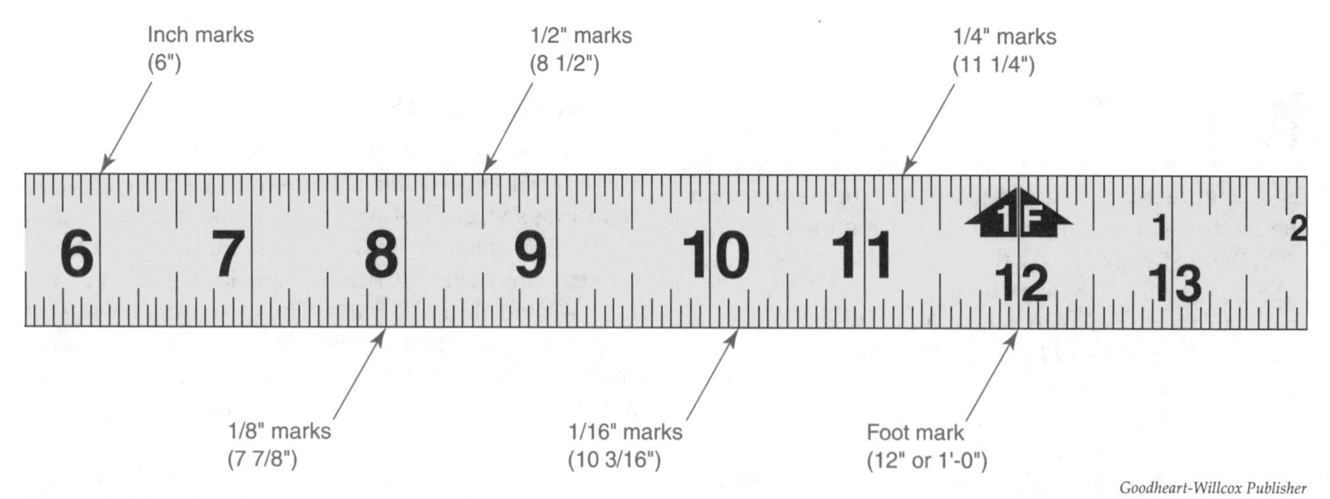

Goodheart-Willcox Publisher

Figure 3-2. Portion of a tape measure used to measure lengths and distances in feet, inches, and fractions of an inch. Shown are the 1″, 1/2″, 1/4″, 1/8″, 1/16″, and 1′ marks. Notice that each has an indicator ("tick mark") of a different length. There are 16 tick marks in each inch; the shortest mark indicates 1/16″ of an inch division.

3. Further study and application of fractional parts will enable you to locate any common fraction that is a multiple of 16ths. For example:

$$\frac{4}{16}=\frac{1}{4} \quad \frac{8}{16}=\frac{1}{2} \quad \frac{10}{16}=\frac{5}{8}$$

Metric Rule

The basic unit of linear measure in the metric system is the ***meter (m)***. Other linear units are either fractions or multiples of a meter. The names of the other units consist of a prefix followed by *meter*. The most common units are the following:

Unit	Abbreviation	Equal to
Millimeter	mm	1/1000 m
Centimeter	cm	1/100 m
Kilometer	km	1000 m

The ***millimeter (mm)*** is normally used for metric drawings in the construction field. The millimeter is equal to 1/1000th of a meter and 1/10th of a ***centimeter (cm)***. Metric dimensions are better to work with because they can be added and subtracted more easily than English units. However, the customary system is used almost exclusively in the United States, because materials are manufactured to customary dimensions. A comparison of the customary and metric systems will help develop an understanding of equivalent values and conversions between the two systems.

1 inch = 25.4 millimeters
12 inches = 304.8 millimeters
1 yard (36 inches) = 914.4 millimeters
39.37 inches = 1000 millimeters (1 meter)

Figure 3-3 shows the relationship between 1″ and the equivalent measurement in millimeters.

To convert inches to millimeters, multiply by 25.4:

Example: Convert 14″ to millimeters.

Solution: 14″ × 25.4 = 355.6 mm

If the measurement is in both feet and inches, first convert the feet to inches, then convert the inches to millimeters. The millimeters can then be converted to meters by dividing by 1000:

Example: Convert 11′-5″ to metric.

Solution: (11′ × 12) + 5″ = 132″ + 5″ = 137″

137″ × 25.4 = 3479.8 mm

$$\frac{3479.8 \text{ mm}}{1000} = 3.48 \text{ m}$$

To convert millimeters to inches, divide by 25.4:

Example: Convert 225 mm to inches.

Solution: 225 ÷ 25.4 = 8.86″ = 8 7/8″

When using other metric units, first convert to millimeters, then to inches, and then to feet:

Example: Convert 23.45 m to customary units.

Solution: 23.45 m = 23450 mm
23450 mm ÷ 25.4 = 923.23″
912″ = 76′
923.23″ – 912″ = 11.23″
923.23″ = 76′-11 1/4″

Some building material sizes are given in millimeters or centimeters. For example, the nominal size of metric brick (including joint thickness) is 225 mm × 112.5 mm × 75 mm. The standard metric modular sheet size is 120 cm × 240 cm. Metric measurements in cabinet work are usually given in centimeters.

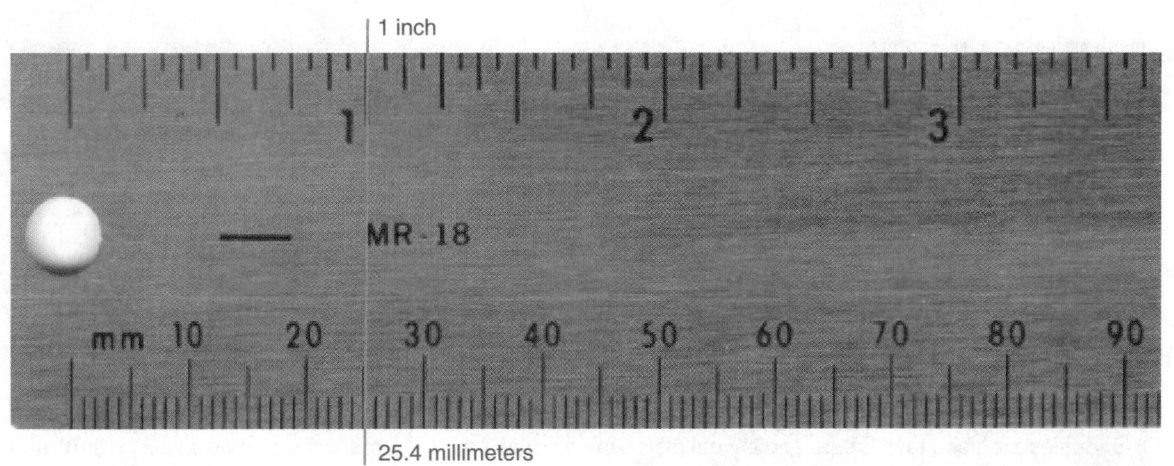

Figure 3-3. Relationship between 1″ and 25.4 mm.

GREEN BUILDING

Using Care with Tools

Best practice is green practice when selecting and using hand and power tools. Select tools with low impact on the environment. Maintain and handle all tools properly. Energy-efficient tools used carelessly create waste by damaging materials and the tools themselves.

In time, building products manufactured to standard metric sizes will become increasingly popular. In addition, many industrial machine installations and parts in global industries are designed to metric sizes. Therefore, the ability to work with metric measurements will be required.

The smallest division on the metric framing square is 1/2 cm. On tapes, the smallest division is the millimeter. To read the metric rule, follow these steps:

1. Study the major divisions marked 1, 2, 3, etc. Each of these represents 1 cm.
2. Each centimeter is divided into 10 millimeters.
3. A meter is 1000 millimeters.
4. For a reading of 3.4 meters (3400 mm), follow the rule to three meters (3 m) and then on to 400 mm (0.4 m), **Figure 3-4.**

Using Scales

The term ***scale*** refers to the relative size at which a drawing has been made. In addition, the term *scale* also refers to the instrument (ruler) used to measure distances on a scaled drawing. The following sections explain how to identify the drawing scale and how to use common scales (instruments) to make measurements.

Identifying the Drawing Scale

Construction drawings are made to a *reduced scale* (smaller than actual size). The scale of a particular floor plan, elevation, or detail is indicated on the sheet, either in the title block or beneath the drawing itself. A note such as *SCALE: 1/4″ = 1′-0″*, for example, identifies the drawing scale for the plan. The drawing scale will determine which scale the user of the plan will have to use to retrieve dimensional information.

Most construction drawings in the United States are made using an ***architect's scale*** or an ***engineer's scale***. See **Figure 3-5**. In most other countries, a metric scale is used. On a drawing made using an architect's scale, the drawing scale is defined so that some fraction of an inch or a whole number quantity in inches equals one foot. For example, typical architect's scales include 1/8″ = 1′-0″, 1/4″ = 1′-0″, and 3/4″ = 1′-0″. These scales mean that every 1/8″, 1/4″, or 3/4″ length on the drawing is equal to one foot in reality. See **Figure 3-6**.

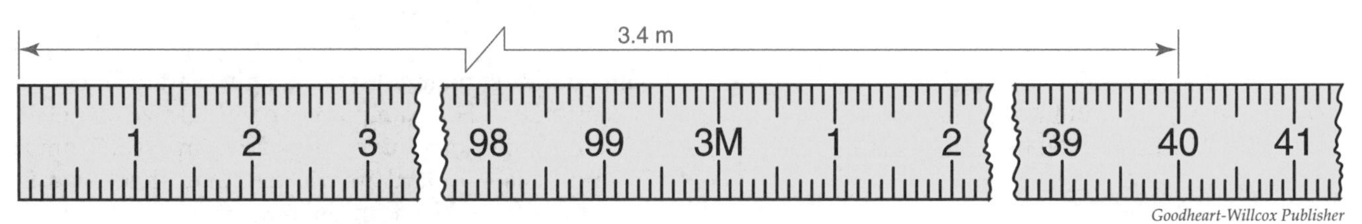

Goodheart-Willcox Publisher

Figure 3-4. A measure of 3.4 meters is shown.

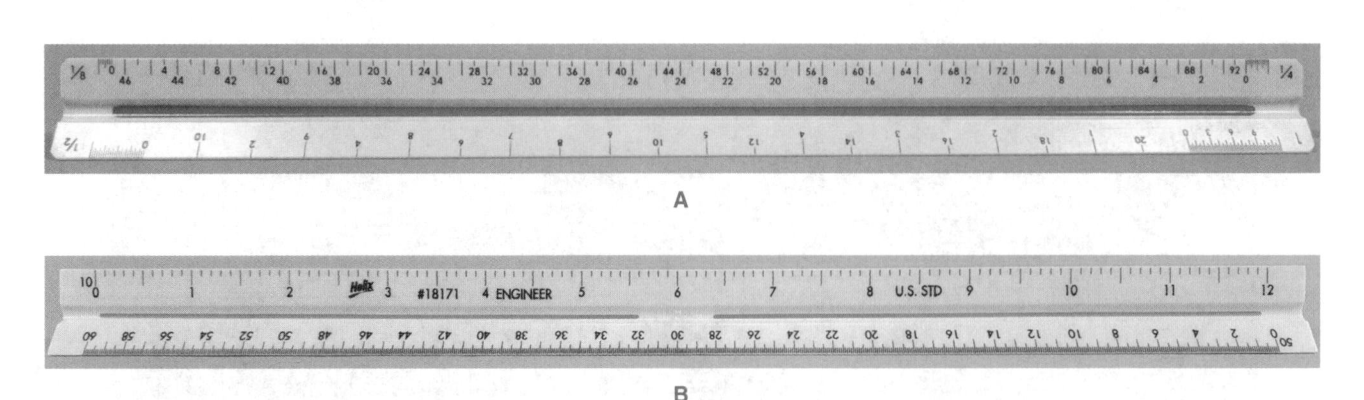

Goodheart-Willcox Publisher

Figure 3-5. Scales are used to measure and lay out distances on drawings. A—Architect's scale with 1/8″ = 1′-0″ and 1/4″ = 1′-0″ scales on the upper face and 1/2″ = 1′-0″ and 1″ = 1′-0″ scales on the lower face. B—Engineer's scale with 1″ = 10′ and 1″ = 50′ scales shown.

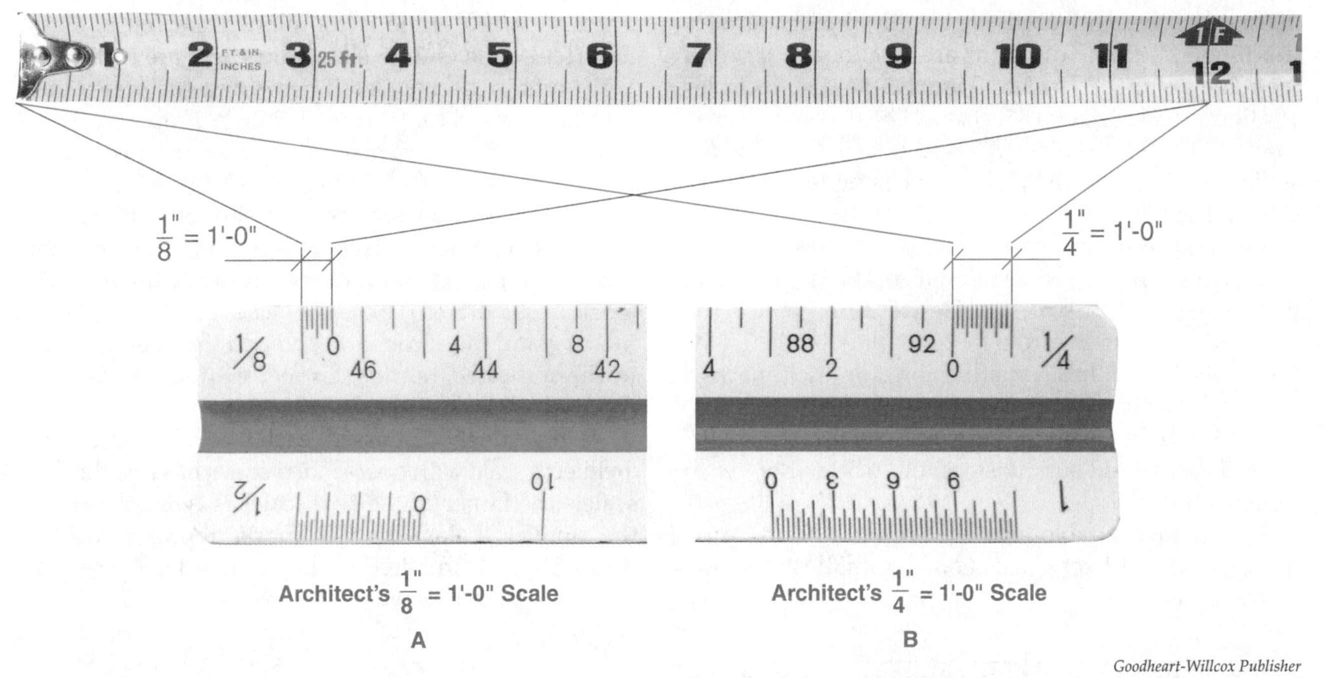

Goodheart-Willcox Publisher

Figure 3-6. Architect's scales. The tape measure is shown for reference. A—On an architect's 1/8″ = 1′-0″ scale, a 1/8″ measurement represents 1′ in actual size. B—On an architect's 1/4″ = 1′-0″ scale, a 1/4″ measurement represents 1′ in actual size.

On a drawing made using an engineer's scale, the drawing scale is defined so that every inch measurement on the plan is equal to a measurement in the designated units (typically feet). For example, typical engineer's scales include 1″ = 10′, 1″ = 20′, and 1″ = 40′. These scales mean that every 1″ length on the drawing is equal to 10′, 20′, or 40′ in reality. See **Figure 3-7**.

The following sections explain how to use scales to read and lay out measurements. Study the procedures closely as you practice using each type of scale.

Using an Architect's Scale

An architect's scale is commonly used to make drawings for buildings, such as floor plans, elevations, and details. An architect's scale can be used to measure distances on a drawing by matching the drawing scale to the appropriate scale listed on the instrument.

Architect's scales are available in both three-sided (triangular) forms, as shown in **Figure 3-5A**, and flat forms, **Figure 3-8**. The architect's scale in **Figure 3-5A** has six faces (two faces on each side) and a total of 11 different scales. On five of the faces, there are two scales reading from opposite ends. The larger of the two scales is twice the size of the smaller scale. Each of these scales is referred to as an ***open-divided scale***. On these scales, the main units are foot divisions numbered along the entire length of the scale. The

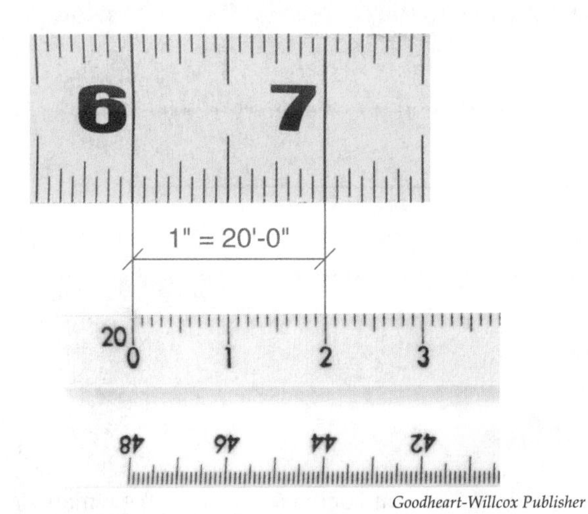

Goodheart-Willcox Publisher

Figure 3-7. An engineer's 20 scale. On this scale, a measurement of 1″ represents 20′ in actual size.

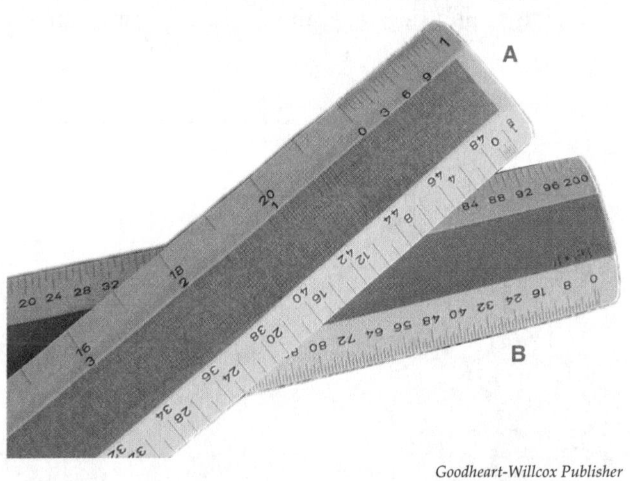

Goodheart-Willcox Publisher

Figure 3-8. Flat architect's scales. A—Architect's scale with 1/8″ = 1′-0″ and 1″ = 1′-0″ scales shown. B—Architect's scale with 1/32″ = 1′-0″ scale shown.

main units are not subdivided. Only the end unit is subdivided and the divisions are fractional parts of a foot. The architect's scale in **Figure 3-5A** also includes a full-size scale on one of its faces. The full-size scale is labeled "16." This scale is referred to as a ***full-divided scale*** because all of the main units along the scale are subdivided.

The end unit on the open-divided scales is subdivided into inches or fractional parts of an inch, depending on the scale. In **Figure 3-6A**, notice the 1′-0″ end unit starting from the zero mark on the 1/8″ = 1′-0″ scale. There are 6 subdivisions, or "tick marks," each representing 2″. In **Figure 3-6B**, notice the 1′-0″ end unit starting from the zero mark on the 1/4″ = 1′-0″ scale. There are 12 subdivisions, or "tick marks," each representing 1″.

The following table lists typical scales found on a three-sided architect's scale and the smallest division on each scale:

Scale	Smallest Tick Mark =
3/32″ = 1′-0″	2″
3/16″ = 1′-0″	1″
1/8″ = 1′-0″	2″
1/4″ = 1′-0″	1″
3/8″ = 1′-0″	1″
1/2″ = 1′-0″	1/2″
3/4″ = 1′-0″	1/2″
1″ = 1′-0″	1/4″
1 1/2″ = 1′-0″	1/4″
3″ = 1′-0″	1/8″

The scales most commonly used for floor plans in the customary (English) measurement system are the 1/8″ = 1′-0″ and 1/4″ = 1′-0″ scales. The 1/8″ = 1′-0″ scale is commonly referred to as the "eighth" scale or 1/96 size scale (there are ninety-six 1/8″ units in 1′). The 1/4″ = 1′-0″ scale is commonly referred to as the "quarter" scale or 1/48 size scale (there are forty-eight 1/4″ units in 1′). For detail drawings, the scales most commonly used range from 1/2″ = 1′-0″ up to full size.

When measuring a distance with an architect's scale, use the method shown in **Figure 3-9**. Place the scale on the drawing so that one end of the line extends past the zero mark (0) into the area with small graduations (the inches area). Next, align the other end of the line with the nearest foot mark. Note the inches and fractions of an inch beyond the zero mark, and add that measurement to the indicated number of feet to find the distance represented by the line.

As previously discussed, each face on a three-sided architect's scale with open-ended scales has two different scales (the larger of the two scales is twice the size of the smaller scale). The tick marks representing foot divisions are numbered to the right and left. The scale selected determines which tick marks to use. The longer tick marks correspond to the larger scale. The shorter tick marks correspond to the smaller scale. For each scale, the tick marks representing foot divisions match the length of the zero tick mark at the end of the scale. When reading the larger scale, only the longer tick marks are used. When reading the smaller scale, both sets of tick marks are used. It is important to note that not every tick mark is labeled. When making measurements, be sure to use the appropriate tick marks on the scale you are reading to obtain the correct measurement.

Referring to **Figure 3-9**, the longer tick marks on the top face of the scale correspond to the 3/4″ = 1′-0″ scale. The foot divisions are numbered left to right. The shorter tick marks on the top face correspond to the 3/8″ = 1′-0″ scale, which is half the size of the 3/4″ = 1′-0″ scale and is read from the right end (not shown). The foot divisions for the 3/8″ = 1′-0″ scale are numbered right to left (they are labeled 16, 18, 20, and so on). When measuring on this scale, the longer tick marks are read along with the shorter tick marks. Read in succession, they represent 1′-0″ increments.

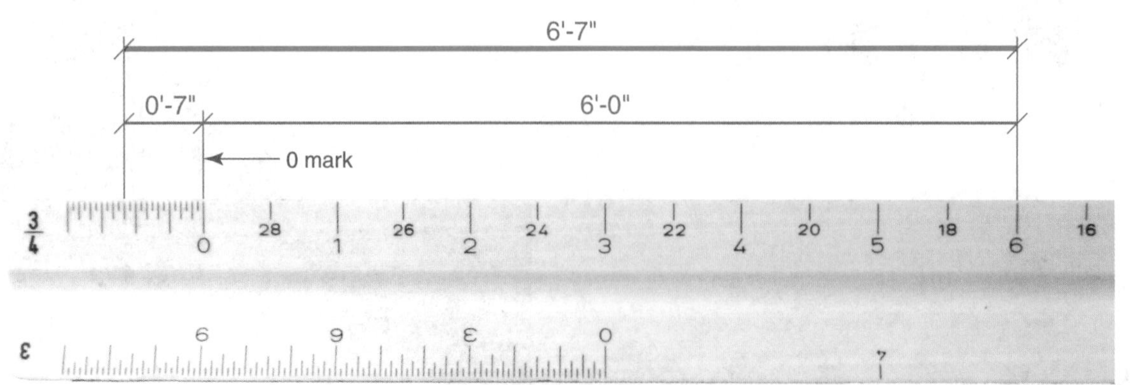

Goodheart-Willcox Publisher

Figure 3-9. Using an architect's scale to measure a line on a drawing. Place one end of the line beyond the 0 mark (in the inches area), then align the other end of the line with the nearest foot mark (in this case, 6′). Determine the number of inches (7″) and add it to the number of feet. The line measures 6′-7″.

> **Note**
>
> It takes practice to become comfortable with reading a scale. Find someone who has had some experience using scales and take turns reading measurements on scaled drawings. Measure lines that are already dimensioned. Then, try drawing some lines on graph paper and measure them for additional practice.

Using an Engineer's Scale

The engineer's scale is typically used on civil drawings, such as site plans, maps, and other drawings depicting anything outside the building. Engineer's scales are referred to in whole numbers and are related to so many feet per inch. For example, a "20 scale" would be noted as 1″ = 20′. When using this scale, every inch on the drawing represents 20 feet.

Notice that engineering scales are much smaller and provide less detail than those used for building drawings. The purpose of the engineer's scale is to be able to lay out larger areas of a project and get the project on one drawing.

Engineer's scales are available in both three-sided forms, as shown in **Figure 3-5B**, and flat forms. Typical scales found on a three-sided engineer's scale include the following:

- 1″ = 10′ (also can represent 100′, 1000′, or even 10,000′)
- 1″ = 20′
- 1″ = 30′
- 1″ = 40′
- 1″ = 50′
- 1″ = 60′

Figure 3-10 shows the measurement of a line using an engineer's scale of 1″ = 10′.

Notice on the 1″ = 10′ scale that each major division is subdivided into 10 parts. The major divisions are whole numbers representing multiples of 10. The other scales found on an engineer's scale are divided in similar fashion. Measurements in feet can be quickly made by adding a zero to the whole number division. See **Figure 3-11**. For example, on the 1″ = 10′ scale, a measurement from the zero mark to the 3 mark represents a distance of 30′. On the 1″ = 20′ scale, a measurement from the zero mark to the 10 mark represents a distance of 100′. If the measurement extends to three additional tick marks, the remainder is in feet and the total measurement is 103′.

Because engineer's scales have major divisions that represent multiples of 10, the units can be expanded to represent any proportional number. For example, the 1″ measurement shown in **Figure 3-10** can represent 10′, 100′, or 1000′.

Metric Scale

The metric system of measurement has seen little use in the United States, primarily because metric construction standards have not been established. Industry groups are at work developing metric standards. Once metric standards have been adopted and metric modular materials become available, metric dimensioning will be used extensively.

The ***metric scale*** most closely representing the customary quarter-inch scale (1/48 size) is the 1:50 scale (1/50 size), in which a two-centimeter length on the drawing equals a one-meter (100 cm) length on the actual object.

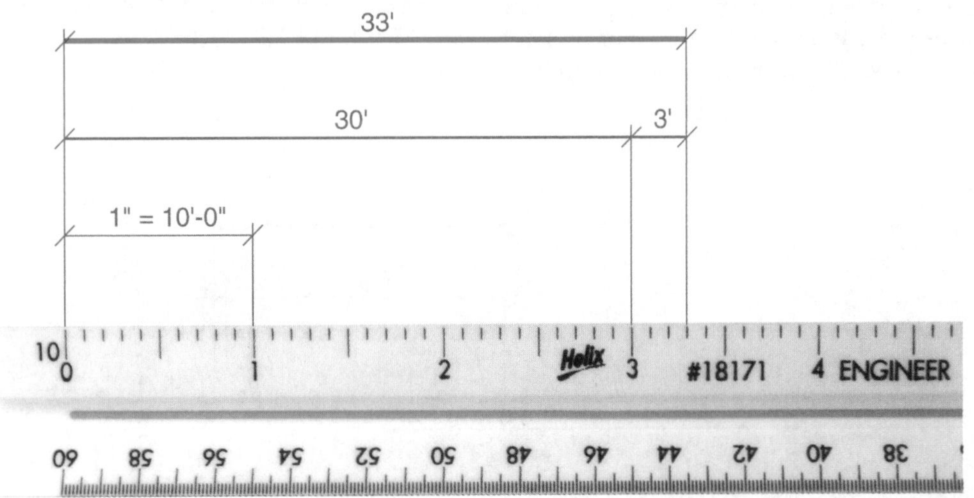

Goodheart-Willcox Publisher

Figure 3-10. Using an engineer's scale to measure a line on a drawing. The "10" at the left end of the scale indicates the number of feet per inch. Each inch is divided into 10 increments as well. Thus, the line measures 3 inches plus three increments, or 33′.

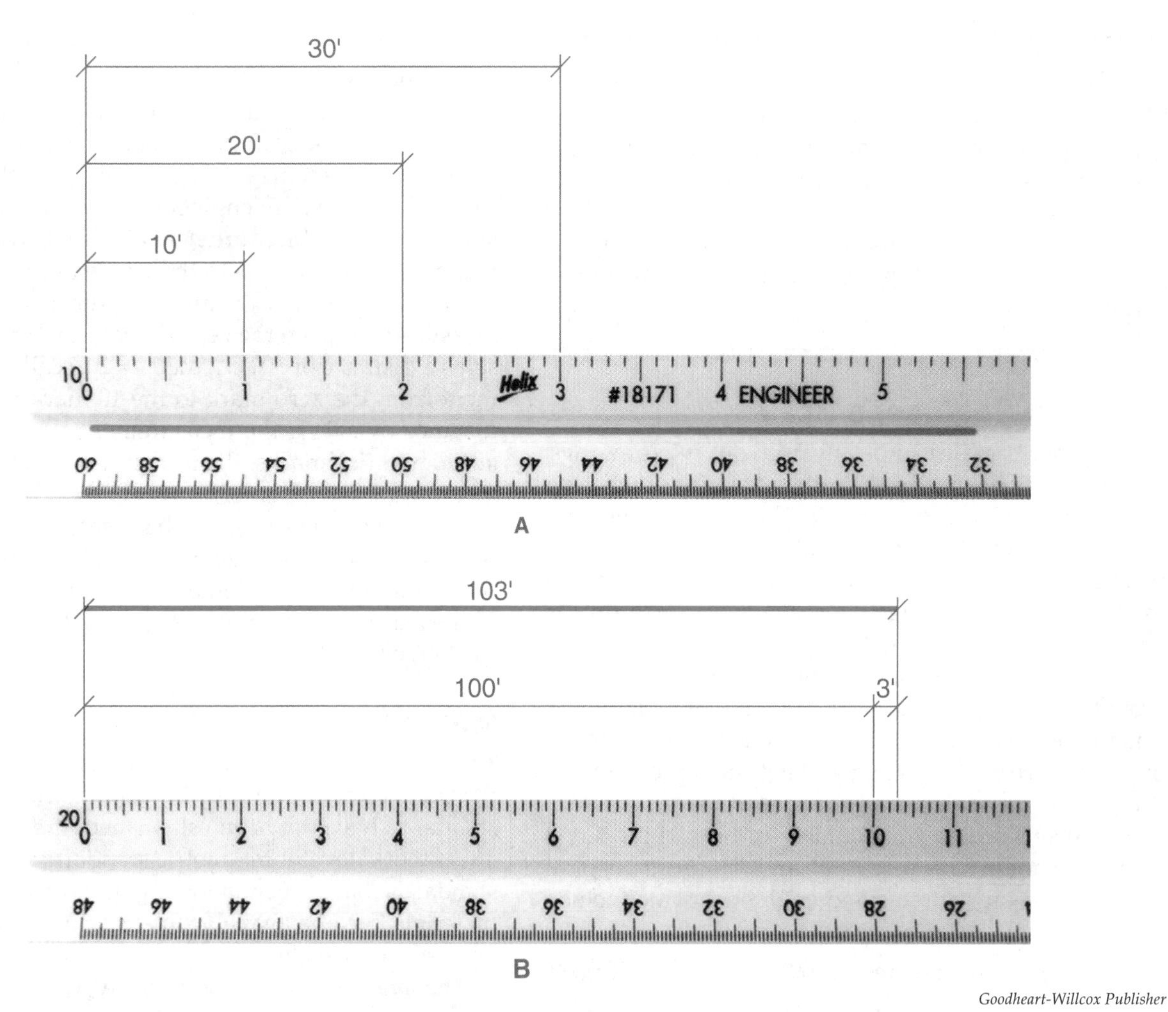

Goodheart-Willcox Publisher

Figure 3-11. The main units on an engineer's scale represent multiples of 10. When measuring distances in feet, add a zero to the whole number division. A—Measurements on a 1″ = 10′ scale. B—Measurements on a 1″ = 20′ scale.

Reading a Scale

As a construction worker, you will be reading dimensions given on a print and laying off measurements on the job. The needed dimensions should be provided on the print. However, there may be times when a particular dimension is desired but has been omitted or was not considered necessary. Also, you may be required to sketch a detail, or lay it out to scale. In these situations, it is necessary to use a scale.

The following procedure describes how to measure a distance on a quarter-inch scale drawing, using an architect's scale. Refer to **Figure 3-12** as you study this procedure.

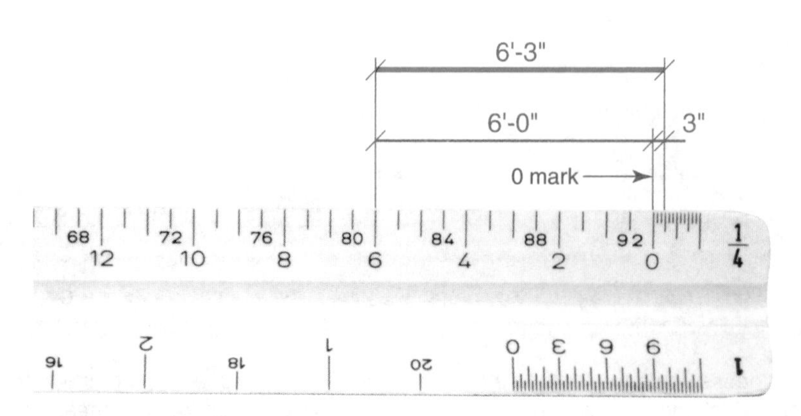

Goodheart-Willcox Publisher

Figure 3-12. A measurement on the architect's 1/4″ = 1′-0″ scale.

Reading an Architect's Scale

1. Locate the scale marked 1/4.
2. Starting at zero, the scaled measurements of feet are numbered to the left, such as 2, 4, 6, 8, and so on. Inches are marked in the section to the right of the zero.
3. To lay off a measurement of 6′-3″, start with the line indicating 6′ and move to the zero, and on to the 3″.

Using the same method, a measurement of 11′-8″ would start at the line indicating 11′ and extend past to 8″.

Architectural details frequently are drawn to larger scales, such as 1 1/2″ = 1′-0″. As shown in **Figure 3-13**, this scale is read in the same manner as the quarter-inch scale.

Readings on the metric scale are made in the same manner as on a customary scale. The unit of metric linear measure is the meter. A reading of 2.5 meters on the 1:50 metric scale is shown in **Figure 3-14**. The unit of metric linear measure in cabinet work is the centimeter (cm). A measurement of 82 centimeters on the 1:20 metric scale is shown in **Figure 3-15**.

Using the Fractional Rule as a Scale

When an architect's scale is not available, a fractional rule can be used to take measurements. For a measurement on the 1/4″ = 1′-0″ scale, let each 1/4 inch on the rule represent one foot of actual measurement on the project. Fractional parts of the 1/4″ would represent the appropriate number of inches on the project. For example, a scaled reading of 5′-6″ would be five quarters of an inch for the 5′ and one eighth of an inch for the 6″, or 1 3/8″. See **Figure 3-16**.

A properly prepared drawing will include all needed dimensions. Use care when scaling a print for measurements not provided. All such measurements should be cross-checked with other dimensions and verified by your supervisor or the architect.

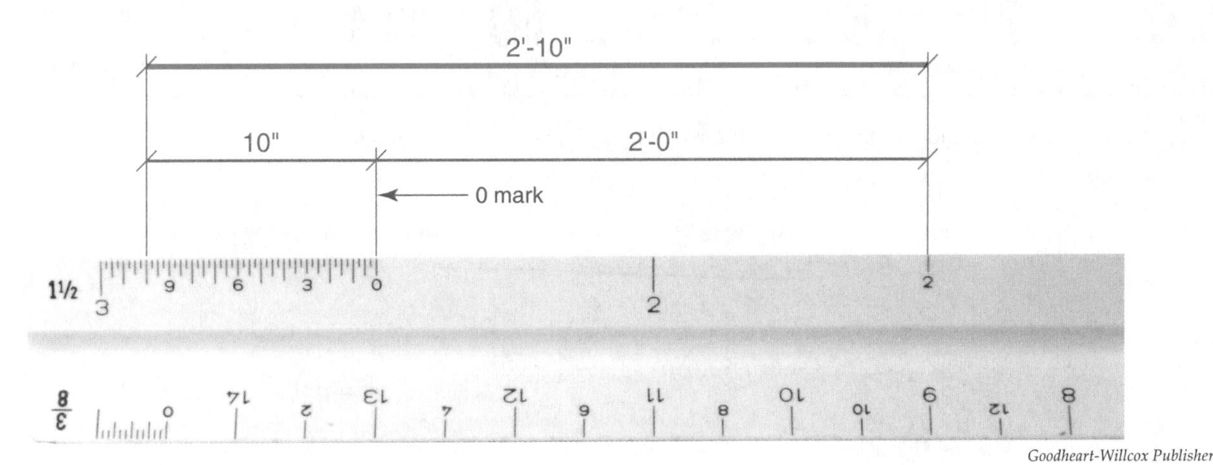

Goodheart-Willcox Publisher

Figure 3-13. A measurement on the 1 1/2″ = 1′-0″ scale.

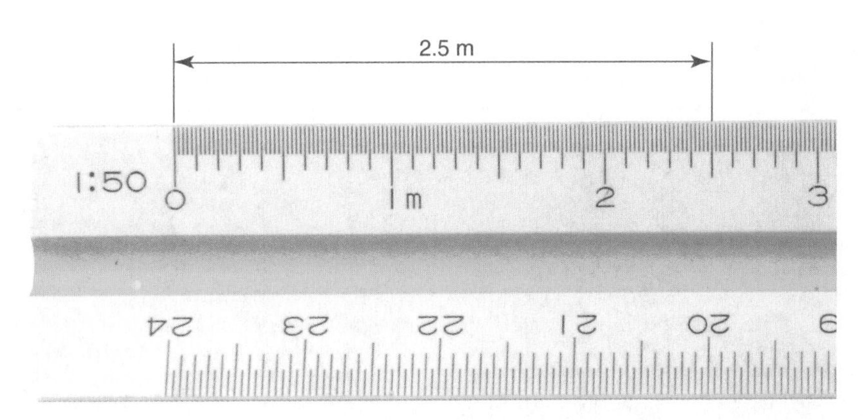

Goodheart-Willcox Publisher

Figure 3-14. A reading of 2.5 meters on the metric 50 scale.

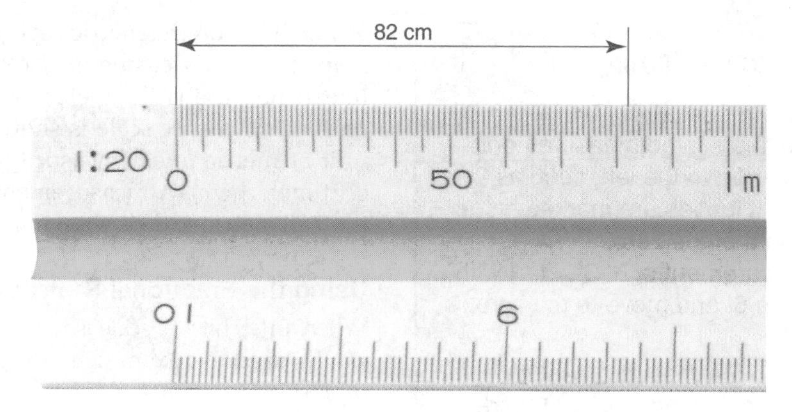

Goodheart-Willcox Publisher

Figure 3-15. A reading of 82 centimeters on the metric 20 scale.

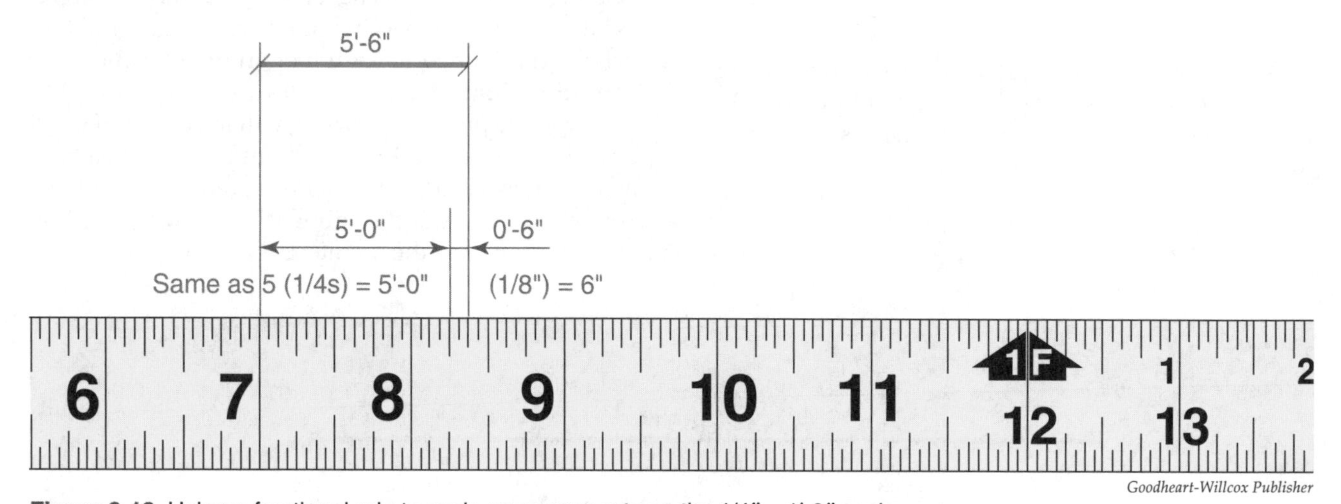

Goodheart-Willcox Publisher

Figure 3-16. Using a fractional rule to scale measurements on the 1/4″ = 1′-0″ scale.

Test Your Knowledge

Name ______________________________

Write your answers in the spaces provided.

____ 1. Convert 25″ to millimeters.
A. 25 mm
B. 635 mm
C. 1 mm
D. 11.8 mm
E. None of these answers is reasonable.

____ 2. Convert 12′-3″ to metric measure.
A. 307.8 mm
B. 5.79 mm
C. 3078 mm
D. 3.73 m
E. None of these answers is reasonable.

____ 3. Convert 38 mm to customary units.
A. 1 1/2″
B. 965.2″
C. 18′
D. 3 1/8″
E. None of these answers is reasonable.

____ 4. Convert 17.42 m to customary units.
A. 57′-1 7/8″
B. 36′-10 1/2″
C. 442 1/2″
D. 36 7/8″
E. None of these answers is reasonable.

____ 5. *True or False?* One inch is longer than 2 cm.

____ 6. Convert 210′-6″ to metric measure.
A. 64.2 m
B. 5346.7 mm
C. 2520″
D. 5.35 m
E. None of these answers is reasonable.

____ 7. Convert 1.42 m to customary units.
A. 3.6′
B. 64″
C. 4′-8″
D. 1/2″
E. None of these answers is reasonable.

____ 8. Convert 3 1/2″ to metric measure.
A. 88.9 cm
B. 0.14 mm
C. 0.14 m
D. 8.89 mm
E. None of these answers is reasonable.

____ 9. *True or False?* One foot is longer than 30 cm.

____ 10. *True or False?* One yard is longer than 90 cm.

____ 11. Which of the following is a typical scale used for floor plans?
A. 1″ = 1′-0″
B. 1/2″ = 1′-0″
C. 1/4″ = 1′-0″
D. 1″ = 60′-0″
E. None of these is the correct scale.

____ 12. The metric system is rarely used in the United States because ____.
A. it is too complicated for people to understand
B. you must speak Spanish, French, or German to understand it
C. it is against the law
D. construction standards using the metric system have not been established
E. All of these reasons are correct.

____ 13. *True or False?* It is often necessary to scale dimensions from drawings, because not all of the required dimensions are normally shown on a drawing.

____ 14. *True or False?* When measuring distances on a drawing, an architect's scale must be used. A normal fractional ruler cannot be used.

____ 15. *True or False?* When measuring the length of a line using a scale, you position the "0" reading on the scale with one end of the line. The other line endpoint corresponds to its reading on the scale.

____ 16. *True or False?* The scale of a drawing is the relationship between the size of the drawing and the size of the actual structure.

_____ 17. What is the smallest tick mark on each of these architect's scales?

A. 3/16" ____________________

B. 1/4" ____________________

C. 1" ____________________

D. 1 1/2" ____________________

_____ 18. A "20 scale" refers to which of the following?

A. 1/4" = 1'-0"

B. 1/2" = 1'-0"

C. 1" = 10'

D. 1" = 20'

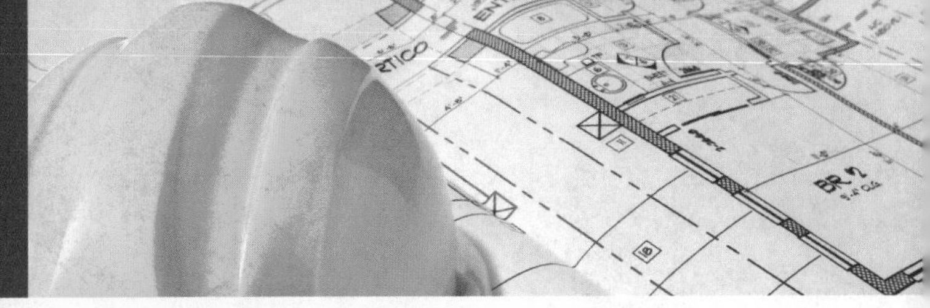

Activity 3-1

Reading a Fractional Rule

Name ________________________________

Fill in the measurements for the six blanks below. Reduce to lowest terms and place your answers in the spaces provided.

A ______ B ______

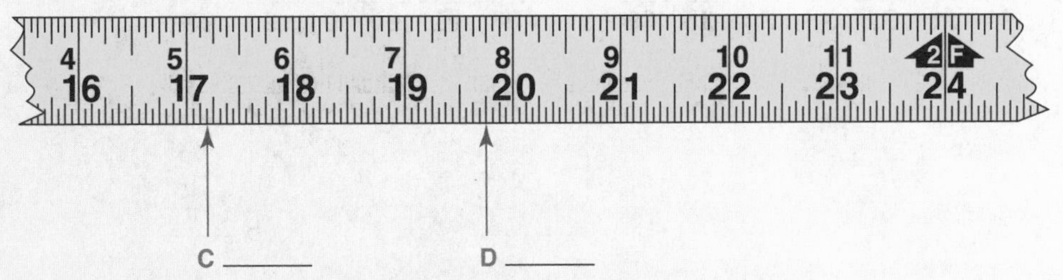

C ______ D ______

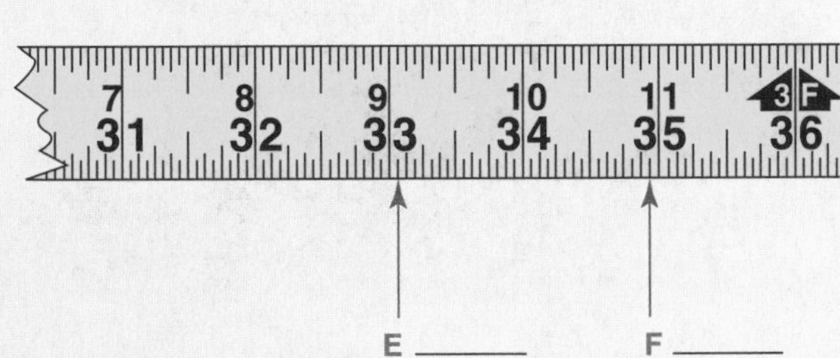

E ______ F ______

Reading a Metric Rule

Name ______________________________

Fill in the measurements below in terms of millimeters. Place your answers in the spaces provided.

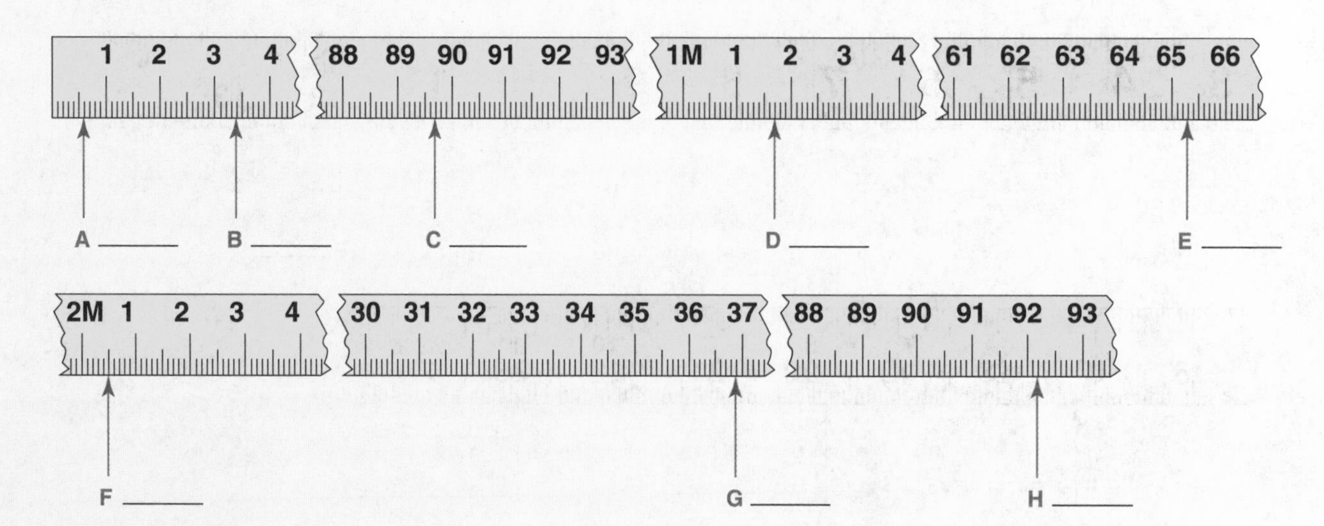

Activity 3-3

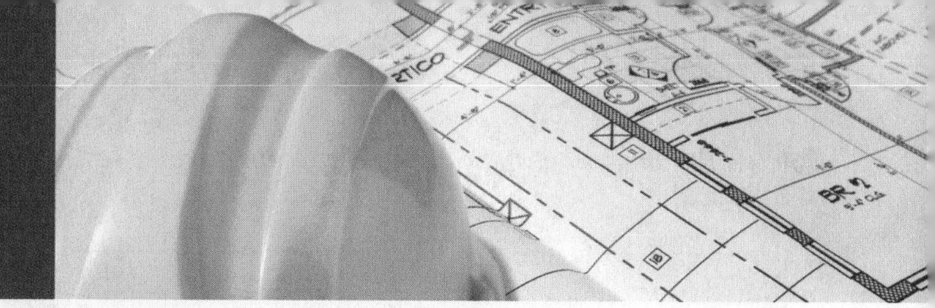

Reading a Scale

Name ______________________________________

Identify the portion of a foot (12″) as indicated by the reading from the zero mark (0) on the scales below. Place your answers in the spaces provided.

1. ________

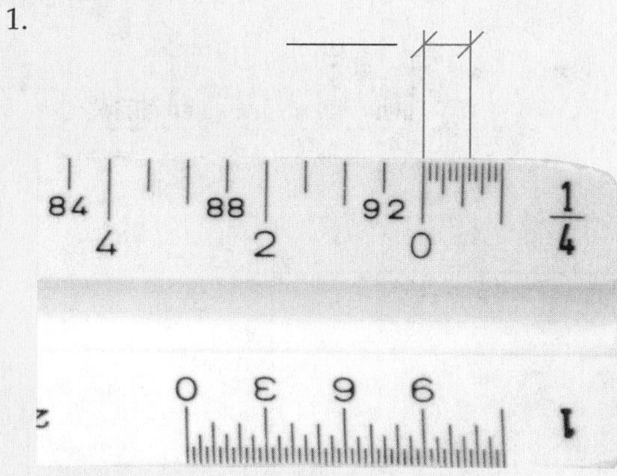

2. ________

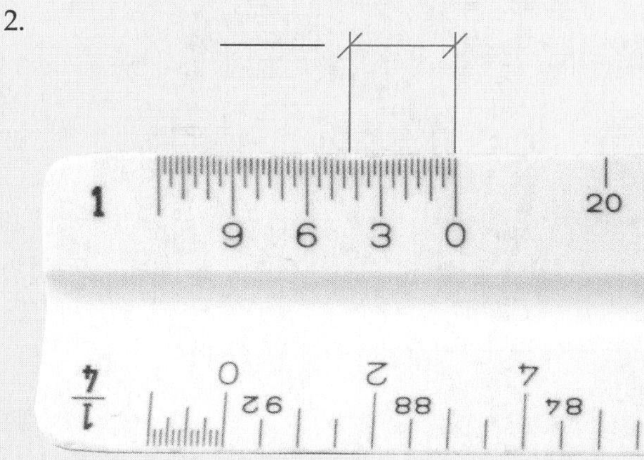

3. ________

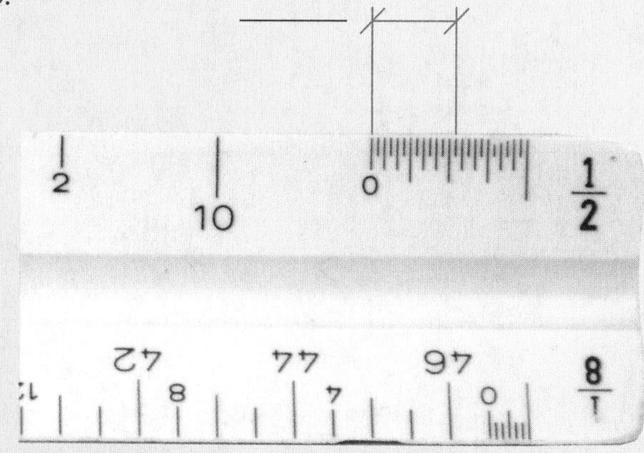

4. ________

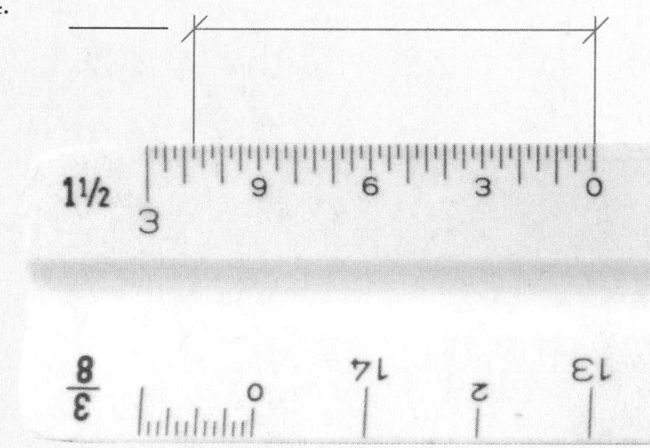

5. ________

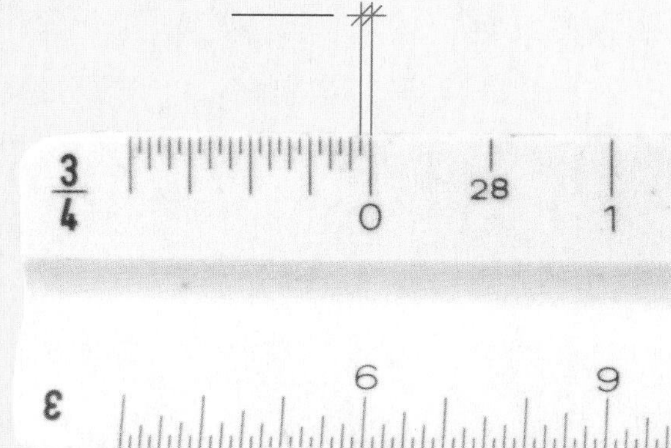

Identify the dimensions indicated by the readings on the scales below. Place your answers in the spaces provided.

6. ________

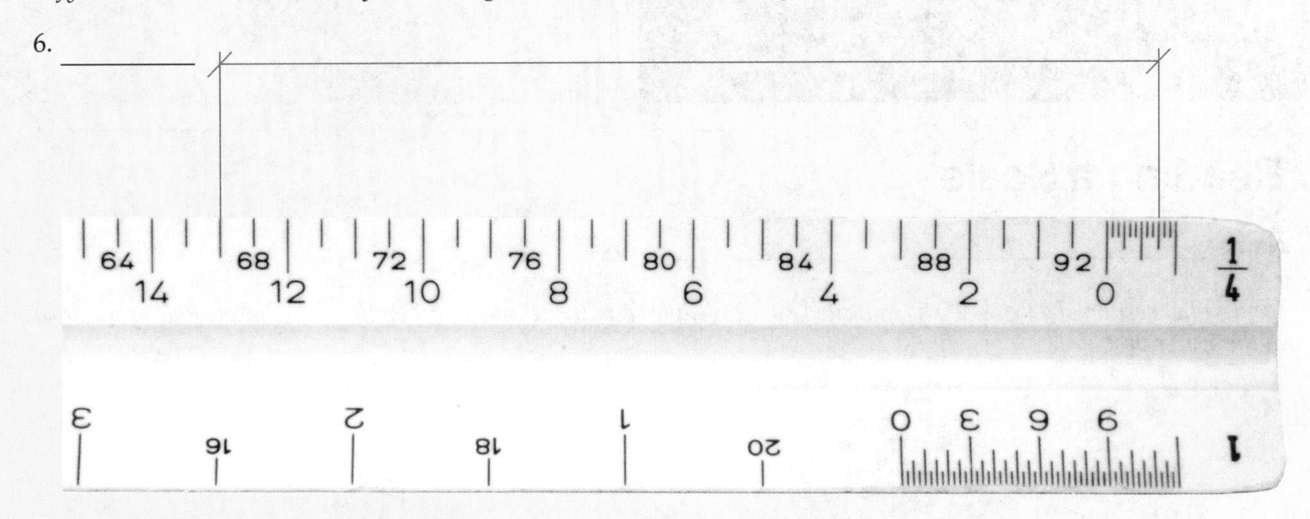

7. ________

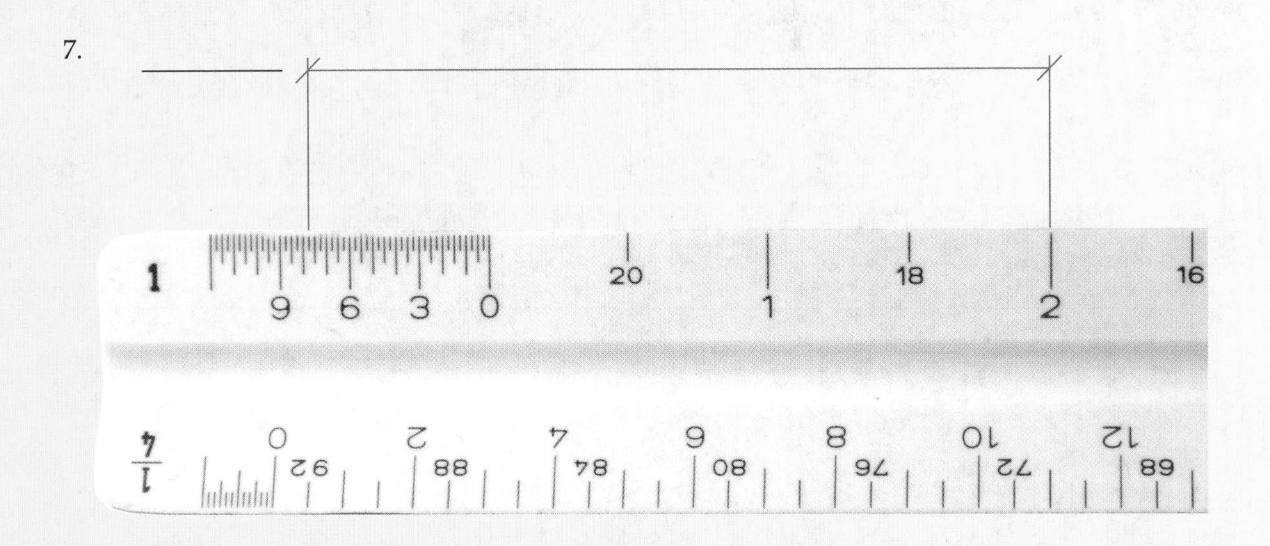

8. ________

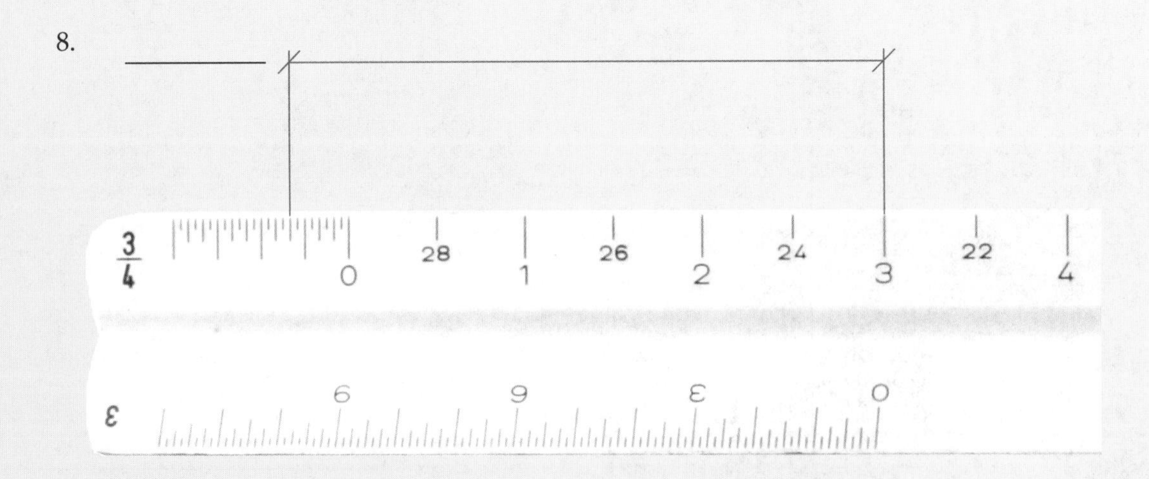

Activity 3-4

Identifying Missing Dimensions

Name ______________________________

Fill in the missing dimensions in the blanks provided by adding and subtracting dimensions (do not use scales). Assume walls are 4" thick unless noted otherwise.

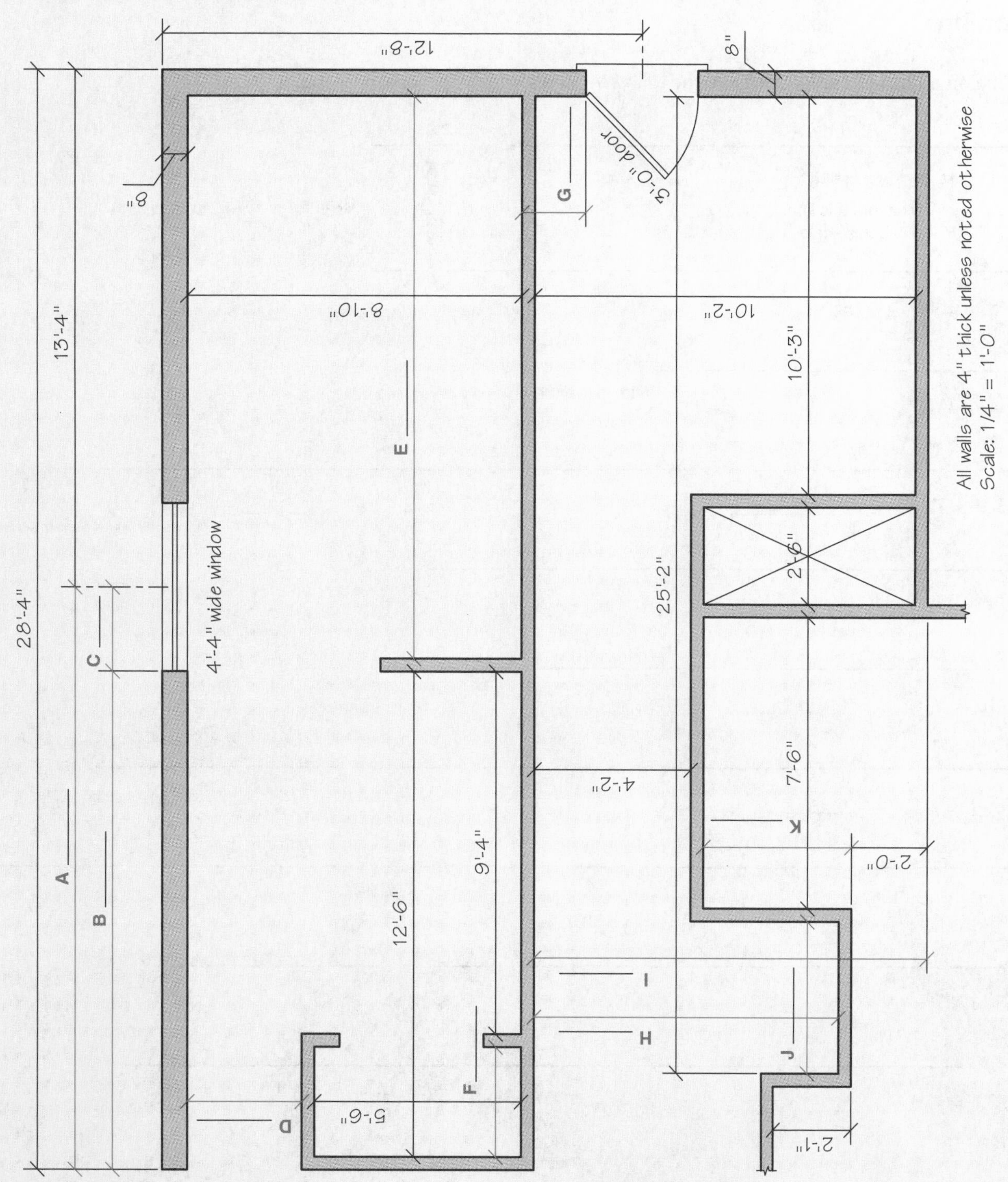

Activity 3-5

Reading and Drawing Using Scales

Name ______________________________

Use an architect's scale or an engineer's scale, as indicated in the activity, to measure or draw the line as indicated.

Scaling I

Using an **architect's scale**, measure the following lines:

3/4″ = 1′-0″ __________ Write answer here

1 1/2″ = 1′-0″ __________

3/8″ = 1′-0″ __________

3/16″ = 1′-0″ __________

3″ = 1′-0″ __________

1/2″ = 1′-0″ __________

1″ = 1′-0″ __________

3/32″ = 1′-0″ __________

Scaling II

Using an **engineer's scale**, measure the following lines:

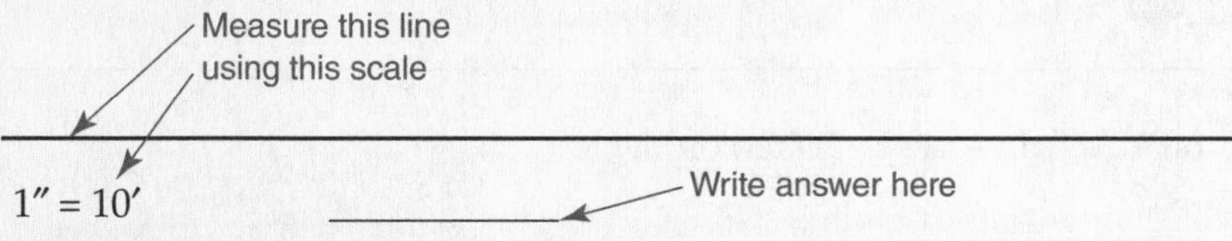
Measure this line
using this scale

1″ = 10′ __________ Write answer here

1″ = 20′ __________

1″ = 60′ __________

1″ = 100′ __________

1″ = 40′ __________

1″ = 50′ __________

1″ = 30′ __________

1″ = 20′ __________

1″ = 10′ __________

1″ = 60′ __________

Dimensioning I

Using an **architect's scale**, draw the following lines:

1/4″ = 1′-0″ Using line above, mark off 12′-8″ (in scale indicated on the left)

1/8″ = 1′-0″ Using line above, mark off 23′-6″

3/4″ = 1′-0″ Using line above, mark off 7′-10″

1 1/2″ = 1′-0″ Using line above, mark off 3′-9 1/2″

3/8″ = 1′-0″ Using line above, mark off 13′-2″

3/16″ = 1′-0″ Using line above, mark off 33′-0″

3″ = 1′-0″ Using line above, mark off 2′-1 3/4″

1/2″ = 1′-0″ Using line above, mark off 7′-3 1/4″

1″ = 1′-0″ Using line above, mark off 5′-5″

3/32″ = 1′-0″ Using line above, mark off 60′-0″

Dimensioning II

Using an **engineer's scale**, draw the following lines:

1″ = 10′ Using line above, mark off 55′-6″ (in scale indicated on the left)

1″ = 20′ Using line above, mark off 105′

1″ = 50′ Using line above, mark off 187′

1″ = 100′ Using line above, mark off 660′

1″ = 60′ Using line above, mark off 50′

1″ = 30′ Using line above, mark off 166′-0″

1″ = 100′ Using line above, mark off 89′

1″ = 40′ Using line above, mark off 222′

1″ = 60′ Using line above, mark off 104′

1″ = 20′ Using line above, mark off 120′-6″

SECTION 2
Print Reading Basics

karamysh/Shutterstock.com

UNIT 4
Lines and Symbols

TECHNICAL TERMS

Alphabet of Lines
arrowheads
border lines
break lines
centerlines
dimension lines
extension lines
hidden lines
leaders
legend
object lines
property lines
section cutting lines
section lines

LEARNING OBJECTIVES

After completing this unit, you will be able to:

- Identify the common types of lines used on prints.
- Identify features from different lines.
- Match drawing symbols with their meanings.

Reading construction prints begins with understanding the types of lines and symbols that appear on drawings, **Figure 4-1**. Architects and engineers use a defined system of lines and symbols. The different types of lines and symbols used on drawings are defined and discussed in this unit.

The Alphabet of Lines

A number of common types of lines are found on construction prints. The different lines used on prints make up a system of line conventions known as the ***Alphabet of Lines***. All lines in this system are normally drawn in the same color, usually black. Some lines vary in width. Some are drawn solid, while others are a combination of broken lines. Each type of line conveys a different meaning. **Figure 4-2** illustrates some common lines.

Lines can represent different features, depending on the type of drawing for which they are used. For instance, a site plan in a set of civil engineering prints may show a train track as a solid line with short, perpendicular lines. On a plumbing drawing, the same type of line might represent an underground utility. In **Figure 4-1**, a dashed line is used for the detail symbol "callout" referencing a footing detail. The same type of line is used for footings. Always refer to the ***legend*** for the appropriate meaning associated with the line. As you will learn later in this unit, a legend is a list of symbols used in a set of prints.

The following types of lines are common on construction drawings. Refer to **Figure 4-2** as you study each type.

- ***Border lines***. Border lines are extra-heavy lines located near the edge of the sheet of drawing paper. When used, they serve as a border for the drawing (not all drawings will show border lines on the print). They are also used to separate the various portions of the drawing, such as the title block, notes, and the revision block.
- ***Property lines***. Property lines are extra-heavy lines made up of long dashes alternating with two short dashes. The length and bearing (direction) of each line usually is identified on the site plan in a set of civil engineering prints.

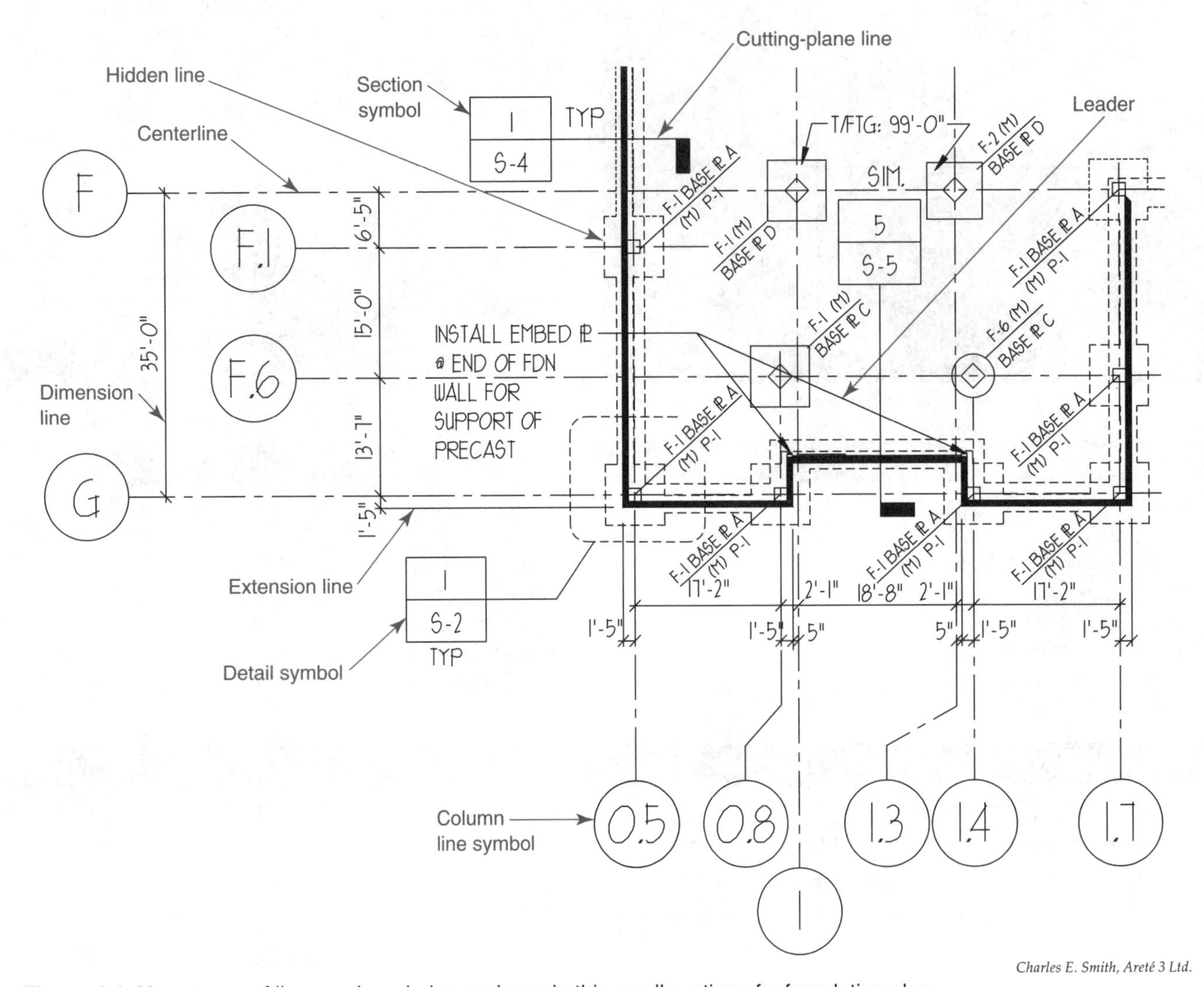

Charles E. Smith, Areté 3 Ltd.

Figure 4-1. Many types of lines and symbols are shown in this small portion of a foundation plan.

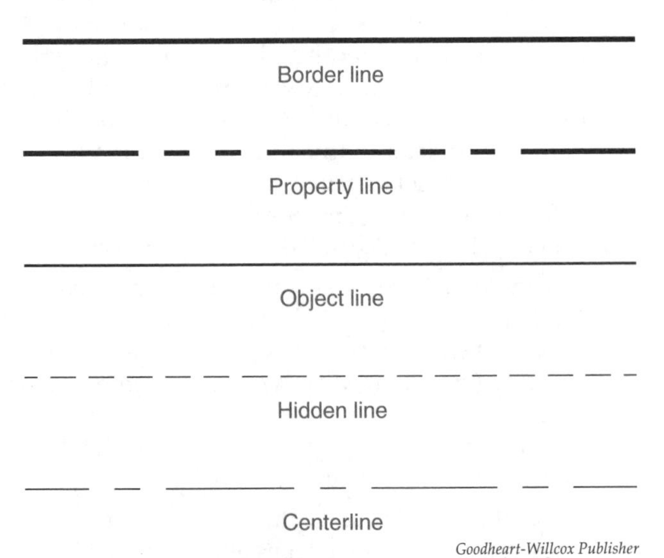

Goodheart-Willcox Publisher

Figure 4-2. Some common lines used on construction drawings.

- ***Object lines.*** Object lines represent the main outline of the features of an object, such as a building, wall, or walkway. An object line is a heavy weight line and is continuous, showing all of the object edges and surfaces. Object lines are the most common lines used in a set of prints.
- ***Hidden lines.*** Hidden lines are composed of a series of short dashes evenly spaced and consistent in length. They are drawn to a lighter weight than object lines. They define edges and surfaces that are behind other objects and are not visible in a particular view, but they provide information needed to interpret the drawings. The print reader must look for another view in the set of drawings to identify where these edges occur and find further detail. Often, these hidden parts will be revealed in an elevation or sectional view. Hidden lines are omitted if they do not add to the clarification of the drawing.

- ***Centerlines***. Centerlines are thin lines used to indicate centers of objects such as footings, columns, fixtures, equipment, and openings. Normally, these objects are located by dimensioning to the center of the intended object. The centerline is also used to indicate a finished floor line. Centerlines are composed of alternating long and short dashes.
- ***Dimension lines*** and ***extension lines***. Dimension lines and extension lines are thin lines that indicate the extent and direction of dimensions, **Figure 4-3**. Dimension lines extend the length of the distance being measured. A marking device—such as an arrow, dot, or tick mark—is placed at each end of the dimension line. Extension lines are drawn perpendicular to the dimension line to specify the features between which the dimension applies.
- ***Leaders***. A leader is also a thin line. It has an arrow at one end and a text note or symbol at the other end, **Figure 4-4**. The arrow points to the feature or detail to which the note or symbol refers. Leaders are the same weight as dimension and extension lines.
- ***Arrowheads***. There are many types of arrowheads; each architect develops a unique style of indication. The purpose is to define the distance between points (extension lines) on the drawing. When space between extension lines is too small, the arrowheads are located on the outside with the dimension in the middle or off to the side and a leader line referring to the location for the dimension. A variety of arrowhead styles is shown in **Figure 4-3**.

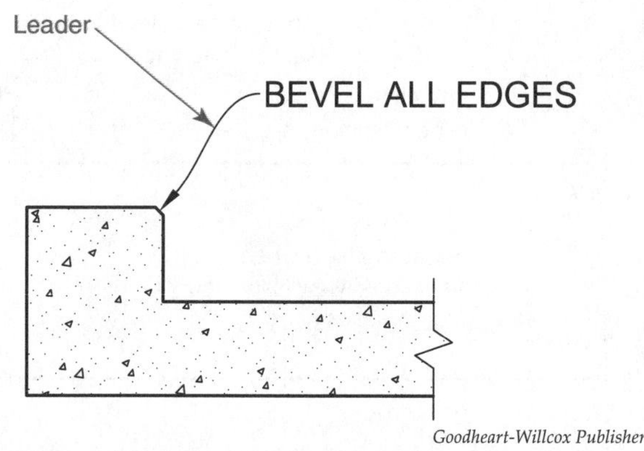

Goodheart-Willcox Publisher

Figure 4-4. A leader is a line from a note to a feature in the drawing.

- ***Break lines***. Break lines, **Figure 4-5**, are used to indicate that only a portion of an object is shown. There are two types of break lines: long break lines and short break lines. Long break lines are used to show long, straight breaks. Short break lines are used to show short, irregular breaks. One common use for long break lines is to indicate that an object (such as a wall) continues but is not shown on the drawing. Refer to **Figure 4-5A**. Another common use for long break lines is to indicate that the object's

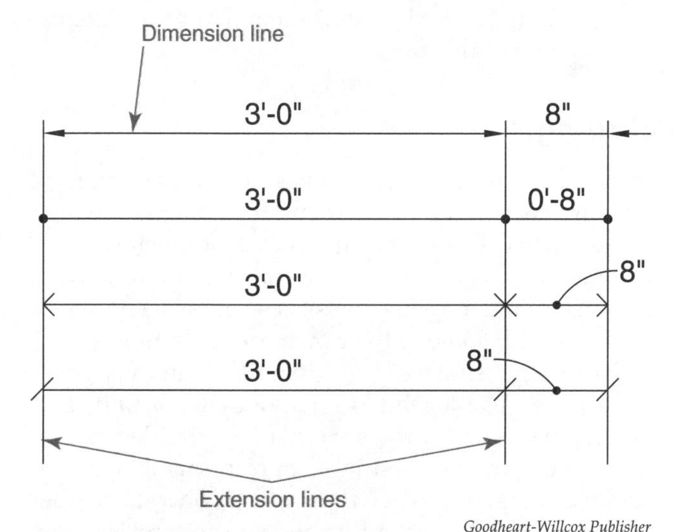

Goodheart-Willcox Publisher

Figure 4-3. Dimension lines and extension lines are thin lines used in dimensioning. Also shown are some of the many ways to express arrowheads.

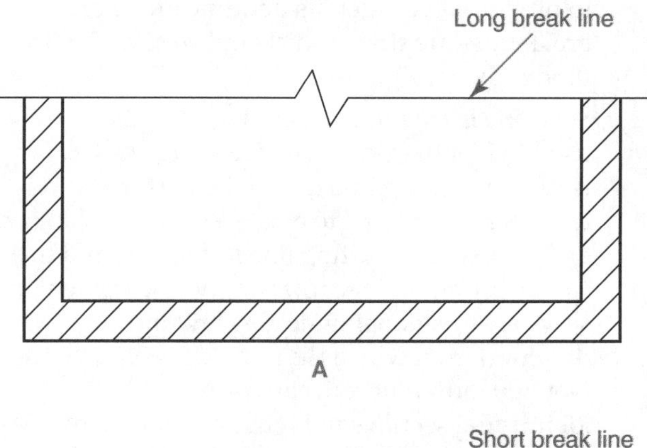

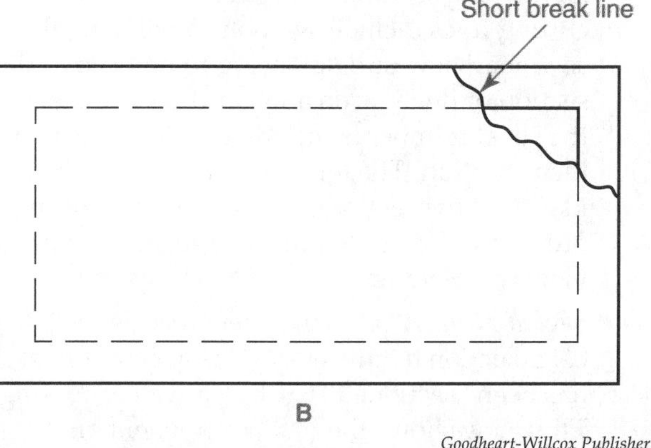

Goodheart-Willcox Publisher

Figure 4-5. Break lines indicate that a portion of the object has been omitted. A—Long break line used to show a portion of a long wall. B—Short break line used to show features "underneath" the break.

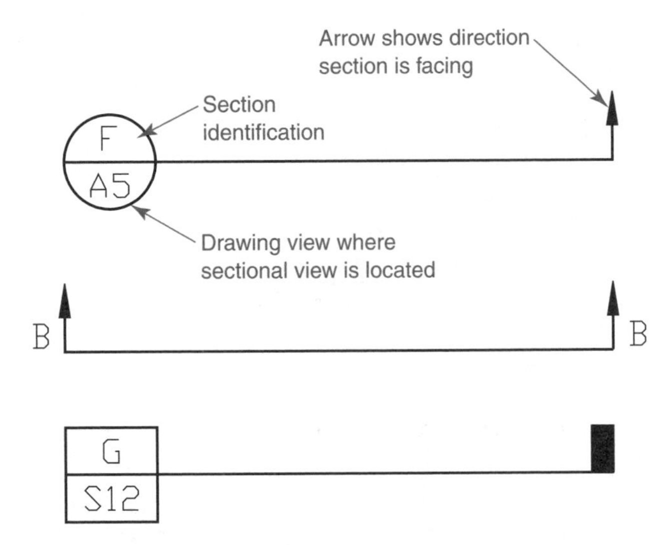

Figure 4-6. A section cutting line marks the point where the drawing has been "cut" to show a sectional view or detail and indicates the viewing direction.

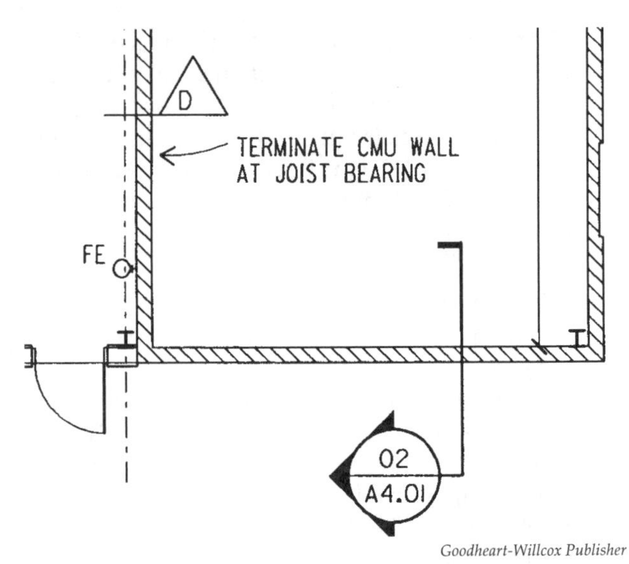

Figure 4-7. The angled section lines are used in this plan view to indicate concrete masonry unit (CMU) walls.

full length is not shown either to save space or because the entire object will not fit on the drawing. Long break lines are drawn thin with freehand "zigzags" to indicate the break. Short break lines, **Figure 4-5B**, are used to show that a portion of the drawing has been removed in order to reveal details underneath. Short break lines are drawn thick and freehand with randomly curving lines.

- ***Section cutting lines.*** Section cutting lines are used to identify sectional views, **Figure 4-6.** A section cutting line marks the part of the drawing being "cut" to create a sectional view or detail. A section cutting line is also referred to as a *cutting-plane line.* Arrows on the ends of the line serve as directional indicators and indicate the direction from which the section is being viewed. Note the different conventions used for identifying sections in **Figure 4-6**. Some section cutting lines include symbols identifying the sectional view or detail being referenced. If the section or detail is on another drawing sheet, the sheet number is included with the section identification. The top number or letter usually identifies the section or detail and the bottom number or letter identifies the sheet on which the view can be found.
- ***Section lines.*** A sectional view reveals the interior construction details of a building component that has been "sectioned" (such as a wall or footing). This view shows the "cut" component and provides a graphic representation of the building materials used in construction. Section lines are thin lines used in a sectional view to represent building materials, **Figure 4-7**. Section lines are typically drawn at a 45° angle and provide a simple way to represent building materials. The section lines in **Figure 4-7** are drawn in this manner to represent concrete masonry unit (CMU) walls (concrete block walls). To provide greater detail, graphic symbols indicating the actual type of building material (such as concrete) may be used in a sectional view. Graphic symbols used to represent building materials are typically referred to as *material rendering symbols, hatching,* or *poche,* depending on whether the drawing has been made manually or with CAD software. Whether section lines or graphic symbols are used, building material symbols help the print reader quickly interpret the types of materials required. For a listing of common building material symbols used on construction drawings, see the *Building Material Symbols* section in the Reference Section of this book.

Symbols

In addition to the various types of lines, a variety of symbols are commonly used on construction drawings. These symbols represent items and products such as different kinds of building materials and fixtures. Most architects and engineers use a standardized set of symbols with some of their own unique symbols.

Normally, symbols are identified in a legend, **Figure 4-8**. The legend is a list of symbols and their corresponding meanings. Sometimes, the legend for all of the symbols used in a set of prints is included on the cover sheet. When the legend is on a sheet other than the one you are reading and you are first learning to read drawings, it is helpful to copy the legend onto a smaller sheet and keep it handy for reference.

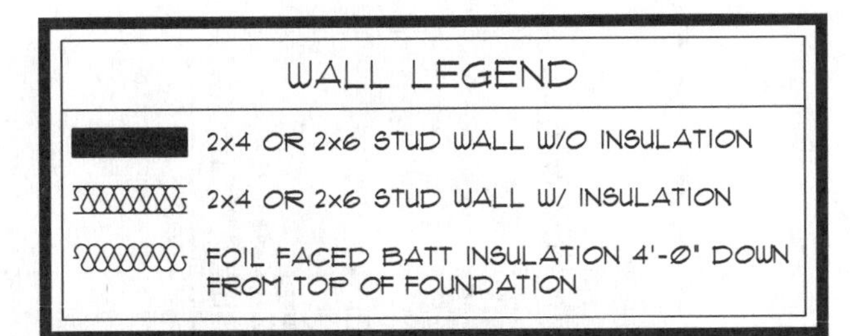

WALL LEGEND
2x4 OR 2x6 STUD WALL W/O INSULATION
2x4 OR 2x6 STUD WALL W/ INSULATION
FOIL FACED BATT INSULATION 4'-0" DOWN FROM TOP OF FOUNDATION

A

Norris & Dierkers Architects/Planners, Inc.

LIGHTING SYMBOL LEGEND

SYMBOLS	DESCRIPTION
	2'-0" x 4'-0" RECESSED LIGHT FIX. W/LENS
	2'-0" x 4'-0" NIGHT LIGHT FIXTURE
	DUAL POWERED EXIT LIGHT
	EXHAUST FAN
	DUAL POWERED EMERGENCY WALL PAK LIGHT
	150 W RECESSED CAN LIGHT
	100 W RECESSED CAN LIGHT
	STAIR NIGHT LIGHT FIXTURE
	2'-0" x 2'-0" RECESSED LIGHT FIX. W/LENS
	2'-0" x 2'-0" NIGHT LIGHT FIXTURE
	1'-0" x 4'-0" RECESSED LIGHT FIX W/ACRYLIC LENS

B

Charles E. Smith, Areté 3 Ltd.

Figure 4-8. Legends are used to identify symbols on a drawing. A—Wall legend. B—Lighting symbol legend.

Many common symbols are shown in the Reference Section of this book. Look at that section now to become more familiar with common symbols on drawings.

Abbreviations

Abbreviations are very important in print reading. They take the place of words in notes and cut down the length of the notes. In a set of prints, a list of abbreviations used in the prints can usually be found on the title page. See **Figure 4-9**.

Abbreviations should be standard, but they can sometimes stand for more than one word. See the Reference Section in this book for a list of abbreviations commonly used on prints.

ABBREVIATIONS

ACT	ACOUSTICAL TILE
AFF	ABOVE FINISHED FLOOR
BLDG	BUILDING
BRG	BEARING
CJ	CONTROL JOINT
CLG	CEILING
CMU	CONCRETE MASONRY UNIT
CONT	CONTINUOUS
DET	DETAIL
DN	DOWN
DS	DOWNSPOUT
EB	EXPANSION BOLT
EJ	EXPANSION JOINT
FDN	FOUNDATION
FIN	FINISH
FLR	FLOOR
FTG	FITTING, FOOTING
GA	GAGE
GALV	GALVANIZED
GYP	GYPSUM
HDW	HARDWARE
INSUL	INSULATION
INT	INTERIOR
JST	JOIST
JT	JOINT
KIT	KITCHEN
LAV	LAVATORY
LIN	LINEAR
MFR	MANUFACTURER
MO	MASONRY OPENING
NOM	NOMINAL
OC	ON CENTER
OD	OUTSIDE DIAMETER
PL	PLATE, PROPERTY LINE
PLWD	PLYWOOD
RA	RETURN AIR
REG	REGISTER
RET	RETURN
RO	ROUGH OPENING
SAN	SANITARY
STRUCT	STRUCTURAL
THRU	THRU
TYP	TYPICAL
UNO	UNLESS NOTED OTHERWISE
WC	WATER CLOSET
WWF	WELDED WIRE FABRIC

Goodheart-Willcox Publisher

Figure 4-9. Examples of abbreviations commonly found on a set of construction drawings. See the Reference Section in this book for a comprehensive list of abbreviations.

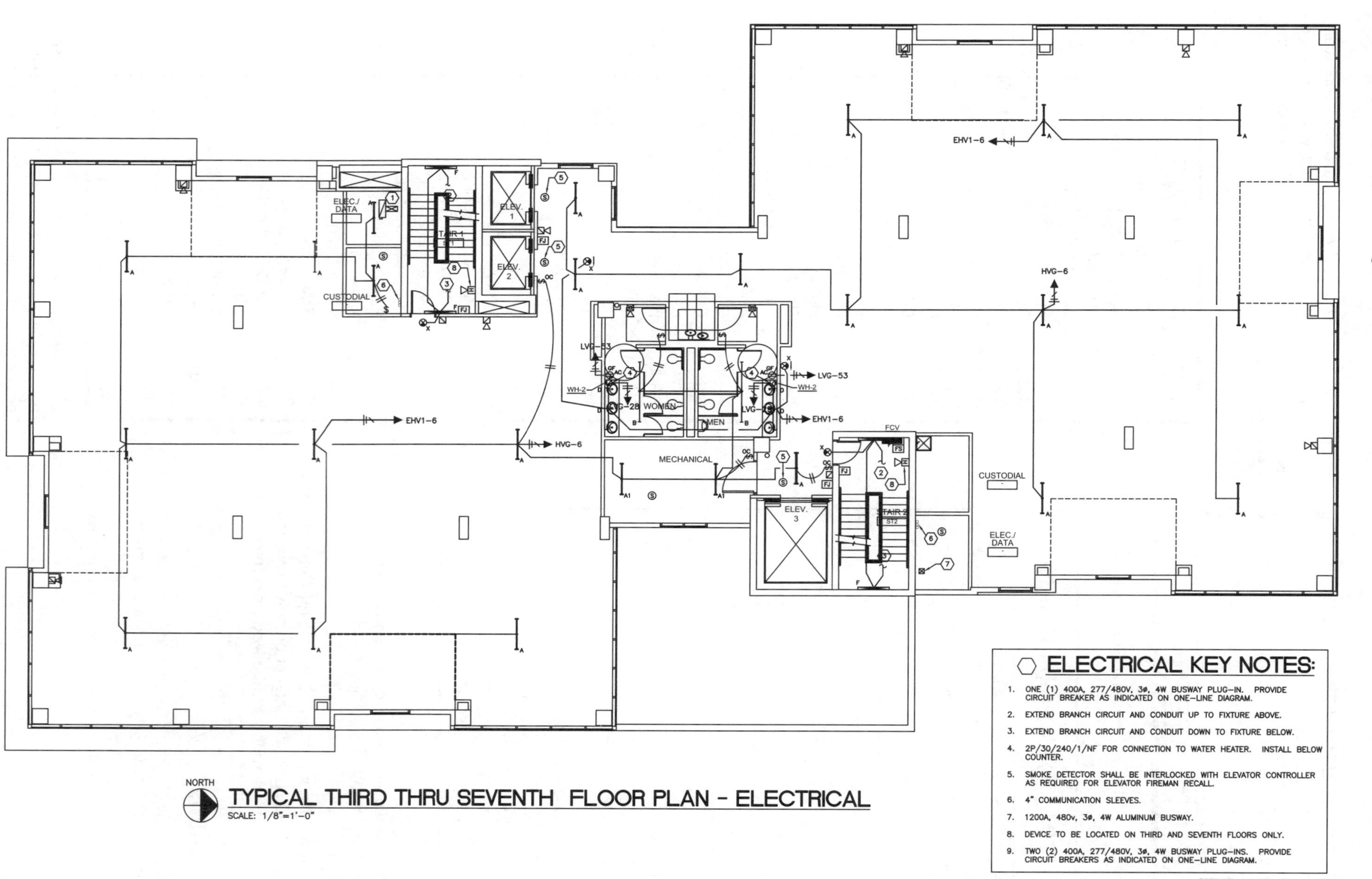
ELECTRICAL KEY NOTES:
1. ONE (1) 400A, 277/480V, 3ø, 4W BUSWAY PLUG-IN. PROVIDE CIRCUIT BREAKER AS INDICATED ON ONE-LINE DIAGRAM.
2. EXTEND BRANCH CIRCUIT AND CONDUIT UP TO FIXTURE ABOVE.
3. EXTEND BRANCH CIRCUIT AND CONDUIT DOWN TO FIXTURE BELOW.
4. 2P/30/240/1/NF FOR CONNECTION TO WATER HEATER. INSTALL BELOW COUNTER.
5. SMOKE DETECTOR SHALL BE INTERLOCKED WITH ELEVATOR CONTROLLER AS REQUIRED FOR ELEVATOR FIREMAN RECALL.
6. 4" COMMUNICATION SLEEVES.
7. 1200A, 480v, 3ø, 4W ALUMINUM BUSWAY.
8. DEVICE TO BE LOCATED ON THIRD AND SEVENTH FLOORS ONLY.
9. TWO (2) 400A, 277/480V, 3ø, 4W BUSWAY PLUG-INS. PROVIDE CIRCUIT BREAKERS AS INDICATED ON ONE-LINE DIAGRAM.
TYPICAL THIRD THRU SEVENTH FLOOR PLAN - ELECTRICAL
SCALE: 1/8"=1'-0"
NORTH
CUSTODIAL
ELEC./ DATA
MECHANICAL
MEN
WOMEN
ELEV. 1
ELEV. 2
ELEV. 3
STAIR-1
STAIR-2
FCV
EHV1-6
HVG-6
LVG-53
WH-2

CITY Properties Group, LLC

Switches, outlets, conductors, electrical paths, lighting fixtures, and other devices are typically represented with symbols on electrical drawings.

Test Your Knowledge

Name ____________________________________

Match the following terms and descriptions. Write your answers in the spaces provided.

____ 1. Illustrates the boundaries of a plot of land.

____ 2. Used to mark the feature for which a dimension is given; drawn perpendicular to the dimension line.

____ 3. Accepted system of line conventions for describing an object or a building.

____ 4. Represents an edge that exists but is blocked from view.

____ 5. A list of symbols included on a print.

____ 6. Shows where material has been "cut" and indicates the viewing direction.

____ 7. Represents a type of fixture or other product used in construction.

____ 8. Used to indicate centers of objects.

____ 9. Defines the drawing limits.

____ 10. Thin line drawn in a section view, typically at a 45° angle.

A. Hidden line
B. Border line
C. Centerline
D. Legend
E. Symbol
F. Alphabet of Lines
G. Property line
H. Object line
I. Dimension line
J. Section cutting line
K. Break line
L. Extension line
M. Section line

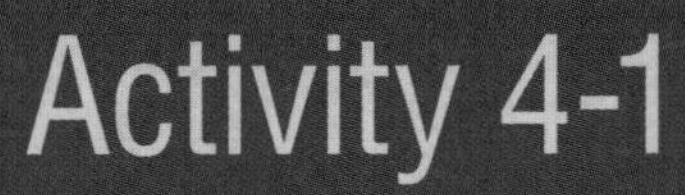

Activity 4-1

Alphabet of Lines

Name ______________________________

In the spaces provided, draw the various lines freehand. Pay special attention to the form and weight of each line.

1. Property line

2. Object line

3. Hidden line

4. Centerline

5. Dimension and extension line

6. Long break line

7. Section line for CMU walls

8. Section cutting line

Activity 4-2

Construction Drawing Symbols

Name ______________________________

In the spaces provided, write the name for the symbol shown. Refer to the Reference Section (Section 7) in this book for the correct symbols.

Electrical Symbols

1. ______________________________

2. ______________________________

3. S D ______________________________

4. ______________________________

5. ______________________________

6. S ______________________________

7. ______________________________

8. ______________________________

9. ______________________________

10. S ______________________________

Plumbing Symbols

11. ______________________________

12. ______________________________

13. ______________________________

14. ______________________________

15. ______________________________

Topographic Symbols

16. ______________________________

17. ______________________________

18. — ELEC — ELEC — ELEC — ______________________________

19. ______________________________

20. ______________________________

21. ______________________________

Building Materials

22. 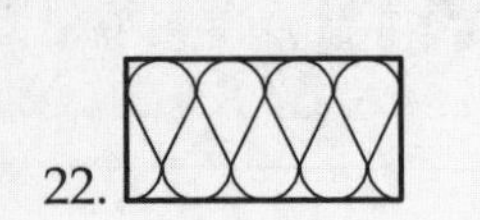 ______________________

23. 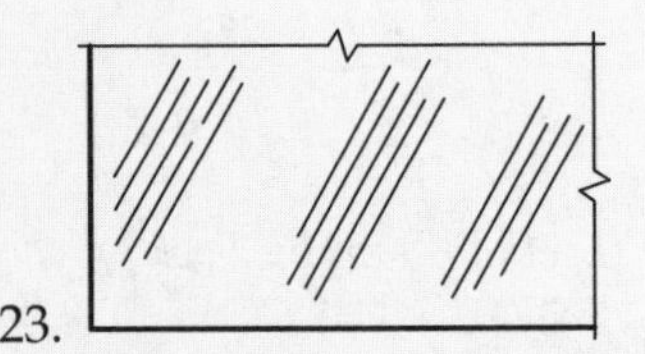 ______________________

24. 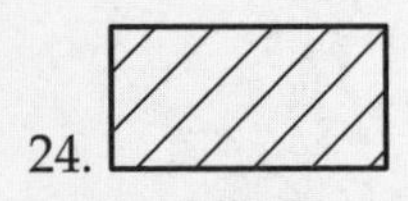 ______________________

25. 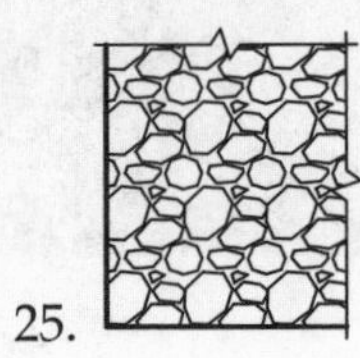 ______________________

26. 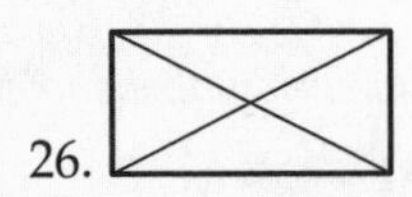 ______________________

27. 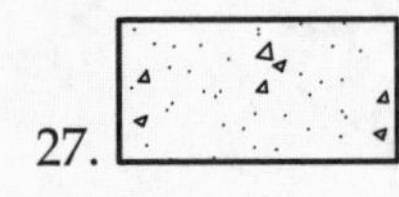 ______________________

UNIT 5
Fundamental Drawing Practices

TECHNICAL TERMS

details
electrical plan
elevations
floor plan
foundation plan
framing plan
graph paper
inclined lines
mechanical plan
orthographic projection
perpendicular projectors
plan view
plumbing plan
proportion
rise
run
section
sheathing
triangle-square method
unit method

LEARNING OBJECTIVES

After completing this unit, you will be able to:

- Use proper techniques to make freehand sketches.
- Sketch lines, circles, arcs, and ellipses.
- Make sketches of objects in correct proportion.
- Understand the principles of orthographic projection.
- Visualize orthographic views of objects and structures.
- Create orthographic views.
- Identify the different types of building views shown in construction drawings.
- Identify standard dimensioning practices used on construction drawings.
- Describe the dimensioning methods used for building features on different drawing types.

Sketching is used by craftspeople, suppliers, architects, and engineers to communicate ideas and construction details. This unit presents the basics of sketching, also known as freehand drawing. By mastering the art of sketching, you will be able to communicate your thoughts and ideas quickly and clearly.

This unit also introduces orthographic projection and fundamental dimensioning practices. You will learn how views are projected to create orthographic drawings. In addition, you will learn how drawings are dimensioned and how to interpret dimensions.

Sketching Technique

When sketching, sharpen your pencil to a conical point, as shown in **Figure 5-1**. When drawing a thin line, use a sharp point. When drawing a thick line, the point can be rounded.

It is important to hold the pencil properly. Grip it firmly enough to control the strokes, but not so rigidly as to stiffen your movements. Your arm and hand should move freely and easily. The point of the pencil should extend approximately 1 1/2″ beyond your fingertips, **Figure 5-2**.

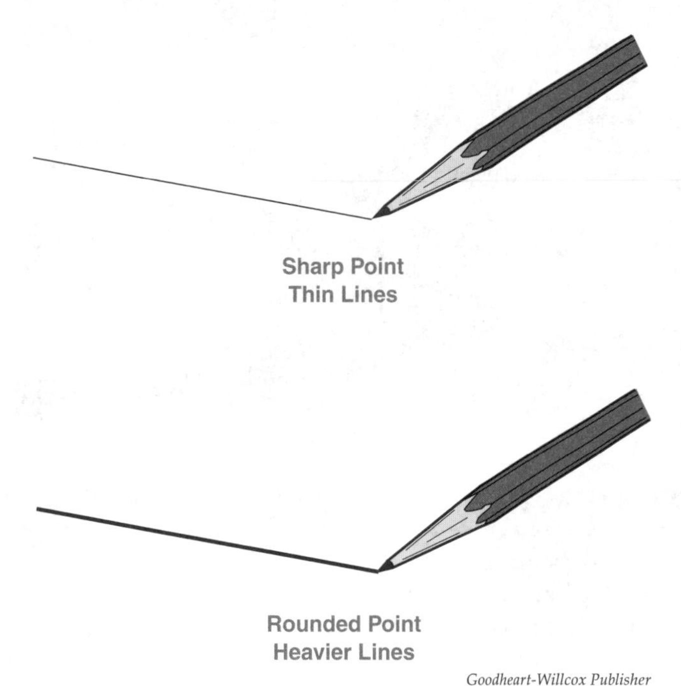

Goodheart-Willcox Publisher

Figure 5-1. Pencils must be sharpened differently for different line thicknesses. To maintain the quality of the line, rotate the pencil while drawing the line.

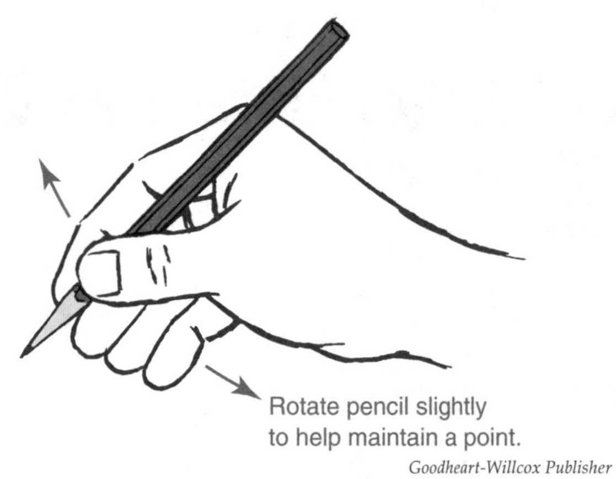

Goodheart-Willcox Publisher

Figure 5-2. The pencil must be held properly—your grip should be firm enough to draw smooth lines and relaxed enough so that your hand remains comfortable.

Rotate your pencil slightly between strokes to maintain the point and line consistency. Initial lines should be firm and light, but not fuzzy. Avoid making grooves in your paper by using too much pressure. In sketching straight lines, your eye should be on the point where the line will terminate, **Figure 5-3**. Make a series of short strokes (lines), rather than one continuous stroke. As you practice and develop these skills, freehand sketching will become more natural and fluid. As you are learning, follow the suggested steps and procedures for the various lines and shapes. They are designed to provide consistent lines and shapes.

There are many types of pencils for sketching and technical drawing. A regular wood and lead pencil

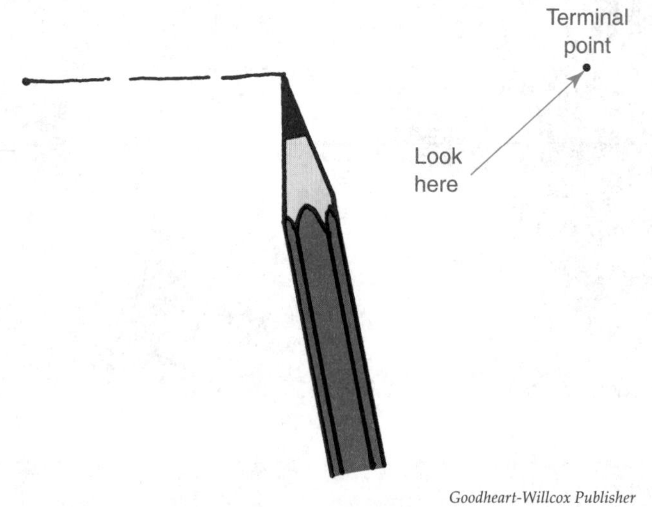

Goodheart-Willcox Publisher

Figure 5-3. Sketch lines as a series of short, faint strokes first. Then go over the strokes to make the line solid.

will work, but mechanical pencils such as the ones shown in **Figure 5-4** are more commonly used. These mechanical pencils come in a variety of lead diameters, with 0.3 mm, 0.5 mm, 0.7 mm, 0.9 mm, and 2 mm the most commonly used. See **Figure 5-5**. Leads for these pencils also come in a number of different hardnesses, depending on the use, a person's hand pressure, and personal preference. See **Figure 5-6**.

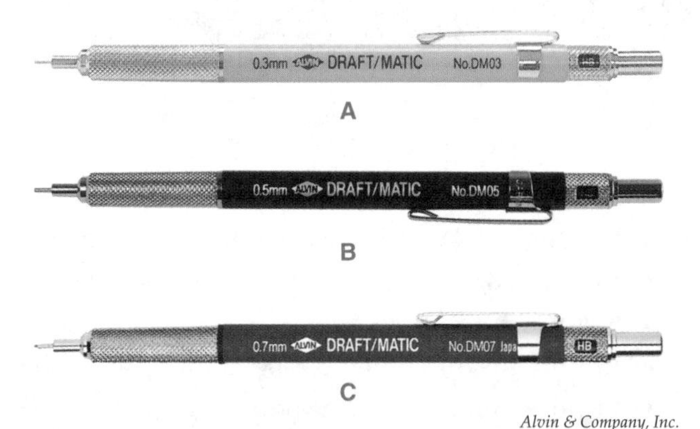

Alvin & Company, Inc.

Figure 5-4. Mechanical pencils are available in a variety of lead diameters. A—0.3 mm. B—0.5 mm. C—0.7 mm.

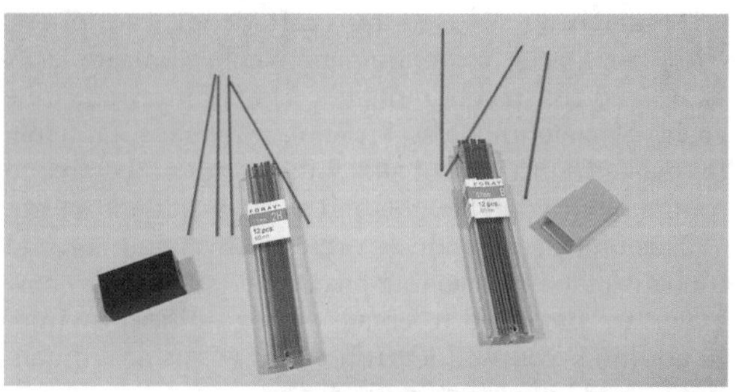

Goodheart-Willcox Publisher

Figure 5-5. Mechanical pencil leads come in a variety of thicknesses. Shown are 0.5 mm 2H and 0.7 mm B.

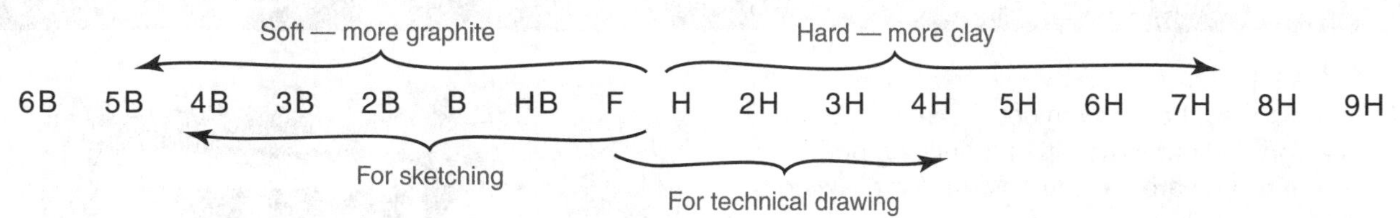

Goodheart-Willcox Publisher

Figure 5-6. Leads are made in a wide range of hardnesses for different applications and drafter preferences.

Sketching Horizontal Lines

Horizontal lines are sketched by positioning your forearm perpendicular to the line being sketched. You then move your arm and hand parallel to the line. While following this procedure, refer to **Figure 5-7**.

Horizontal Lines

1. Locate and mark the end points of the line to be sketched.
2. Position your arm by making trial movements from left to right (if you are left-handed, from right to left), without marking the paper.
3. Sketch short, light lines between the points. While sketching the line, look at the point where the line is to end.
4. Darken the line to form a continuous line with uniform weight. Your eye should lead the pencil along the lightly sketched line as you darken it.

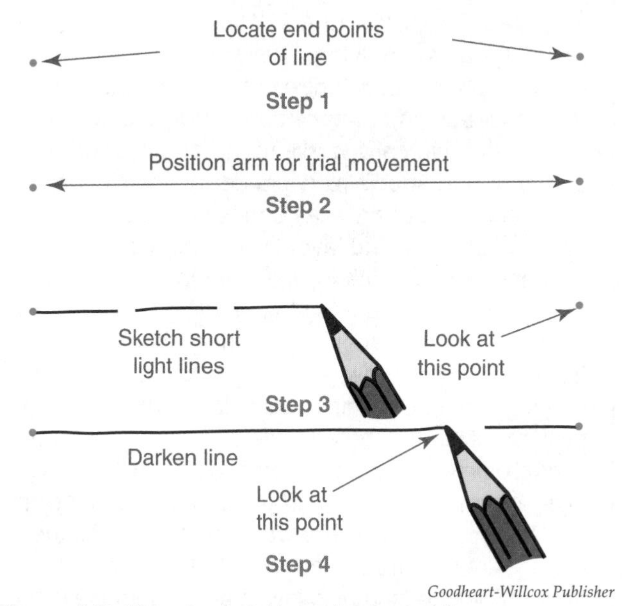

Goodheart-Willcox Publisher

Figure 5-7. Steps in sketching a horizontal line.

Sketching Vertical Lines

Vertical lines are sketched from top to bottom, using the same short strokes as with horizontal lines. When making the strokes, position your arm comfortably, approximately 15° from vertical. While following this procedure, refer to **Figure 5-8**.

Vertical Lines

1. Locate and mark the end points of the line to be sketched.
2. Position your arm by making trial movements from top to bottom.
3. Sketch short, light lines between the points. Keep your eye on the point where the line ends.
4. Darken the line to form one continuous line of uniform weight. Your eye should lead the pencil along the lightly sketched line.

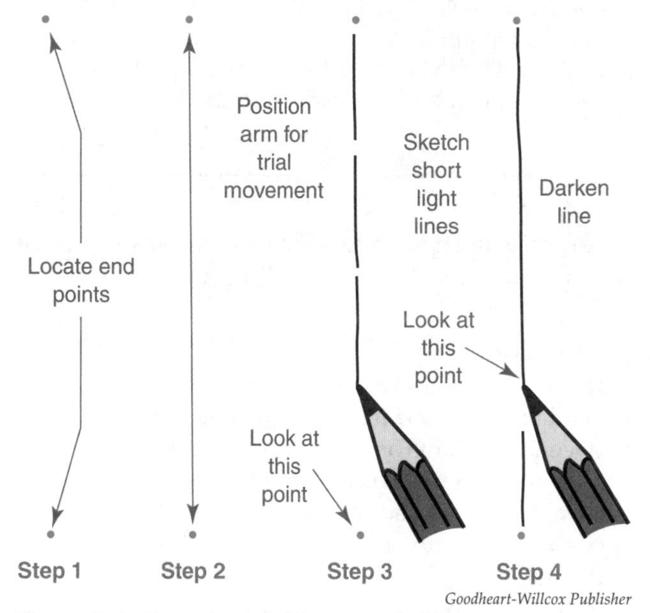

Goodheart-Willcox Publisher

Figure 5-8. Steps in sketching a vertical line.

CAREERS IN CONSTRUCTION

Drafter

Drafters create the drawing documents used in the design and construction of a building project. These drawings are called *construction drawings*. A drafter typically works with a computer-aided design and drafting (CADD) software system to produce construction drawings. A drafter using CADD software to make original designs is known as a *CADD designer* or *CADD detailer*. In addition to being trained in using CADD software, drafters have a working knowledge of building construction methods, building materials, and building codes.

A drafter typically works for an architect or an engineer and specializes in a certain discipline. Architectural drafters make drawings to show the architectural building layout. Structural drafters reference the architectural drawings to overlay the structural components of the building. This process continues as the other engineering disciplines work from the project drawing files to add the required items for the mechanical, electrical, and plumbing systems. The drafting and design work of each trade is coordinated to maintain accuracy and prevent interference between building components in the field. For example, the mechanical contractor uses the project drawing files to coordinate the air distribution system with the plumbing layout and determine if there are conflicts between the air ducts and plumbing lines. The design teams can determine if any of these components need to be rerouted before being installed in the field. Eliminating issues such as these is vital to keeping the project on schedule.

Depending on the project and the type of software used, drafters may prepare drawings in two-dimensional (2D) or three-dimensional (3D) form. In a project that utilizes building information modeling (BIM) processes, drafters use specialized software to create the building as a 3D parametric model. In this type of project, all design data is defined in the 3D model and the 2D drawing documentation is generated from the model automatically.

Making a career as a drafter requires ongoing education to stay current with advancements in technology. To be successful, drafters must be able to adapt to new developments in computer technology, design software, and construction methods and materials.

FutroZen/Shutterstock.com

Drafters use design software to create drawing documents used in construction.

You may find it easier to sketch horizontal and vertical lines if you rotate the paper slightly counterclockwise.

Sketching Inclined Lines and Angles

Straight lines that are neither horizontal nor vertical are called ***inclined lines***. You can sketch inclined lines between two points or at a designated angle. The same strokes and techniques used for sketching horizontal and vertical lines are used for inclined lines. If you prefer, rotate the paper to sketch these lines horizontally or vertically.

Angles can be estimated by first sketching a right (90°) angle, then subdividing its arc to get the desired angle. **Figure 5-9** illustrates how to establish a 30° angle.

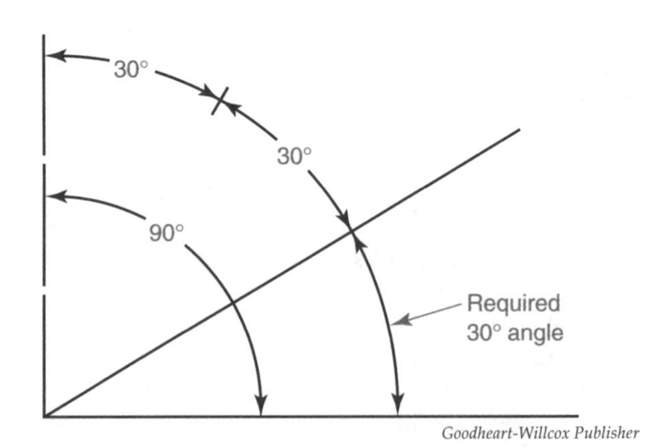

Goodheart-Willcox Publisher

Figure 5-9. Estimating angle sizes.

Sketching Arcs and Circles

There are several methods of sketching arcs and circles. One of the most commonly used is the ***triangle-square method***. This method is used in the following procedures.

Sketching Arcs between Lines

Use the following procedure to sketch an arc between two lines. See **Figure 5-10A**.

Arcs

1. Project the two lines until they intersect using light construction lines.
2. Lay out the arc radius from the point of intersection of the lines. Project two light construction lines to locate the center point of the radius.
3. Form a triangle by connecting the end points and locate the center point of the triangle.
4. Sketch short, light strokes from the point where the arc is to start on the vertical line through the center point to the point on the horizontal line where the arc ends.
5. Darken the line to form one continuous arc, which should join smoothly with each straight line. Erase the construction lines.

Sketching Circles

When sketching circles, use the following procedure. See **Figure 5-10B**.

Circles

1. Locate the center of the circle and sketch the center lines. Then, approximate half the diameter on each side of the center.
2. Sketch a square lightly at the diameter ends.
3. Across each corner, sketch a diagonal line to form a triangle. Then, locate the center point of each triangle.
4. Sketch short, light strokes through each quarter of the circle, making sure the arc passes through the triangle center point and joins smoothly with the square at the diameter ends.
5. Darken the line to make a smooth, well-formed circle. Erase construction lines.

Sketching Ellipses

Sketching an ellipse is similar to sketching a circle. Refer to **Figure 5-11** as you follow the procedure.

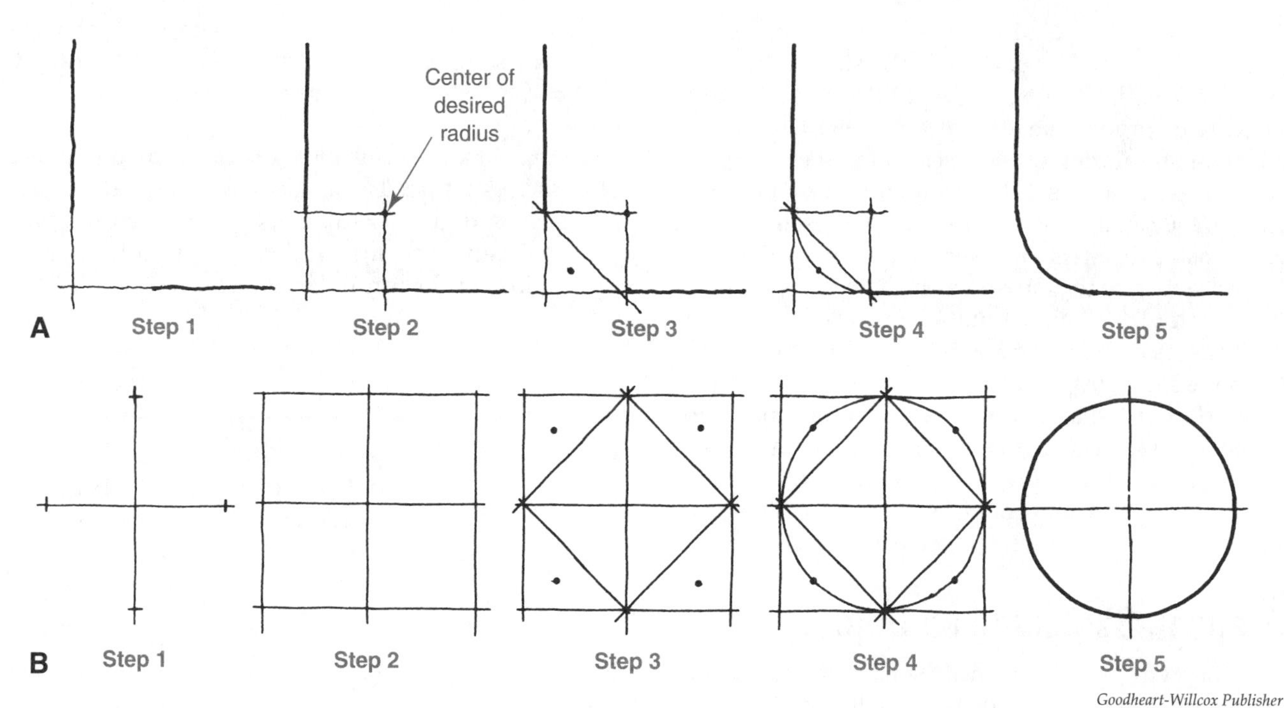

Figure 5-10. Using the triangle-square method. A—Sketching an arc. B—Sketching a circle.

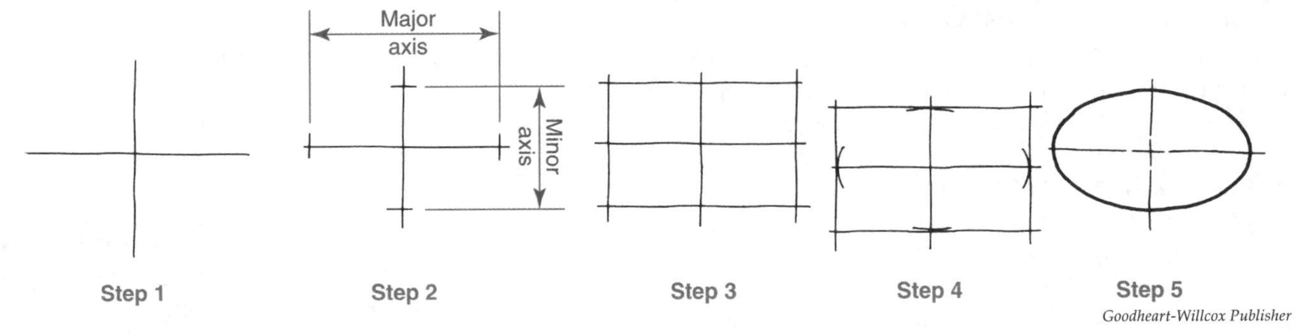

Figure 5-11. Steps in sketching an ellipse.

Ellipses

1. Locate the center of the ellipse and sketch center lines.
2. Lay out the major axis of the ellipse on the horizontal center line and the minor axis on the vertical center line.
3. Sketch a rectangle through the points on the axes.
4. Sketch tangent arcs at the points where the center lines cross the rectangle.
5. Complete the ellipse and darken the line, then erase the construction lines.

Proportion in Sketching

Proportion is the relationship of the size of one part to the size of another part, or to the size of the entire object. The width, height, and depth of your sketch must be kept in proportion so that the sketch conveys an accurate description of the object being sketched.

One technique useful in estimating proportions is the ***unit method***. This method involves establishing a relationship between measurements on an object by breaking each of the measurements into units. Compare the width to the height, and select a unit that will fit each measurement, **Figure 5-12**. Unit lengths should be in the same proportion.

Proportion is a matter of estimating lengths on a part or assembly, then setting these down on your sketch in the same ratio of units. Practice this method of establishing proportion in sketches. It will help you to represent the objects you sketch accurately.

Steps in Sketching an Object

The following steps will help you in laying out and completing your freehand sketches. **Figure 5-13** illustrates these steps for sketching an ellipse containing three circles.

Sketching an Object

1. Sketch a rectangle or square of the correct proportion.
2. Sketch major subdivisions and features of the object.
3. Erase unnecessary lines.
4. Darken lines to correct weight.

Sketching Aids

Many aids can be used to make your sketching more efficient and effective. ***Graph paper*** has a grid of light lines on it and is available in different grid sizes. The grid helps to keep lines straight and makes proportioning easier. Graph paper can be used for setting scale when drawing details to scale. For example, graph paper with four squares per inch can be used for drawing to 1/4″ = 1′-0″ scale. Using drawing scales is introduced in Unit 3.

Other aids are available to help prepare sketches. *Straightedges* are used to sketch lines. *Rules* help to keep proportions. If a rule or straightedge is not handy, a piece of folded paper or cardboard can be used to measure proportions or determine distances.

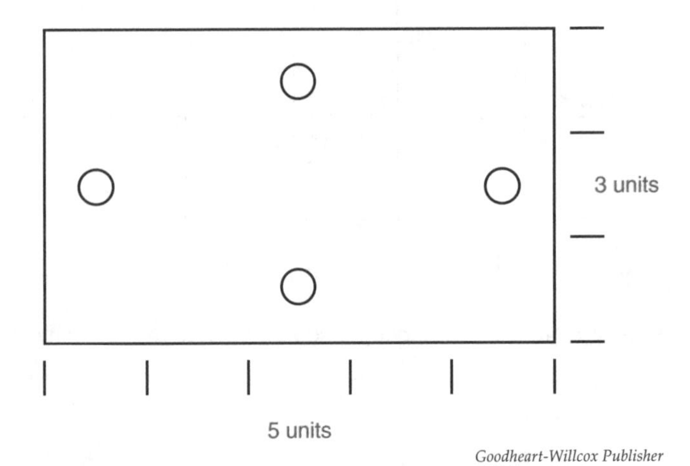

Figure 5-12. Proportioning an object using the unit method.

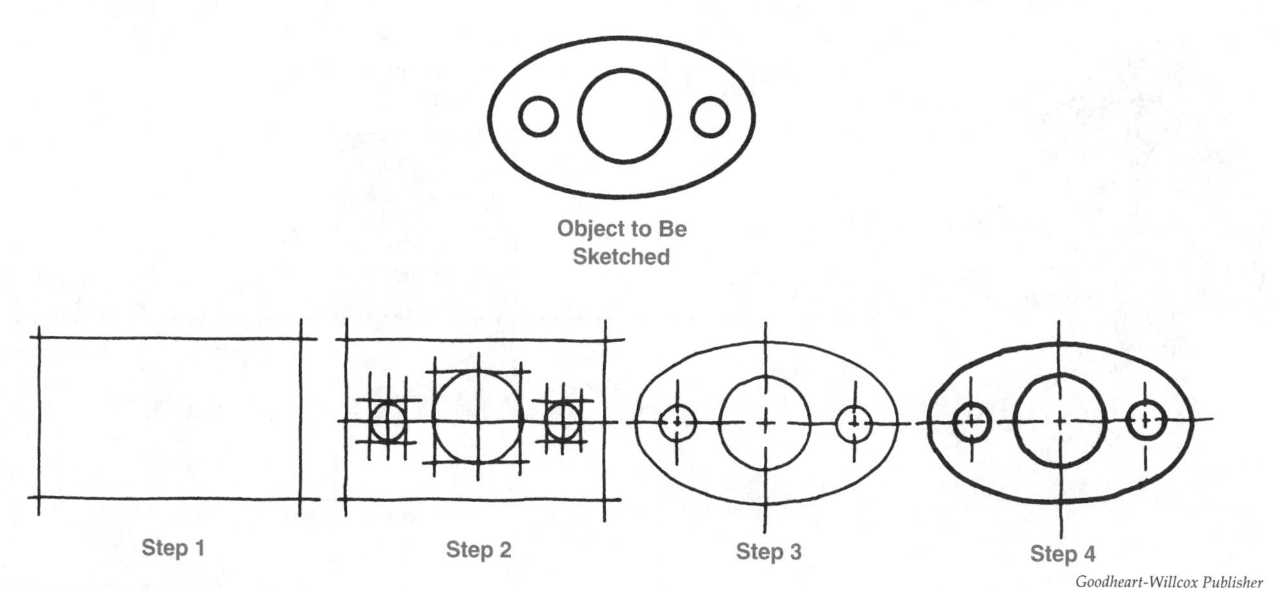

Figure 5-13. Steps in sketching an object.

You should eventually develop your sketching skill to a point where aids are no longer needed. Continue practicing sketching until you are comfortable with your technique.

The ability to sketch is an important skill that will help you visualize a design and communicate it to others. It will also give you a basic understanding of how views are constructed in orthographic drawings, discussed next.

> **Note**
>
> Making accurate sketches is a skill that takes practice. To improve your sketching ability, practice making sketches daily. You can sketch any object (it does not have to be a construction detail). Do this again and again and in time, you will be impressed with how you improve in using this valuable technique.

Introduction to Orthographic Projection

Nearly all drawings used on a construction project are orthographic drawings. These drawings are created using ***orthographic projection***, a process by which an object or structure is described using various views. Each view defines one face, or side, of the subject.

The views of an orthographic drawing are projected at a right angle (90°) to each other. The best way to visualize this is by cutting and unfolding a cardboard box, as shown in **Figure 5-14**. The front view remains in position. The four adjoining views revolve 90° around the "folds," bringing them into the same plane as the front view. The rear view is shown next to the left-side view, but it could be shown in several alternate positions, as indicated.

Imagine placing an object inside a glass box. If you then viewed the object through any of the six sides of the box, you would see only one face of the object. Each view through a side of the box would create one orthographic view, **Figure 5-15**.

It is easier to form a three-dimensional mental picture of an object or structure from a pictorial drawing. However, orthographic drawings are preferred because more details can be shown. Eventually, you will learn to visualize using orthographic drawings as easily as you would with pictorial drawings.

Creating Orthographic Drawings

Orthographic drawings are not difficult to create. Each view should be carefully drawn. Begin with the front view, then move to the top and side views. Lines parallel to the plane of the view being drawn are drawn at true size.

Figure 5-16 shows a block with a notch cut into one corner. The orthographic drawing of the block is shown in **Figure 5-17**. Refer to Figure 5-17 as you work through the following procedure.

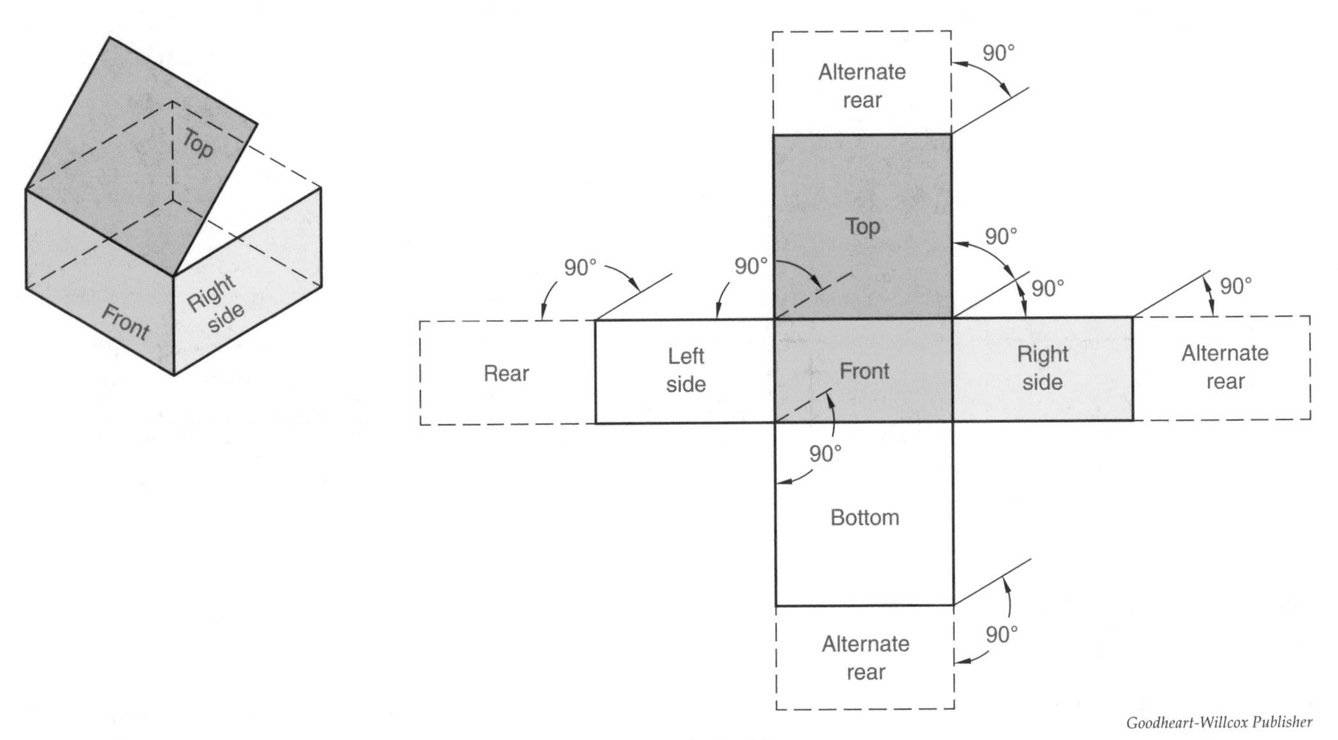

Goodheart-Willcox Publisher

Figure 5-14. Projection of orthographic views is shown by an unfolded box.

Drawing Orthographic Views

1. Begin by drawing the front view. All views should be drawn to scale. In an orthographic drawing, the object is oriented so that most of its features are located in the front, right side, and top views.
2. At every edge and feature shown on the front view, ***perpendicular projectors*** are drawn in the vertical and horizontal directions. These construction lines are drawn lightly, and erased when the drawing is complete.
3. Draw the top and side views. Start with the front edges. The projection lines connect common features between views.
4. From the front edge of the top view, draw a horizontal projection line. Draw a vertical projection line from the front edge of the side view.
5. At the intersection of these lines, draw a line at a 45° angle. Projection lines for features common to the top and side views will intersect at this line.

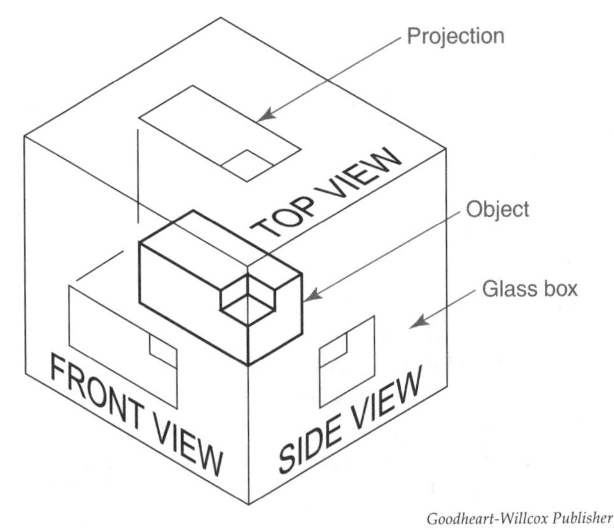

Goodheart-Willcox Publisher

Figure 5-15. Imagining an object within a glass box can help you visualize orthographic views and the projection of object lines.

Construction Drawings

Construction drawings that show different views of a building (such as floor plans and elevations) are obtained using orthographic projection. Imagine a building enclosed in a glass box, **Figure 5-18A**. Each view is projected toward its viewing plane, **Figure 5-18B**, and then unfolded and brought into the same plane as the front view, **Figure 5-18C**.

Due to the size of most buildings and the amount of information that must be shown, the different views usually are separated and placed on individual sheets. Prints are made from these separate drawings and fastened together to form the set of prints for a particular job. The bottom view is not shown; the top of the structure, or *roof plan*, is shown only for complex structures.

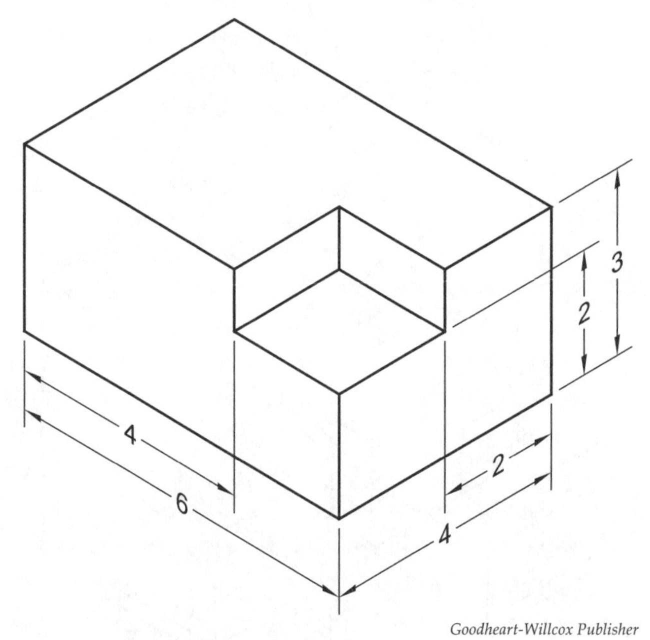

Goodheart-Willcox Publisher

Figure 5-16. This simple notched block can be easily described using orthographic projection.

Plan Views

A ***plan view*** is a drawing showing a view of the building from directly above. Plan views are taken at different levels throughout the building. For example, one plan view may show only the third story, while another would show the basement. In more complicated buildings, each floor may require multiple plan views to illustrate all construction details.

There are several different classifications of plan drawings (often referred to simply as *plans*). Only the most complicated buildings use all types of plans. Some basic buildings can fit all of the required information on a single plan. The common types of plan drawings are as follows:

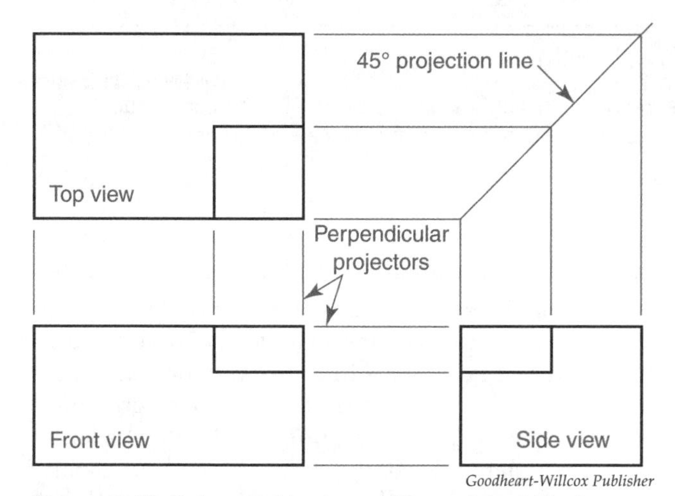

Goodheart-Willcox Publisher

Figure 5-17. Orthographic views of the notched block shown in Figure 5-16.

- ***Floor plan***. The floor plan shows the layout of the building, **Figure 5-19**. Walls, doors, windows, rooms, and stairs are shown. When space is available, other materials (such as plumbing and electrical fixtures) can also be shown on the floor plan. The floor plan is normally the first drawing in a set of prints. Many other drawings can be referenced from the floor plan, including section and detail drawings. (Floor plans are covered in detail in Unit 9.)
- ***Foundation plan***. The foundation plan is similar to the floor plan, except it shows the foundation for the building. Foundation walls, slabs, piers, and footings are shown. The basement is also shown on the foundation plan. (Foundation plans are covered in detail in Unit 10.)
- ***Framing plan***. A framing plan shows the layout of the structural members supporting a floor or roof. A framing plan is often included for each floor. If there is room, detail drawings of the connections between members may be included. (Framing plans are covered in detail in Unit 12.)
- ***Electrical plan***. The electrical plan is often created by the electrical contractor. Locations of receptacles, switches, and fixtures are included on the drawing. (Electrical plans are covered in detail in Unit 15.)
- ***Plumbing plan***. The plumbing plan shows heating and circulating equipment, supply and waste systems, plumbing fixtures, and locations where piping enters the building. (Plumbing plans are covered in detail in Unit 13.)
- ***Mechanical plan***. A mechanical plan shows the heating, ventilating, and air-conditioning (HVAC) system and any mechanical equipment and systems located in the building. Piping systems are also sometimes shown on the mechanical plan. (Mechanical HVAC plans are covered in detail in Unit 14.)

Elevations

Elevations are exterior views of a building, **Figure 5-20**. Elevations are orthographic views showing the vertical layout of a structure. Exterior elevations show features such as the style of the building, building materials, doors, windows, chimneys, and moldings.

Interior elevations may be provided to show the construction of a particular interior wall or area. Elevations are covered in detail in Unit 9.

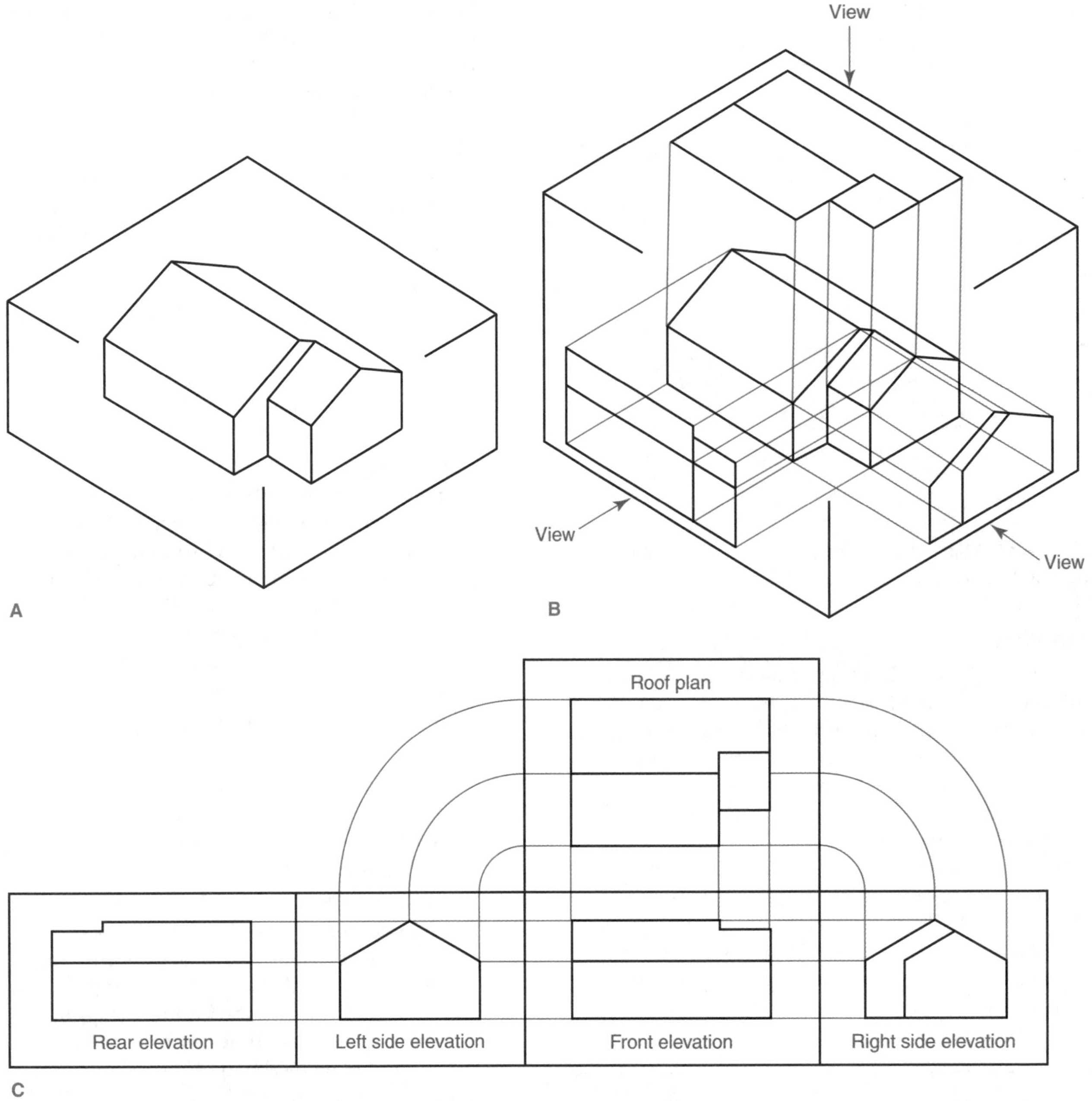

Goodheart-Willcox Publisher

Figure 5-18. Orthographic projection of a building. A—Imagine the building within a glass box. B—Each building face is projected onto a side of the box. C—The box and projections are unfolded into one plane. A view of the bottom of the structure is not needed.

Sections

The purpose of a set of prints is to show the construction details of a building. Plans and elevations give most of the information needed. Sometimes, however, it is necessary to show the "inside" of a wall, cabinet, or roof structure to clarify construction materials, connections, and procedures. When the drawing is an imaginary "cut" through a wall or other feature, it is known as a sectional view, or ***section***. Sections are covered in detail in Unit 9.

Details

Due to the scale at which construction drawings are usually made, certain features are not clearly shown on the plan, elevation, or sectional views. These features will require a large-scale illustration to provide information necessary for construction. In these situations, a detail drawing is used. ***Details*** are drawn at a larger scale than plans, elevations, and sections, and usually take precedence over drawings shown in less detail. Detail drawings are covered in Unit 9.

Dimensioning Practices

Dimensions are an essential part of a drawing. The person reading a print must understand various dimensioning techniques. You will find that drafting standards vary greatly from company to company, even though a considerable effort is being made to unify standard practices.

Standard practices for dimensioning are discussed in the following sections. A careful study of these practices will enable you to properly interpret construction drawings.

Dimensioning Technique

Drafting and dimensioning styles vary greatly from country to country, company to company, and drafter to drafter. However, certain practices remain fairly common. Although drafting is quite standardized, personal preference and style still exist.

Individual style is often used when drawing dimensions. A dimension line can terminate in an arrowhead, dot, or tick mark. The dimension can be placed above, below, or within the dimension line. It is common practice to use tick marks and show the dimension above the dimension line on construction drawings.

However, the selection of objects to be dimensioned and the techniques used to dimension them are fairly standard. Any dimension that may be needed during construction should be included on the drawing. Unnecessary dimensions should not be included.

Dimensioning Floor Plans

Floor plans show the arrangement of rooms, location of walls, and placement of windows. The floor plan should identify the materials and construction methods needed. The dimensions on the floor plan must be correct because other drawings will use the floor plan as their basis. Dimensions of walls, windows, and doors are included. Locations of unusual features should also be shown.

When dimensioning walls, different types of walls are dimensioned differently. Masonry walls are dimensioned to their exterior surface, **Figure 5-21**. Dimensions for exterior walls of frame and brick-veneer buildings usually start at the exterior surface of the stud wall, **Figure 5-22**. Some architects show the dimension to the outside of the masonry on brick-veneer buildings as well. This provides the workers with the dimensions necessary for laying out the sole plates over subfloors and in locating window and door openings before the ***sheathing*** (structural covering over studs) is added.

Some architects dimension exterior walls of single-story frame construction from the surface of the sheathing. This surface should align with the foundation wall. If the drawing scale is too small to show the dimension clearly, a note should be added, stating the surface or point to which the dimensions are referring. In the absence of a note or clarity of dimensions, you should calculate the exact location of the exterior dimension by referring to other prints—floor plans, elevations, or details.

Interior walls are usually dimensioned to the center or side of partitions. Some architects dimension to partition surfaces, then dimension the thickness of each partition.

Window and door openings are located by their centerlines for frame construction. For masonry construction, these openings are dimensioned to the edges of the masonry surface openings. Doors or windows in narrow areas may not be dimensioned for location if it is obvious they would be centered in the space available.

Dimensioning Elevations

Dimensions provided on elevation drawings are those related to the vertical plane, since most horizontal dimensions are included on plan drawings. The footing thickness, depth of footing below grade, floor and ceiling heights, window and door heights, and chimney height are provided on elevation drawings. Refer to **Figure 5-20**. In addition to vertical dimensions, elevations supply information through notes. Notes are typically given for grade information, materials, and special details.

Roof slope usually is given on a drawing as a slope triangle. The diagram represents the ratio between the ***rise*** (the change in elevation from the top to the bottom of the roof) and ***run*** (one-half the entire span of the building). A typical slope would be 4:12 or 4 units of rise for 12 units of run. The roof shown in **Figure 5-20** has a rather steep slope of 12:12.

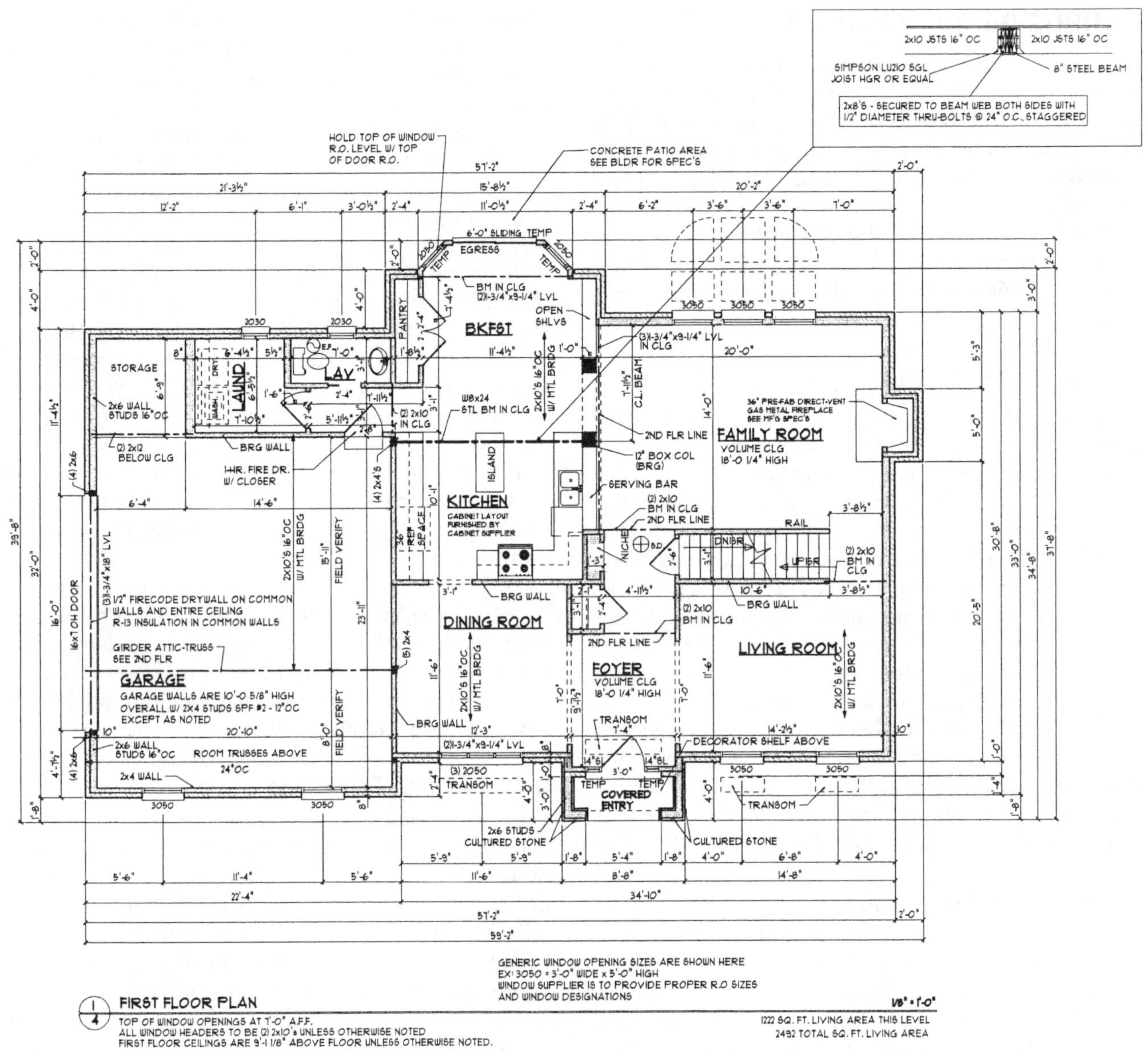

Vanderbuilt Homes, Inc.

Figure 5-19. The floor plan of a building.

Vanderbuilt Homes, Inc.

Figure 5-20. A front elevation of a building. Compare to Figure 5-19 and note how the floor plan relates to the elevation.

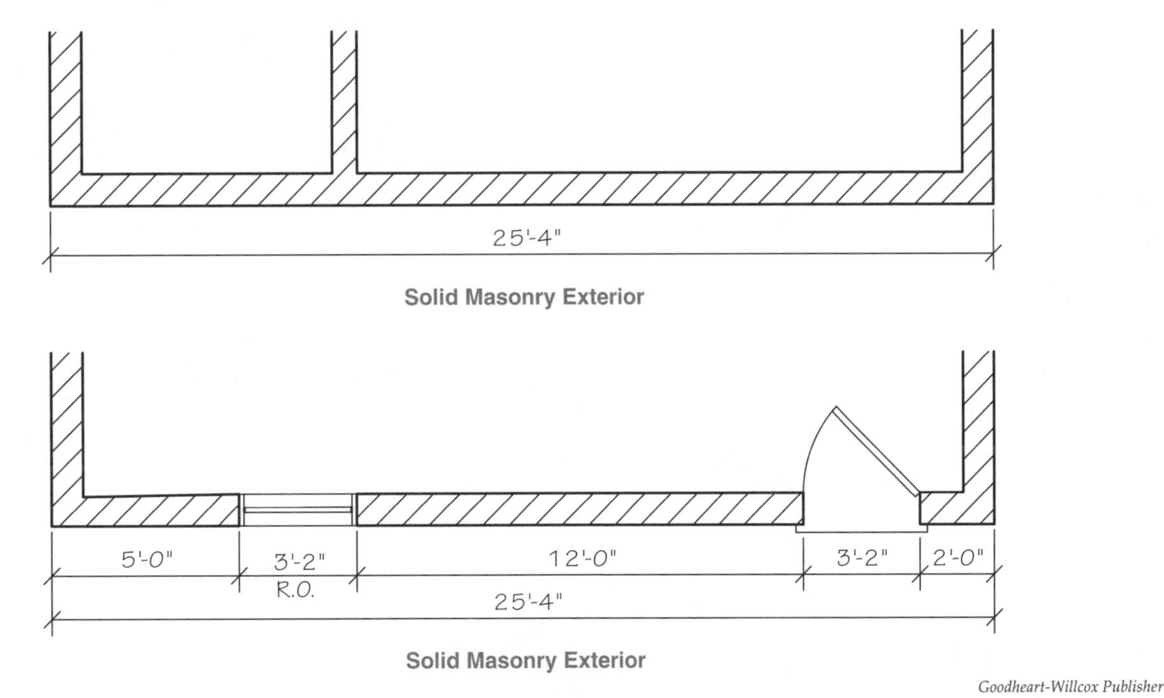

Goodheart-Willcox Publisher

Figure 5-21. Examples of dimensioning practices for masonry walls.

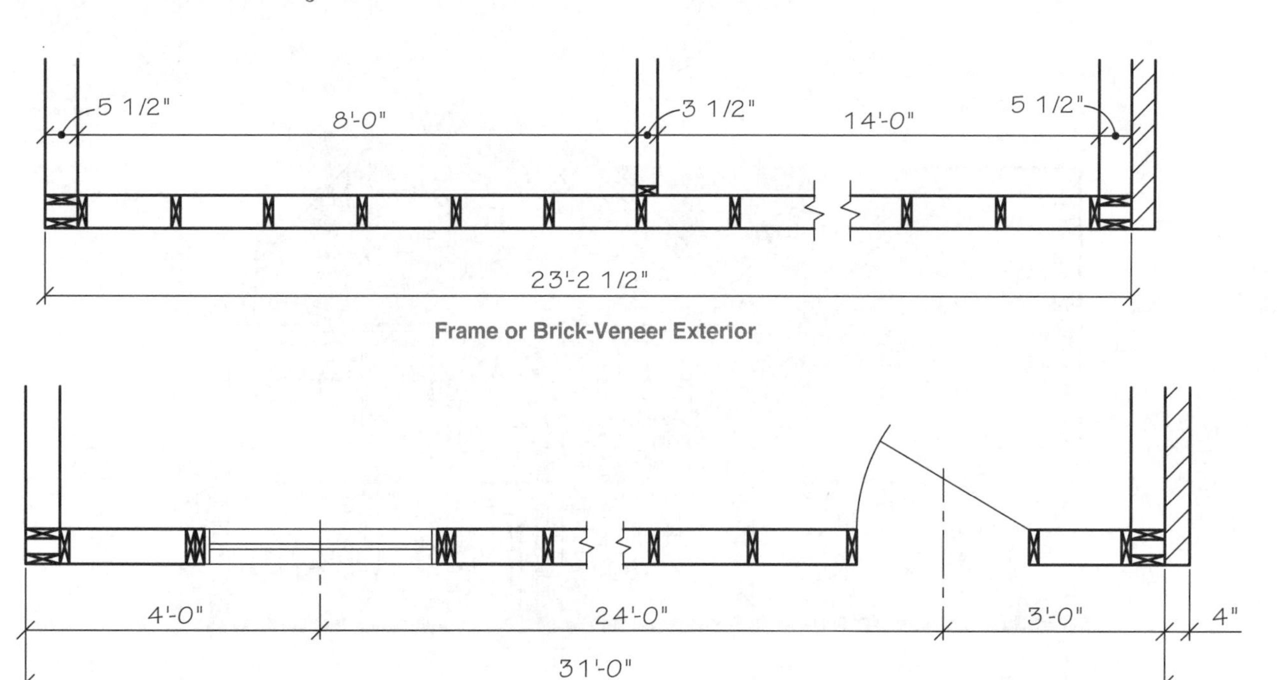

Goodheart-Willcox Publisher

Figure 5-22. Examples of dimensioning walls and openings in wood-framed and masonry buildings.

Test Your Knowledge

Name ______________________

Write your answers in the spaces provided.

_____ 1. It is important to have good sketching skills in order to _____.

A. impress people with your artistic ability

B. communicate ideas

C. read prints properly

D. not waste drafting supplies

E. All of the above.

_____ 2. When sketching a horizontal line, you begin by sketching short, light lines. When drawing these initial construction lines, your eyes should focus on _____.

A. the endpoint of the line

B. the tip of the pencil

C. the beginning of the line

D. your thumb

E. None of the above.

_____ 3. *True or False?* In the unit method of proportioning, the longest dimension of an object is set equal to one unit. The other dimensions are then estimated as fractions of that unit.

_____ 4. *True or False?* The more experience you gain sketching, the more tools and aids (such as rules and graph paper) you should use.

_____ 5. *True or False?* The point of the pencil should be sharpened differently for different weights of lines.

_____ 6. Which of the leads listed below would be used for technical drawing?

A. 4B

B. 6B

C. H

D. All of the above.

_____ 7. Why are orthographic drawings, rather than pictorial drawings, used for construction?

A. Orthographic drawings are easier to create.

B. Pictorial drawings are too large to fit on the drawing sheets.

C. More details can be shown on orthographic drawings.

D. Most standards require orthographic drawings.

E. Pictorial drawings are more commonly used.

_____ 8. The maximum number of orthographic views of an object is _____.

A. two

B. three

C. four

D. five

E. None of the above.

_____ 9. Which type of drawing is a view taken from directly above the building?

A. Sectional view

B. Elevation

C. Plan view

D. Detail

E. None of the above.

_____ 10. Which type of drawing shows the building as if part of it were removed?

A. Sectional view

B. Elevation

C. Plan view

D. Detail

E. None of the above.

_____ 11. Which type of drawing is used when there isn't enough room on another drawing to illustrate something?

A. Sectional view

B. Elevation

C. Plan view

D. Detail

E. None of the above.

_____ 12. Which type of drawing illustrates the building as if you were standing on the ground looking at it?

A. Sectional view

B. Elevation

C. Plan view

D. Detail

E. None of the above.

_____ 13. The location of light switches would be found on the _____.

A. foundation plan

B. framing plan

C. electrical plan

D. plumbing plan

E. None of the above.

_____ 14. The type of floor joists would be found on the _____.

A. foundation plan

B. framing plan

C. electrical plan

D. plumbing plan

E. None of the above.

_____ 15. The size of the basement wall would be found on the _____.

A. foundation plan

B. framing plan

C. electrical plan

D. plumbing plan

E. None of the above.

_____ 16. The location of hose bibbs would be found on the _____.

A. foundation plan

B. framing plan

C. electrical plan

D. plumbing plan

E. None of the above.

_____ 17. Which of the following dimensions wouldn't appear in an elevation drawing?

A. Elevation of the second-floor joists

B. Depth of the footing

C. Thickness of a wall

D. Height of windows

E. None of the above.

_____ 18. A typical roof slope (rise to run) is _____.

A. 1:12

B. 4:12

C. 12:4

D. 12:1

E. None of these are common roof slopes.

_____ 19. *True or False?* Masonry walls are dimensioned to their exterior surfaces.

_____ 20. *True or False?* Dimensions for exterior walls of brick-veneer buildings usually start at the exterior surface of the stud wall.

Activity 5-1

Sketching Lines

Name ____________________________________

Horizontal Lines

Sketch horizontal lines between points A–A′ through J–J′.

A • • A′ B • • B′

C • • C′ D • • D′

E • • E′

F • • F′

G • • G′

H • • H′

I • • I′

J • • J′

Vertical Lines

Sketch vertical lines between points K–K′ through T–T′.

K M O P Q R S T

K′ M′

L N

L′ N′ O′ P′ Q′ R′ S′ T′

Activity 5-2

Angles

Name ______________________________

Start at the indicated point and sketch the required angles.

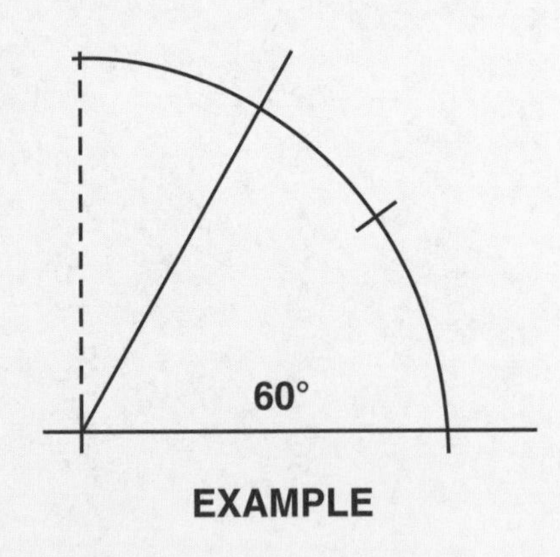

EXAMPLE

45°

A

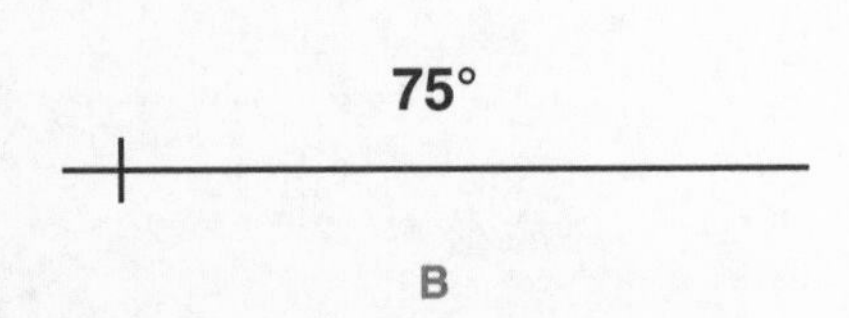

B

20°

C

100°

D

120°

E

22½°

F

85°

G

50°

H

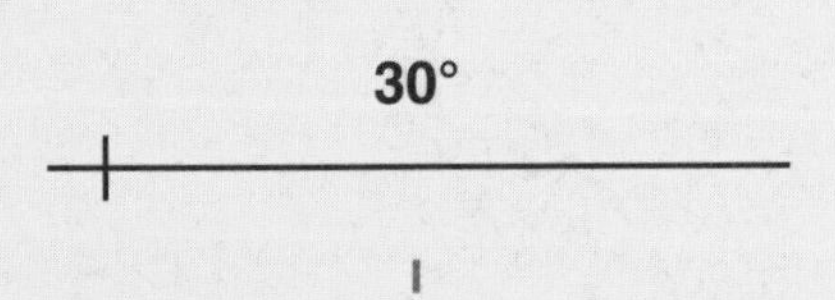

I

15°

J

K

Activity 5-3

Arcs and Circles

Name ________________________________

Sketch arcs joining the sets of lines for A through F. Show construction lines for A through C. Erase construction lines for D through F. Sketch circles for G through L. Show construction lines for G through I. Erase construction lines for J through L.

A

B

C

D

E

F

G

H

I

J

K

L

Matching Views

Name ____________________

Study the pictorial views and match each orthographic drawing by inserting the correct letter in the space provided.

A

B

C

D

E

F

1 ____

2 ____

3 ____

4 ____

5 ____

6 ____

Activity 5-5

Floor Plan

Name ______________________________

On a separate sheet of paper, sketch the layout for three adjoining rooms in the floor plan. Make the sketch at a larger scale than shown. Do not measure or dimension the sketch.

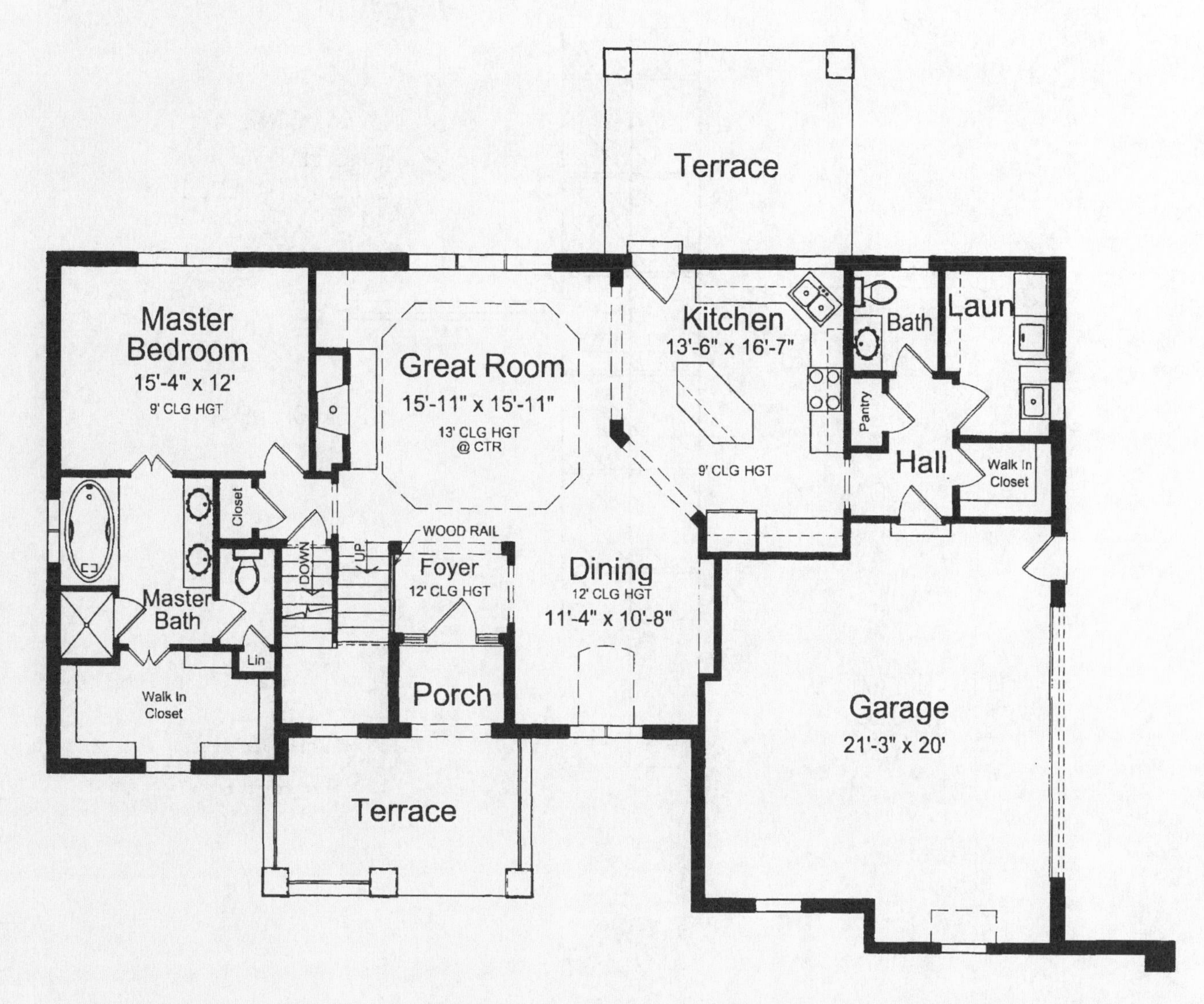

Activity 5-6

Commercial Wall Section

Name ______________________________

On a separate sheet of paper, make a sketch of the wall section. Do not include notes. Do not measure or dimension the sketch.

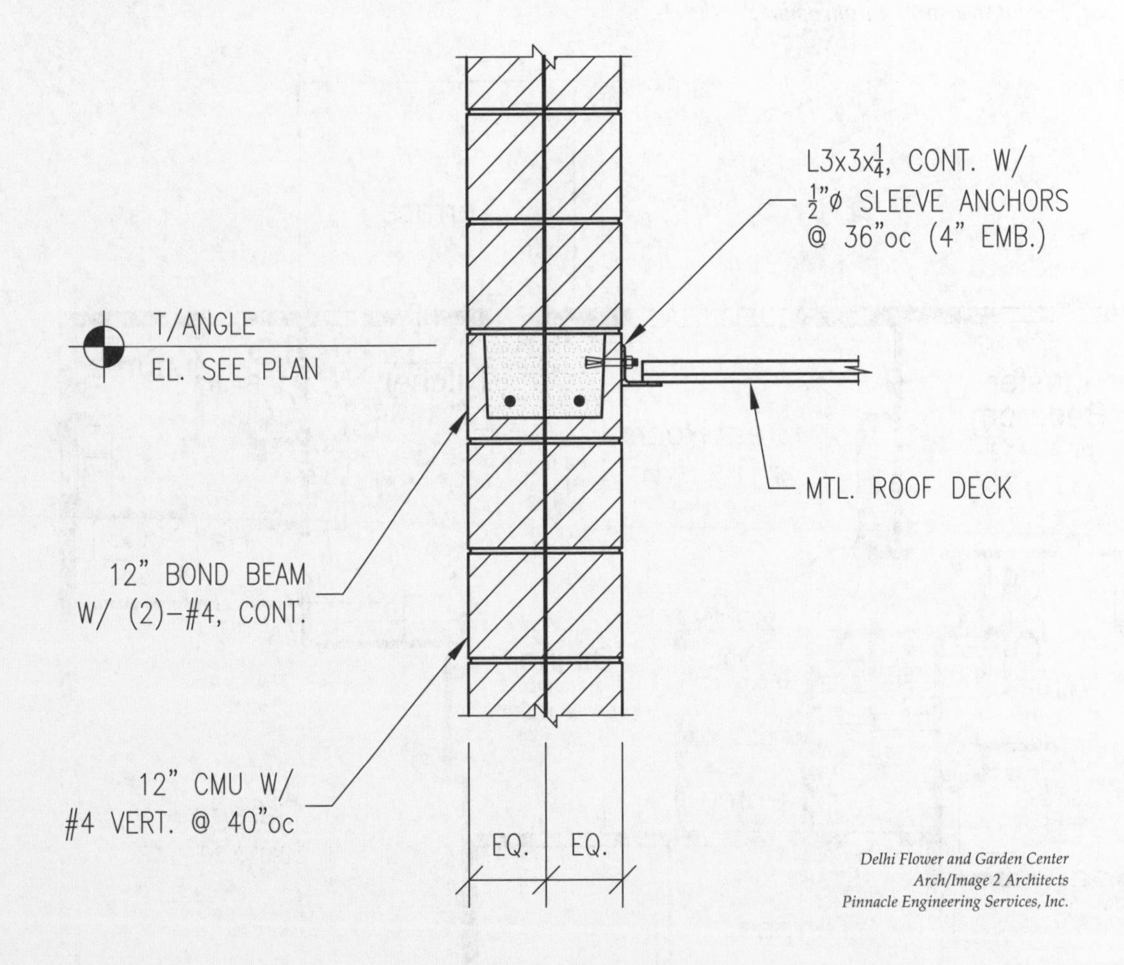

Delhi Flower and Garden Center
Arch/Image 2 Architects
Pinnacle Engineering Services, Inc.

Activity 5-7

Parking Lot

Name ___________________________________

On a separate sheet of paper, sketch the Visitor Parking area. Estimate the proportions. It is not necessary to show the dimensions.

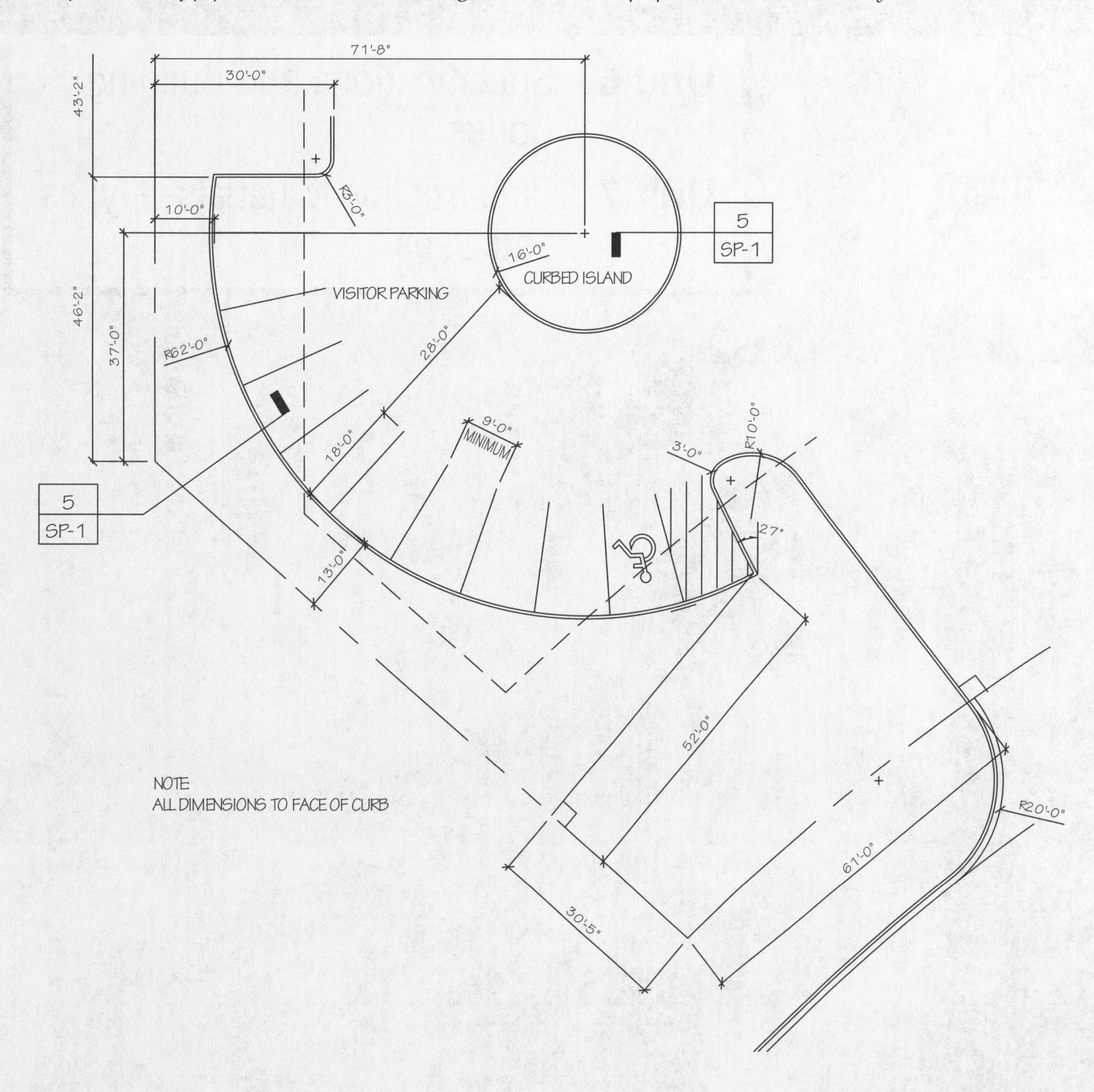

SECTION 3
Specifications and Materials

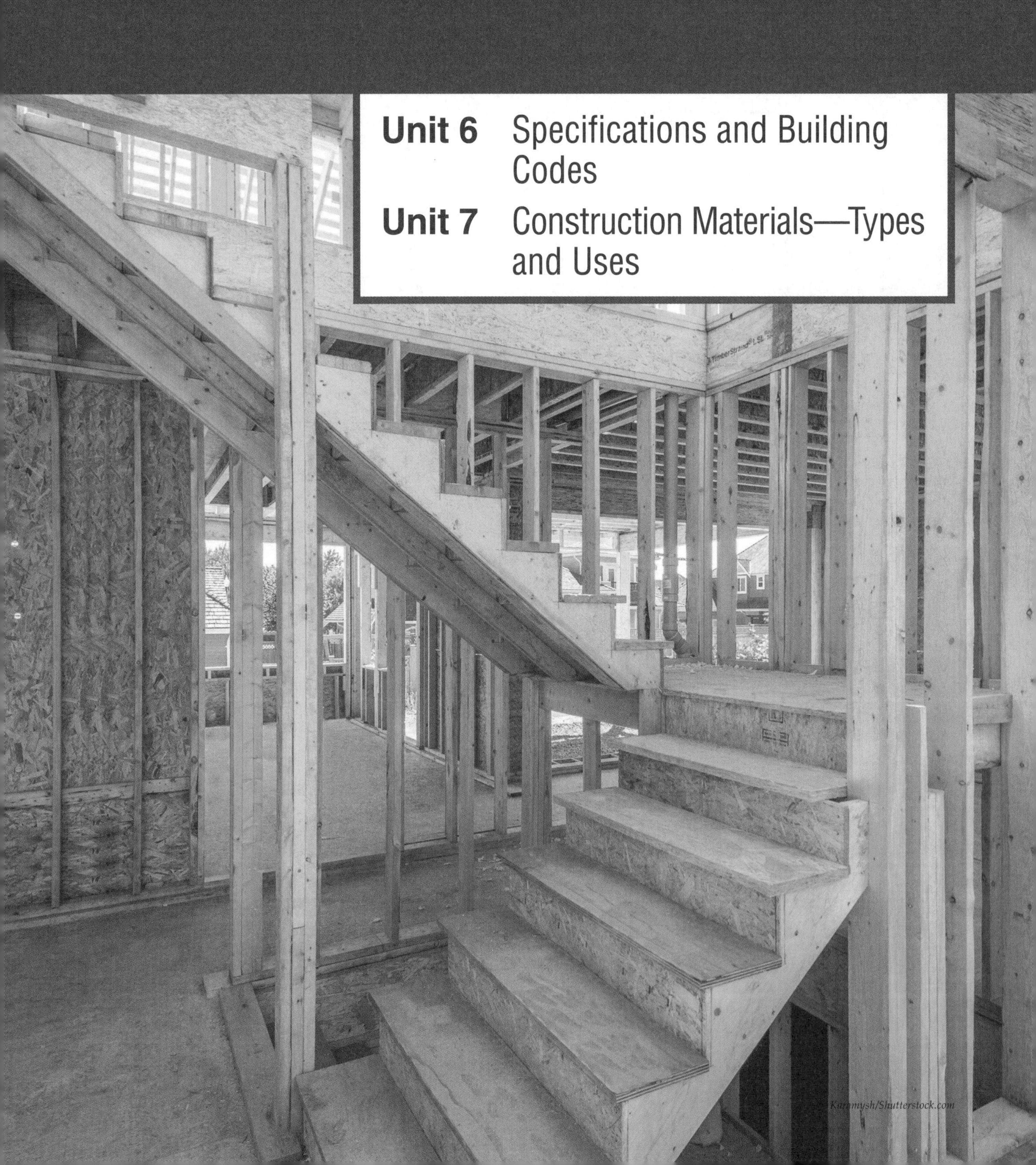

Kuramysh/Shutterstock.com

UNIT 6 Specifications and Building Codes

TECHNICAL TERMS

ASTM International
building codes
Construction Specifications Institute (CSI)
divisions
MasterFormat®
specifications

LEARNING OBJECTIVES

After completing this unit, you will be able to:

- Explain what is included in specifications.
- Explain why specifications are needed.
- Find a particular topic within specifications.
- Understand how building code requirements impact the building process.

Specifications are written statements that define the extent and quality of work to be done and the materials to be used. They supplement the drawings which, in turn, describe the physical location, size, and shape of the project. The specifications for a project are often referred to as the *specs*.

In this unit, specifications are explained. Also, a small sample set of specifications is presented to acquaint you with how specifications are organized and to assist you in locating information concerning a particular phase of construction.

Specifications vary in length from a few pages to several hundred pages. They are normally prepared by a person who is thoroughly familiar with construction procedures, building materials, and building codes. Building codes are introduced in this unit. Building materials are introduced in Unit 7.

Specifications and drawings supplement each other, so that all information needed for a construction project is included. An example would be the hardware for use on doors. The drawings would show location and direction of door swing; the specification would indicate quality or brand name, style, finish, and any other information to ensure that the proper hardware is purchased and installed.

Overview of Specifications and Scope of Use

The owner, architect, engineer, general contractor, subcontractors, material suppliers, building departments, and attorneys all use the specifications, as do the craftspeople engaged in the construction.

The specifications become the measuring device. The owner uses the specs to define the quality of work for the construction. The architect uses the specifications to define materials, processes, quality standards, industry standards, installation procedures, who is responsible for what portion of work, and anything that cannot be illustrated on drawings. Civil, mechanical, and electrical engineers use specs to define quality and industry standards, provide schedules for equipment, and define intricate portions of the project that may be difficult to illustrate on the drawings.

Contractors commonly joke about specifications, calling them "comics," and stating that the price of the job is a direct relationship to the weight of the specification book. For the contractor, the specifications define the scope of work to be performed by his or her own workforce and each subcontractor. Without them, the bids might be difficult to compare for the general contractor and the owner.

Building officials need specifications to determine compliance to local and state codes. Attorneys need specifications to assist in the resolution of disputes. Good clear specs written with fairness to all parties will keep the project moving and keep the attorneys out of the building process. The specifications, along with the drawings, ensure that the construction project will be completed in the manner that was intended. See **Figure 6-1**.

Purpose of Specifications

Specifications provide detailed information that cannot be shown on the drawing due to space constraints. Together, the specifications and drawings make up the *drawing documentation*, also referred to as the *construction documents* or *contract documents*. Both components provide information on details of construction.

Specifications serve several purposes:

- As a written guide for contractors bidding on a construction job. The specifications tell the contractor the exact materials to be used. This makes the estimate more accurate than if the contractor had to make assumptions.
- As a standard for quality of material and work to be completed on the project.
- As a guide for the building inspector in checking for compliance with building codes and zoning ordinances.
- As the basis of agreement between the owner, architect, contractors, and subcontractors in preparing scope-of-work contracts, so that disputes may be avoided during the construction of the project. Specifications can also aid in settling any disputes that may arise.

General Requirements:

Work shall include all items (building and site) indicated on these drawings unless otherwise noted. The American Institute of Architects (AIA) General Conditions, latest edition, are hereby a part of this contract. General contractor shall remove all construction debris from the job site and leave the building broom-clean.

Structural Specifications:

All footing shall be carried to undisturbed medium sand or rock, and to at least the depth shown on the drawings.

All concrete to attain a minimum ultimate compressive strength of 3000 psi in 28 days. Aggregates to be clean and well-graded, maximum size of 1″. Concrete slump 3″ minimum to 5″ maximum.

Goodheart-Willcox Publisher

Figure 6-1. Excerpts from a typical set of specifications.

Contents of Specifications

Specifications, and the information contained in them, vary from a simple outline to elaborate specification books.

The contract requirements, or *documents*, are contained within the specifications. These cover nontechnical aspects of the project, such as the terms of the contract agreement and responsibilities for inspection, insurance, permits, utilities, and supervision. It is in these areas that misunderstandings can often occur. Therefore, these requirements are more detailed on complex projects.

The technical aspects of the project are listed by divisions, usually in the order the work is performed on the job. Materials are specified by standard numbers. ***ASTM International***, formerly known as the American Society for Testing and Materials (ASTM), has standards identifying specification of materials and testing procedures.

In addition to specifying materials, the specifications identify standards for the quality of work to be done. Inspection and testing procedures and quality criteria are stated. Provisions may be included for determining responsibility and corrective measures if the quality standards are not met.

Specification Divisions

Specifications differ for various construction projects. However, there are certain elements that are common to any construction job. The ***Construction Specifications Institute (CSI)*** has developed the ***MasterFormat®*** standard for identifying requirements, products, and activities used in the construction industry. The MasterFormat standard is made up of a system of numbered ***divisions*** and numbered titles within each division. The current MasterFormat standard includes 50 divisions, although approximately one-third of those are reserved for future expansion. The MasterFormat standard uses a 6-digit numbering system to organize information. This system provides a standard convention for organizing content in specifications and other types of documentation used in the construction industry.

In the list that follows, only the MasterFormat divisions within the scope of this textbook have been included. For a complete list of MasterFormat divisions, see the Reference Section.

- Division 01—General Requirements
- Division 02—Existing Conditions
- Division 03—Concrete
- Division 04—Masonry
- Division 05—Metals
- Division 06—Wood, Plastics, and Composites
- Division 07—Thermal and Moisture Protection
- Division 08—Openings

- Division 09—Finishes
- Division 10—Specialties
- Division 11—Equipment
- Division 12—Furnishings
- Division 13—Special Construction
- Division 14—Conveying Equipment
- Division 21—Fire Suppression
- Division 22—Plumbing
- Division 23—Heating, Ventilating, and Air Conditioning (HVAC)
- Division 25—Integrated Automation
- Division 26—Electrical
- Division 27—Communications
- Division 31—Earthwork
- Division 32—Exterior Improvements
- Division 33—Utilities
- Division 34—Transportation

Organization of Information in Specifications

All specification documents are divided into sections of information. This organization helps the user find the information that is required to either estimate or construct the building. Section headings are used to organize the information and identify content. The following section headings are typical in a set of specifications:

- **Related Documents.** This section lists other documents with information that is of interest to the given topic but does not need to be repeated. The documents referenced may include a list of drawings or information about pay requests, retainage, insurance requirements, prevailing wages, and other requirements.
- **Summary.** A summary provides an overview of the type of work included under the given specification, but is not necessarily all-inclusive.
- **Related Sections.** This section identifies other sections that are of interest in the project specification book, such as General Conditions, Contract Documents, Insurance Requirements, etc.
- **Definitions.** This section lists special definitions that may not be typically understood by the reader. For example, the following definition might be included for clarification:

  Concrete Stain: Permanent coloration of existing cured concrete caused by chemical reaction.

  While this definition might be obvious to a decorative concrete contractor, the uninformed person reading this might interpret the term *concrete stain* to mean a simple stain on the concrete.
- **References.** This section is where the specification can make referenced items part of the contract, such as other reference material and items not specifically included within the specification. For example, standards or documents developed by the American Concrete Institute (ACI) or ASTM International might be referenced. These documents can be made part of a specification by simply listing them as a reference. An individual or company working from a specification must be familiar with this referenced information.
- **Description of Work.** This section provides a more detailed description of the work included and covered within the given specification.
- **Quality Assurance.** This section covers testing requirements for the project. This section may also list contractor qualifications, such as certifications or years of experience. In addition, mock-ups and samples of the work might be provided in this section, or they could be covered under Submittals.
- **Submittals.** This section provides samples of the work and product information, such as color charts and technical data information. This is where the owner's representative reviews, without obligation, the products or material that a contractor intends to use on a job.
- **Special Inspections.** This section describes information about independent lab testing, site visits, and preexisting conditions to the project.
- **Project Conditions.** Surrounding conditions and other considerations related to the project are described in this section. For example, there may be information related to traffic control, weather conditions, and protection of the work or surrounding areas.
- **Products.** This section lists all approved manufacturers or contractors for this scope of work. The term "or equal" is commonly used while referring to a list of products. This simply means that if a product being used by a contractor is not the one listed, the alternative product must meet the minimum requirements of the listed product and the contractor must obtain prior approval by the design team before use on the project.
- **Execution.** This section describes how the product is to be installed and/or perform.
- **Protection.** More specific information about protection of the work is covered under this section. This information identifies who is responsible and what is required at the time of completion when the work is turned over to the owner.
- **Maintenance.** Most owners want to know what is required to continue proper performance of the work. Maintenance information is addressed under this section.

- **Warranties.** This section addresses the number of years the product is under warranty against material and installation defects. Most warranties are one year, but the warranty period can be listed as longer under the given specification or under General Requirements. The warranty information explains what constitutes a defect and how it should be repaired.
- **Cleaning and Repairs.** As a project is turned over to an owner at the end of construction, sometimes special cleaning and repairs may be required. This section contains information addressing cleanup and repair requirements.

Reading Specifications

When reading specifications, first review the table of contents to become familiar with the type of information included. This will also give you an overview of what is included in the project.

Information of a general nature, such as insurance, supervision, and inspection, is found in *Division 01—General Requirements.*

When searching for a particular topic, determine under which division it would be included. For example, if you needed to know the required concrete strength, you would look under *Division 03—Concrete.*

> **Note**
> Whether you are bidding a specified material item, specification section, or the entire project, it is recommended that you read and become familiar with information included under General Requirements and Project Conditions in the specifications. It is common to hear material suppliers, subcontractors, and contractors say they were required to perform something they did not include in the bid that cost significant amounts of money. Become familiar with the entire set of documents. This is also helpful when preparing a proposal and statement of qualifications.

Building Codes

Building codes are laws and standards specifying requirements for building construction. Building codes establish standard construction practices and materials and are enforced by local building officials to ensure that required standards are met. Building codes impact all aspects of the building design and construction process. Becoming familiar with the requirements of building codes and careful inspection of them is important in all building projects.

There are different types of building codes in existence. *Model codes* are national building codes that promote uniformity nationwide. Examples of model codes include the International Building Code, the National Electrical Code, and the International Plumbing Code.

Local building codes are ordinances stating minimum standards for construction requirements and practices in a given area or community. Local building codes are developed to protect the health, safety, and welfare of the public. Many municipalities adopt model codes in lieu of writing and updating their own codes.

Safety and health standards are laws and standards stating safety and health requirements. Examples of safety and health standards are those developed by the Occupational Safety and Health Administration (OSHA) to protect the welfare of employees in all occupation types. These standards establish rules for proper ventilation, noise control, hazard protection around dangerous equipment, and personal body protection. These standards affect many aspects of the construction process, from the design of a safe workplace environment to the use of safety equipment by construction workers.

Building codes and standards are constantly monitored and revised with the advice of experts in the area of interest. To find out what codes are applicable in your area or an area where you plan to design and build, you can usually visit the community website. Look for information organized under the heading or title referring to *effective codes.* These codes are minimum requirements. Building officials will usually allow some variations as long as the variations comply with the basic principles and standards of the code.

Building code requirements are enforced by city, county, or local inspectors. These inspectors will examine work and sign off on select portions of a structure. For example, inspections are made to check items such as footings, framing of a residential structure, and plumbing and electrical rough-in.

Always be familiar with what inspections are required in the area where you are building. Visiting the local building official's office and developing a working relationship with inspection officials is a good way to learn about local building codes.

> **GREEN BUILDING**
> **Green Building Codes**
> Practices that started as 'green' are becoming 'code' across the country. Structures built with green materials are healthier and safer for their occupants. The International Green Construction Code, published by the International Code Council, establishes regulations that govern the impact of buildings on the environment and promote sustainable construction practices.

Test Your Knowledge

Name ____________________________________

Match each of the subjects in the left column with the MasterFormat division in which it would be located by writing its letter in the blank. Some of the divisions in the right column will be used more than once; others will not be used at all.

____ 1. Ceramic tile
____ 2. Steel beams
____ 3. Pressure-treated lumber
____ 4. Piping material
____ 5. Curtains
____ 6. Concrete curbs
____ 7. Ceiling paint
____ 8. Closet bypass doors
____ 9. Conveyors
____ 10. Bricks
____ 11. Light switches
____ 12. Fire sprinkler heads
____ 13. Grading
____ 14. Stone retaining wall
____ 15. Acoustical paneling
____ 16. Laundry sink
____ 17. Thermostat
____ 18. Exercise equipment
____ 19. Reinforcing bars
____ 20. Batt insulation

A. Division 01—General Requirements
B. Division 02—Existing Conditions
C. Division 03—Concrete
D. Division 04—Masonry
E. Division 05—Metals
F. Division 06—Wood, Plastics, and Composites
G. Division 07—Thermal and Moisture Protection
H. Division 08—Openings
I. Division 09—Finishes
J. Division 10—Specialties
K. Division 11—Equipment
L. Division 12—Furnishings
M. Division 13—Special Construction
N. Division 14—Conveying Equipment
O. Division 21—Fire Suppression
P. Division 22—Plumbing
Q. Division 23—Heating, Ventilating, and Air Conditioning (HVAC)
R. Division 26—Electrical
S. Division 27—Communications
T. Division 31—Earthwork
U. Division 32—Exterior Improvements
V. Division 33—Utilities

21. List the effective codes applicable in your community.

____

____

____

____

22. List five on-site inspections applicable in your community that require a building official inspector to sign off prior to continuing the construction.

____

____

____

____

____

Activity 6-1

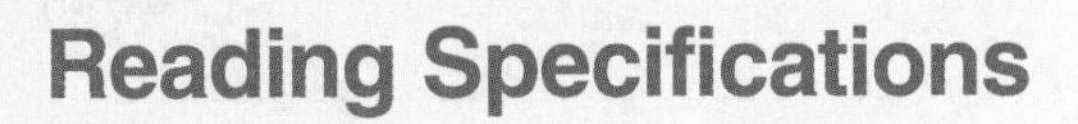

Reading Specifications

Name ______________________________

Use the partial set of specifications at the end of this activity and the list of MasterFormat divisions in the Reference Section to answer the following questions. Write your answers in the spaces provided.

1. Under which division would you locate

   A. Roofing materials and processes? 1. A. ______________________

   B. Interior wood trim? B. ______________________

   C. Interior finishing of walls? C. ______________________

2. What must all bidders do before submitting a bid? 2. ______________________

3. If an item is mentioned in the specifications and not in the drawings, must the bidder supply it? 3. ______________________

4. Who has the final decision as to the interpretation of the drawings and specifications? 4. ______________________

5. To whom does the term *contractor* refer? 5. ______________________

6. Who is responsible for providing and paying for temporary electrical service? 6. ______________________

7. Who shall provide the temporary heat needed to accomplish the work? 7. ______________________

8. If the permanent heating plant is used for temporary heat, who will pay for the cost of its operation? 8. ______________________

9. Who may provide and maintain temporary storage sheds for storage of all materials that may be damaged by the weather? 9. ______________________

10. How much of an overlap must be made between the new and existing roof? 10. ______________________

11. If a dimension is shown as 20′-0″ on the drawings, but is scaled as 18′-6″, which dimension is used? 11. ______________________

12. What thickness of rigid insulation is used on flat roofs? 12. ______________________

13. Where can a copy of the General Conditions of the Contract be viewed? 13. ______________________

    ____

14. These specifications are for what type of construction? 14. ______________________

15. If a detail drawing shows a 3′-4″ wide sidewalk and the plan view shows the same sidewalk as 3′-6″ wide, how wide should the sidewalk be? 15. ______________________

16. What are the following concrete material specification requirements for the Architectural walls?

A. Specified strength? 16. A. ____________

B. Max W/C ratio? B. ____________

C. What are the fly ash requirements? C. ____________

D. Why is retarder required for the Architectural concrete? D. ____________

____

17. What is the cement content for the Backfill concrete? 17. ____________

SPECIFICATIONS FOR AN OFFICE ADDITION and EXCERPTS FROM OTHER STRUCTURES

INDEX

SUPPLEMENTARY GENERAL CONDITIONS

General:

These Supplementary General Conditions and the Specifications bound herewith shall be subject to all the requirements of the "General Conditions of the Contract for the Construction of Building," latest edition, Standard Form of the A.I.A., except that these Supplementary General Conditions shall take precedence over and modify any pages or statements of the General Conditions of the Contract and shall be used in conjunction with them as a part of the Contract Documents. The General Conditions of the Contract are hereby, except as same may be inconsistent herewith, made a part of this Specification, to the same extent as if herein written in full.

Copies of the General Conditions of the Contract are on file and may be referred to at the Office of the Architects.

Scope of Work:

The work involved and outlined by these Specifications is for the construction work for the completion of the Office Addition for ________________________, as further illustrated, indicated, or shown on the accompanying drawings.

Examination of Site:

Before submitting a proposal for this work, each bidder will be held to have examined the site, satisfied himself fully as to existing conditions under which he will be obligated to operate in performing his part of the work, or which will in any manner affect the work under this contract. He shall include in his proposal any and all sums required to execute his work under existing conditions. No allowance for additional compensation will be subsequently in this connection, in behalf of any contractor, or for any error or negligence on his part.

Drawings and Specifications:

These Specifications are intended to supplement the Drawings, the two being considered complementary, therefore, it will not be the province of these Specifications to mention any portion of the construction which the drawings are competent to explain and such omission will not relieve the contractor from carrying out such portions as are only indicated on the drawings. Should the items be required by these Specifications which are not indicated on the drawings, they are to be supplied, even if of such nature that they could have been indicated thereon.

Any items which may not be indicated on the drawings or mentioned herein, but are necessary to complete the entire work, as shown and intended, shall be implied and must be furnished in place.

The decision of the architects as to the proper interpretation of the Drawings and these Specifications shall be final and shall require compliance by the contractor in executing the work.

Figured dimensions shall have precedence over scale measurements, and details over smaller scale general drawings. Should any...

Principles and Definitions:

Where the words "approved," "satisfactory," "equal," "proper," "ordered," "as directed," etc., are used, approval, etc., by architects is understood.

It is understood that when the word "Contractor" is used in these Supplementary General Conditions, the work described in the paragraph may apply to all Contractors and Subcontractors involved with the work.

Temporary Facilities:

Temporary Heat: Each contractor shall provide the necessary temporary heat needed for materials, water, etc., or any other heat required to accomplish his work. If the permanent heating plant is used for temporary heat, the owner will pay for its operation.

Temporary Light & Power: The general contractor shall arrange and pay for a temporary electrical service taken from the existing building for use by all trades, include 60 amp, 2 pole, 4 outlet fused panel mounted on pole or wall at the site. The electrical contractor shall provide temporary light within the structure as necessary and directed. General contractor will pay for all temporary electrical service until such time when permanent meter shall be installed.

Temporary Sheds for Storage: The general contractor may provide and maintain on the premises, where directed, watertight storage shed, or sheds, for storage of all material which may be damaged by the weather. These sheds shall have wood floors raised above the grounds.

Division 03

CONCRETE

Mix Type	Location	Specified Strength (at days)	Min. Portland Cement #/C.Y.(1)	% Max. Chloride by Weight of Cement	Max W/C Ratio	% AE(2)	Agg. Size(3)
A	Foundations: footings	3000 @ 28	500	0.30	0.60	–	#57, 1"
B	Exterior concrete (including topping slab in loading dock)	4500 @ 28	565	0.15	0.45	5–7	#57, 1"
C	Typical concrete (U.N.O.)	4000 @ 28	520	0.30	0.50	–	#57, 1"
D	Architectural concrete walls	4000 @ 28	565 (no fly ash)	0.30	0.40	(6)	#57, 1"
E	Architectural concrete columns	6000 @ 56	565 (no fly ash)	0.30	0.35	–	#8, crushed
F	Stair pan fills, masonry grout	3000 @ 28	500	0.30	0.50	–	#8, 3/8"
G	Concrete on composite metal deck (noted CWT– see arch. dwg.)	4000 @ 28	520 + 120 fly ash + steel fibers (4)	0.30	0.50	–	#57, 1"
H	Backfill concrete	1500 @ 28	280	1.0	–	–	#57, 1"

NOTES:

1. Including fly ash in mixes where permitted. Not applicable if a specified amount of fly ash is listed with the mix. The minimum cement requirement may be met by substituting 1.33 lb. of fly ash for each 1.0 lb. of Portland cement replaced, to the maximum allowed ratio of fly ash to cement. The maximum ratio of fly ash to Portland cement shall be limited to 20%.
2. Tolerance on air content shall be as delivered.
3. Normal weight.
4. Steel fibers are required in floor slabs designated "CWT" on Architectural finish schedule.
5. The Architectural concrete mix shall be designed to minimize shrinkage. A retarder shall be used to minimize heat gain and maximize long-term strength gain. A high range water-reducing admixture shall be used to limit the W/C ratio. The retarder may cause a delay in removal of the formwork. Strength development of the concrete may be critical in cycling formwork, depending on form system.

Division 07

THERMAL AND MOISTURE PROTECTION

General Conditions:

The contractor shall read the General Conditions and Supplementary General Conditions which are a part of these Specifications.

Scope of Work:

Furnish all materials and labor necessary to complete the entire roofing as shown on the drawings or hereinafter specified.

Roofing:

Flat roofing shall be 4 ply tar and gravel Spec. 102 of the Chicago Roofing Contractors Association, except single pour on gravel.

Carry new roofing onto existing roof a minimum of 3′-0″ and roof new saddles on existing.

Rigid Insulation:

Cover all flat roof surfaces over steel deck with 2″ of rigid insulation.

Insulation board shall be Fesco or equal.

Insulation shall be installed according to Spec. 102 of the Chicago Roofing Contractors Association. Form saddles on existing roof of rigid insulation so pitch is to new roof drain.

Activity 6-2

General Notes for a Residential Building Project

Name ____________________________________

Refer to Sheet 1 from the Sullivan residential building plans in the Large Prints supplement to answer the following questions.

1. What is the design load for the first floor?

    ____

2. Give the specified psi rating for the following items:

    A. Interior slab-on-grade ____________________________________

    B. Foundation walls ____

    C. Garage slab ____

3. What is specified for the framing lumber?

    ____

    ____

4. What is the size of the air space specified for masonry veneer walls?

    ____

5. What manufacturer of windows is specified?

    ____

Specifications for various aspects of concrete work, including reinforcement, formwork, and cast-in-place material, are found in the Construction Specification Institute's Division 03.

UNIT 7
Construction Materials—Types and Uses

TECHNICAL TERMS

admixture
adobe brick
aggregates
angles
ashlar
bent glass
bond
bond beams
brick masonry
building brick
cement
ceramic tiles
concrete
concrete brick
concrete masonry unit (CMU)
cut stones
face brick
ferrous metals
firebrick
fired-clay tile
flexible insulation
float glass
gage system
glass
glass block
glazed brick
gravel
green building
grout
header course
hydration
insulating glass
kiln-burned brick
laminating
lintels
loose-fill insulation
lumber
mortar
mosaic tiles
nonferrous metals
open-web steel joists
patterned glass
paving brick
pervious concrete
plywood
quarry tile
reflective insulation
reinforcing bars
rigid insulation
rough-sawn lumber
rubble
R-value
safety glass
sand-lime brick
single-wythe walls
stained glass
stone masonry
stretcher
structural clay tile
structural steel
surfaced lumber
sustainability
sustainable design
terra cotta
thermal insulation
welded wire fabric (WWF)

LEARNING OBJECTIVES

After completing this unit, you will be able to:

- Identify a variety of basic materials used in construction.
- Identify the basic components of concrete.
- Describe different types of masonry brick, block, and mortar.
- Classify wood as hardwood or softwood.
- Recognize different structural steel shapes.
- Describe various types of glass, plastics, and insulation.
- Identify symbols representing materials on a drawing.
- Explain the fundamentals of green building construction.

It takes many different kinds of materials to construct a building project. Most of these materials are carefully detailed in the project construction documents. Information about materials may also be included in the drawing notes and the project specifications.

The purpose of this unit is to provide you with an overview of the most common materials used in building construction. The materials will be discussed as they relate to numbered divisions in the MasterFormat standard published by the Construction Specifications Institute (CSI). Also discussed are material symbols used on construction drawings and the ways in which materials may be identified on drawings.

Division 31—Earthwork

Aggregates (also referred to as ***gravel***) are used for many different construction applications. In building construction, aggregates are typically used for subbase support under concrete slabs, foundation backfill, underground drainage, and fill around pipes, **Figure 7-1**. Aggregates also serve as one of the three basic ingredients of concrete.

Aggregates start out as solid rock in the earth's crust, and over many thousands of years, are broken down by freeze/thaw cycles and other natural action into pieces of various sizes. The larger pieces are called coarse aggregates or gravel; the smaller pieces are usually referred to as sand or fine aggregates.

There are various kinds of aggregates, such as:

Pea gravel, small, rounded pieces ranging in diameter from ¼" to 3/8".

River gravel, larger aggregates dredged from river bottoms, banks, and flood plains.

Crushed stone, mined stone that has been mechanically fractured to a specified size.

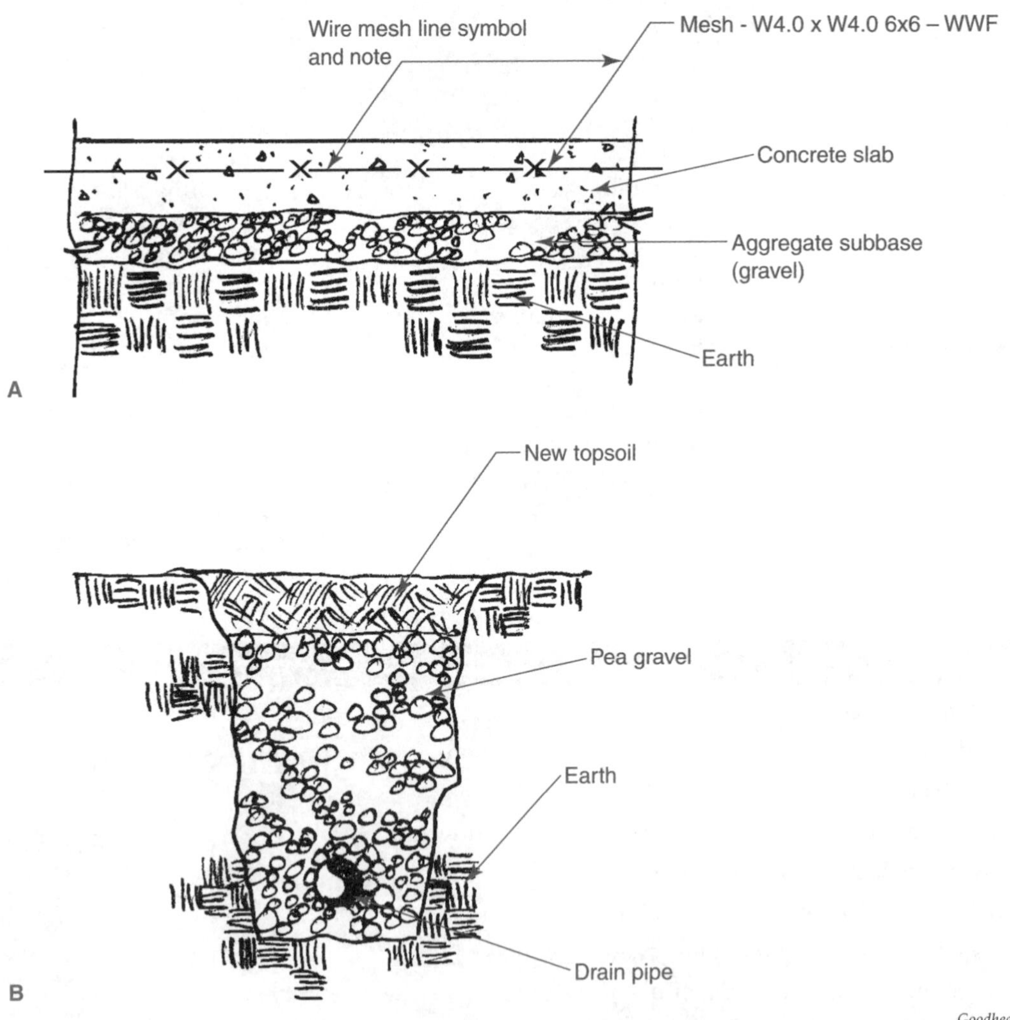

Goodheart-Willcox Publisher

Figure 7-1. Typical uses for aggregates. A—Gravel used as a subbase support for a concrete slab. Typically, this type of gravel is crushed with some fines in it so it can be compacted. Note the different symbols used for the various materials. Wire mesh, used for reinforcement, is represented by a line symbol consisting of a series of short line segments and a repeating "X" symbol. B—Pea gravel used to fill a curtain or trench drain. This type of gravel has a rounded shape and aids in water drainage. Note that both types of gravel shown in A and B have the same material symbol. Information about the required type is determined by referring to a general note or the project specifications.

Characteristics of aggregates are:

- Soundness
- Chemical stability
- Abrasion resistance
- Grading and sieve analysis
- Percentage of crushed particles
- Particle shape
- Surface texture
- Specific gravity
- Absorption
- Moisture content
- Volume stability

Some gravel drawing symbols are shown in **Figure 7-2**.

Division 03—Concrete

Concrete is one of the oldest building materials, having been used by the Romans as early as 100 B.C. Concrete is a mixture of cement, sand, coarse and fine aggregates, admixtures, and water. When first mixed, it is *plastic* (able to flow and be shaped) and can be cast to take the shape of the formwork provided.

Hardening of the concrete is caused by a chemical reaction between the cement and water called ***hydration***. Most mixtures of concrete set within 4 hours but can take up to 12 hours, depending on the temperature, the volume of the pour, type of cement, and admixtures. When the temperature is below 70°F (20°C), the chemical reaction slows. Very little chemical reaction takes place below 40°F (4°C), and almost none occurs at 32°F (0°C). The rule of thumb is: if you are comfortable, the concrete is comfortable. Concrete continues to harden for months after the initial set, but most placements reach their compressive or design strength within 28 days. Forms can be removed after one to several days or when the concrete can support itself. This should be determined by an engineer and testing lab.

Types of Cement

Cement binds the concrete mixture together. There are five basic types of cement. The most common, used for general construction, is called *Type I Normal Portland cement*. Another variation used in construction is *white Portland cement*. It is light-colored and used chiefly for architectural effects. White Portland cement is made from carefully selected raw materials and develops the same strength as the normal gray-colored Portland cement.

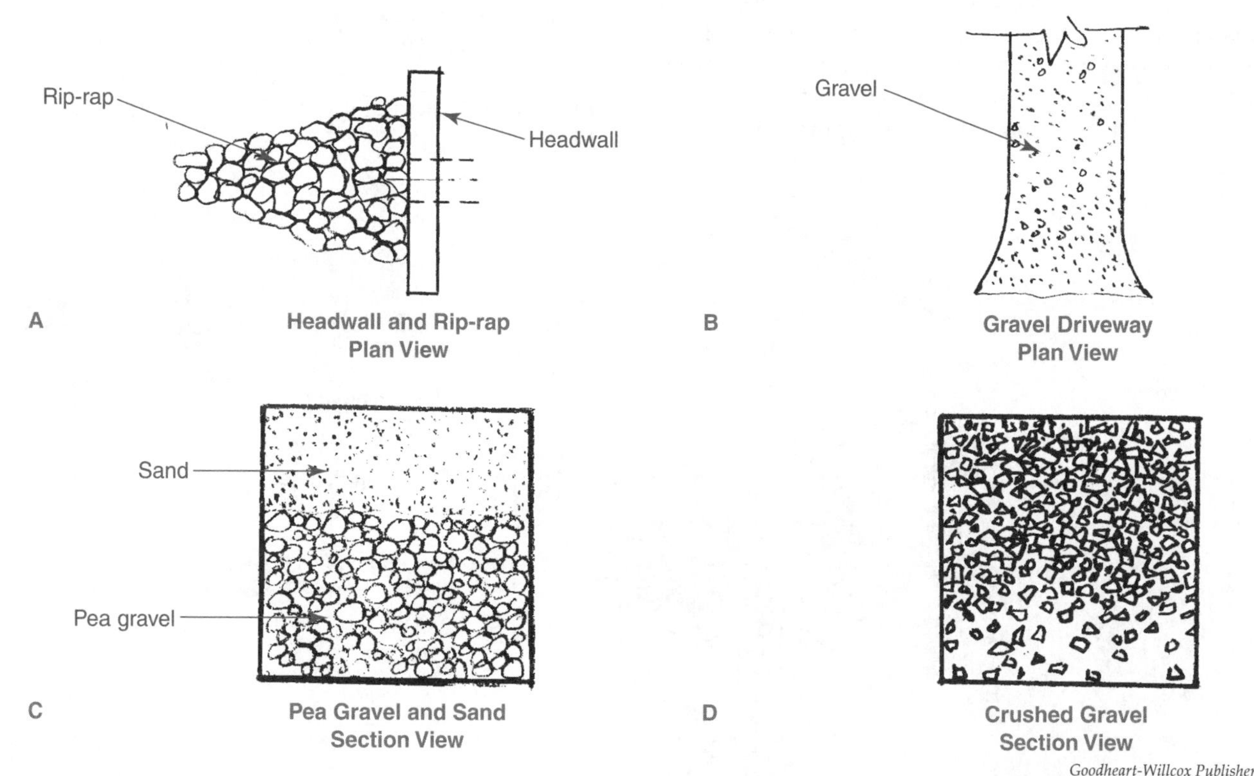

Goodheart-Willcox Publisher

Figure 7-2. Samples of drawing symbols for gravel. A—Plan view of a headwall with rip-rap used to prevent erosion from drainage water outflow. This type of structure would be found on a site plan. B—Plan view of a gravel driveway. C—Section view of pea gravel with a sand layer above the gravel. D—Section view of crushed gravel.

The five basic types of cement include:

- Type I, Normal Cement (most common)
- Type II, Moderate Sulfate Resistance (slow-reacting)
- Type III, High Early Strength (fast-setting)
- Type IV, Low Heat of Hydration (low heat generation)
- Type V, High Sulfate Resistance

These cements, along with aggregates and admixtures, are available to produce special types of concrete. For example, Type IV cement is for low heat generation for large construction building foundation projects, such as dams. Other cements have high early strength to produce concrete that gains strength faster than normal, Type I or II cements, permitting earlier form removal and thus speeding cold weather construction. Still others are more resistant to deterioration caused by sulfates and alkalis in the soil.

Concrete Mixes

A concrete mix should be designed to produce the desired result. Characteristics and properties of concrete depend on the materials, and their proportions, that make up the mixture. This will determine the workability, strength, durability, economy, volume stability, and appearance of the finished hardened concrete. Enough water is added to make the mixture plastic, so that it will flow into the forms. Too much water, however, will reduce the strength and durability of the concrete, so the contractor needs to be careful. A typical mix would consist of 10% cement, 15% water, 25% fine aggregates, 45% coarse aggregates, and 1% to 5% entrained or entrapped air. See **Figure 7-3**. Concrete mixes can vary from job to job and area to area depending on engineering and structural requirements.

> **GREEN BUILDING**
>
> **Fly Ash**
>
> Fly ash is produced when coal is burned to produce electricity. This by-product can be used in concrete as a replacement for part of the Portland cement. Environmental benefits of reusing fly ash to replace Portland cement include conservation of resources and energy, reduction of greenhouse gas emissions, and reduced costs. In addition, fly ash improves the workability of the fresh concrete; requires less water, which reduces cracking; and makes the final product more durable because permeability is reduced.

Any material added to the concrete mix—other than cement, sand, aggregate, and water—is known as an ***admixture***. Admixtures are used to make the mix more workable, retard or speed up hardening, increase freeze resistance, or increase chemical resistance. Common admixtures to concrete include *air entrainment*, used to improve durability in freeze/thaw environments; *retarders*, used to slow down the initial set of fresh concrete, especially in hot weather; *accelerators*, used to speed up the initial set of fresh concrete in cold weather; *water reducers*, used to reduce the amount of water required for a desired workability and water-to-cement ratio for strength; and *coloring agents*, used for altering the color of the concrete mixture. Concrete is typically transported to the job site in a ready-mix truck. See **Figure 7-4**.

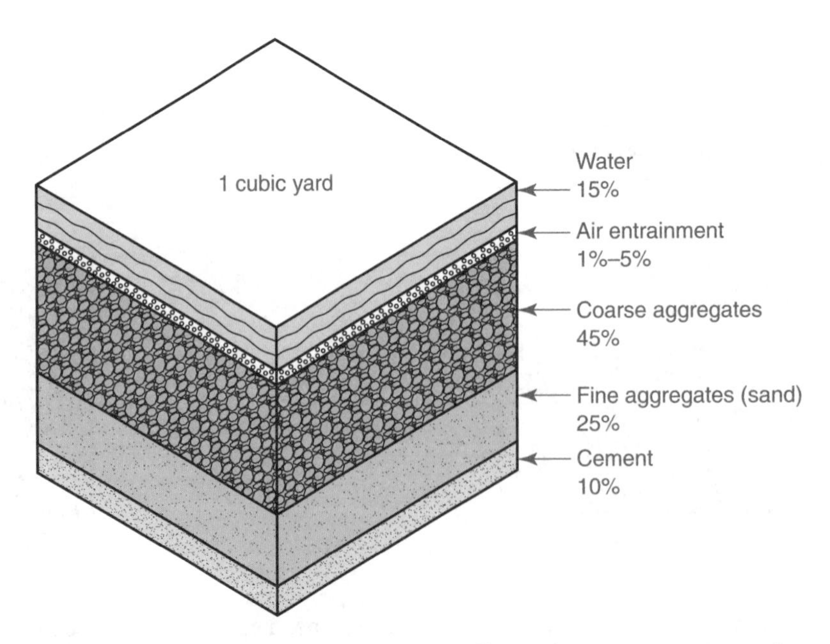

Goodheart-Willcox Publisher

Figure 7-3. Common proportions of materials making up concrete. Proportions can vary depending on local materials, design mix and strength requirements, and air entrainment.

Goodheart-Willcox Publisher

Figure 7-4. A ready-mix truck is typically used to transport concrete to the construction site.

Reinforced Concrete

Concrete has great compressive strength, but very little tensile (pulling) strength. To overcome this weakness, concrete is cast around steel ***reinforcing bars***. These bars (commonly referred to as "rebar") have high tensile strength. As the concrete hardens, it grips the steel to form a bond. The size of the bar is indicated by the bar number, which is a multiple of 1/8″. For example, a #4 bar is 1/2″ in diameter (4 × 1/8″ = 1/2″). See **Figure 7-5**. Refer to Unit 10 for more information.

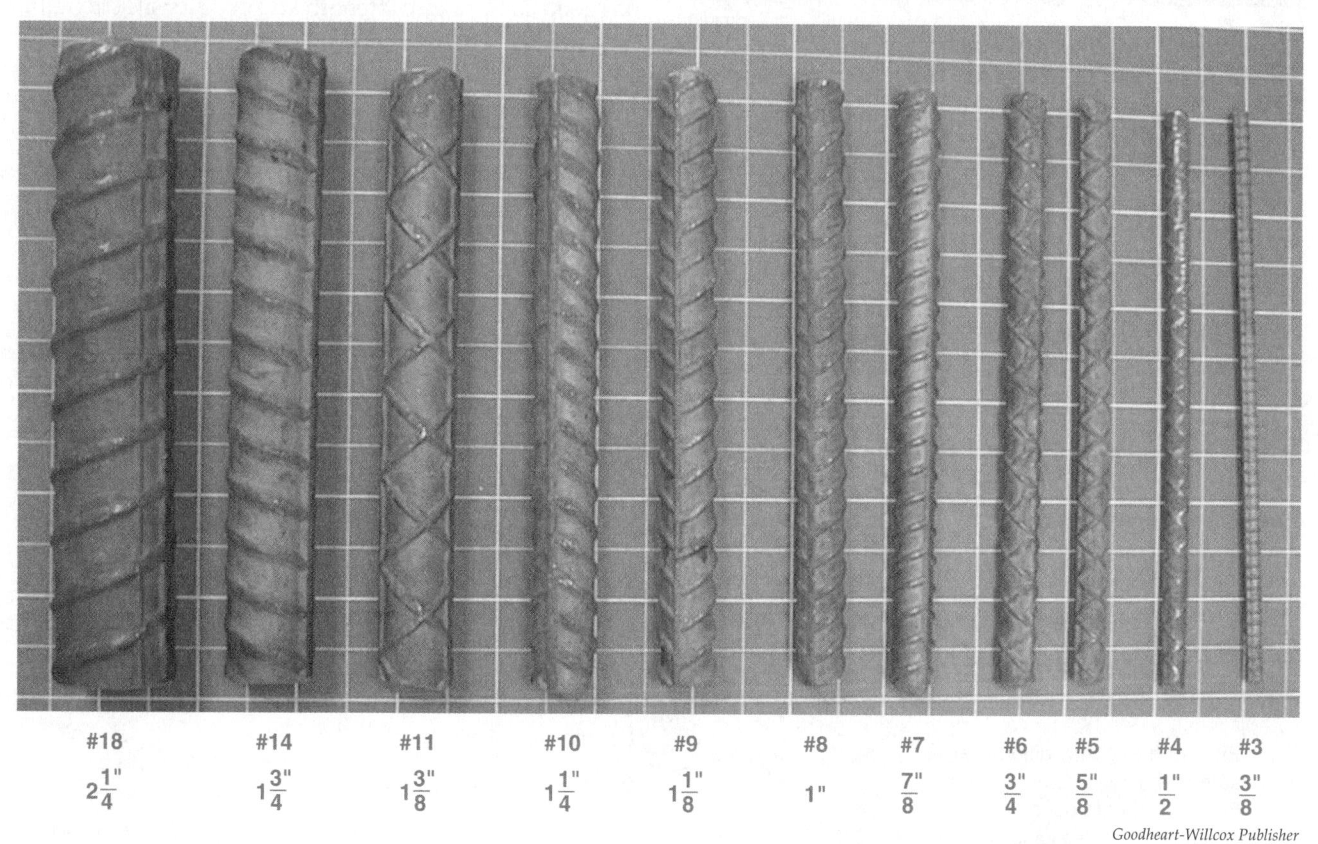

Goodheart-Willcox Publisher

Figure 7-5. Reinforcing steel bars ("rebar"). Bar diameters are identified by a number that is a multiple of 1/8″. The #18 bar is 2 1/4″ in diameter; the #3 bar is 3/8″.

Reinforcing bars are round in shape, with surface projections (called *deformations*) formed in the rolling process to strengthen bonding with the concrete. Bars are placed after the forms are constructed, **Figure 7-6**. The concrete is then cast around the bars.

Sheets of wire mesh are also used for reinforcement. ***Welded wire fabric (WWF)*** is a prefabricated material used to reinforce concrete slabs, floors, and pipe. It consists of a mesh of steel wires welded together, **Figure 7-7**. It is available in sheets and rolls and many wire sizes.

There are two types of welded wire fabric: *smooth* (or plain), designated by a *W*; and *deformed*, designated *D*.

Dan Dorfmueller

Figure 7-6. This grid of reinforcing steel bars will add strength to the completed concrete foundation and structure it supports above.

The "D" fabric has deformations along the wire to better develop anchorage in the concrete. Previously, the fabric was specified by gage number, and some drawings still use this system.

Welded wire fabric is further designated by numbers. An example is 6×8–W8.0×W4.0. The first number (6) gives the spacing of the longitudinal wire in inches. The second number (8) gives the spacing of the transverse wire in inches. The first letter-number combination (W8.0) gives the type and size of the longitudinal wire. The second combination (W4.0) gives information on the transverse wire.

In the example given, the longitudinal wires are 6″ apart. The transverse wires are 8″ apart. The longitudinal wire is smooth and has a cross-sectional area of 0.08 in^2. The transverse wire is also smooth with an area of 0.04 in^2. **Figure 7-8** lists some of the common stock styles of welded wire fabric. **Figure 7-9** shows how mesh is placed in a concrete slab.

Division 04—Masonry

Masonry structures are made from a number of smaller units held together with a bonding material known as *mortar*. Masonry units are manufactured as brick, concrete block, stone, and clay tile. ***Mortar*** is a cementitious material that bonds the individual units together.

Almost all masonry construction must be strengthened with reinforcing materials. Like concrete, masonry has good compressive strength and poor tensile strength.

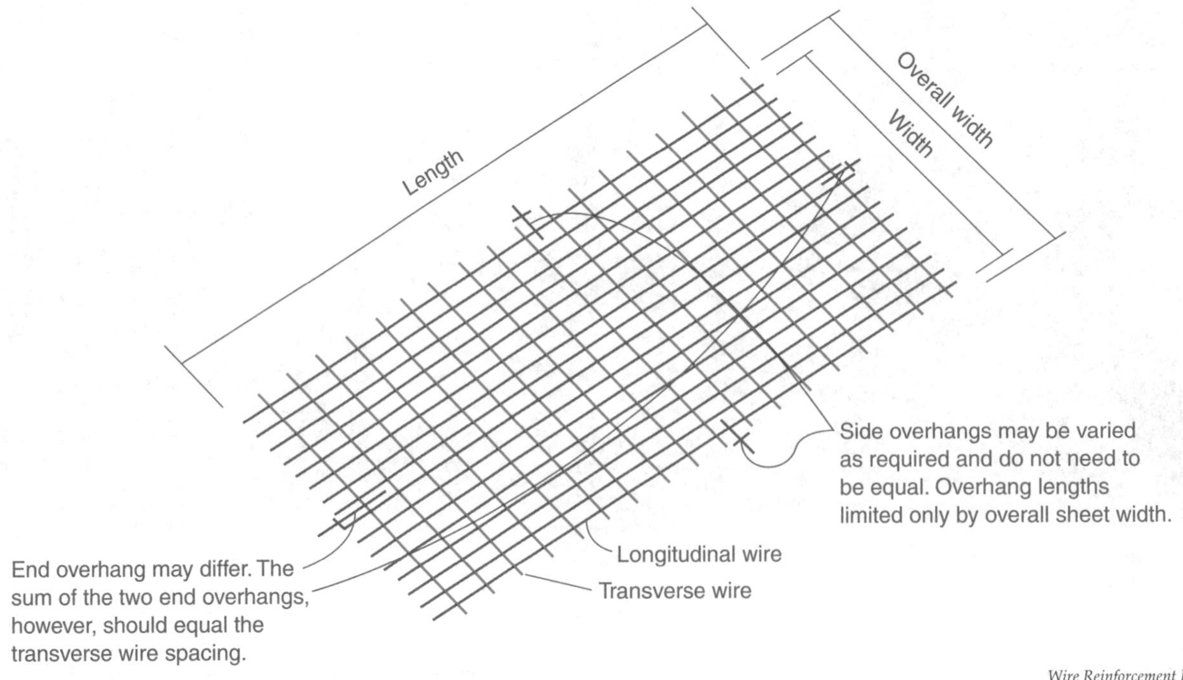
Length
Overall width
Width
Side overhangs may be varied as required and do not need to be equal. Overhang lengths limited only by overall sheet width.
End overhang may differ. The sum of the two end overhangs, however, should equal the transverse wire spacing.
Longitudinal wire
Transverse wire
Wire Reinforcement Institute

Figure 7-7. Nomenclature used for welded wire fabric.

Common Stock Styles of Welded Wire Fabric				
Style Designation		Steel Area (in²/ft)		Weight (lb/100 ft²)
New designation (by W-number)	Old designation (by steel wire gage)	Longit.	Trans.	
Roll				
6×6–W1.4×W1.4	6×6–10×10	.028	.028	21
6×6–W2.0×W2.0	6×6–8×8*	.040	.040	29
6×6–W2.9×W2.9	6×6–6×6	.058	.058	42
6×6–W4.0×W4.0	6×6–4×4	.080	.080	58
4×4–W1.4×W1.4	4×4–10×10	.042	.042	31
4×4–W2.0×W2.0	4×4–8×8*	.060	.060	43
4×4–W2.9×W2.9	4×4–6×6	.087	.087	62
4×4–W4.0×W4.0	4×4–4×4	.120	.120	85
Sheets				
6×6–W2.9×W2.9	6×6–6×6	.058	.058	42
6×6–W4.0×W4.0	6×6–4×4	.080	.080	58
6×6–W5.5×W5.5	6×6–2×2**	.110	.110	80
4×4–W4.0×W4.0	4×4–4×4	.120	.120	85

*Exact W-number size for 8 gage is W2.1
**Exact W-number size for 2 gage is W5.4

Wire Reinforcement Institute

Figure 7-8. Welded wire fabric is available in a number of stock sizes, in either roll or sheet form.

Brick Masonry

Brick masonry uses units (*bricks*) that are manufactured, rather than removed from quarries. There are many types:

- ***Adobe brick***. Made from natural sun-dried clays or earth and a binder.
- ***Kiln-burned brick***. Made from natural clays or shales (sometimes with other materials added, such as coloring) and molded to shape, dried, and fired for hardness.
- ***Sand-lime brick***. Made from a mixture of sand and lime, molded into shape and hardened under steam pressure and heat.
- ***Concrete brick***. Made from a mixture of Portland cement and aggregates, molded into solid or cored units and hardened chemically by the hydration of cement.

There are many types and sizes of brick. Most brick is either building brick or face brick. Some special types are used to a lesser extent.

- ***Building brick***. Usually called "common brick," this is the most commonly used type. It is used for interior and exterior walls, backing, and other applications where appearance is not important.
- ***Face brick***. This material is manufactured under more controlled conditions to produce bricks of specific dimensions, colors, and structural qualities. Face bricks are more expensive than building bricks, because of the care going into their manufacture. Face bricks with defects are often sold as common bricks.
- ***Glazed brick***. Finished with a hard, smooth coating, this type of brick is used for decorative and special service applications.
- ***Firebrick***. Used where masonry units are subjected to extreme heat, such as fireplaces, incinerators, and industrial furnaces.
- ***Paving brick***. Used in driveways or areas where abrasion is a concern.

PhotoByToR/Shutterstock.com

Figure 7-9. Wire mesh in a concrete slab.

Special bricks are also available in unusual shapes for window sills, rounded corners, and other nonstandard applications.

Brick Symbols

Brick is indicated on plan and section drawings with 45° crosshatch lines. For common brick, the lines are widely spaced; for face brick, the spacing is narrower. See **Figure 7-10**. Firebrick is shown on plan drawings with the usual 45° lines indicating brick, plus vertical lines that designate it as firebrick.

On elevation drawings, brick is normally indicated by horizontal lines. A note identifies the type of brick. Some architects only draw horizontal lines around the outer surface of brick walls.

Brick Bonds

There are several types of brick bonds, so construction workers should be familiar with those used most widely. A ***bond*** is the bricklaying pattern. Bonds are designed to improve appearance, add strength, or tie a wythe wall to a backing wall.

The bonds used most widely are common bond, running bond, English bond, English cross bond, Flemish bond, and stacked bond. See **Figure 7-11**.

Brick Positions

Bricks can be positioned in different ways. These positions are used by the architect to develop a design or style in the building, as well as to add to the structural strength of the brickwork. Each position has a name that identifies it. For example, the most common position is the ***stretcher***, **Figure 7-12**. It is laid in a flat position, lengthwise with the wall. Bricks in the stretcher position make up a large portion of most walls.

In some bonds, such as the common bond (also called American bond), every sixth or seventh course is turned 90°. This is done to improve appearance or to tie the face brick with the backing wall. Bricks laid in this manner are called *headers*, and the course is referred to as a ***header course***.

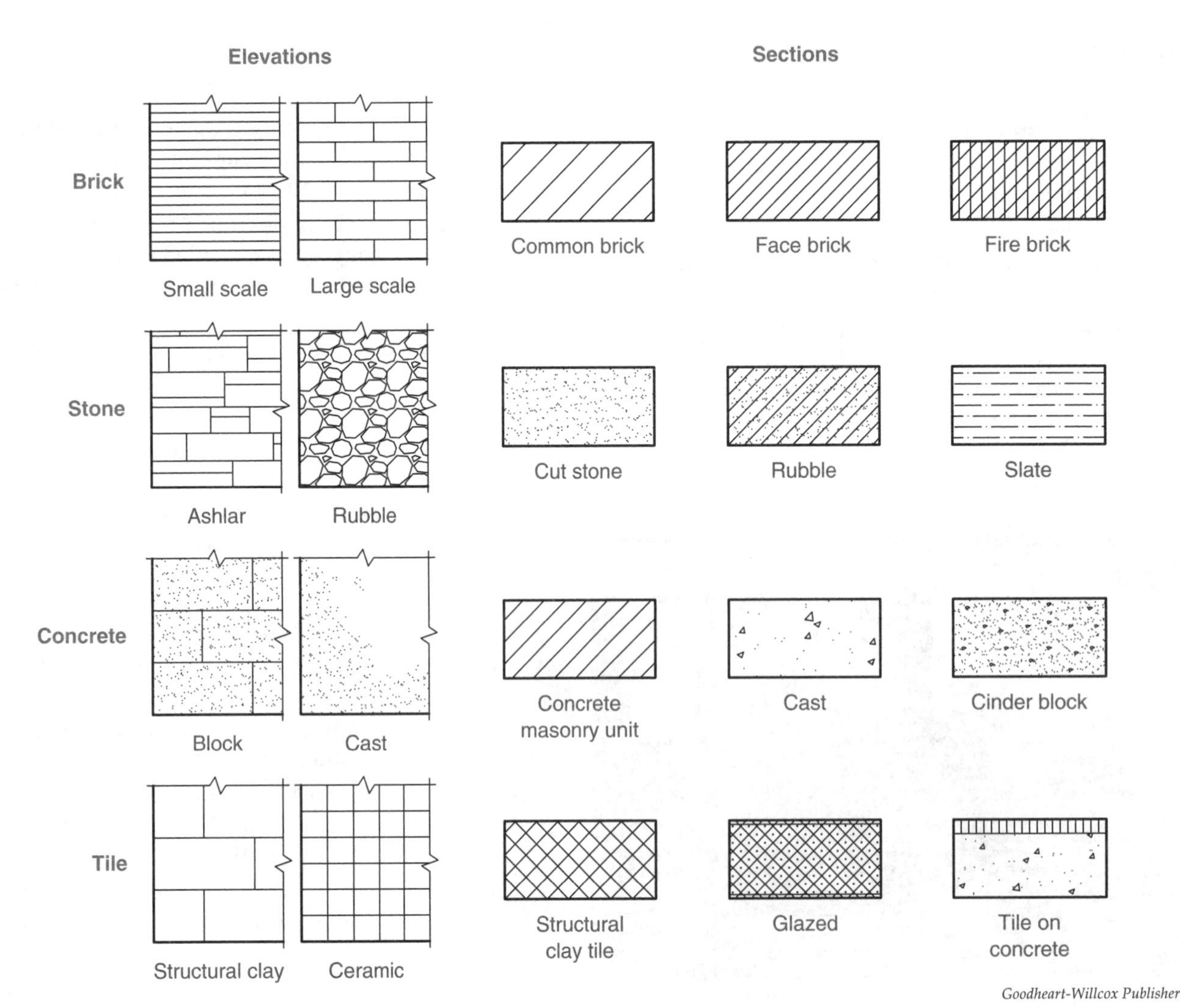

Figure 7-10. Symbols for common masonry materials.

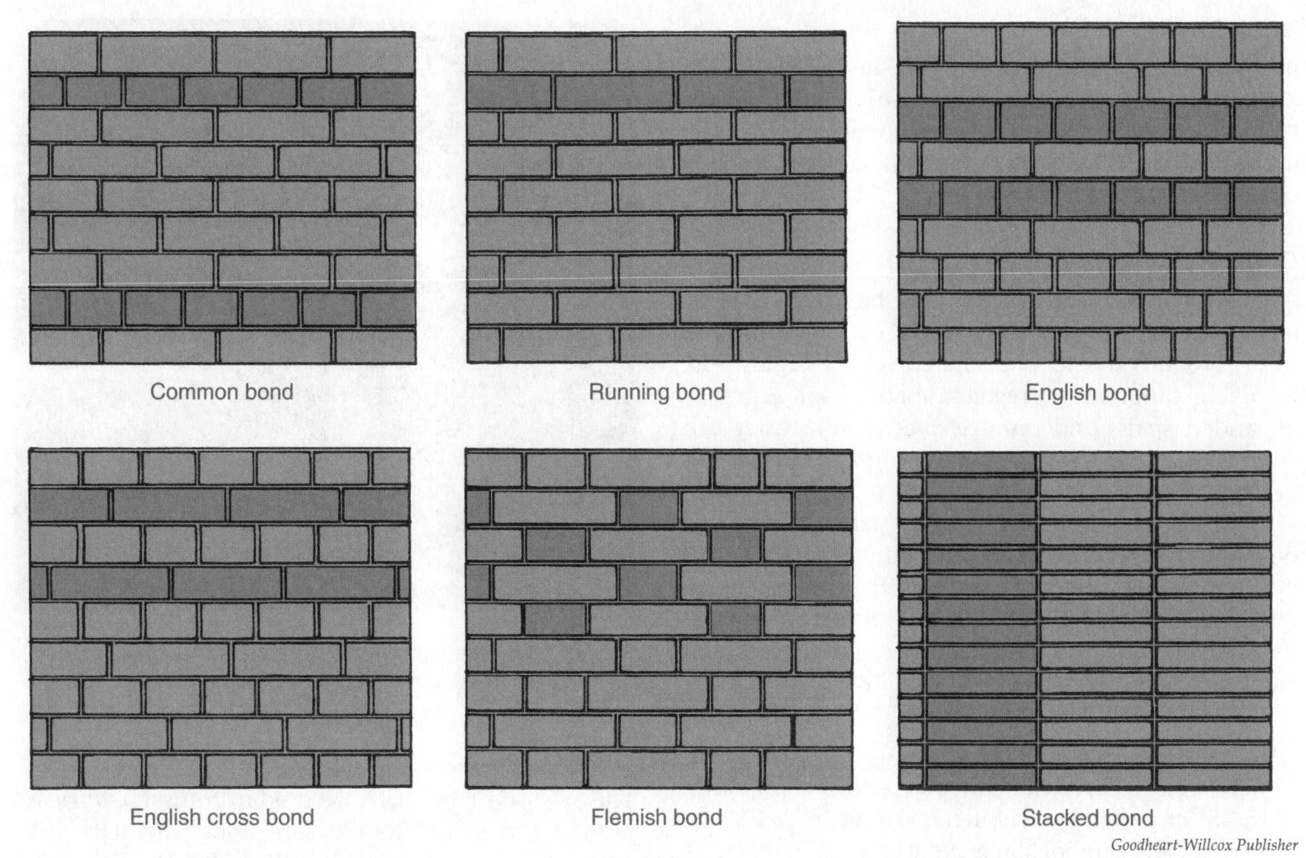

Figure 7-11. Different bond patterns vary the appearance of brick structures. Note: The red highlighting is used to help illustrate how the patterns are formed.

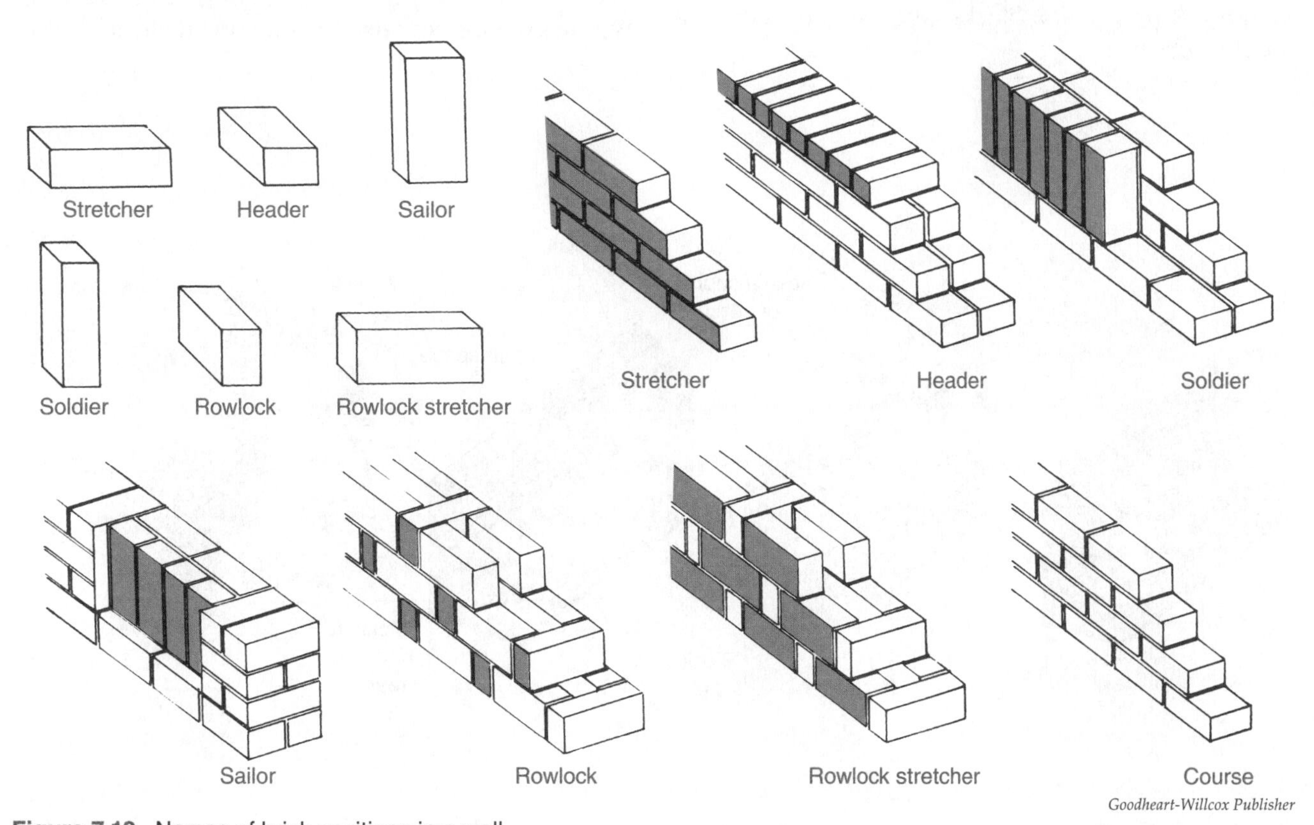

Figure 7-12. Names of brick positions in a wall.

Brick Lay-Up

Brick walls can be laid up as ***single-wythe walls*** that are 8″ or more in thickness. These can be solid or have two or more cells (spaces) in them. Sometimes the cells of the brick are filled with granular insulation. Various masonry walls are shown in **Figure 7-13**.

Concrete Masonry Units (CMUs)

Another popular and widely used building material is the ***concrete masonry unit (CMU)***, which is formed from Portland cement, sand, and gravel. See **Figure 7-14**. By using different aggregates, such as sand, gravel, expanded shale, and pumice, the manufacturer can control the weight, strength, and acoustical properties of the CMU.

CMUs are made in a variety of sizes, shapes, and densities to meet specific construction needs. The most popular is the standard block, with dimensions of 7 5/8″ × 7 5/8″ × 15 5/8″. When 3/8″ of mortar is used, this becomes an 8″ × 8″ × 16″ module. Another common block size is the 3 5/8″ × 7 5/8″ × 15 5/8″.

Concrete Block Symbols

Concrete block is indicated in plan and section views with 45° crosshatch lines. Refer to **Figure 7-10**. The elevation view symbol for concrete block is the same as that for poured concrete with lines added to represent the block pattern.

Concrete block can have colored surfaces and special design features. Slump blocks give the appearance of rough adobe brick.

Dan Dorfmueller

Figure 7-14. The standard concrete block (CMU) is the most widely used type of concrete masonry unit.

Stone Masonry

The most common materials used in ***stone masonry*** are granite, limestone, marble, sandstone, and slate. Like concrete, stone has been used as a building material for many centuries. In the past, stones were used for structural members, roofing, and finishing. Due to the development of new materials and methods of construction, stones are now used mainly for their decorative value.

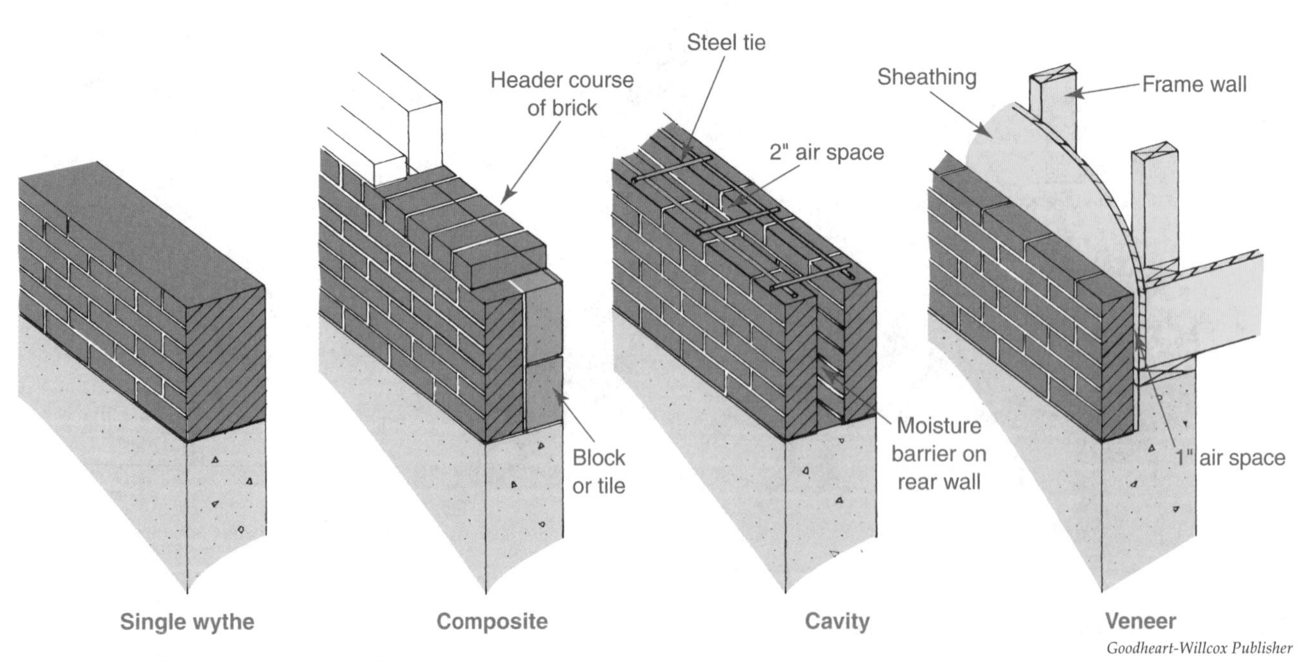

Goodheart-Willcox Publisher

Figure 7-13. Types of masonry walls.

Most stones are mined from a quarry and sent to a finishing mill for final dressing. Some stones are used in their original shapes and surface finishes. Others are cut to a specific shape, size, and finish. These are known as *cut stones.*

Stone masonry can be laid as solid walls of stone or as composite walls backed with concrete block or tile. It is also used as a veneer. Stone walls are classified according to the shape and surface finish of the stone, such as rubble, ashlar, and cut stone. Examples of rubble and ashlar masonry are shown in **Figure 7-15**.

Rubble consists of stones as they come from the quarry or are gathered from a field or stream. Such stones may be smooth with rounded edges, or they may be rough and angular. The random rubble wall consists of stones laid in an irregular pattern with varying sizes and shapes. Other rubble patterns are coursed, mosaic, and strip.

Ashlar stones are squared stones that have been laid in a pattern but not cut to dimensions. There are several ashlar patterns:

- **Regular.** Constructed with a uniform continuous height.
- **Stacked.** Tends to form columns.
- **Broken range.** Consists of squared stones of different sizes laid in uniform courses, but range is broken within a course.
- **Random range.** Neither course nor range remain uniform.
- **Random ashlar.** Course is not uniform and ends are broken, not square.

Cut stones, also known as *dimensional stones*, are cut and finished at the mill to meet the specifications of a particular construction job. Each stone is numbered for

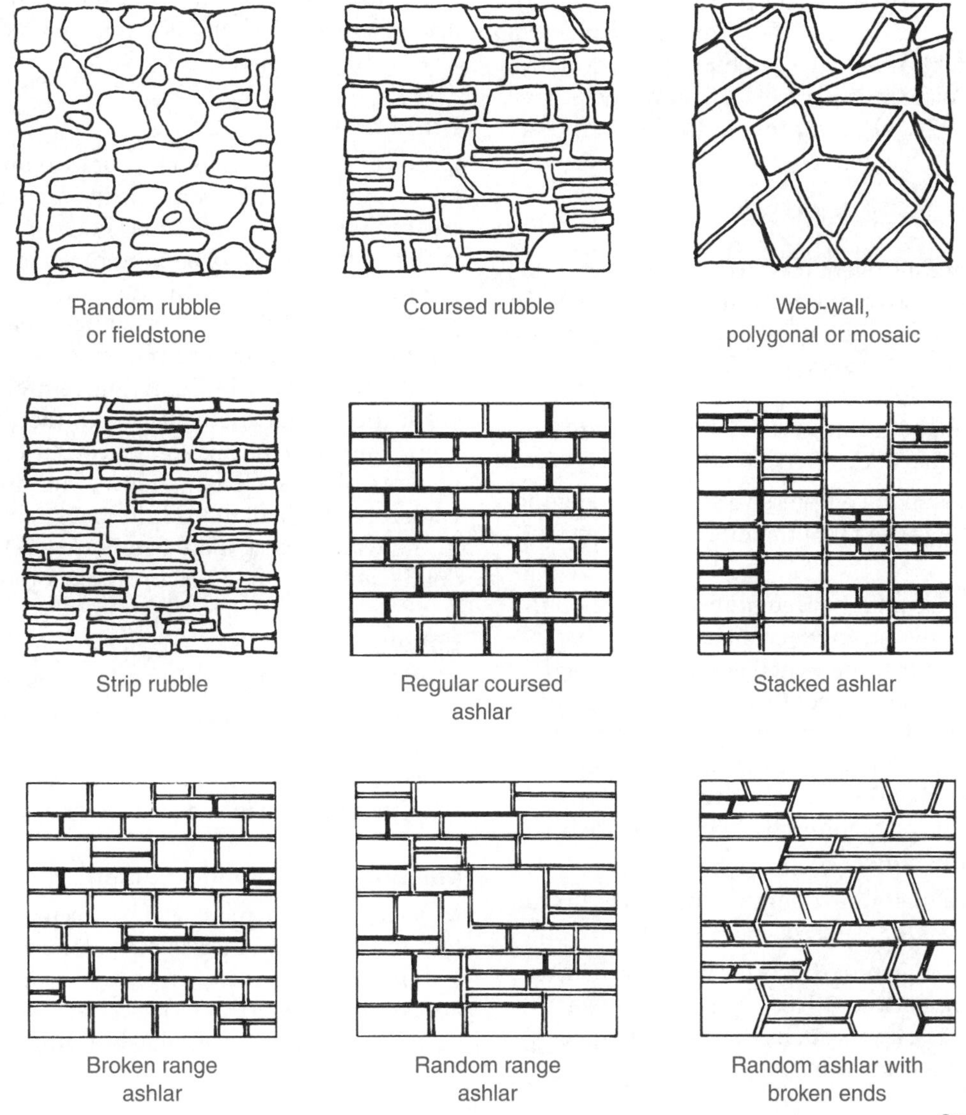

Figure 7-15. Various stone masonry walls, shown in elevation.

location. Unlike ashlar masonry, which is laid largely at the design of the mason, cut stones are laid according to the design of the architect.

Structural Clay Tile

Structural clay tile is made of materials similar to those used in brick, but it is a larger building unit. It has many uses in construction—as load-bearing walls, backup for curtain walls, and fireproofing around structural steel. Rectangular open cells pass through each unit, and tile comes in a variety of shapes and sizes.

Clay tile has largely been replaced with hollow brick and concrete masonry units. Most masonry walls today are composite walls of a finish surface material and a less expensive backup material.

Terra cotta is a type of structural clay tile principally used for nonbearing ornamental and decorative effects.

Gypsum Blocks

Gypsum masonry blocks are used primarily for interior nonbearing walls, fire-resistant partitions, and enclosures around structural steel. Made from gypsum and a binder of vegetable fiber or wood chips, gypsum blocks can be given a plaster finish coat. Gypsum blocks have a face size of 12″ × 30″ and come in thicknesses of 2″, 3″, 4″, and 6″.

Mortar

Mortar is the binding agent used to hold masonry units together. Mortar also compensates for the differences in brick and stone sizes. Metal ties and reinforcement are secured in mortar. Mortar consists of cement, hydrated lime, sand, and water.

ASTM International has established standards for mortar. There are five different standardized types:

- **Type M** has high compressive strength and good durability. It is used for unreinforced underground masonry.
- **Type S** is also a high-strength mortar. Although its compressive strength is not as high as that associated with Type M, it has a stronger bond and greater lateral strength. It also has the greatest tensile strength of any mortar type. Type S is used for reinforced masonry, unreinforced masonry subjected to bending, and in situations where mortar is the only connection between face brick and backing brick.
- **Type N** is a medium-strength, general-use mortar. It is best used for exposed, above-ground masonry.
- **Type O** is low-strength mortar used for interior, nonbearing masonry. This type of mortar should not be used in applications that will be exposed to freezing temperatures.
- **Type K** has very low strength and should only be used for interior applications when strength is not a concern.

Stone work usually requires a special type of mortar consisting of white Portland cement, hydrated lime, and sand. This mortar prevents stains caused by ordinary cements. Mortar in the joints is normally raked back from the surface as the stone is set. Later, the joint is pointed, using the same mortar or a colored sealant.

Trade associations recommend proportions of cement, hydrated lime, and sand for mortar. Local building codes also set allowable limits. A typical mixture would be 1 part Portland cement by volume, one-fourth part hydrated lime, and 3 parts sand. Water is added as needed to make the mortar workable. Many brick masons use a special masonry cement containing plasticizing agents to make the mortar more workable.

Masonry Accessories

Besides the masonry units and mortar, additional components are needed in masonry construction. Bond beams, joints, lintels, and flashing are needed to complete a wall.

Bond Beams

Concrete masonry walls are usually reinforced horizontally and vertically by constructing a reinforced beam or column within the wall. This is done by pouring ***grout*** (a flowable mixture of sand and cement) around reinforcing steel inserted in the units. Special channel blocks are used to form horizontal ***bond beams***, using reinforcing steel and mortar or grout. Vertical bond beams are formed by inserting reinforcing bars in a vertical cell after the wall is laid, then filling the cell with grout. Bond beams are typically used over masonry wall openings such as windows and doors, the top of walls, or anywhere that additional strength is required in the wall construction. See **Figure 7-16** for a typical detail showing a bond beam at the top of a wall.

Lintels

Lintels are members placed in masonry walls above door and window openings. Lintels are supported on either side of the opening. They can be made of precast concrete, steel, or other materials. The door and window schedule will often include the size and type of lintel used.

Masonry Joint Reinforcement

Masonry walls are reinforced by placing various anchors, ties, and rods in the mortar joints. These reinforcing units are produced in many sizes and shapes for different applications. Reinforcing devices are called out directly on the prints or in applicable building codes.

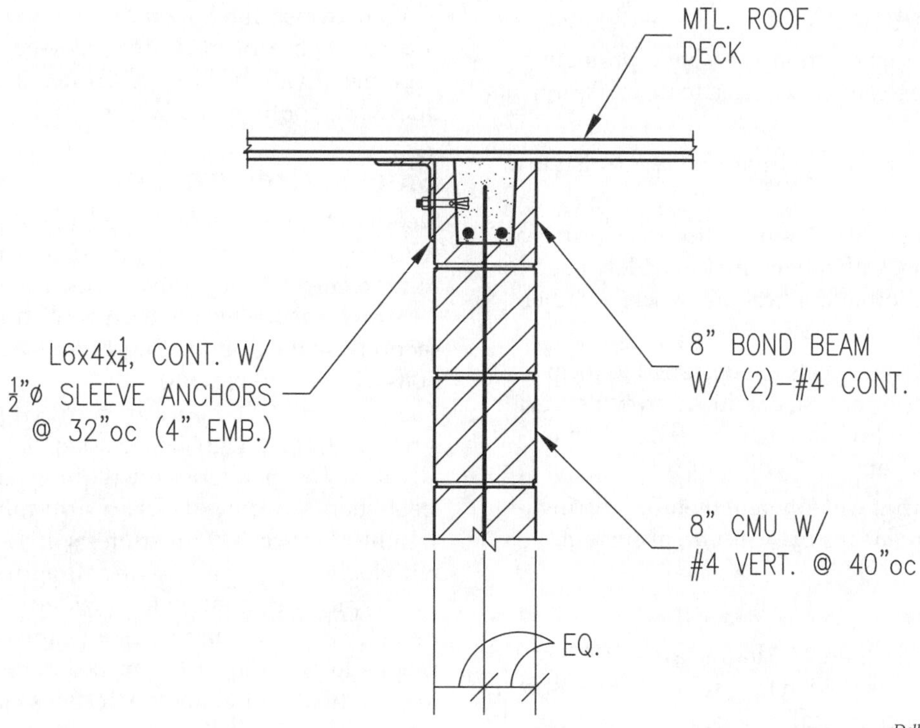

Delhi Flower and Garden Center
Arch/Image 2 Architects
Pinnacle Engineering Services, Inc.

Figure 7-16. Detail showing a bond beam in a concrete masonry unit wall.

Division 05—Metals

Metal is used extensively in the construction industry. Large commercial buildings use structural steel. Construction jobs make use of metal windows, doors, studs, beams, joists, wall facings, roofing, plumbing, and hardware.

Kinds of Metals

Metals can be divided into two categories, ferrous and nonferrous. ***Ferrous metals*** contain iron as a principal element, and usually have magnetic properties. They typically have more strength than nonferrous metals. ***Nonferrous metals*** contain no iron and do not have magnetic properties. They are typically lighter and less strong than ferrous metals.

Ferrous Metals

Iron. Iron is malleable, ductile, magnetic, and silver white in color. There are many kinds of iron, such as pig iron, cast iron, and wrought iron. Iron is one of the principal ingredients of steel.

Steel. Steel is an iron-based alloy with a carbon content ranging from 0.2% to 2.0%. Depending on the intended use, steel has many chemical compositions. Carbon steel, alloy steel, mild steel, medium steel, tool steel, spring steel, and stainless steel are a few of the varieties available. By varying the carbon content, steel can be made harder and stronger, but it also becomes more brittle. Adding chromium, nickel, and magnesium will make the metal highly resistant to corrosion.

Nonferrous Metals

Aluminum. Pure aluminum and aluminum alloys are used in many building applications where resistance to corrosion is important. Aluminum that is 99% pure is soft and ductile, and therefore weak. Aluminum is typically used in such building applications as flashing, downspouts, some kinds of roofing, doors and trim, and mullions for windows. Aluminum also is made as an alloy with copper added for strength and casting uniformity. Aluminum alloys are made in structural shapes for use as H-beams, I-beams, and angles. Aluminum is also used for some ductwork, screens, and electrical wiring.

Copper. Copper and copper alloys have a high electrical conductivity and are resistant to corrosion. This metal is typically used in construction for water distribution, electrical wiring, and flashing. The particular metal would be indicated on the drawing as a note and most likely detailed in the specifications.

Brass. This metal is widely used for door and window hardware, trim, grilles, and railings. Brass is copper with zinc as its principal alloy element.

Bronze. A copper-tin alloy, bronze also can contain various other elements, such as aluminum or silicon. Bronze resists corrosion and is widely used for ornamental architectural products.

Surface Treatments

Metals used in construction may have protective or decorative surface treatments applied to them. Treatments include:

- **Chroming.** Applying chromium as a finish to a metal surface.
- **Galvanizing.** Coating steel or iron with zinc to resist rusting. Galvanized iron is widely used for flashing and other applications where weather tends to corrode metals.
- **Rusting.** Certain steels are designed to form a protective layer of *rust*, a reddish-brown surface coating formed when the metal is exposed to moisture and air.
- **Electroplating.** An adherent metallic coating deposited by means of an electrical current.

Metal Gages

Metal materials less than 1/4″ in thickness (often called "sheet metal") are classified in the ***gage system***. As shown in **Figure 7-17**, gages typically range from 4 (thickest) to 38 (thinnest). For example, 4-gage metal has a thickness of 0.2242″, a little less than 1/4″, while 22-gage metal is 0.0299″ (approximately 1/32″).

Steel manufacturer's standard gage number	Weight, psf	Equivalent sheet thickness (in.)
4	9.3750	0.2242
6	8.1250	0.1943
8	6.8750	0.1644
10	5.6250	0.1345
12	4.3750	0.1046
14	3.1250	0.0747
16	2.5000	0.0598
18	2.0000	0.0478
20	1.5000	0.0359
22	1.2500	0.0299
24	1.0000	0.0239
26	0.7500	0.0179
28	0.6250	0.0149
30	0.5000	0.0120
32	0.40625	0.0097
34	0.34375	0.0082
36	0.28125	0.0067
38	0.25000	0.0060

Goodheart-Willcox Publisher

Figure 7-17. Gages, weights, and thicknesses of sheet metal.

Wall studs, lintels, window and door frames, and floor joists are made from heavy-gage metals. Thin-gage metals are used for such items as roof flashing, ductwork, roofing, and wall siding.

Steel in Construction

Steel is the most widely used metal in construction, with applications ranging from structural support to reinforcement to decorative uses.

Structural steel is the term applied to hot-rolled steel sections, shapes, and plates, **Figure 7-18**. This includes bolts, rivets, and bracing.

Structural steel shapes are formed by passing heated strips of steel through a succession of rollers that gradually form the metal into the required shape. Structural steel shapes are available in a number of sizes and weights. **Figure 7-19** illustrates some standard shapes, with identifying symbols and designations.

A typical designation for a wide-flange beam would be W12×16, which indicates a beam 12″ in depth that weighs 16 pounds per linear foot. A typical designation for a lightweight beam, sometimes called an I-beam, would be S8×23. This indicates that the beam is 8″ in depth and weighs 23 pounds per linear foot.

Steel members are connected to form building frames, **Figure 7-20**. The frame is usually then hidden behind other materials, such as a masonry wall, precast panels, or sheet metal siding.

Steel ***angles*** (sometimes called *angle iron*) are used as bracing in steel framing and to construct ***open-web steel joists*** and many other products, **Figure 7-21**. Angles are designated with the letter "L," followed by the lengths and thickness of the "legs." For an angle identified as L3x3x1/2, both legs are 3″ long and 1/2″ thick.

Goodheart-Willcox Publisher

Figure 7-18. Structural steel beams, columns, web joists, and other shapes are widely used in construction projects. These workers are installing a column onto a concrete footing.

Descriptive Name	Shape	Identifying Shape	Typical Designation Depth Wt/Ft in Lb.	Nominal Size Depth Width
Wide flange shapes	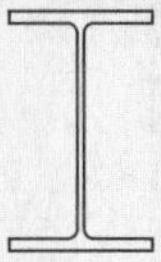	W	W21 × 147	21 × 13
Miscellaneous shapes	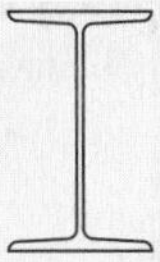	M	M8 × 6.5	8 × 2¼
American Standard beams	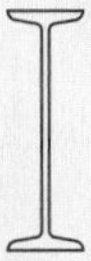	S	S8 × 23	8 × 4
American Standard channels		C	C6 × 13	6 × 2
Miscellaneous channels		MC	MC8 × 20	8 × 3
Angles — equal legs	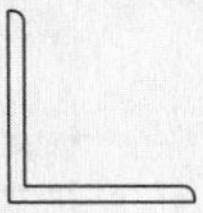	L	L6 × 6 × ½*	6 × 6
Angles — unequal legs		L	L8 × 6 × ½*	8 × 6
Structural tees (cut from wide flange)	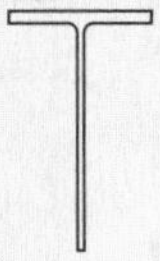	WT	WT12 × 60	12
Structural tees (cut from miscellaneous shapes)	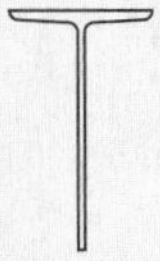	MT	MT5 × 4.5	5
Structural tees (cut from Am. Std. beams)	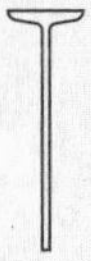	ST	ST9 × 35	9
Zees	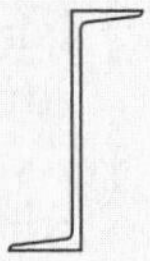	Z	Z6 × 15.7	6 × 3½

*Leg thickness

Goodheart-Willcox Publisher

Figure 7-19. Some examples of structural steel shapes.

A

Goodheart-Willcox Publisher

B

Jim Parkin/Shutterstock.com

Figure 7-20. A—This steel frame is designed to support the entire weight of the building. The exterior walls of this building are precast concrete panels. B—Steel beams and columns are common members used to support a building. Connections between the structural members are specified on detail drawings.

Division 06—Wood, Plastics, and Composites

Wood continues to be one of the chief building materials, **Figure 7-22**. It is used for structural framing (rough carpentry), trim, floors, walls, and cabinetry (finish carpentry and architectural woodwork). Relative to its weight, wood has high strength in compression, tension, and bending. It also has excellent impact resistance. While steps have been taken to substitute other materials, wood remains a valuable and widely used residential construction material.

Wood Classification

Woods are broadly classified as either hardwoods or softwoods. There are many varieties used for construction. These classifications are not an exact measure of hardness or softness (because this varies) but a general classification based on type of tree. In addition to hardness or softness, woods vary in strength, weight, texture, workability, and cost. Building specifications usually indicate the type and grade of lumber to be used in different parts of the construction.

Wood Classifications	
Hardwoods	**Softwoods**
Ash	Cypress
Ash, White	Fir, Douglas
Beech	Fir, White
Birch	Hemlock
Cherry	Pine, Ponderosa
Elm	Pine, Southern
Gum	Pine, White
Hickory	Poplar, Yellow
Mahogany	Red Cedar, Eastern
Maple	Red Cedar, Western
Oak	Redwood
Walnut	Spruce

Lumber

When wood is cut into pieces of specific thickness, width, and length, it is called ***lumber***. Lumber products include rough framing members (at least 2″ thick), such as beams, headers, and posts; finished lumber, such as flooring, door and window trim, paneling, and moldings; and specialty items, such as decorative panels, carved doors, ornamental overlay designs, and turned balusters (stair-rail posts).

Lumber is classified as rough-sawn or surfaced to size. ***Rough-sawn lumber*** has been cut to size but not dressed or surfaced. ***Surfaced lumber*** has been dressed or finished to size by running it through a planer. The designation "S2S" is used for lumber dressed on two sides, and "S4S" for lumber that is surfaced or planed on all four sides.

Plywood is a wood product made of several layers of lumber arranged with the grain at right angles in each successive layer and bonded with an adhesive. An odd number of layers is used, so that the grain of the face and back are running in the same direction. The panels are

Goodheart-Willcox Publisher

Figure 7-21. Open-web steel joists combine strength with light weight.

usually 4′ × 8′ in size, and are available in finished thicknesses ranging from 1/8″ to over 1″. Because of its modular size and uniformity, plywood speeds construction and is considered an economical building material.

Interior plywood is bonded with an adhesive that is water-resistant. It is used for cabinetry, rough flooring, and finished walls. *Exterior* or *structural plywood* is bonded with a waterproof adhesive. It is used for wall sheathing, finished walls, roof sheathing, and concrete forms.

Goodheart-Willcox Publisher

Figure 7-22. Wood continues to be the primary building material for residences and for many larger structures, such as this low-rise hotel.

Glue-Laminated Timber

The process of ***laminating*** (bonding layers of lumber together with adhesive) has made it possible to span larger distances and change traditional construction techniques. Wood beams, arches, and other members of nearly any size and shape can be fabricated. These laminated products are made of kiln-dried lumber and prepared for interior and exterior use. These members are usually prefinished at the factory and delivered to the job with protective wrapping.

Division 09—Finishes

Finishes are the visible surfaces of a building, such as gypsum wallboard, paint or wallpaper, ceramic tile, carpeting, wood flooring, and acoustical ceilings. Finishes are, in part, delineated on the drawings and specified in a room finish schedule.

GREEN BUILDING

Reducing Lumber Waste

Precut, custom, and fabricated components can be cut to specification at the lumber mill and shipped to the job site. This reduces job site waste due to material misuse, poor storage, carpenter error, and other mishaps.

Gypsum Wallboard and Plaster

Gypsum wallboard and plaster products consist of a core of air-entrained gypsum between two layers of treated paper. Wallboard comes in standard 4′ × 8′ sheets, but is available in 4′ × 9′ and 4′ × 12′. If ordered in truckload quantities, almost any special size desired can be produced. Wallboard is fastened to wood or metal studs with nails or screws, and varies in thickness from 1/4″ to 1″. Joints are sealed with a joint compound and paper tape to provide a smooth, even surface. The wall can be painted, papered, or given a surface texture to enhance its appearance.

Lath for gypsum plaster walls is available in 16″ × 48″ sheets with thicknesses of 3/8″ and 1/2″. The lath is fastened to the studs and a three-coat gypsum plaster finish—consisting of a scratch coat, brown coat, and finish coat—is applied.

Building papers are available in a variety of types suitable for sheathing walls and roofs. Some papers are reinforced for strength and tear resistance. Building papers are treated with asphalt, plastic, tar, or other materials. The building paper is usually specified on the print or in the specifications.

Ceramic Tile

There are many kinds of ceramic tiles, commonly referred to merely as "tile," for use on floors, walls, and other surfaces. Ceramic tile is made from nonmetallic minerals, fired at a very high temperature.

- ***Ceramic tiles*** come in sizes from 3/8″ square to 16″ × 18″ units. Popular wall sizes are 4 1/4″ × 4 1/4″, 4 1/4″ × 6″, and 6″ × 6″. Hexagonal and octagonal tiles are also available. The tiles can be glazed or unglazed. Glazed tiles are usually 5/16″ thick. Unglazed tiles vary from 7/16″ to 3/4″ thick.
- ***Mosaic tiles*** can be laid to form a design or pattern. They are normally smaller than 6″ (typically 1″) square, and can be glazed or unglazed.
- ***Quarry tile*** is used for floor coverings and is produced from clays that provide a wear-resistant surface.
- ***Fired-clay tile*** is used primarily for floor coverings. It is produced from clays and is fired in a kiln to harden the surface.

Tile-Setting Materials

There is a wide variety of tile-setting materials available for use, based on flooring types and final use of the tiling products. Typically, ceramic tile is set in Portland cement, latex adhesive, or epoxy mortar. **Figure 7-23** shows two applications: tile set on plywood using a cementitious backer board with a bond coat, and tile set directly on a concrete slab surface.

The joints between the tiles are grouted, with different types of *grout* available to meet varying kinds of joint and tile exposure. The grout consists of Portland cement modified to meet various joint conditions.

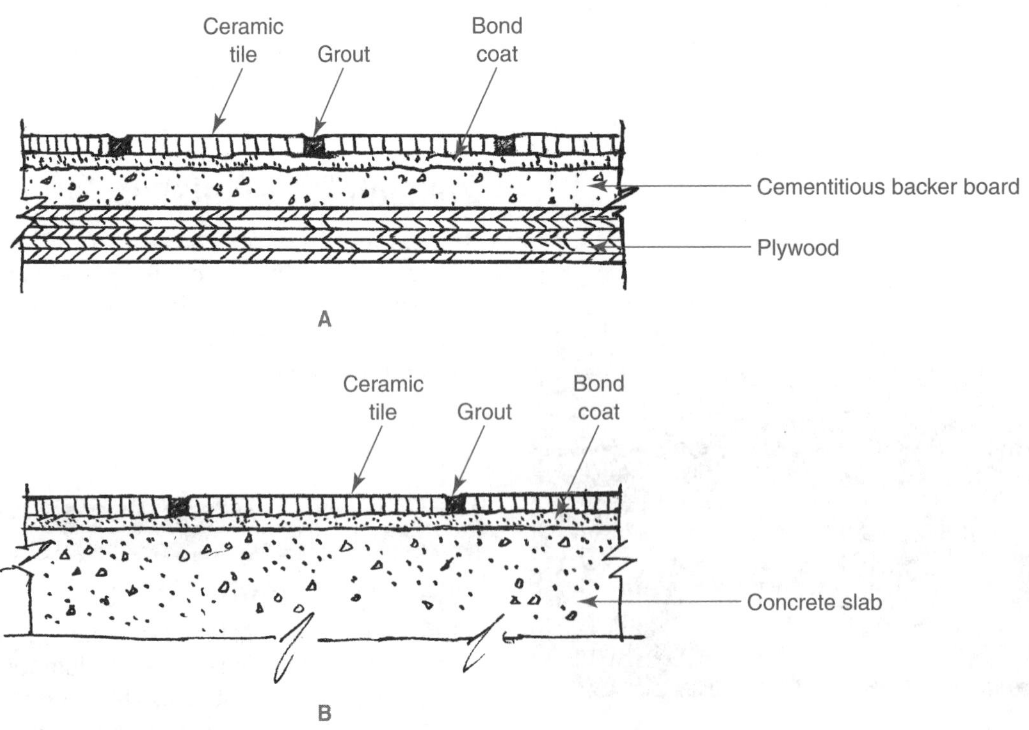

Figure 7-23. Ceramic tile installation. A—Section view of ceramic tile installed on a plywood floor with a cementitious backer board. B—Section view of ceramic tile installed on a concrete slab.

Glass

Glass is a ceramic material formed at temperatures above 2300°F (1260°C). It is made from sand (silica), soda (sodium oxide), and lime (calcium oxide). Other chemicals can be added to change its characteristics.

Float glass is the most common type of glass. A continuous ribbon of molten glass flows out of a furnace and floats on a bath of molten tin. Irregularities melt out and the glass becomes flat. The ribbon of glass is fire-polished and annealed, without grinding or polishing. Over 90% of the world's flat glass is made by the float process. Flat glass produced by the float process is commonly used for windows in thicknesses of 3/32″ (single strength, or "SS") and 1/8″ (double strength, or "DS").

After the float process, other processes can further modify the properties of the glass, producing several types:

- ***Bent glass*** is produced by heating annealed glass to the point where it softens so it can be pressed over a form.
- ***Safety glass*** was developed to overcome the hazards of flat glass in large, exposed, or public areas. Three types of safety glass are available: tempered, laminated, and wired glass.
  - *Tempered glass* is developed by heating annealed glass to near its melting point, then chilling it rapidly. This creates high compression on the exterior surfaces and high tension internally, making the piece of glass three to five times as strong as annealed glass. Tempered glass can be broken, but it shatters into small, pebble-like pieces rather than sharp slivers. Tempered glass must be ordered to the exact size needed before tempering, because it cannot be cut, drilled, or ground after it has been tempered.
  - *Laminated glass* consists of a layer of vinyl between sheets of glass. The layers are bonded together with heat and pressure. This glass can be broken, but the plastic layer holds the small, sharp pieces in place.
  - *Wired glass* has a wire mesh molded into its center. Wired glass can be broken, but the wire holds the pieces together. Wired glass can be obtained with an etched finish, a sandblasted finish, or a patterned finish.
- ***Insulating glass*** is a unit of two or more sheets of glass separated by an air space that is dehydrated and sealed. Insulating glass units serve as a good insulator for heat and sound transfer. A typical insulating glass installation in a window sash is shown in **Figure 7-24**.

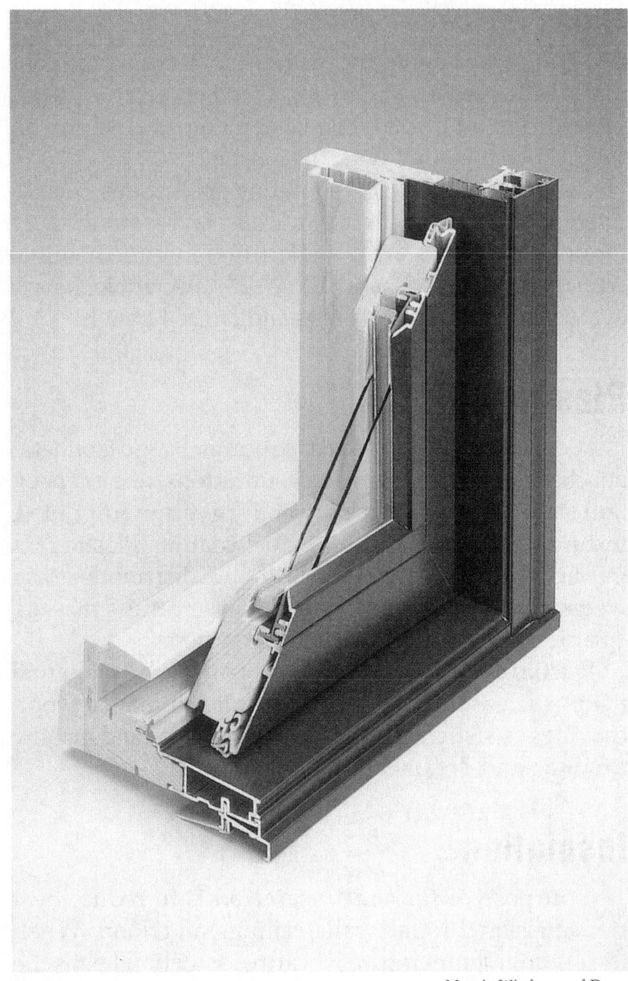

Marvin Windows and Doors

Figure 7-24. Window with insulating glass.

- ***Patterned glass*** is flat glass with a pattern rolled into one or both sides to diffuse the light and provide privacy.
- ***Stained glass***, sometimes called *art* or *cathedral glass*, is produced by adding metallic oxides in the molten state. This glass can be used in sheets or cut into smaller pieces and made into leaded glass for windows and decorative pieces.

GREEN BUILDING

High-Efficiency Windows

Windows and glass doors can leak a significant amount of heat in winter. Installing high-efficiency, triple-glazed windows in cold climates can reduce heat loss by 15%–30%. Double-glazed windows will meet weather needs and code specifications in warmer, more moderate parts of the country.

Glass Block

Glass block is made by fusing two sections of glass together. A partial vacuum is created between the pieces, providing good insulating qualities. Edges are left rough to improve the bond with the mortar.

Glass block is used in nonbearing applications, such as interior walls, screens, curtain walls, and windows. See **Figure 7-25**. Blocks are manufactured in three nominal sizes: 6″ × 6″, 8″ × 8″, and 12″ × 12″. Special blocks are available for forming corners and curved panels.

Plastics

Plastics have many uses in construction. Plastic laminates attached to plywood serve as countertops, door veneer, and wall surfacing. Panels of wood or gypsum are printed, textured, and given a plastic vinyl coating. Plastic rain gutters and downspouts collect and distribute storm water. Plastic pipes are used for water-transmission, sprinkling, drainage, and sewage systems.

Plastics are also used for many trim and ornamental items, such as moldings on doors and simulated wood carvings. Plastic materials are usually noted on the drawing and detailed in the specifications.

Insulation

The purpose of ***thermal insulation*** is to reduce heat transmission through walls, ceilings, and floors. When the outdoor temperature is warm, insulation keeps the heat from entering the structure. When the outdoor temperature is cool, insulation helps keep the warm air indoors.

Insulation is manufactured in a variety of forms and types to meet specific construction requirements. Each type of material or composite system will have an ***R-value*** (resistance to heat transfer), depending on the manner of application and amount of material. A high R-value number means good insulation qualities. Insulation materials are classified as:

- Flexible (blanket or batt).
- Loose-fill.
- Reflective.
- Rigid (structural and nonstructural).

Flexible insulation is available in blanket and batt form. Blankets come in widths suitable to fit 16″ and 24″ stud and joist spacing and in thicknesses from 1″ to 14″. The body of the blanket is made of mineral or vegetable fiber, such as rock wool, glass wool, wood fiber, or cotton. Organic materials are treated to resist fire, decay, insects, and vermin. Blankets are covered with a paper sheet on one or both sides. The vapor barrier is installed facing the warm side of the wall. Batt insulation is made of the same material as blankets, in thicknesses from 3 1/2″ to 14″ and lengths of 48″, 93″, and 96″.

Loose-fill insulation, **Figure 7-26**, is available in bags or bales. It is either poured, blown, or packed in place by hand. Loose-fill insulation is made from materials such as rock wool, glass wool, and cellulose. This type of insulation is suited for insulating sidewalls and attics of buildings. It is also used to fill cells in block walls.

Reflective insulation is designed to reflect radiant heat. It is made of a material surfaced with metal foil (usually aluminum foil). To be effective, the reflective surface must face an air space of at least 3/4″. When the reflective surface contacts another material, such as a wall or ceiling, the reflective properties are lost along with the insulating value. This material is often used on the back of gypsum lath and blanket insulation.

Rigid insulation is made of a foam material in sheet form. Common types are expanded polystyrene foam (EPS), extruded polystyrene foam (XPS), and polyisocyanurate foam (PIR). These produce a lightweight,

Artazum/Shutterstock.com

Figure 7-25. Glass block has many applications in construction. The shower enclosure in this home is constructed to form a radial wall section. Enclosures of this type can be constructed from prefabricated sections or individual block units.

CertainTeed Corporation

Figure 7-26. Loose-fill insulation is commonly used for insulating open attic areas. This insulation is designed to be installed with special pneumatic equipment. R-values ranging from R-11 to R-60 are achieved based on minimum installed thicknesses specified by the manufacturer.

low-density product with good heat and acoustical insulating qualities. Rigid insulation is used as sheathing for walls and roof decks when additional insulation is needed.

Material Symbols on Drawings

Besides showing the location and sizes of different construction components, drawings also identify the materials that are used. Materials are identified in several ways:

- A section view on a drawing may contain a pattern that is unique to a specific material.
- Often, materials are specified in the notes included on a drawing. This allows for easy reference.
- Materials are also included in the project specifications with additional information regarding quality standards.

Materials may be shown differently in plan, elevation, and section views. For example, in **Figure 7-27**, concrete block is represented in the section view with diagonal crosshatch lines. On plan views, concrete block is shown with diagonal crosshatch lines or with the concrete hatch pattern.

Section views that cut across structural framing members show these pieces with an "X" within each member. Finished lumber (such as trim, fascia boards, and moldings) is shown in section with the wood end grain. For a wood frame wall on the plan view, the usual practice is to leave the wall blank, **Figure 7-27**. Some architects shade this area lightly to better outline the building and its partitions.

When shown in small scale, plywood is represented with the same symbol as lumber. In section (if the scale permits), lines may be drawn to indicate the plies (not necessarily the exact number). In elevation views, wood siding and panels are represented as shown in **Figure 7-27**.

Symbols for glass consist of a single line on plan drawings and section drawings. The symbol may consist of several lines on large-scale drawings. Glass areas in elevation views are left plain or consist of a series of random diagonal lines.

Introduction to Green Building Technology

Green building refers to design strategies and construction methods used to build structures that make efficient use of resources and energy. Green building is also known as ***sustainable design***. ***Sustainability*** is the ability of a structure to maintain operational efficiency and have minimal impact on the environment throughout its lifetime. Green building has both environmental and economic benefits. These benefits include the following:

- Protecting the ecosystem
- Improving air and water quality
- Reducing solid and material waste
- Reducing building operating costs
- Improving employee productivity and job satisfaction
- Contributing to the overall quality of life

What does it mean to build green? Simply put, it means to design and construct energy-efficient buildings that are made from sustainable materials and engineered to have minimal environmental impact. See **Figure 7-28**. Green building technology is not only intended to help protect the environment, it is designed to improve on older construction methods and save money. For these reasons, green building technology is becoming more widespread in residential and commercial construction. The following sections serve as an introduction to the primary design considerations and types of products used in green building construction.

Site Selection

Green building starts with the selection of the site, orientation of the building, and proper utilization of the natural terrain. During site development, considerations are made to remove as few trees as possible, control erosion, and control detrimental water runoff. Building materials and products may also be considered during site development. For example, ***pervious concrete*** (highly porous concrete that allows water to pass through it) can be used to reduce water runoff. Landscaping is another design consideration during site development. For example, trees are located to provide shade where needed. The trees selected should be appropriate for the local climate. In addition, when selecting plants and shrubs for landscaping, it is important to consider local climate patterns. If an area is subject to drought, the landscaping should include plants and shrubs that do not require an abundance of water.

Energy Efficiency

Energy consumption is a primary design consideration in green building. The shape of the building and orientation to the sun are important factors in controlling heating and cooling costs. Passive energy strategies can be used to provide solar heat in the winter and shade in the summer. The sun orientation and building shape are also analyzed with respect to other functions, such as lighting the interior spaces of the building. Special construction features can be added to take advantage of natural light. For example, interior windows allow light to penetrate deeper into a building. This saves lighting costs and also has a positive impact on the productivity of the workforce. Additional strategies can be used to reduce lighting costs. Efficient lighting systems with sensor controls allow lights to be powered off automatically when they are not needed. See **Figure 7-29**.

Material	Plan	Elevation	Section
Earth	None	None	
Concrete			Same as plan view
Concrete block			
Gravel fill	Same as section	None	
Wood	Floor areas left blank	Siding Panel	Finish Framing
Brick	Face Common	Face or Common	Same as plan view
Stone	Cut Rubble	Cut Rubble	Cut Rubble
Structural steel		Indicate by note	Specify
Sheet metal flashing	Indicate by note		Show contour
Insulation	Same as section	Insulation	Loose fill or batt Board
Plaster	Same as section	Plaster	Stud Lath and plaster
Glass			Large scale Small scale
Tile			

Goodheart-Willcox Publisher

Figure 7-27. Symbols used for construction materials.

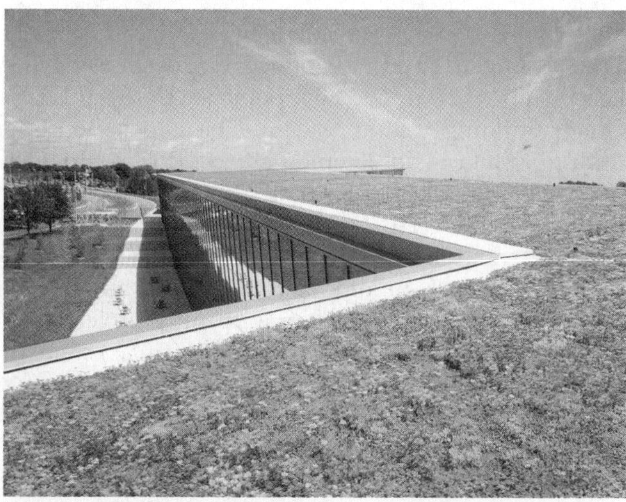

LiveRoof® Hybrid Green Roof System; courtesy of LiveRoof, LLC

Figure 7-28. The green roof system installed on this commercial building uses 1′ × 2′ modules with mature plants. Green roofs have both environmental and economic benefits, including reduction of roof temperature, capture and absorption of storm water, and runoff reduction.

In three-dimensional (3D) design and modeling, simulation and analysis software is commonly used to test energy-efficient structures that utilize green building principles. When used with a 3D building model in building information modeling (BIM), simulation and analysis programs can be used to optimize electrical and mechanical systems. The software evaluates the design data of the building and generates information related to energy performance, including estimated usage and costs.

Material Selection

Material selection plays a significant role in green building construction. The use of recycled and *renewable* (recyclable) materials helps conserve resources and directly impacts the sustainability of the structure. Sustainable materials such as steel and engineered lumber are used in framing to maintain long-term building performance. When selecting materials, the entire life cycle of the structure is considered. For example, products such as steel, aluminum, and plastic are recyclable and can be reclaimed when a building is demolished. Cost is another factor in material selection. Using sustainable building products may come at greater expense, but it may prove more economical. A sustainable building product may cost more than a less efficient product, but it may last twice as long.

Costs associated with transporting materials and building products to the site are also important considerations. When possible, local or regional resources and materials are used to save fuel and other transportation costs. Environmental concerns are also weighed when selecting materials. Certain materials and construction processes release toxins and may be harmful to the environment. For example, some adhesives and finishes emit volatile organic compounds (VOCs) when applied. Water-based products may be available and may offer a safer alternative.

Thermal efficiency is another important consideration in material selection. Products with good insulating characteristics help conserve energy. Well-sealed windows with low thermal transmission, such as low-e and argon-filled units, have high R-values and are highly efficient. These types of windows are designed to reduce heat loss or gain but permit light gain.

Water Efficiency

A clean and plentiful water supply is a valuable resource. In green building projects, water conservation is an important priority. There are many ways water can be used more efficiently. Roof runoff water (recycled water) can be used for toilet flushing, garden irrigation, and other purposes. Many products used in plumbing work are designed for water efficiency, including low-volume fixtures, faucets, and shower heads. Energy-efficient appliances, such as point-of-use water heaters, help conserve water and provide other advantages, such as reduced heat loss.

Goodheart-Willcox Publisher

Figure 7-29. The lighting system in this warehouse utilizes motion-activated lights to reduce energy use.

GREEN BUILDING

Volatile Organic Compounds (VOCs)

Volatile organic compounds (VOCs) are toxic substances that evaporate into the atmosphere. The evaporation of VOCs contributes to the development of environmental hazards such as smog, but VOCs can also reduce the quality of air indoors. Not only are VOCs harmful to the environment, but they can also be harmful to your health. Continued exposure to these toxins can produce symptoms such as headaches and nausea or more severe damage, such as organ damage and cancer. Products that contain VOCs include substances such as paint and cleaning supplies, but VOCs are also found in fabrics or carpets.

In selecting materials for a construction project, it is possible to reduce the owner's exposure to VOCs being released into the building. Many VOC-containing products are offered in more organic alternatives. These products use plant-based materials rather than chemicals such as benzene. Plan to use zero-VOC paints, and choose carpets and other flooring materials carefully. Take the time to research the materials. In general, choose materials with natural finishes or fabric that is made from organic cotton.

LEED Building Certification

The US Green Building Council has established the Leadership in Energy and Environmental Design (LEED®) Green Building Rating System. The Green Building Rating System is an accepted standard that provides construction guidelines for builders and municipalities. LEED building certification, available for commercial and residential construction projects, is based on implementation of green building objectives. LEED certification recognizes successful performance of a project in a series of categories in a specific LEED rating system. In the certification process, points are awarded for meeting criteria in each category. The categories addressed under the New Construction and Major Renovation rating system include the following:

- Location and transportation
- Sustainable sites
- Water efficiency
- Energy and atmosphere
- Materials and resources
- Indoor environmental quality (IEQ)
- Innovation
- Regional priority

To earn LEED certification, the construction project must score the required number of points in the specific LEED rating system. Four different certification levels are available: Certified, Silver, Gold, and Platinum. The LEED Platinum certification is the highest level attainable.

Test Your Knowledge

Name ______________________________

Write your answers in the spaces provided.

_____ 1. *True or False?* Concrete is stronger when it is in tension (pulled) than when it is in compression (pushed).

_____ 2. When preparing a batch of concrete, which of the following would not be included?
A. Water
B. Cement
C. Aggregate (sand, gravel)
D. Chemical admixtures
E. All of these items could be included in the mix.

_____ 3. Which of the following types of brick is made from a mixture of Portland cement and aggregates?
A. Sand-lime brick
B. Adobe brick
C. Kiln-burned brick
D. Concrete brick
E. All of the above.

_____ 4. *True or False?* Firebrick is referred to as "common brick" because it is the most commonly used type of brick.

_____ 5. Squared stones that have been laid in a pattern but not cut to dimensions are called _____.
A. ashlar stones
B. pea gravel
C. rubble
D. concrete masonry units (CMUs)
E. cut stones

_____ 6. Which mortar type is best-suited for situations requiring high lateral strength?
A. Type M
B. Type S
C. Type N
D. Type O
E. None of the above.

_____ 7. *True or False?* Lumber classified as rough-sawn lumber has been dressed or finished to size by running it through a planer.

_____ 8. What is the difference between interior and exterior plywood?
A. The type of wood used
B. The size of the sheets
C. The type of adhesive used
D. The thickness of the sheets
E. Interior and exterior plywood are identical.

_____ 9. The steel beam identification W8×15 represents a wide-flange beam _____.
A. 8″ deep and weighing 15 pounds per linear foot
B. 8″ deep and 15″ wide
C. 8″ wide and 15″ deep
D. 8″ deep and 15′ long
E. None of the above is correct.

_____ 10. *True or False?* Nonferrous metals contain little or no aluminum.

_____ 11. *True or False?* Concrete completely hardens within 48 hours after it is poured.

_____ 12. Which of the following is *not* considered to be a type of safety glass?
A. Wired glass
B. Laminated glass
C. Tempered glass
D. All of these are types of safety glass.

_____ 13. *True or False?* An air space between two sealed panes of glass serves as good insulation.

_____ 14. What type of cement would you use in the winter to make the concrete set up faster?
A. Type I, Normal Cement
B. Type III, High Early Strength
C. You wouldn't make concrete in the winter because it might freeze.
D. Use a retarder admixture.

15. Give the actual diameter of the reinforcing steel bar listed below:

A. #9

B. #4

C. #8

D. #5

E. #18

A. ______________________________

B. ______________________________

C. ______________________________

D. ______________________________

E. ______________________________

16. Name the brick positions in the walls indicated below.

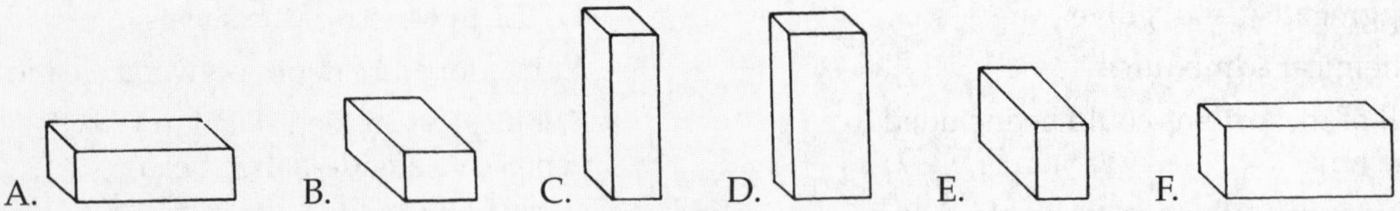

A. B. C. D. E. F.

A. ______________________________

B. ______________________________

C. ______________________________

D. ______________________________

E. ______________________________

F. ______________________________

_____ 17. What are the two basic kinds of metal?

A. Ferrous and nonferrous

B. Steel and copper

C. Steel and aluminum

D. Brittle and malleable

18. What shape does each of the following designations for structural steel indicate?

A. W

B. S

C. C

D. L

A. ______________________________

B. ______________________________

C. ______________________________

D. ______________________________

_____ 19. What causes concrete to harden?

A. Dehydration

B. Evaporation

C. Consolidation

D. Hydration

20. What is the system used to measure the thickness of sheet metal products?

____

_____ 21. *True or False?* A sheet metal product made from 12-gage metal weighs more than the same product made from 6-gage metal.

_____ 22. *True or False?* Loose-fill insulation is an insulating material available in blanket and batt form.

_____ 23. Which of the following is a benefit of green building?

A. Reducing solid and material waste

B. Improving employee productivity and job satisfaction

C. Improving air and water quality

D. All of the above.

_____ 24. *True or False?* Thermally efficient, argon-filled window units have low R-values.

_____ 25. *True or False?* Storm water can be collected from a roof and used for garden irrigation or toilet flushing.

SECTION 4
Reading Prints

Marykit/Shutterstock.com

UNIT 8
Site Plans

TECHNICAL TERMS

assumed benchmark
bearing
benchmark
contour lines
delta
easement
legend
minute
north arrow
Plan North
plot plan
property lines
second
setback
site plan
topography
True North
zoning

LEARNING OBJECTIVES

After completing this unit, you will be able to:

- Recognize common features of site plans.
- Identify property line descriptions.
- Explain the difference between True North and Plan North.
- Read contour lines on a site plan.
- Plot topography sections.

A building can be greatly enhanced by its location on the plot of land. Therefore, architects take advantage of the topography of the land, surrounding trees, view from the street, and other features to improve the appearance of a structure.

Features of Site Plans

A ***site plan*** (also referred to as a ***plot plan***) is a view from above the property that shows the outline of the property and the location of the building or structures on the lot. See **Figure 8-1**. Many features may be shown on the site plan:

- Lot number, block number, or property address
- Bearing (direction) and length of property lines
- North arrow
- Dimensions of front, rear, and side yards
- Location of other accessory buildings (garage, carport, etc.)
- Location of walks, drives, fences, and patios
- Location of utility easement and property setbacks
- Location of utilities (gas, electric, water, and sewage lines)
- Elevations at the various locations
- Trees and shrubs to be removed or retained
- Grade elevations and topography of the site

North Arrow

The ***north arrow*** indicates the north direction. When you begin looking at a site plan, first find the north arrow. It will help you to visualize sun tracking and orientation relative to the structure. This is particularly true if you are familiar with the plot of land on which the building will be constructed.

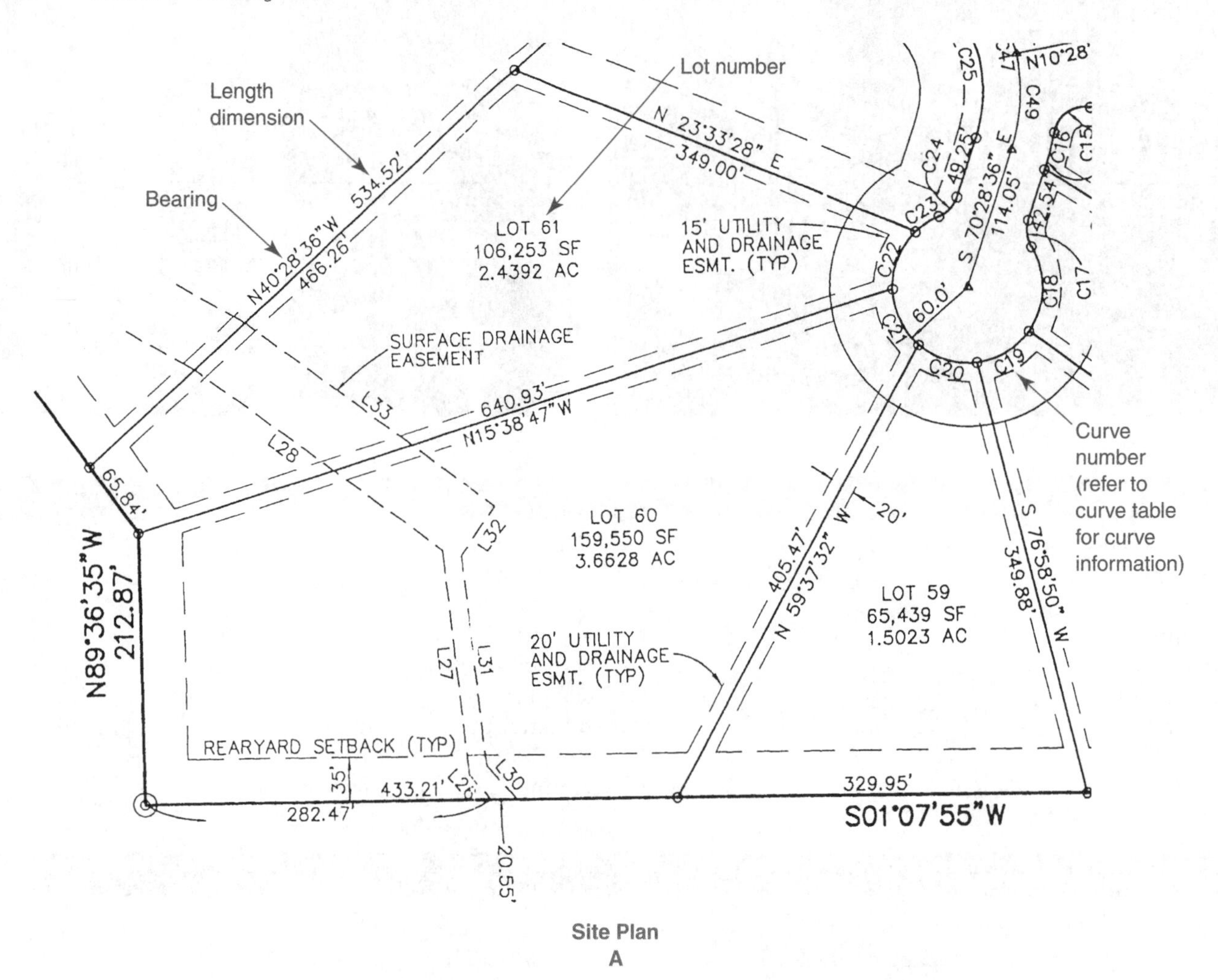

Site Plan
A

CRV NO.	RADIUS	ARC	CHORD	CHORD BRG.	DELTA	TANGENT
C10	25.00'	39.29'	35.37'	S 11°58'59" W	90°03'21"	25.02'
C11	60.00'	43.70'	42.74'	S 12°10'55" E	41°43'33"	22.87'
C12	60.00'	65.95'	62.68'	S 40°10'13" W	62°58'42"	36.75'
C13	60.00'	50.00'	48.57'	N 84°28'02" W	47°44'47"	26.55'
C14	60.00'	76.12'	71.11'	N 24°15'04" W	72°41'09"	44.14'
C15	25.00'	39.29'	35.37'	N 32°56'10" W	90°03'21"	25.02'
C16	180.00'	23.52'	23.51'	N 74°13'13" W	07°29'16"	11.78'
C17	25.00'	21.68'	21.00'	S 84°41'01" W	49°40'47"	11.57'
C18	60.00'	70.25'	66.31'	N 86°36'45" W	67°05'16"	39.78'
C19	60.00'	50.00'	48.57'	N 29°11'43" W	47°44'47"	26.55'
C20	60.00'	50.00'	48.57'	N 18°33'05" E	47°44'47"	26.55'
C21	60.00'	50.00'	48.57'	N 66°17'52" E	47°44'47"	26.55'
C22	60.00'	50.00'	48.57'	S 65°57'21" E	47°44'47"	26.55'
C23	60.00'	22.29'	22.16'	S 31°26'23" E	21°17'09"	11.28'
C24	25.00'	21.68'	21.00'	S 45°38'12" E	49°40'47"	11.57'
C25	120.00'	125.66'	120.00'	N 79°31'24" E	60°00'00"	69.28'

Curve Table
B

CESO Engineers & Architects

Figure 8-1. A site plan locates the building on the property. A—Site plan showing the location of lots, easements, and lengths and bearings of property lines. B—A curve table listing information for curves referenced in the site plan.

If the walls of the building are not parallel to the compass directions, a ***Plan North*** may be designated. The Plan North will typically be slightly different from ***True North***, **Figure 8-2**. A Plan North is provided so that there is a reference orientation aligned with the building. This also simplifies the description of building elevation views.

Property Lines

Lines defining the limits of a building plot or a piece of property are called ***property lines***. The property line is defined by a length and a ***bearing*** (direction of the line). Each property line is identified on the site plan. Bearings are expressed as degrees east or west of north or south, **Figure 8-3**. Bearings are expressed in degrees, minutes, and seconds. A circle is divided into 360 degrees (360°); one degree (1°) is 1/360th of a circle. A ***minute*** is 1/60th of a degree; a ***second*** is 1/60th of a minute.

When the property line is a curve rather than a straight line, it is identified by a radius, length, and angle of tangency. A ***delta*** (Δ) is the central angle formed by the radii meeting the curve at the points of tangency. In **Figure 8-1**, a curve table is used to list data for the cul-de-sac curves and other curves representing property lines in the site plan.

Contour Lines

Contour lines are lines that identify the ground elevation. All of the points along a single contour line are at the same elevation. The elevation of the line is listed, **Figure 8-4**. Contour lines are drawn on the site plan to indicate the changing elevation of the land; as a group, they show the topography of the land.

The interval between contour lines (the change in vertical distance) can be any convenient distance, such as 1′, 2′, 5′, or 10′. If the interval is too small, there will be too many contour lines and the drawing will become crowded and hard to interpret. If the interval is too large, some detail will be lost. For example, if a 10′ interval is used, a 7′ high mound of dirt wouldn't be shown if it was located between contour lines. This could affect the excavation estimating.

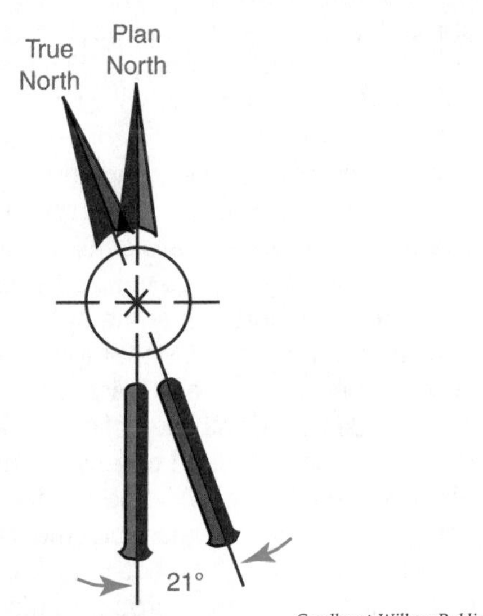

Goodheart-Willcox Publisher

Figure 8-2. The north arrow indicates the difference between True North and Plan North.

N
45°
20°
65°
N 45° W
N 20° E
N 65° E
W
E
S 80° W
S 30° E
80°
30°
S
Bearing Directions

Goodheart-Willcox Publisher

Figure 8-3. Expressing directions as bearings related to north and south.

GREEN BUILDING

Working with Existing Vegetation

Architects take many factors into account when planning the location of a building on a piece of property. One of the most important factors is the existing vegetation on the property. This includes trees, bushes, and any other growing plants. When possible, a building is constructed around existing trees and vegetation to reduce impact on the natural surroundings. This has the additional benefit of saving the owner money on purchasing new trees, shrubs, and plants.

When it is not possible to build around one or more trees, the architect may consider relocating them to another place on the property. Even mature trees can be moved successfully. If this is done, it is important to take precautions to prevent damage to the trees or the surrounding structures and nearby equipment, such as electrical wires and other utility equipment.

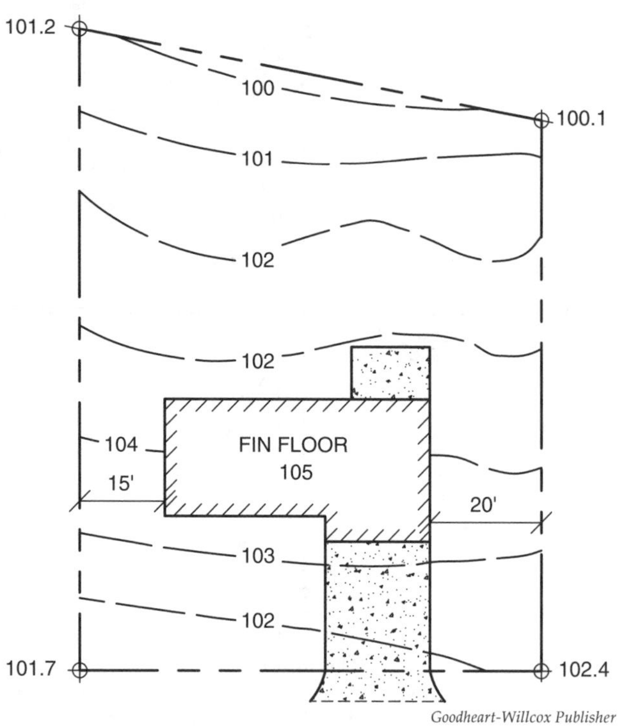

Goodheart-Willcox Publisher

Figure 8-4. Contour lines show the elevation and the general lay of the plot of land.

The distance between contour lines indicates slope. Contour lines that are far apart indicate a gradual slope of the land. Lines that are close together indicate a steep slope. See **Figure 8-5**, Notes B and C.

Drawing sections through contours is a good way to understand or determine the change in grade. In **Figure 8-5**, section A_1-A_2 shows a gradual slope down to a slight valley and back up again. The section was plotted by marking off horizontal grade lines at intervals of 2′ and plotting the intersection of each contour line along the section cut. The intersection is projected perpendicularly down to the proper grade indicated on the horizontal grid. A line drawn from mark to mark will indicate the slope of the section.

Figure 8-6 indicates the topography and section as a pond or valley with the numbers decreasing toward the middle. **Figure 8-7** indicates the exact same configuration with the elevation numbers increasing toward the middle, indicating a mound or hill in plan.

Contour lines are long, solid or dashed lines drawn freehand. When it is desired to show both the original grade and a finish grade of contour, the original grade is shown with short, dashed lines or grayed-out lines. The finish grade is shown with solid lines, **Figure 8-8**.

The elevations on a particular plot are referenced to a local permanent marker of known elevation, such as a survey marker plate, a fire hydrant, or a manhole cover. The marker is called a ***benchmark***. To simplify the reading and laying out of elevations on building structures, some architects set the elevation of a particular point to 0.0′ or 100.00′. All elevations are then taken with respect to this point. This elevation is sometimes called an ***assumed benchmark*** or building elevation. This makes the changes in elevation on higher and lower floors easier for the construction team to interpret.

Topographic Features

The ***topography*** (location and elevation of features) is often displayed on the site plan. Topographic features include natural objects, such as trees or shrubs, and

GREEN BUILDING

Rainwater Harvesting

You may not think about it, but buildings consume a considerable amount of resources. Buildings consume many raw materials used in construction and a great deal of electricity, natural gas, and water. Because water is a valuable natural resource, many homeowners and building managers take steps to make water conservation a priority. For example, installing low-volume fixtures helps reduce the amount of water required for the building. However, beyond limiting water use indoors, there are additional ways to conserve water. A common practice is to collect rainwater and use it for irrigating landscaping and gardens. Rain barrels are popular for rainwater harvesting. These are designed to collect water from a roof downspout and usually have a spigot for attaching a garden hose. Rain barrels typically hold 55 gallons of water, but larger containers with storage capacities up to several hundred gallons are available.

More complex rainwater catchment systems are available for harvesting larger quantities of water. Rainwater can be collected in one or more large storage tanks or a cistern capable of storing several thousands of gallons. Rainwater is collected from the building roof or an approved catchment surface above ground. Large storage tanks typically include a pump and can be installed above or below ground. Tanks come in different shapes and sizes and are available in materials such as polyethylene, galvanized steel, fiberglass, wood, and concrete. In some systems, collected rainwater can be recycled for indoor purposes, such as toilet flushing and washing clothes. Special systems designed with water filtration and treatment equipment can be used to recycle rainwater as potable water (water suitable for drinking). The construction of a rainwater catchment system must meet local building code requirements. Underground installations must have proper reinforcement and access for cleaning and maintenance.

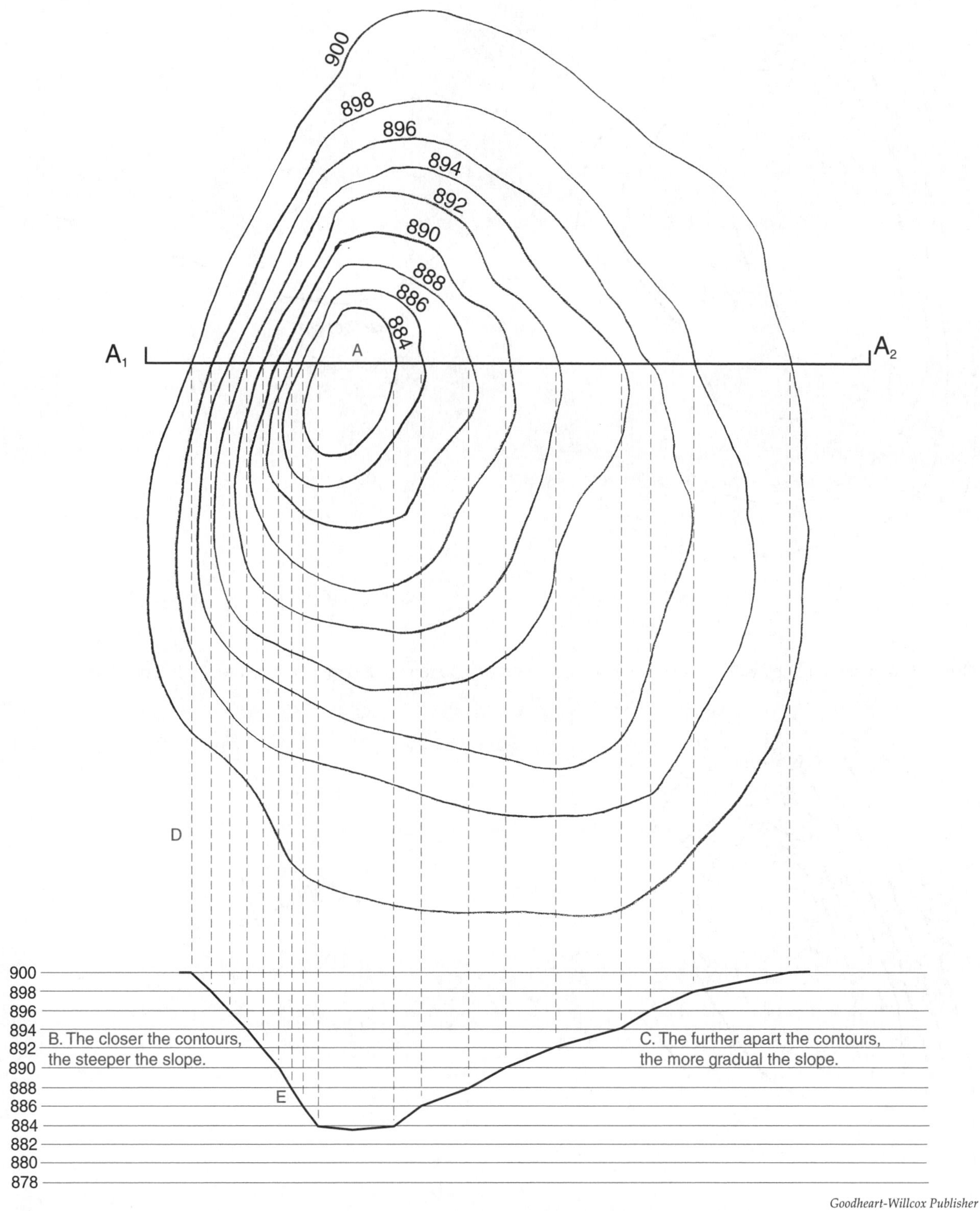

Goodheart-Willcox Publisher

Figure 8-5. The contours show the grades decreasing toward the middle (A), which would indicate a valley or pond. Notice that the closer the contour lines (B), the steeper the slope; the farther apart the contour lines (C), the more gradual the slope. A section (A_1-A_2) is plotted by extending lines downward from each contour intersection to a horizontal grid (D). Connecting the intersections of the horizontal and vertical lines on the grid provides a cross section of the slope (E).

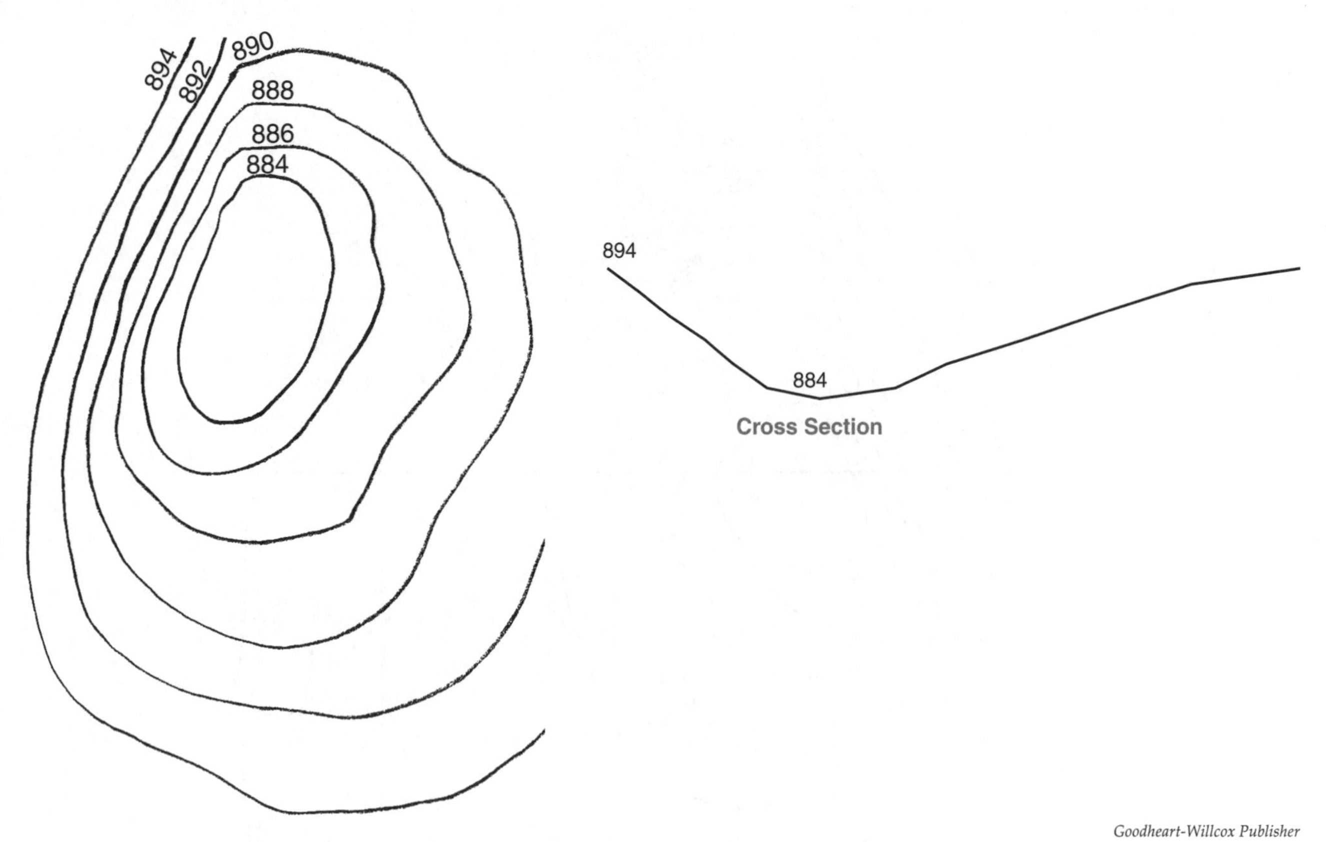

Goodheart-Willcox Publisher

Figure 8-6. Contour lines with the elevation numbers decreasing toward the middle indicate a pond or valley.

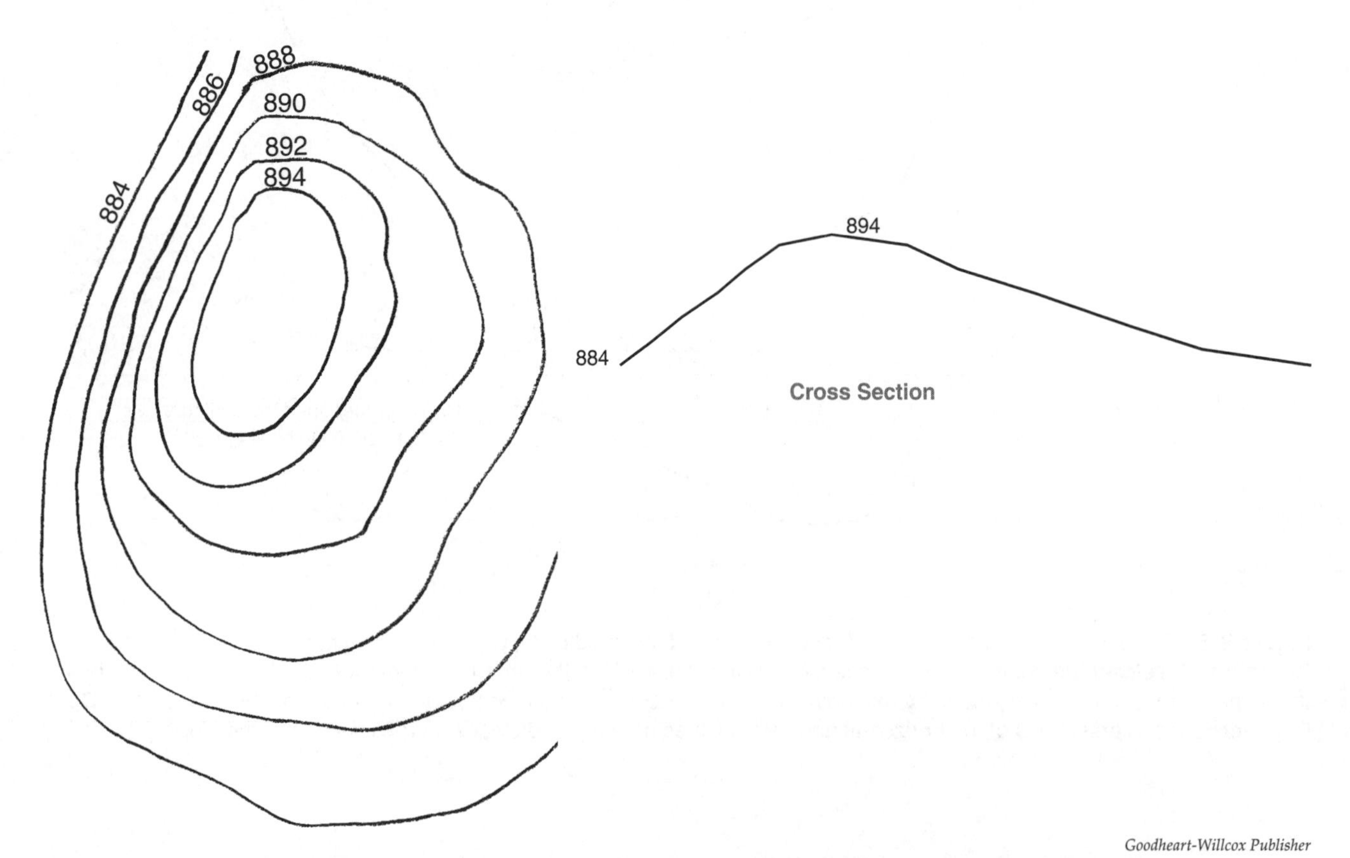

Goodheart-Willcox Publisher

Figure 8-7. Contour lines with the elevation numbers increasing toward the middle indicate a mound or hill.

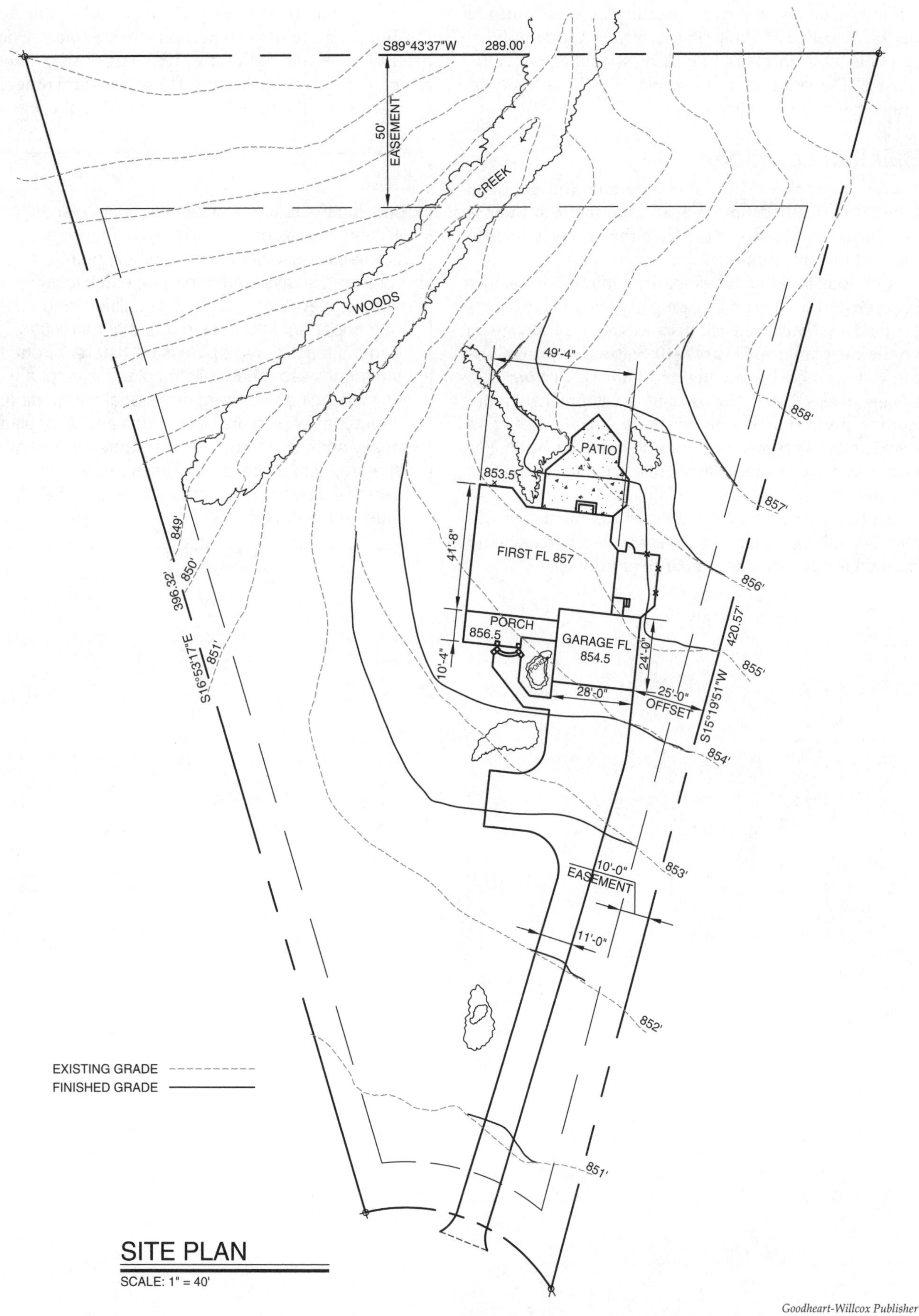

Goodheart-Willcox Publisher

Figure 8-8. The contour lines on this residential site plan show the original and finished grade. The original grade is shown with grayed-out dashed lines. The finish grade is shown with solid lines.

human-made objects, such as utility poles or railroad tracks. **Figure 8-9** illustrates common topographical symbols used on plans. Site plans should also include a list of the symbols used to identify these features. This list is called a ***legend***.

Building Location

An outline of the structure is shown on the site plan. Often, the finished floor elevation of the first floor is also included. The distances from the property lines to the building are shown.

Most local building codes specify a minimum distance between the building and the property lines. This distance is called a ***setback***. Setback distances are typically given on the site plan. See **Figure 8-10**. Setback requirements are determined by zoning ordinances. ***Zoning*** is a system of classifying the use and development of land. Zoning divides land into zones for specific uses and restricts what types of structures can be built. For example, a given zone may be restricted to residential, commercial, or industrial use.

Required minimum distances for easements are also shown on the site plan. An ***easement*** is an area of land marked for use and maintenance of utilities.

The connections between the main utility lines and the building are also shown on the site plan. Underground pipes and cables are shown as dashed lines or as noted in a site plan legend. These lines and other line types used on the drawing are defined in the legend.

Note

Studying the features shown on a site plan will help you become familiar with important information associated with a building project. A site plan shows land contours, established points of elevation, locations of utilities, and other important information about construction. Learning to read a site plan is no different from learning to read other building plans, except a site plan includes three-dimensional information. Other types of plan drawings, such as a floor plan, usually show all features at the same elevation. When reading a site plan, always try to form an overall bird's-eye view of the site layout, then focus on the details.

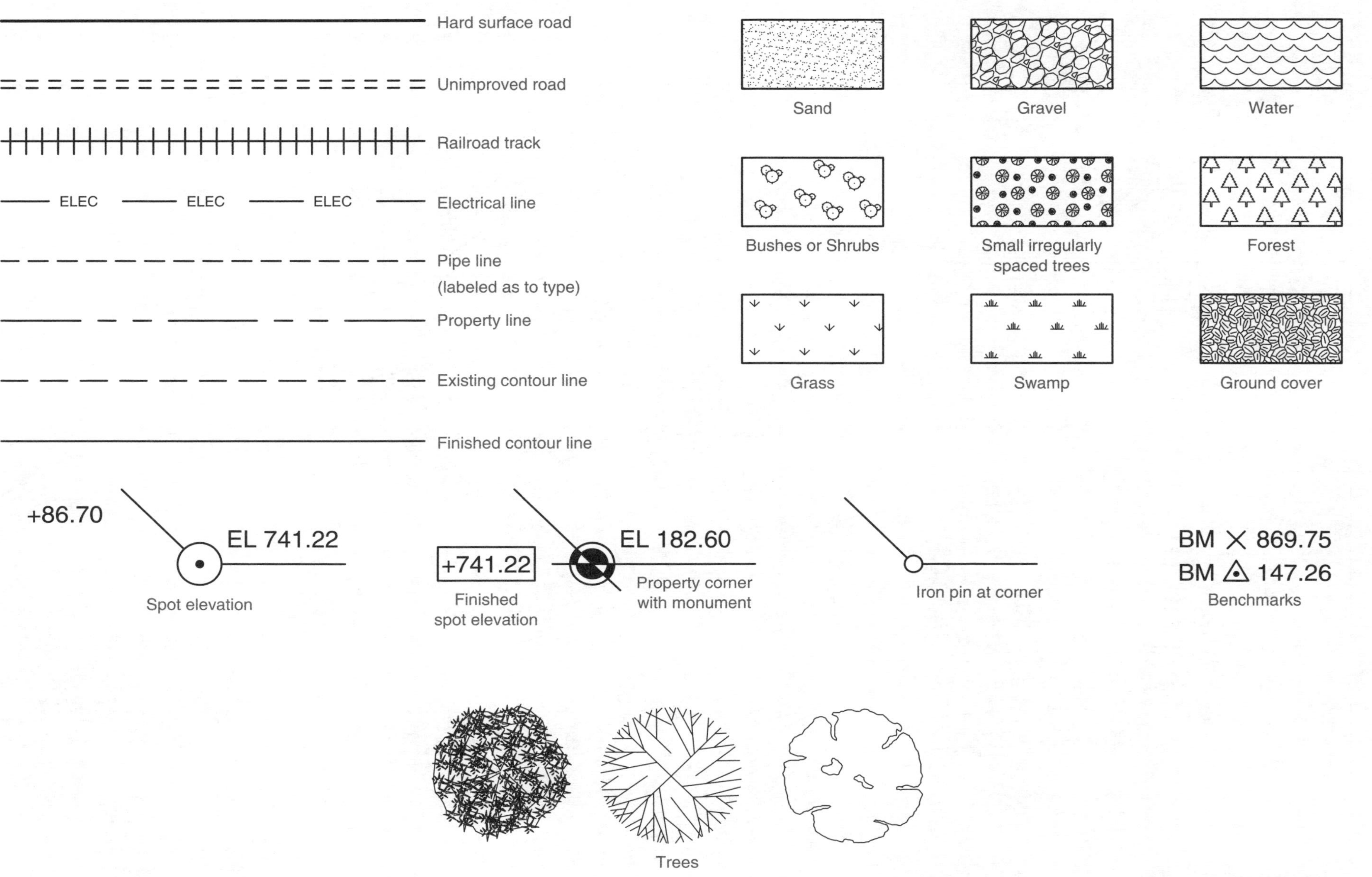

Figure 8-9. Common symbols used on site plans. Always refer to the plan legend for proper interpretation.

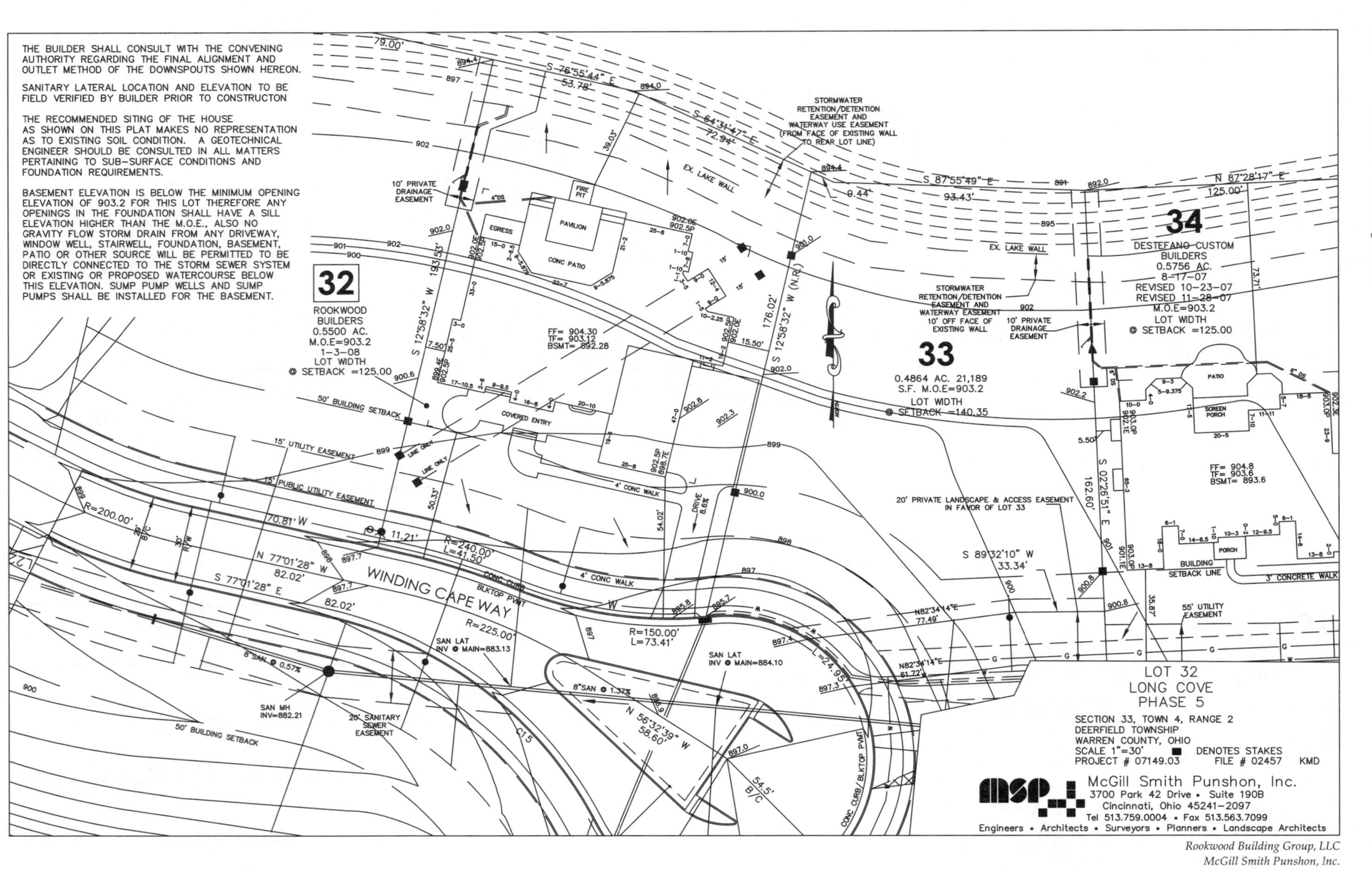

Rookwood Building Group, LLC
McGill Smith Punshon, Inc.

Figure 8-10. A site plan showing setbacks and easement requirements.

Test Your Knowledge

Name ______________________________

Write your answers in the spaces provided.

____ 1. A site plan is also known as a ____ plan.
- A. mechanical
- B. plot
- C. floor
- D. foundation

____ 2. A benchmark is ____.
- A. used to mark the location of lawn furniture
- B. a portable device used to measure elevations
- C. a line drawn with chalk on the ground
- D. a permanent object of known elevation used to measure other elevations
- E. None of these things.

____ 3. *True or False?* Plan North is designated by the creator of the drawing.

____ 4. Which of the following is *not* used to define a curved property line?
- A. Radius
- B. Curve length
- C. Elevation change
- D. Central angle
- E. All of these are needed to define a curved property line.

____ 5. *True or False?* If the interval between contour lines is too large, the contour lines will be crowded too closely.

____ 6. Which of the following would *not* be shown on a site plan?
- A. Location of a garage
- B. Trees
- C. Utility lines
- D. Location of the chimney on a building
- E. All of these would be shown on a site plan.

____ 7. A minimum distance required between the building and the property lines is called a(n) ____.
- A. assumed benchmark
- B. delta
- C. setback
- D. bearing

8. Using the bearings shown below, draw on the compass the approximate location and direction that you would be turning.
- A. N 25° E
- B. S 65° E
- C. S 45° W

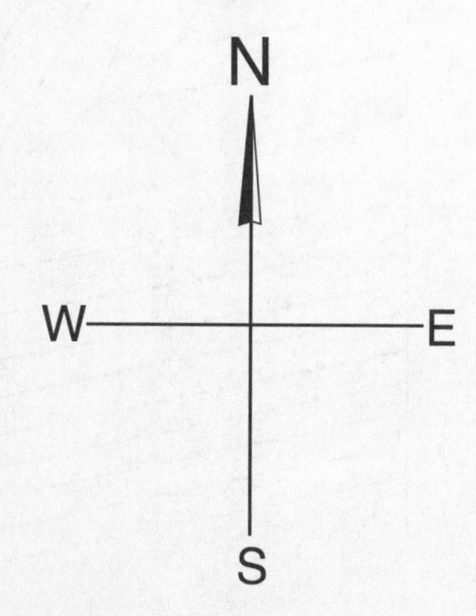

9. Using the site plan shown, answer the following questions. A 2′ interval is used for contour lines.
   A. Is the point indicated in cut (where material is removed) or fill (where material is added)? What is the change in elevation?

   ____

   B. Is the point indicated in cut (where material is removed) or fill (where material is added)? What is the change in elevation?

   ____

   C. What is the elevation of the point indicated?

   ____

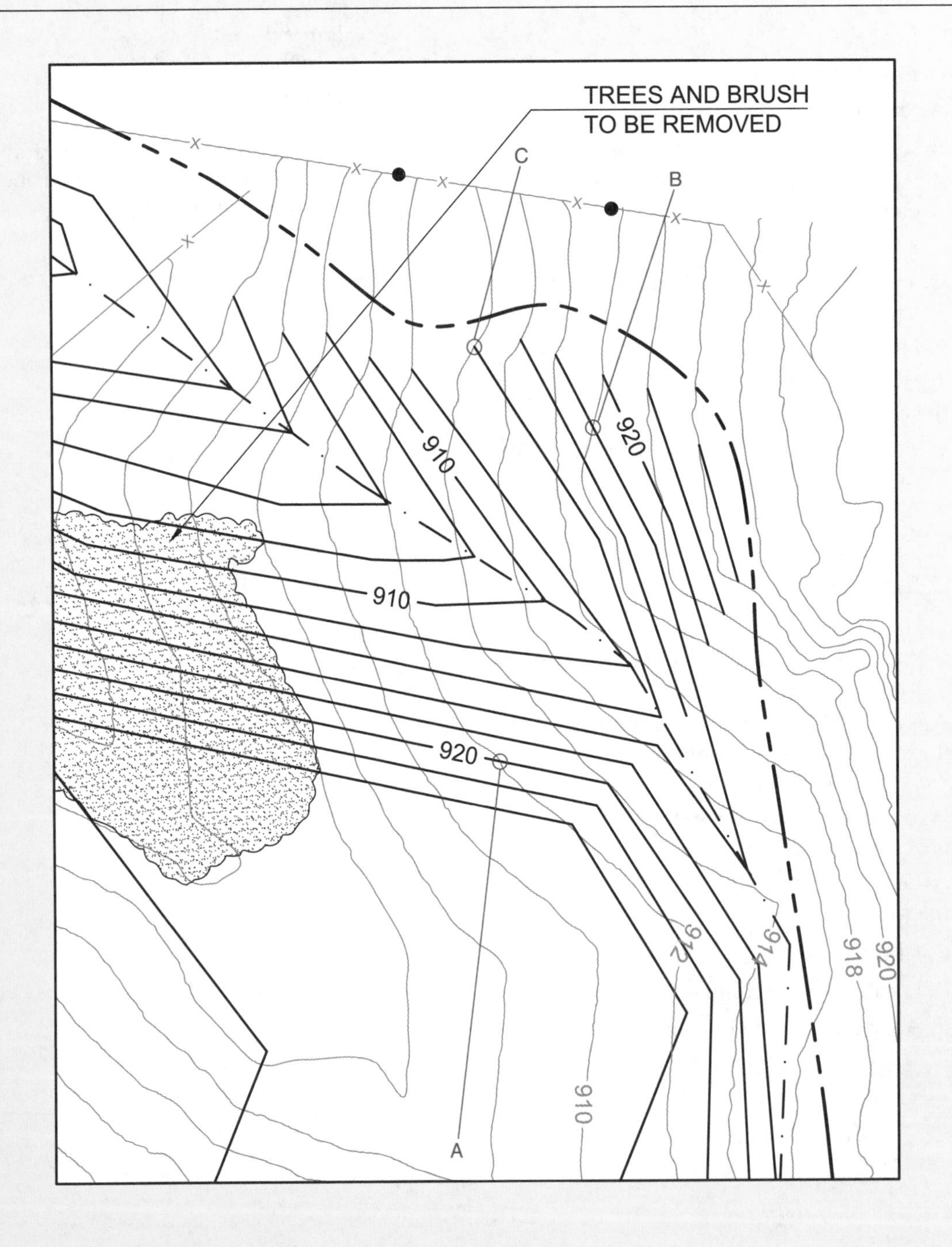

Activity 8-1

Site Plan for a Residence

Name ________________________________

Refer to **Figure 8-10** *to answer the following questions.*

1. What is the scale of the site plan?

   ____

2. For which lot has the site plan been prepared?

   ____

3. What is the acreage of the lot?

   ____

4. What is the bearing and length of the property line along the east side of the property?

   ____

5. Which part of the property generally faces True North?

   ____

6. Give the distance the house is set back from the following locations.

   A. The street ________________________________

   B. West side ________________________________

   C. East side ________________________________

7. What is the FF elevation and what does FF stand for?

   ____

8. How wide is the sidewalk?

   ____

9. What is specified for the slope of the drive extending toward the street?

   ____

10. What line symbol is used to indicate the existing lake wall? Sketch the symbol freehand.

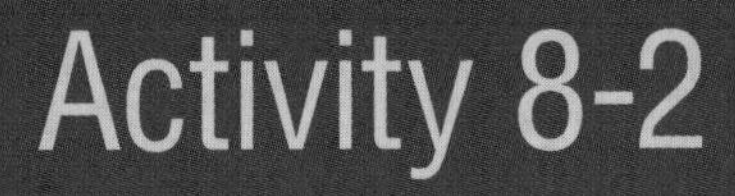

Activity 8-2

Plans for a Commercial Building Project

Name ______________________________

Refer to Sheets 1, 4, and 5 from the Delhi Flower and Garden Centers commercial building plans in the Large Prints supplement to answer the following questions.

1. What is the scale of the grading plan? 1. ______________________

2. How many sheets are there in this set of plans? 2. ______________________

3. Give the bearing and length for each of the following property lines.
   A. North side of Lot 56
   B. East side of Lot 56
   C. The curve of Lot 55

   3. A. ______________________
   B. ______________________
   C. ______________________

4. Based on the north arrow, on which side of the Delhi Flower and Garden Centers building is the detention pond? 4. ______________________

5. What do the large "X" symbols indicate? 5. ______________________

6. What is the size of the Type 6 concrete curb? 6. ______________________

7. What is the lowest elevation of the detention pond? 7. ______________________

8. Where are the construction entrances located? 8. ______________________
   ____

9. What is the right-of-way (R/W) dimension? 9. ______________________

10. What is the existing grade and finished floor level for the covered sales area?
    A. Existing grade
    B. Finished floor level

    10. A. ______________________
    B. ______________________

11. On the grading plan, are there any other buildings indicated in addition to the Delhi Flower and Garden Centers building? 11. ______________________

12. What is the temporary bench mark elevation and where is it located?
    A. Elevation
    B. Location

    12. A. ______________________
    B. ______________________

UNIT 9
Architectural Drawings

TECHNICAL TERMS

architectural drawings
detail
elevation
floor plan
schedule
section

LEARNING OBJECTIVES

After completing this unit, you will be able to:

- Identify the types of drawings classified as architectural drawings.
- List different types of floor plans used in construction projects.
- Describe the purpose of elevations, sections, and details.
- Identify materials specified on architectural drawings.
- Interpret construction requirements specified on architectural drawings.

Introduction

As you learned in Unit 1, construction drawings are organized into categories and trade disciplines as they relate to a construction project. ***Architectural drawings*** show the materials and construction processes that define the structure and create the final space for the building. Architectural drawings represent the core of the drawing set and serve to pull the entire project together. As discussed in Unit 1, drawings in this classification are typically prefixed with the letter *A*. Architectural drawings normally compose one of the largest sections in the set of construction documents, **Figure 9-1**.

Information on architectural drawings includes materials used for exterior and interior wall coverings, such as brick, metal siding, and drywall. Architectural drawings specify requirements for items such as roof systems, ceiling layouts, finish floor materials, and walls and partitions. Architectural drawings provide information for "hidden" materials behind walls or above ceilings, such as the insulation behind a brick exterior or above the acoustical ceiling. Architectural drawings are where windows, doors, and interior finishes are defined. The architectural drawings fill in the "gap" defining the accessory building materials and products that are added to complete the building structure.

Architectural Drawing Organization

Architectural drawings are organized in a manner that simplifies finding information. A typical organization consists of the following:

- Floor plans
- Architectural elevations
- Architectural sections
- Architectural details
- Schedules

Floor Plans

A ***floor plan*** shows the building layout and serves as a reference for making many other types of drawings, details, and schedules. See **Figure 9-2**. The floor plan is typically referenced when making framing plans, mechanical plans, electrical plans, and plumbing plans. In larger residential projects and commercial projects, a number of specialized floor plans may be used. These

ARCHITECTURAL

A0.01 GENERAL NOTES, PARTITION TYPES, & LEGENDS
A0.02 MASTER MATERIAL NOTES LIST
AD.01 LOWER FLOOR DEMOLITION PLAN-SECTORS L-1, L-2
AD.02 FIRST FLOOR DEMOLITION PLAN-SECTOR 1-1
AD.03 FIRST FLOOR DEMOLITION PLAN-SECTOR 1-2
AD.04 FIRST FLOOR DEMOLITION PLAN-SECTOR 1-3
AD.05 SECOND FLOOR DEMOLITION PLAN-SECTOR 2-2
AD.06 SECOND FLOOR DEMOLITION PLAN-SECTOR 2-3
A1.01 FIRST FLOOR MASTER/CODE PLAN
A1.02 LOWER FLOOR & SECOND FLOOR MASTER/CODE PLAN
A1.03 ROOF PLAN
A2.01 LOWER FLOOR PLAN - SECTORS L-1, L-2
A2.02 FIRST FLOOR PLAN-SECTOR 1-1
A2.03 FIRST FLOOR PLAN-SECTOR 1-2
A2.04 FIRST FLOOR PLAN-SECTOR 1-3
A2.05 SECOND FLOOR PLAN-SECTORS 2-1, 2-2
A2.06 SECOND FLOOR PLAN-SECTOR 2-3
A3.01 ENLARGED PLANS AND ELEVATIONS
A4.01 EXTERIOR ELEVATIONS-EAST AND SOUTH
A4.02 EXTERIOR ELEVATIONS-WEST AND NORTH
A4.05 BUILDING SECTIONS
A5.01 WALL SECTIONS
A5.02 DETAILS
A6.01 DOOR, WINDOW, AND STOREFRONT ELEVATIONS
A6.02 OPENING DETAILS
A7.01 LOWER LEVEL ENLARGED EQUIPMENT PLAN-SECTORS L-1, L-2
A7.02 FIRST FLOOR ENLARGED EQUIPMENT PLAN-SECTOR 1-1
A7.03 FIRST FLOOR ENLARGED EQUIPMENT PLAN- SECTOR 1-2
A7.04 FIRST FLOOR ENLARGED EQUIPMENT PLAN-SECTOR 1-3
A7.05 SECOND FLOOR ENLARGED EQUIPMENT PLAN-SECTOR 2-1, 2-2
A7.06 SECOND FLOOR ENLARGED EQUIPMENT PLAN-SECTOR 2-3
A9.01 FIRST FLOOR REFLECTED CEILING PLAN
A9.02 LOWER FLOOR & SECOND FLOOR REFLECTED CEILING PLAN

STRUCTURAL

S2.01 FOUNDATION PLAN AND GENERAL NOTES-SECTOR 1-1
S2.02 LOWER LEVEL FOUNDATION PLAN; SECTIONS & DETAILS

CR architecture + design

Figure 9-1. A portion of a drawing index for a commercial building project. The architectural drawings typically make up one of the largest sections in a set of construction documents.

plans are used to show information about existing construction or additional information that would not be shown on a basic floor plan. Floor plans provide fundamental information about the building layout and are intended to help the builder quickly understand the construction requirements.

Your study of a floor plan should start with a review of the general layout. Get an idea of the room sizes, halls, and storage before studying details of construction. This will help you to understand the complete set of plans.

Floor plans are drawn to scale (usually 1/48 size, or 1/4″ = 1′-0″). However, you should rely on the dimensions shown and not scale the drawing.

The most common type of floor plan is generated by making a horizontal cut about 42″ to 48″ above the finished floor to show walls, doors, and windows. A separate drawing is made for each floor level of the building. In addition, there are various types of plan drawings that fall under the classification of floor plans. The following are a few of the most common.

- **Demolition plans.** On a renovation or remodeling project, plans that show existing walls, doors, and windows that are to be removed or changed. See **Figure 9-3**.
- **Enlarged plans.** Plans that show more detail in a larger scale in order to show more information in complex areas, such as stairways, bathrooms, entry lobbies, and kitchens. See **Figure 9-4**.
- **Reflected ceiling plans.** Plans that show items in the ceiling layout, such as structural members, light fixtures, and air supply registers. Looking at a reflected ceiling plan is like looking at a mirror image of the ceiling, where the floor acts as the mirror.
- **Floor finish plans.** Plans that show floor finishes and layouts in more complex designs.
- **Overview plans.** Plans that show a vast layout at a small scale and sections of the layout at a larger scale on the same sheet. An overview plan is helpful in understanding the entire project.

Architectural Elevations

Whereas plan drawings show the horizontal layout of a building, elevations show the vertical layout. An ***elevation*** shows an exterior view of a building or an interior view, such as a kitchen or hospital cabinet layout. Architects use elevations to communicate the beauty of a building as well as define construction materials, vertical elevation marks, grading and foundation lines, and section and detail cuts. There is usually one elevation drawing for each face of a building. On a four-sided building, for example, there would be four elevations. Elevations are typically identified in one of two ways. Elevations may be designated as *Front*, *Right Side*, *Left Side*, and *Rear*. More commonly, they are identified by the compass direction or designated direction that the elevation faces—*North*, *South*, *East*, and *West*. See **Figure 9-5**. When the building faces east, the front elevation would be identified as the *East Elevation*.

Types of Elevations

Exterior elevations are used to show exterior features exposed to view, such as doors, windows, gutters, and roof lines. Exterior elevations indicate the building materials used for construction, such as siding and trim. Underground features of the building, such as the basement or foundation walls and footings, are shown with hidden lines on elevations.

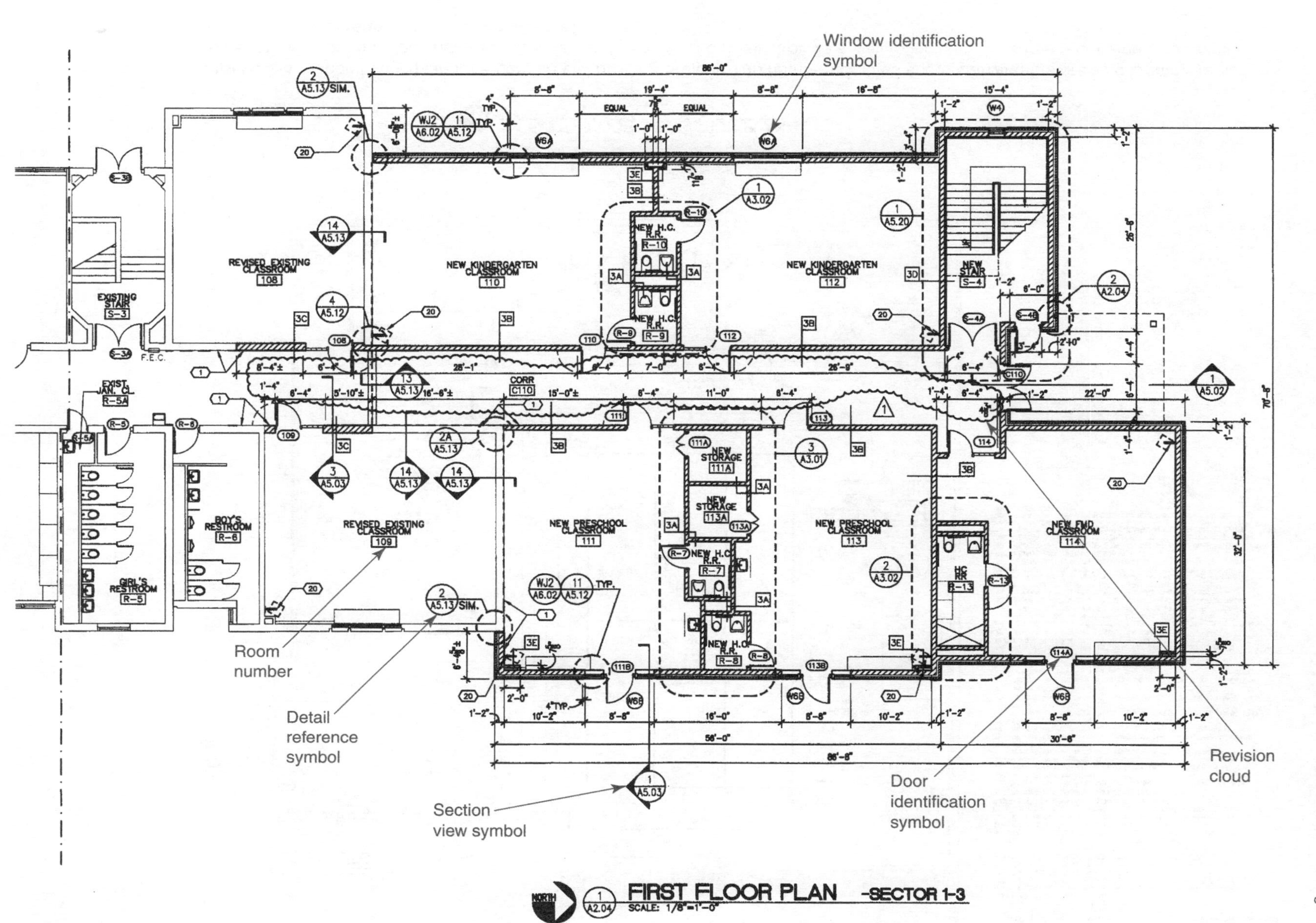

CR architecture + design

Figure 9-2. Floor plans show the building layout and are frequently used as a reference in making other drawings. The callout symbols on this floor plan refer to section views and detail drawings located on other sheets in the construction documents.

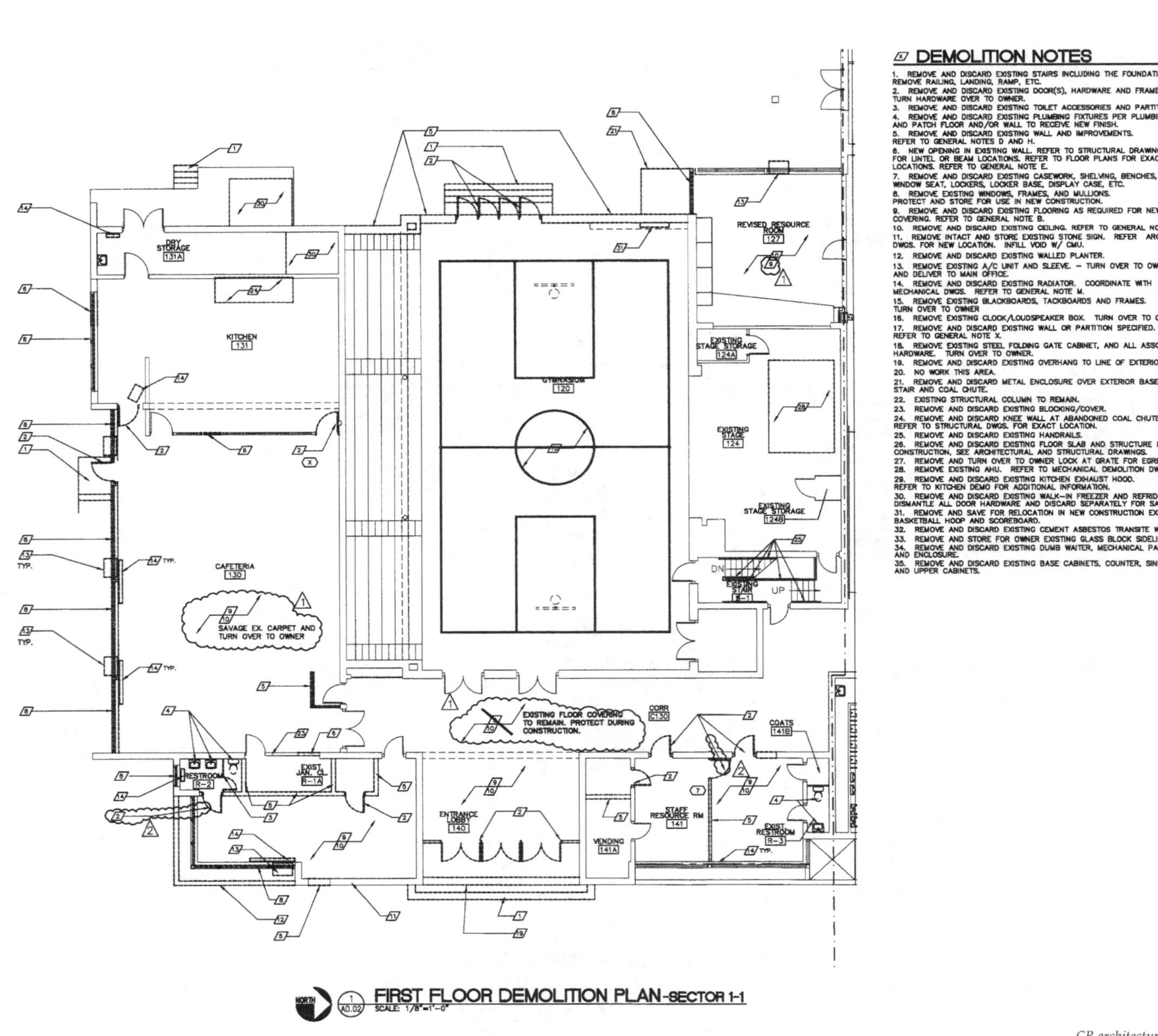

DEMOLITION NOTES
1. REMOVE AND DISCARD EXISTING STAIRS INCLUDING THE FOUNDATION. REMOVE RAILING, LANDING, RAMP, ETC.
2. REMOVE AND DISCARD EXISTING DOOR(S), HARDWARE AND FRAME. TURN HARDWARE OVER TO OWNER.
3. REMOVE AND DISCARD EXISTING TOILET ACCESSORIES AND PARTITIONS.
4. REMOVE AND DISCARD EXISTING PLUMBING FIXTURES PER PLUMBING DRAWINGS AND PATCH FLOOR AND/OR WALL TO RECEIVE NEW FINISH.
5. REMOVE AND DISCARD EXISTING WALL AND IMPROVEMENTS. REFER TO GENERAL NOTES D AND H.
6. NEW OPENING IN EXISTING WALL. REFER TO STRUCTURAL DRAWINGS FOR LINTEL OR BEAM LOCATIONS. REFER TO FLOOR PLANS FOR EXACT LOCATIONS. REFER TO GENERAL NOTE E.
7. REMOVE AND DISCARD EXISTING CASEWORK, SHELVING, BENCHES, WINDOW SEAT, LOCKERS, LOCKER BASE, DISPLAY CASE, ETC.
8. REMOVE EXISTING WINDOWS, FRAMES, AND MULLIONS. PROTECT AND STORE FOR USE IN NEW CONSTRUCTION.
9. REMOVE AND DISCARD EXISTING FLOORING AS REQUIRED FOR NEW FLOOR COVERING. REFER TO GENERAL NOTE B.
10. REMOVE AND DISCARD EXISTING CEILING. REFER TO GENERAL NOTE C.
11. REMOVE INTACT AND STORE EXISTING STONE SIGN. REFER ARCH. ELEV. DWGS. FOR NEW LOCATION. INFILL VOID W/ CMU.
12. REMOVE AND DISCARD EXISTING WALLED PLANTER.
13. REMOVE EXISTING A/C UNIT AND SLEEVE. – TURN OVER TO OWNER AND DELIVER TO MAIN OFFICE.
14. REMOVE AND DISCARD EXISTING RADIATOR. COORDINATE WITH MECHANICAL DWGS. REFER TO GENERAL NOTE M.
15. REMOVE EXISTING BLACKBOARDS, TACKBOARDS AND FRAMES. TURN OVER TO OWNER
16. REMOVE EXISTING CLOCK/LOUDSPEAKER BOX. TURN OVER TO OWNER
17. REMOVE AND DISCARD EXISTING WALL OR PARTITION SPECIFIED. REFER TO GENERAL NOTE X.
18. REMOVE EXISTING STEEL FOLDING GATE CABINET, AND ALL ASSOCIATED HARDWARE. TURN OVER TO OWNER.
19. REMOVE AND DISCARD EXISTING OVERHANG TO LINE OF EXTERIOR WALL.
20. NO WORK THIS AREA.
21. REMOVE AND DISCARD METAL ENCLOSURE OVER EXTERIOR BASEMENT STAIR AND COAL CHUTE.
22. EXISTING STRUCTURAL COLUMN TO REMAIN.
23. REMOVE AND DISCARD EXISTING BLOCKING/COVER.
24. REMOVE AND DISCARD KNEE WALL AT ABANDONED COAL CHUTE. REFER TO STRUCTURAL DWGS. FOR EXACT LOCATION.
25. REMOVE AND DISCARD EXISTING HANDRAILS.
26. REMOVE AND DISCARD EXISTING FLOOR SLAB AND STRUCTURE FOR NEW CONSTRUCTION, SEE ARCHITECTURAL AND STRUCTURAL DRAWINGS.
27. REMOVE AND TURN OVER TO OWNER LOCK AT GRATE FOR EGRESS.
28. REMOVE EXISTING AHU. REFER TO MECHANICAL DEMOLITION DWGS.
29. REMOVE AND DISCARD EXISTING KITCHEN EXHAUST HOOD. REFER TO KITCHEN DEMO FOR ADDITIONAL INFORMATION.
30. REMOVE AND DISCARD EXISTING WALK-IN FREEZER AND REFRIDGERATOR. DISMANTLE ALL DOOR HARDWARE AND DISCARD SEPARATELY FOR SAFETY.
31. REMOVE AND SAVE FOR RELOCATION IN NEW CONSTRUCTION EXISTING BASKETBALL HOOP AND SCOREBOARD.
32. REMOVE AND DISCARD EXISTING CEMENT ASBESTOS TRANSITE WALL PANELS.
33. REMOVE AND STORE FOR OWNER EXISTING GLASS BLOCK SIDELIGHTS.
34. REMOVE AND DISCARD EXISTING DUMB WAITER, MECHANICAL PARTS, AND ENCLOSURE.
35. REMOVE AND DISCARD EXISTING BASE CABINETS, COUNTER, SINK, AND UPPER CABINETS.
DRY STORAGE 131A
KITCHEN 131
CAFETERIA 130
SAVAGE EX. CARPET AND TURN OVER TO OWNER
GYMNASIUM 120
REVISED RESOURCE ROOM 127
EXISTING STAGE STORAGE 124A
EXISTING STAGE 124
EXISTING STAGE STORAGE 124B
EXISTING STAIR
DN
UP
EXISTING FLOOR COVERING TO REMAIN. PROTECT DURING CONSTRUCTION.
CORR C130
COATS 141B
RESTROOM R-2
EXIST. JAN. CL. R-1A
ENTRANCE LOBBY 140
VENDING 141A
STAFF RESOURCE RM 141
EXIST. RESTROOM R-3
TYP.
FIRST FLOOR DEMOLITION PLAN-SECTOR 1-1
SCALE: 1/8"=1'-0"
NORTH
1
A0.02
CR architecture + design

Figure 9-3. A demolition plan for a commercial building project. The numbered notes accompanying the drawing identify items to be removed or changed and are referenced on the plan. The notes provide a way to indicate the same requirements in multiple locations without repeating the entire information.

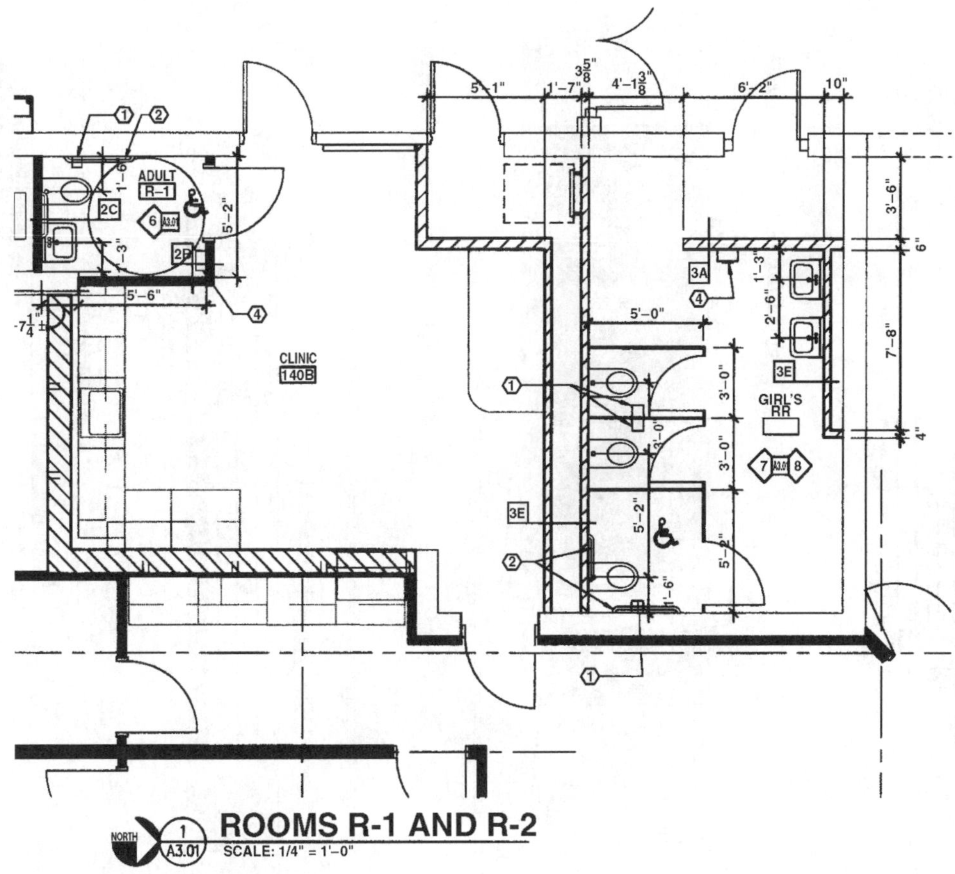

CR architecture + design

Figure 9-4. An enlarged floor plan showing details for two rooms in the structure.

Interior elevations are used to show the vertical surfaces of an interior feature when the feature has complex details or special construction requirements. Interior elevations are typically used for drawing cabinet layouts, bathroom layouts, architectural window walls, and any other special or unique vertical surfaces. See **Figure 9-6**.

Architectural Sections

A ***section*** is a view representing a "cut" through the building to clarify construction of a particular item, such as a wall or roof. Sections are typically drawn at a larger scale to show complete information related to building materials and construction methods. Sections are taken from a plan or elevation drawing and identified by locating a section cutting plane on the drawing being "sectioned." Sections are provided for walls, cabinets, chimneys, stairs, and other features whose construction is not shown clearly on the plan or elevation. Sections show how the various components are fastened and assembled. In general, a different section is provided for each location where the typical construction requirements change from requirements at other locations.

Common types of section drawings are full building sections, wall sections, and detail sections. A *full building section* shows a cut through the entire structure, **Figure 9-7**. A *wall section* shows the entire wall of a building from floor to floor from the lowest level to the roof. See **Figure 9-8**. A wall section defines all materials and dimensions pertaining to the particular wall. A *detail section* is simply a portion of a building drawn at a larger scale to clarify the construction assembly requirements. See **Figure 9-9**.

Sometimes, building materials indicated on a section drawing are identified in a legend located elsewhere on the plan or on a different sheet, such as in the General Notes. This helps organize the information and reduces crowding. See **Figure 9-10**.

Architectural Details

A ***detail*** shows a small portion of the building to clarify the materials and type of construction used. See **Figure 9-11**. Details are made at a larger scale than plan drawings, elevations, and sections. A detail drawing may be placed on the same sheet as the corresponding plan or elevation view, but most often the detail is found on a separate sheet and referenced by detail numbers or letters and sheet location numbers.

Schedules

Schedules are one of the most useful tools in a set of construction documents. A ***schedule*** is a listing of materials or products required in the structure. See

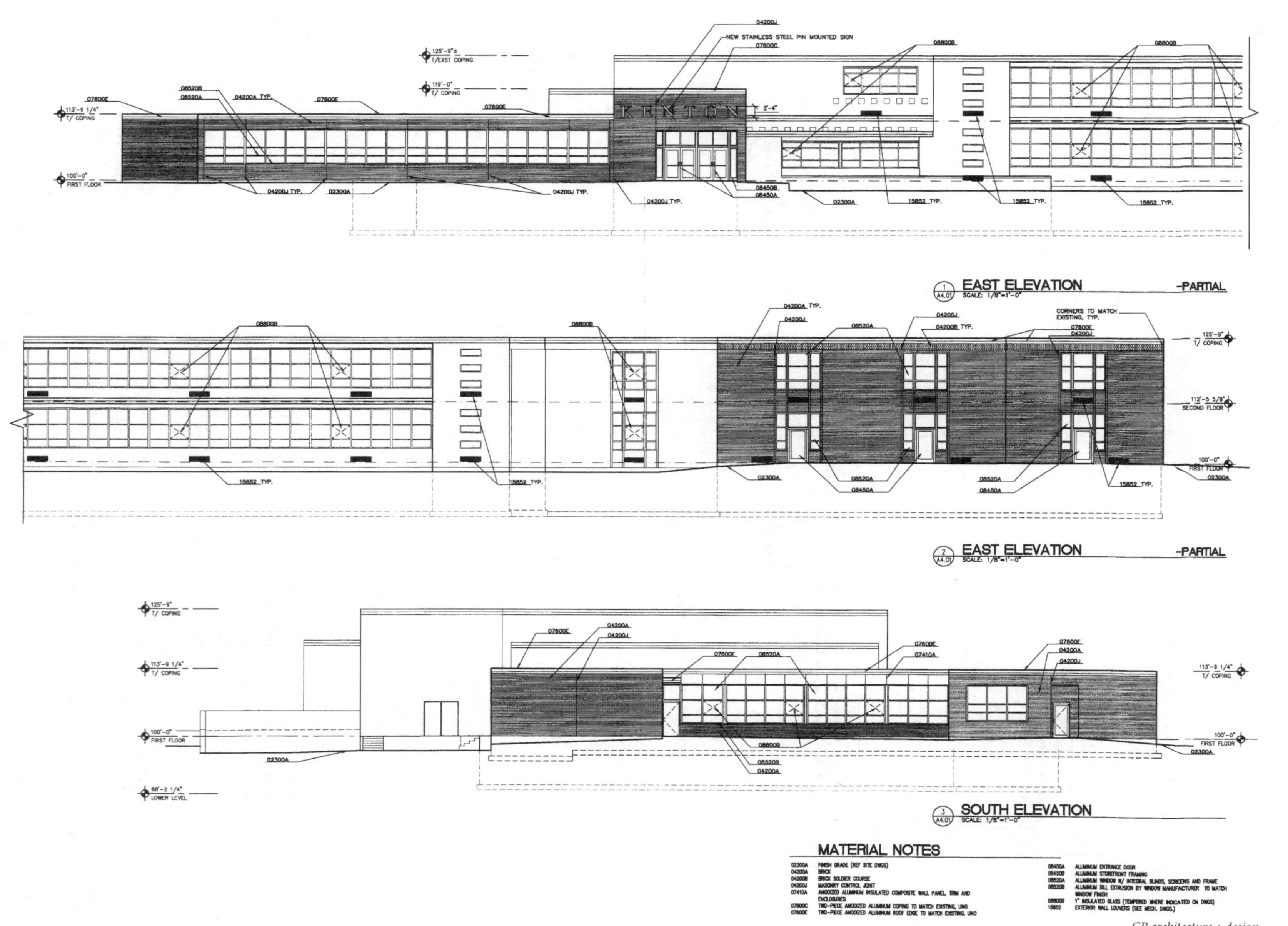

Figure 9-5. Shown are the south and east exterior elevations of a building. A break line is used in the east elevation to show the entire view of the elevation.

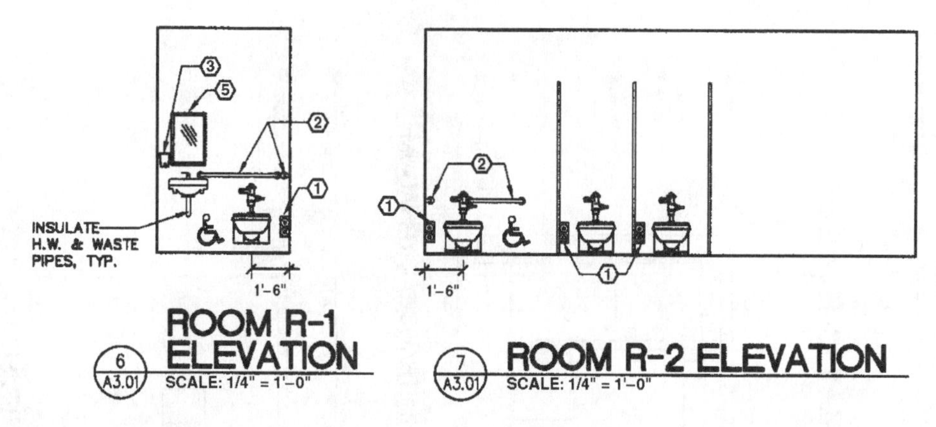

CR architecture + design

Figure 9-6. Interior elevations showing layouts for bathrooms. Compare these views to Figure 9-4.

Figure 9-12. Typical schedules used on architectural drawings include door schedules, window schedules, and room finish schedules. Schedules make it convenient for the designer to provide a large amount of information in one location. Typically, schedules specify product types and sizes, number of items required, and manufacturer information. A symbol or tag identifies each item in the schedule. The same identification is used on the drawing where the item is referenced.

A *door schedule* specifies information for all doors shown on the floor plan and elevation drawings. Each door is identified with a numbered symbol or tag. The number designation typically includes the room number where the door is to be installed. Additional data includes the door type, size, finish, fire rating, and required hardware, such as hinges, knobs, and locks. If any special door treatments are required, they are also listed.

CAREERS IN CONSTRUCTION

Architect

Architects plan and design homes, office buildings, factories, and many other types of structures. An architect is a licensed professional who works with clients, engineers, and contractors throughout a construction project.

An architect designs a building for a client by determining the client's objectives and requirements for the structure whether it is a residential home or a major commercial building. The architectural team develops preliminary design drawings and budgets for the owner's approval. As the design firms up, the construction documents are developed. These include floor plans, elevations, sections, and detail drawings, along with specifications. Today, most construction drawings are created using computer-aided design and drafting (CADD) software.

The architect also plays a vital role during the construction phase by visiting the job site, approving submittals, answering document questions to make sure quality standards are met, and assisting the owner. The architect works with many other professionals, such as civil, mechanical, and electrical engineers, landscape architects, urban planners, and interior designers.

Becoming an architect requires earning a bachelor's or master's degree, gaining experience through paid internships at an architectural firm, and passing a state licensing examination. A bachelor's degree normally requires completion of a five-year college or university program.

A career in architecture offers many positive rewards. One of the most valuable benefits is the satisfaction of seeing the development of a design that becomes a reality when built.

Syda Productions/Shutterstock.com

Architects plan and design buildings based on the needs and objectives of a client.

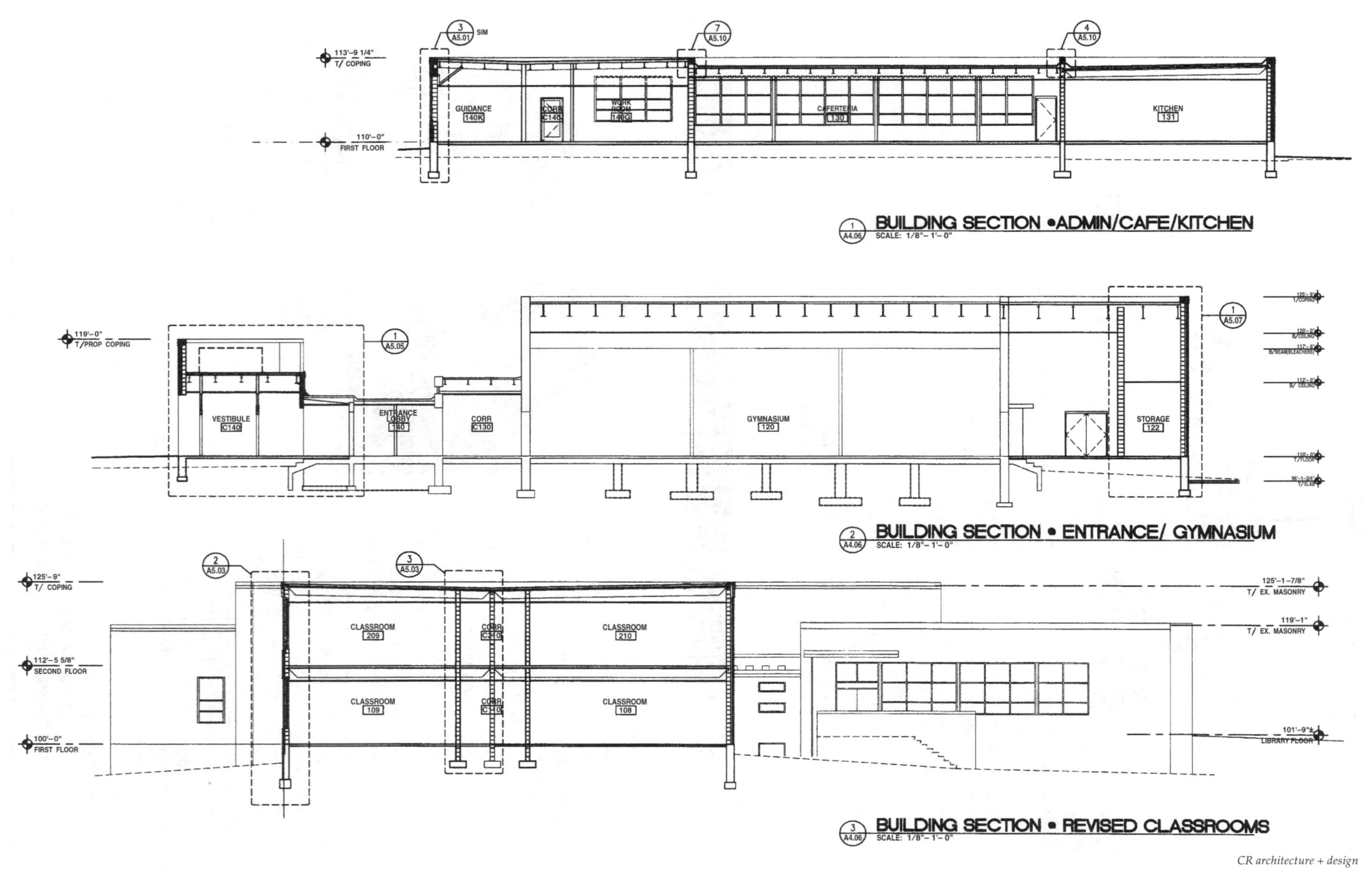

Figure 9-7. Full building sections represent a "cut" through the entire structure.

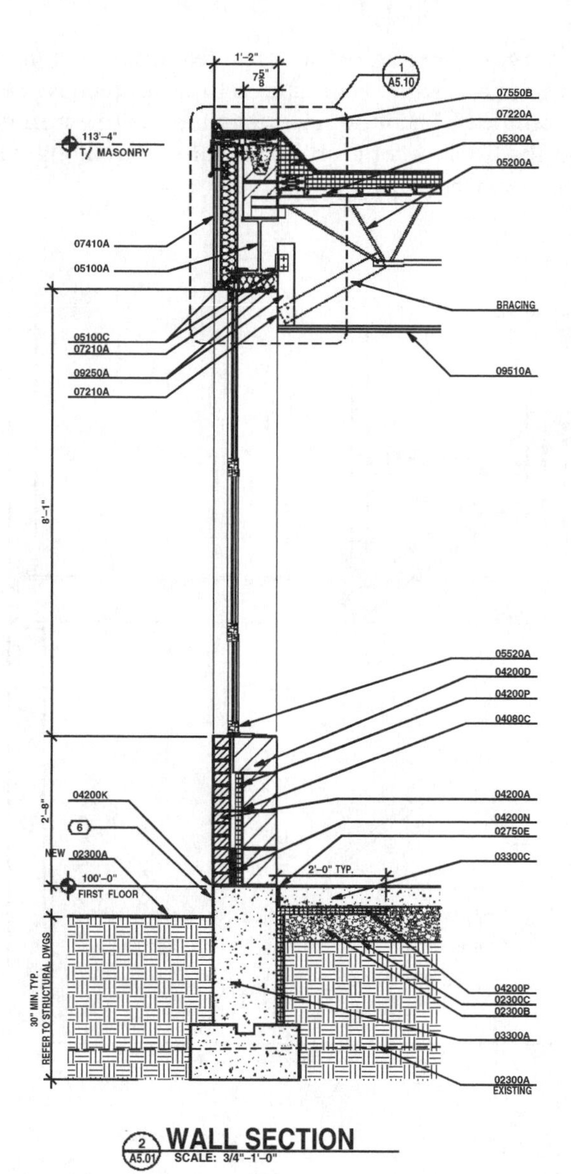

CR architecture + design

Figure 9-8. A wall section shows all dimensions and construction requirements for a particular wall. This section uses reference numbers to identify building materials and construction requirements. The corresponding information is noted in a legend. This helps reduce crowding on the drawing, but requires the print reader to become familiar with the information in the legend.

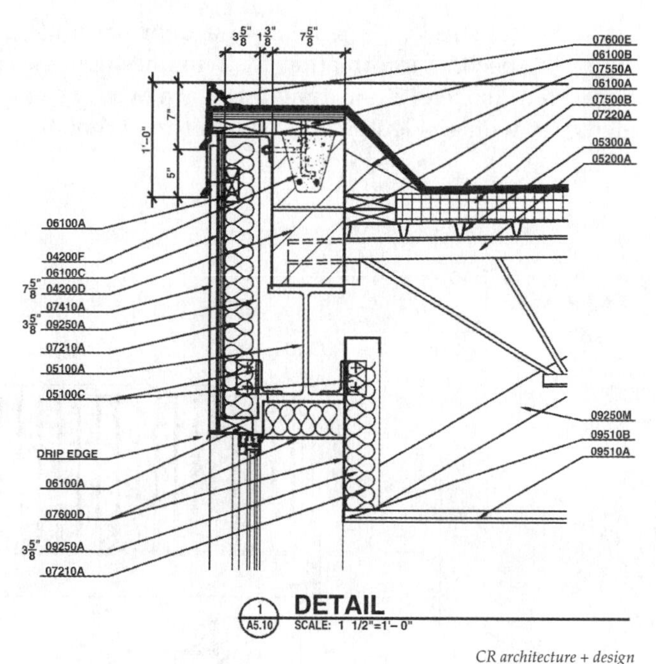

CR architecture + design

Figure 9-9. A detail section referenced from the wall section shown in Figure 9-8.

MATERIAL NOTES

02300A	FINISH GRADE (REF SITE DWGS)
02300B	COMPACT GRANULAR DRAINAGE FILL
02300C	VAPOR BARRIER
02750E	EXPANSION-JOINT MATERIAL
03300A	CONCRETE FOOTING/FOUNDATION (REF 'S' DWGS)
03300C	ELEVATED CONCRETE FLOOR SLAB (REF 'S' DWGS)
03300G	PERIMETER INSULATION
04070A	GROUT SOLID (REF 'S' DWGS)
04080B	HORIZONTAL JOINT REINFORCING
04080C	MASONRY ANCHOR
04200A	BRICK
04200D	CONCRETE MASONRY UNITS (CMU)
04200E	SOLID CMU
04200F	CMU BOND BEAM
04200G	CMU BULLNOSE UNIT
04200K	THRU-WALL MASONRY FLASHING W/ WEEPS @ 16" O.C.
04200N	CAVITY MORTAR PROTECTION
04200P	CAVITY WALL INSULATION
05100A	STRUCTURAL STEEL FRAMING (REF 'S' DWGS)
05100C	STEEL ANGLE OR PLATE (REF 'S' DWGS)
05100E	STEEL LINTEL (REF 'S' DWGS)
05200A	STEEL JOIST (REF 'S' DWGS)
05300A	STEEL DECK (REF 'S' DWGS)
07210A	GLASS FIBER BATT INSULATION - FOIL FACED
07220A	ROOF INSULATION BOARD
07410A	ANODIZED ALUMINUM INSULATED COMPOSITE WALL PANEL, TRIM AND ENCLOSURES
07550B	MODIFIED BITUMINOUS ROOFING (FELTS + CAP SHEET)
08520A	ALUMINUM WINDOW W/ INTEGRAL BLINDS, SCREENS AND FRAME
09250A	STEEL STUD FRAMING, 16" OC UNO
09510A	SUSPENDED ACOUSTICAL CEILING (REF RM FIN SCHED)
09910A	PAINT (REF RM FIN SCHED)

DRAWING NOTES

1. THE PROJECT MANUAL IS AN INTEGRAL PART OF THESE DOCUMENTS
2. ALL DIMENSIONING IS TO FACE OF MASONRY/FACE OF FRAMING UNLESS NOTED OTHERWISE.
3. DOOR, HARDWARE, AND FINISH SCHEDULES ARE LOCATED IN THE SPECIFICATIONS.
4. TOP OF FIRST FLOOR ELEVATION IS 100'-0"; MEDIA CENTER AT 101'-9"
5. UNLESS NOTED OTHERWISE ALL MASONRY OPENINGS BEGIN 4" FROM FACE OF ADJACENT WALL SURFACE.
6. REFER TO STRUCTURAL DRAWINGS FOR VERTICAL MASONRY REINFORCING.
7. REFER TO ENLARGED PLANS ON A3.02 FOR TOILET ACCESSORIES AND LOCATIONS.
8. ALL WALLS TO BE PARTITION TYPE 3A U.N.O.

CR architecture + design

Figure 9-10. A legend identifying building materials and general drawing notes for the wall section shown in Figure 9-8.

A *window schedule* is similar to a door schedule, except it specifies information for windows. On the floor plan and elevation drawings, each window is identified with an identification symbol and number.

A *room finish schedule* specifies final treatments required for floors, walls, and ceilings in each room of the building. Each room is designated with a number, and the finishes for each surface are listed along with any special treatments required.

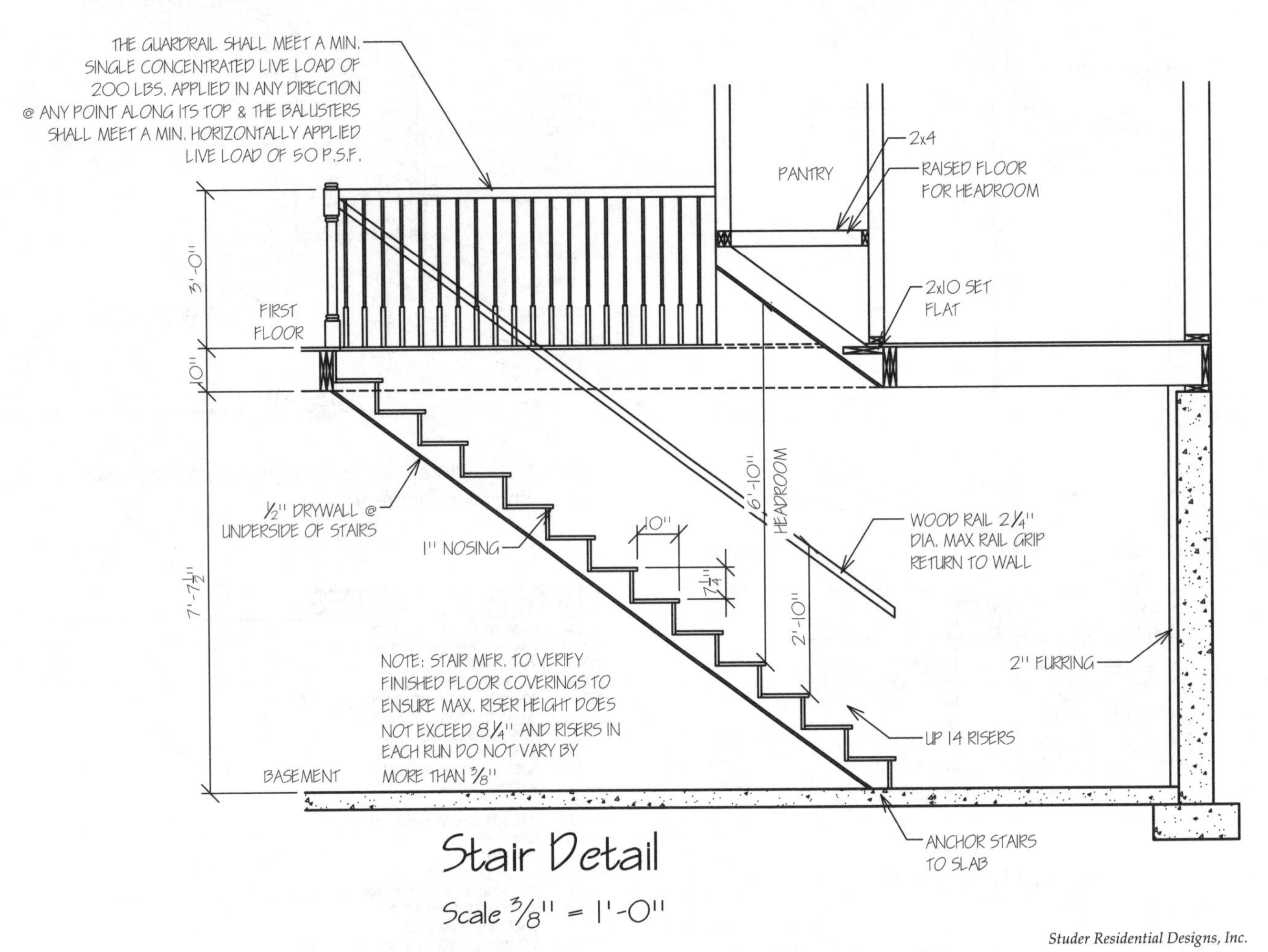

Figure 9-11. A stair detail showing dimensions and construction requirements for a set of stairs.

DOOR AND FRAME SCHEDULE						
No.	DESCRIPTION		DOOR		FRAME	
	LOCATION	SIZE	TYPE	MAT'L	TYPE	MAT'L
101	MECHANICAL ROOM	3'-0" x 7'-0" x 1 3/4"	A	HM	1	HM
102	BREAK ROOM	3'-0" x 7'-0" x 1 3/4"	A	HM	1	HM
103	LOCKED STORAGE	3'-0" x 7'-0" x 1 3/4"	A	HM	1	HM
104	LOCKER ROOM	3'-0" x 7'-0" x 1 3/4"	A	HM	1	HM
105A	EXTERIOR DOOR./STOR. ROOM	4'-0" x 7'-0" x 1 3/4"	A	HM	2	HM
105B	STOR. ROOM INTO HARDLINES	6'-0" x 10'-0" (IMPACT DOORS)	B			
105C	NOT USED					
106A	LOADING DOCK	12'-0" x 14'-0" (SECT OVERHEAD)	C			
106B	EXTERIOR DOOR/SHIPPING	3'-0" x 7'-0" x 1 3/4"	A	HM	2	HM
106C	RECESSED LOADING DOCK	7'-0" x 9'-0" (SECT. OVERHEAD)	C			
106D	SHIPPING INTO GREENHOUSE	10'-0" x 12'-0" (SLIDER)				
107A	SHIPPING INTO OFFICE	3'-0" x 7'-0" x 1 3/4"	A	HM	1	HM
107B	HALLWAY INTO OFFICE	3'-0" x 7'-0" x 1 3/4"	A	HM	1	HM
108	OFFICE	3'-0" x 7'-0" x 1 3/4"	A	HM	1	HM
109	SHIPPING INTO HALLWAY	3'-0" x 7'-0" x 1 3/4"	A	HM	1	HM
110	MEN'S RESTROOM	3'-0" x 7'-0" x 1 3/4"	A	HM	1	HM
111	WOMEN'S RESTROOM	3'-0" x 7'-0" x 1 3/4"	A	HM	1	HM
112A	EXTERIOR DOOR/HARDLINES	3'-0" x 7'-0" x 1 3/4"	A	HM	2	HM
112B	HARDLINES TO ATRIUM	11'-2" x 10'-0" (AUTO-SLIDER)	D			
112C	HARDLINES TO ATRIUM	11'-2" x 10'-0" (AUTO-SLIDER)	D			
112D	HARDLINES TO COV. SALES	11'-2" x 10'-0" (AUTO-SLIDER)	D			
113	HARDLINES TO MEZZ. STAIRS	3'-0" x 7'-0" x 1 3/4"	A	HM	1	HM
114A	ATRIUM TO GREENHOUSE	11'-8" x 10'-0" (AUTO-SLIDER)	D			
114B	ATRIUM TO GREENHOUSE	11'-8" x 10'-0" (AUTO-SLIDER)	D			
114C	ATRIUM TO OUTSIDE SALES	11'-8" x 10'-0" (AUTO-SLIDER)	D			
114D	ATRIUM MAIN ENTRANCE	11'-8" x 10'-0" (AUTO-SLIDER)	D			
115A	GREENHOUSE TO OUT. SALES	12'-0" x 12'-0" (SECT. OVERHEAD)	C			
115B	GREENHOUSE TO OUT. SALES	11'-8" x 10'-0" (AUTO-SLIDER)	D			
115C	GREENHOUSE TO OUT. SALES	12'-0" x 12'-0" (SECT. OVERHEAD)	C			
115D	GREENHOUSE TO OUTSIDE	3'-6" x 7'-0" x 1 3/4"	E			
116	COUNTING ROOM	3'-0" x 7'-0" x 1 3/4"	A	HM	1	HM
117	COVERED SALES	10'-0" x 10'-0" (SLIDER)				
118A	CHECK-OUT MAIN ENTRANCE	11'-8" x 10'-0" (AUTO-SLIDER)	D			
118B	CHECK-OUT/ COVERED SALES	10'-0" x 10'-0" (SECT. OVERHEAD)	C			
118C	CHECK-OUT/ COVERED SALES	11'-8" x 10'-0" (AUTO-SLIDER)	D			

A

Delhi Flower and Garden Center
Arch/Image 2 Architects

ACCESSORY/FIXTURE SCHEDULE				
TAG NO.	MODEL NO.	DESCRIPTION	MANUFACTURER/MODEL	REMARKS
1	B-2730	TOILET TISSUE HOLDER	BOBRICK	SURFACE MOUNTED
2	B-5837	TOILET GRAB BAR	BOBRICK	LENGTH AS INDICATED
3	B-147	SOAP DISPENSER	BOBRICK	SURFACE MOUNTED
4	-	HAND DRYER	SEE M.E.P. DWGS.	
5	B-292-2436	MIRROR	BOBRICK	
6		SHOWER GRAB BAR		
7		FOLDING SHOWER SEAT		

B

CR architecture + design

Figure 9-12. A schedule identifies building items and additional information related to each item, such as size, type, model number, and manufacturer name. A—A door and frame schedule. Each door is identified with a number. B—An accessory/fixture schedule. A symbol (or "tag") is used to identify each item.

Henry Bros. Co. Photography by Larry Morris

Large projects, such as the one being built in the background, require complex sets of prints with many sheets. The prints are often fastened together at one edge to keep sheets together and in proper order.

Test Your Knowledge

Name ______________________________

Write your answers in the spaces provided.

____ 1. Where would the size and type of a window be found?

A. Room finish schedule

B. Floor plan

C. Reflected ceiling plan

D. Window schedule

E. Both C and D.

____ 2. *True or False?* Floor plans show all of the information you need to build the building.

____ 3. Which of the following items would *not* be found in the architectural drawings?

A. Types of paint

B. Types of insulation

C. Property line lengths

D. Layout of light switches

E. Toilet partitions

____ 4. *True or False?* Details are typically made at a smaller scale than floor plans.

____ 5. Where would batt insulation typically be detailed?

A. Floor plan

B. Structural drawing

C. Grading plan

D. Room finish schedule

E. Wall section

____ 6. What is a typical scale used to make a floor plan?

A. 1/4″ = 1′-0″

B. 3/4″ = 1′-0″

C. 1″ = 1′-0″

D. 1″ = 10′-0″

E. 1″ = 30′-0″

____ 7. On what type of drawing would you find a ceiling layout?

A. Roof section

B. Wall section

C. Reflected ceiling plan

D. Interior elevation

E. Window schedule

____ 8. Where would ceiling heights be indicated?

A. Full building section

B. Floor plan

C. Door schedule

D. Interior elevation

E. Both A and D.

____ 9. *True or False?* Elevations show the materials used on the exterior of the building.

____ 10. *True or False?* On an exterior elevation, underground footings are shown with solid, continuous lines.

Activity 9-1

Plan Drawings for a Residential Building Project

Name ______________________

Refer to Sheets 1, 3, and 5 from the Sullivan residential building plans in the Large Prints supplement to answer the following questions.

1. On which sheet is the first floor plan?

   ____

2. What is the scale of the first floor plan?

   ____

3. On which sheets are the elevations found?

   ____

4. What is the overall size of the following rooms?

   A. Garage ______________________

   B. Kitchen ______________________

   C. Master Bedroom ______________________

5. How many bedrooms are in this house?

   ____

6. Which two bathrooms back up to each other?

   ____

7. What kind of ceiling is in the great room?

   ____

8. What are the materials specified for the exterior covering?

   ____

   ____

9. What is the insulation required in the roof?

   ____

10. What material will be used to wrap the house?

    ____

11. From the left-hand side of the house, what is the distance to the center of the second window on the front of the house?

____

12. What is the distance from the first floor to the truss bearing?

____

13. What are the overall dimensions of the house?

____

14. How thick is the typical exterior wall without the siding?

____

15. What is the dimension from the top of the foundation to the first floor?

____

16. What is the cabinet size at the kitchen sink?

____

17. What is the extension of the countertop overhang beyond the rear of the kitchen sink?

____

18. What is noted differently about the kitchen pantry floor?

____

19. What is the height of the stair guardrail and what is the load it must resist?

____

20. What is the specification for the fireplace?

____

____

Plan Drawings for a Commercial Building Project

Name ______________________________

Refer to Sheets A0.1, A1.1, A1.2, A1.3, A2.1, and A3.1 from the Delhi Flower and Garden Centers commercial building plans in the Large Prints supplement to answer the following questions.

1. What sheet are the supplemental notes on?

____

2. What typical size are the metal studs to be for interior walls unless noted differently on the drawings? What spacing is to be used?

____

____

3. Wood that is in contact with concrete should be what type?

____

4. What is the cabinet contractor to do prior to construction?

____

____

5. What scale is the first floor plan?

____

6. What is the overall size of the loading/shipping area?

____

7. What do the diagonal lines in the walls indicate?

____

8. Identify the flooring material in the following locations:

A. Check-out area ______________________________

B. Greenhouse sales ______________________________

C. Covered sales ______________________________

D. Men's and women's restrooms ______________________________

9. What kind of glass is used in the count room?

____

10. Where is the roof access hatch located? What detail provides additional information?

11. What details provide additional information for the paver floors?

12. Referring to Question 11, what is the difference between the greenhouse sales floor and the outside sales floor?

13. Answer the following questions about the planters at the covered entry.

A. What detail is referenced?

B. How thick is the concrete wall?

C. What is to be added to the concrete wall inside of the planter?

D. What is to be installed over the concrete wall outside of the planter?

E. How high is the wall above the paver floor?

14. What type of structure is above the checkout counters? What detail is referenced?

15. Which way is Plan North facing, to the right or left side of the drawings?

16. Where are the stairs to the mezzanine located?

17. What is the riser and tread information for the stairs to the mezzanine?

18. What sheet is the roof plan on?

19. What is the slope of the rafter framing above the gate into the outside sales area?

20. What type of roof is used above the covered entry?

21. What is to be done to the bar joists in the hardlines room?

____

22. What is the symbol for ceiling fans? Draw the symbol freehand.

23. Where are the compact fluorescent recessed can fixtures located?

____

____

Activity 9-3

Elevations, Details, and Sections for a Commercial Building Project

Name ______________________________

Refer to Sheets A4.1, A4.2, A5.1, A6.1, A9.1, and A9.2 from the Delhi Flower and Garden Centers commercial building plans in the Large Prints supplement to answer the following questions.

1. What is to be done to the concrete block on the side elevation?

____

2. What is the width of the gate opening into the outside sales area?

____

3. How high are the block columns on either side of the gate?

____

4. How high are the ceilings in the men's and women's restrooms? What is the ceiling material?

____

____

5. Referring to the trellis structure above the checkout counters, how big are the columns? What is the overall height of the structure?

____

____

6. Answer the following questions about the overhead door for the recessed loading dock.

   A. What is the size of the door? ______________________________

   B. What is the door reference number? ______________________________

   C. Where is the door located? ______________________________

   D. What type of door is it? ______________________________

7. Refer to the wall section for the north wall through the hardlines room.

   A. What is the number of the section referenced? ______________________________

   B. How high is the wall to the top of the bond beam? ______________________________

   C. What does the dashed vertical line indicate? ______________________________

   D. What are the roof materials? ______________________________

8. Refer to the wall section for the wall between the office and the atrium.

   A. What is the number of the section referenced? ____________________

   B. How far is the brick veneer offset from the masonry wall? ____________________

   C. What is used to tie the masonry and the brick veneer together? ____________________

9. What is used to insulate the foundation wall?

____

UNIT 10
Foundation Prints

TECHNICAL TERMS

auger cast pile
basement plan
caissons
concrete masonry units
elevation
expansion joints
floating slab construction
footings
foundation plan
friction pile
frost line
grade beam
keyway
monolithic slab
slab-on-grade
steel pile
waterproofing
waterstop

LEARNING OBJECTIVES

After completing this unit, you will be able to:

- Identify footings on a foundation plan.
- Describe different types of foundation support systems.
- Identify various components of a foundation system.

Once the building has been located on the plot and the necessary site clearing and excavation is complete, the work begins on constructing the building structure. The building structure consists of two major components: the foundation and the above-grade framing system. The foundation consists of the concrete footings, grade beams, and foundation walls. On residential prints, the footings, beams, and other structural components for the foundation system are shown on a ***foundation plan*** or ***basement plan***. Sections and details providing additional information are referenced on the foundation or basement plan and shown on additional structural prints.

The footings and foundation walls must be carefully laid out. The entire structure is built upon the footings and walls and alignment of the above structure depends on the accuracy of their construction. To make print reading easier, this unit explains footings, foundations, and slab-on-grade construction.

Footings

Footings are the "feet" upon which the entire building rests, **Figure 10-1**. The sizes of the footings are shown on the foundation plan or on a detail of the foundation wall. The footing size is determined by engineers based on the type of soil to resist loads (determined by soil tests) and the weight of the building. The load of the building is transferred through the upper portion of the structure down to the foundation walls onto the footings and into the ground. There are many types of foundation systems; this unit will cover those listed below.

- Footings and walls
- Grade beams
- Auger cast piles
- Caissons
- Steel piles

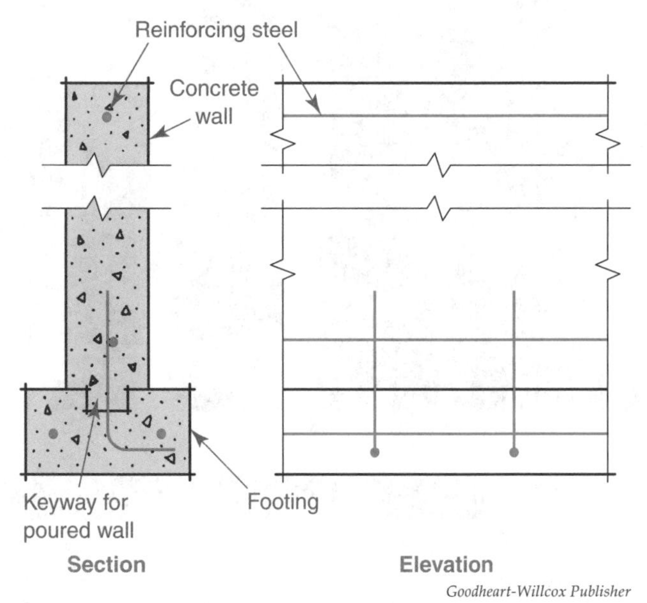

Goodheart-Willcox Publisher

Figure 10-1. Footings are designed to carry the weight of the entire building and to transfer that weight to the earth below.

Footings are also required under columns. These footings frequently are wider and thicker than footings for foundation walls, because the column loads are concentrated in one spot. See **Figure 10-2**. Fireplace chimneys and similar concentrations of weight also require larger footings.

Footings must rest on undisturbed earth below the region's ***frost line***, the deepest point to which the ground will freeze in a given area. The local building code will give the depth of the frost line, and how far below it the bottoms of the footings must be placed.

Steel reinforcing rods (commonly called "rebar") are placed in the footings. This is especially important when footings must pass over earth previously disturbed due to an earlier excavation. When a poured concrete foundation wall is to be erected on the footing, the drawing may call for a ***keyway*** to be cast in the footing to help anchor the wall. Refer to **Figure 10-1**.

On a residential foundation plan, footings are shown as hidden lines, **Figure 10-3**. The width of the footing under the foundation walls and columns is shown. On drawings of commercial buildings, it is common to

CAREERS IN CONSTRUCTION

Concrete Construction

Concrete construction offers many employment opportunities. Concrete is a widely used material in building construction. In residential construction, concrete is typically used to construct foundations and basements. In commercial construction, concrete is often used for both structural and architectural purposes. The number of individuals involved in concrete work on a construction project depends on the size and complexity of the project.

On larger projects, workers involved in concrete construction include laborers, carpenters, cement masons, and reinforcing steel installers. A laborer is a worker who enters the trade without experience and learns and advances with training. The laborer assists the other workers on the job by helping the carpenter move materials, the cement mason place concrete, and the reinforcing steel installer move rebar. Carpenters build formwork for footings and foundation walls. Cement masons work with fresh concrete or lay concrete block. Cement masons are also responsible for concrete finishing. A concrete finisher screeds and trowels the placed concrete and must have a working knowledge of the material properties of concrete. Reinforcing steel installers, commonly called rebar installers, build the reinforcing inside the formwork before the concrete is placed. On large projects, reinforcing steel assemblies and rebar cages can be very substantial.

serato/Shutterstock.com

Concrete construction is performed by skilled workers. These workers are using a metal screed to level the concrete as it is placed.

Workers entering concrete construction typically learn skills on the job by working with experienced tradeworkers. They may also receive training from a trade school or through an apprenticeship program. Work in the concrete industry is fast paced and requires physical strength. Concrete construction workers must be skilled in knowing how to work with concrete and must be able to work in different weather conditions because most work occurs outdoors.

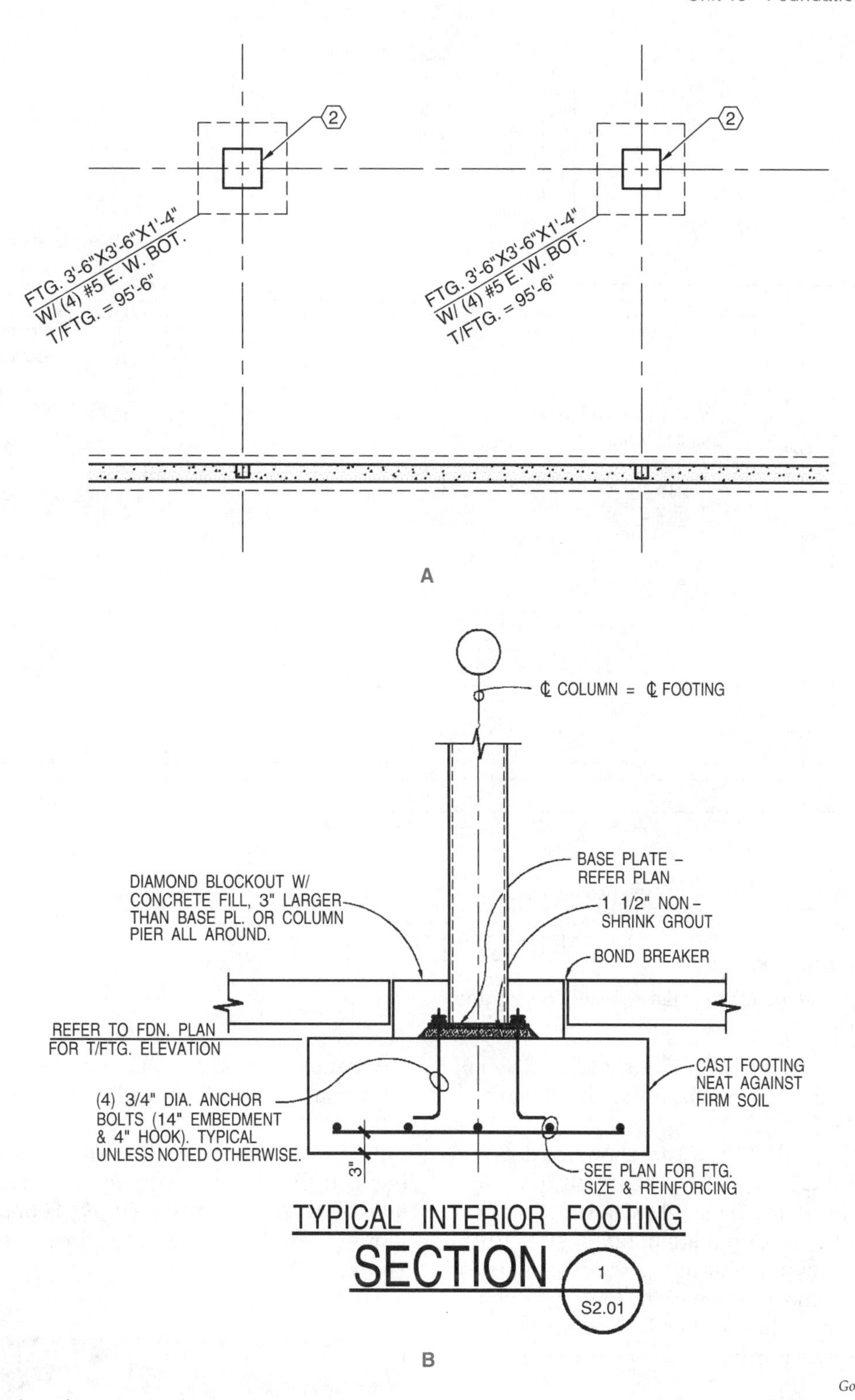

Goodheart-Willcox Publisher

Figure 10-2. Footings for columns have a larger cross section than wall footings. A square shape is normally used. A—Portion of a plan view drawing showing the layout for columns and footings. Note the dimensions given for the footings. B—Section drawing showing column and footing details.

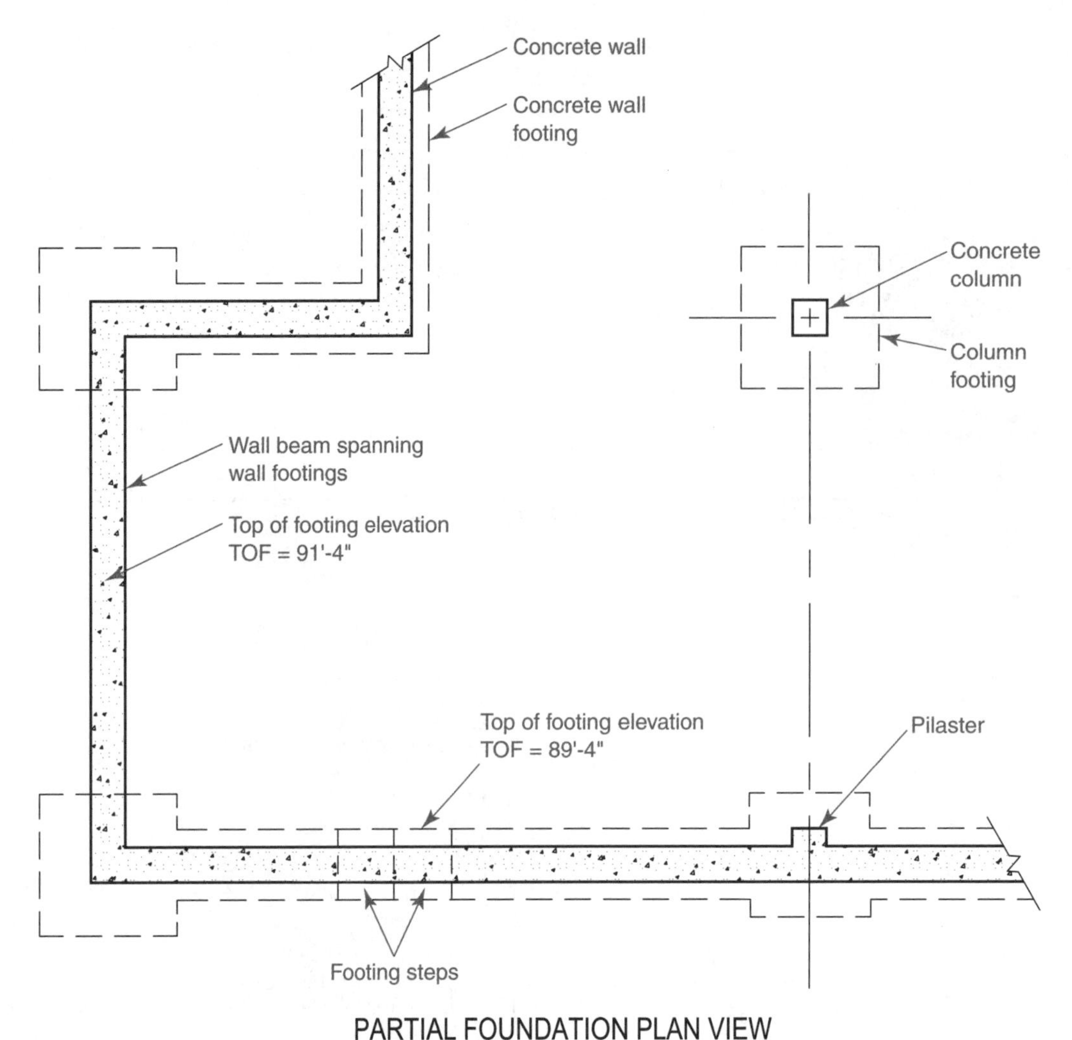

Figure 10-3. A partial foundation plan showing footing and wall locations.

show footings as solid lines. See **Figure 10-4**. On some commercial drawings, the engineer may choose to show footings as dashed lines.

Steel reinforcing rods ("rebar") are shown as dots and solid lines in sectional views. On elevation drawings, these rods are indicated by solid lines showing how the rods are laid out. Special attention must be given to notes on the drawing and in the specifications for details relating to construction of the footings, such as the strength of the concrete specified in psi (pounds per square inch) and minimum coverage of steel rebar.

Foundation Walls

Foundation walls are the base of the building. They transfer the weight of the building to the footings and then to the ground below. Foundation walls can be poured-in-place concrete or ***concrete masonry units*** (concrete block). Poured-in-place concrete is used where soil and weather conditions exert considerable side pressure on the walls. Where feasible, the use of concrete masonry units is an efficient way of constructing a foundation wall, because no forms are required. Usually, both poured and block walls are reinforced with steel rebar.

Foundation walls and columns are shown as solid lines on the foundation plan. The space between the lines represents the material used. Foundation walls and footings are shown as hidden lines in elevation views, **Figure 10-5**.

GREEN BUILDING

Material Storage

Damaged and lost materials create unnecessary waste on a project. Store materials in a secure location to minimize theft. Protect stored materials from damaging weather. Locate materials away from worker pathways to reduce the chance of accidental damage. Keep materials neatly organized to ensure that the materials can be located when they are needed.

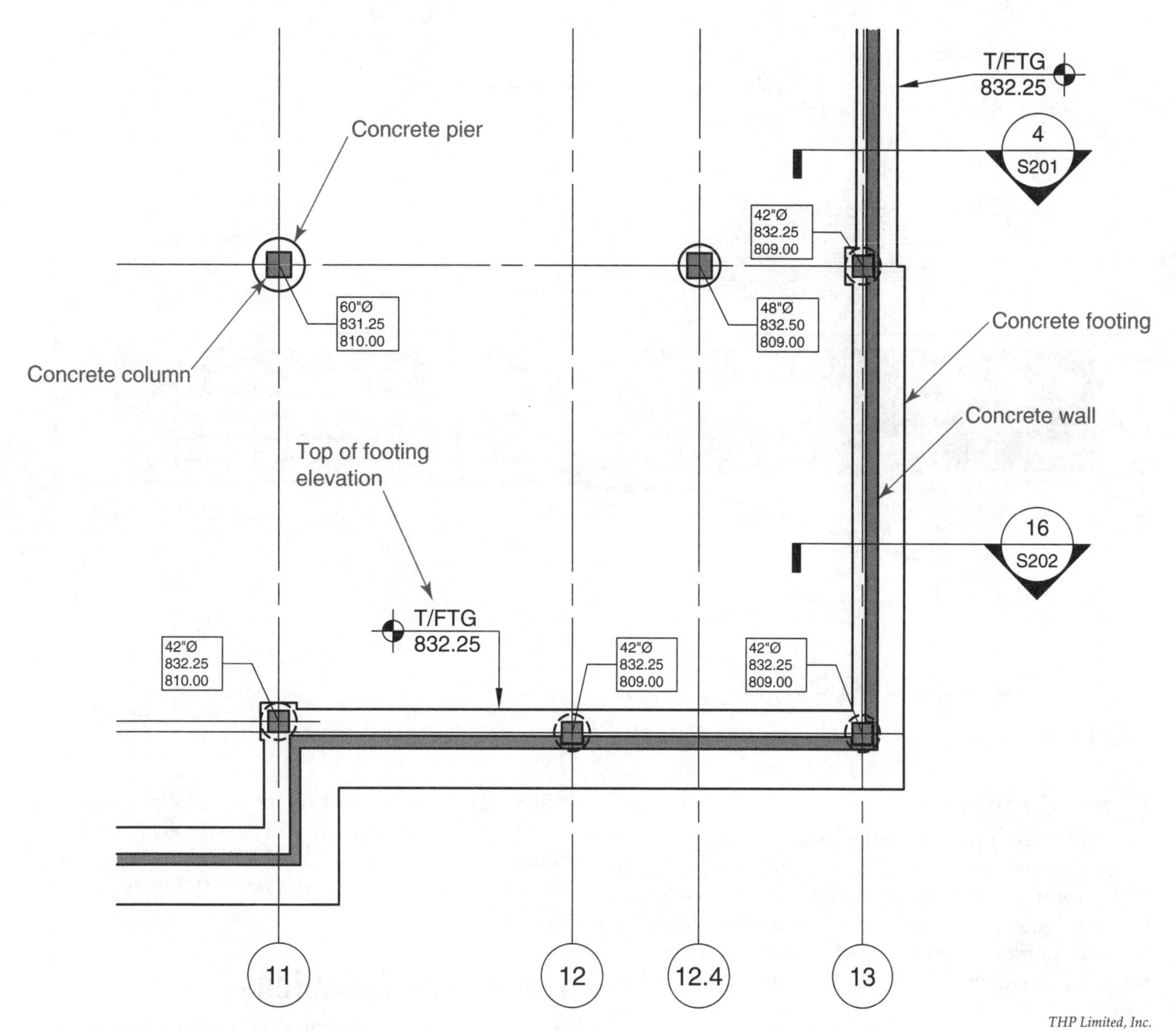

THP Limited, Inc.

Figure 10-4. A portion of a foundation plan for a commercial building. Note the solid lines used to represent footings.

A foundation wall section is shown in **Figure 10-6**. The symbols used in the section indicate the general types of materials used. However, these materials will be further detailed in notes on the drawings or in the specifications.

Residential drawings are usually fairly simple and very few detail drawings are required. Details are used to show more complex features, such as fireplaces and chimneys, exterior stairs, and (in certain cases) a retaining wall. When details of a fireplace are required, additional views of the fireplace support are drawn to provide a better understanding of the construction, **Figure 10-7**.

Commercial foundation drawings are part of the (S) Structural prints. A commercial foundation plan shows the foundation system of the building. It shows the basic layout of the column centerlines, footing support system, and columns and walls. The foundation plan is where the section cuts and details are indicated. Refer to **Figure 10-4**. The section drawings and details are shown on other referenced sheets in the set of plans.

Grade Beams

A reinforced concrete beam that spans from footing to footing is called a ***grade beam***. The beam is formed on the ground, or formed by a trench in the earth. It is typically used to span over weak disturbed spots in the soil support system. See **Figure 10-8**.

Piles and Caissons

Auger cast piles, caissons, and steel piles are all used to transfer loads down through unsuitable soil to more appropriate load-carrying materials.

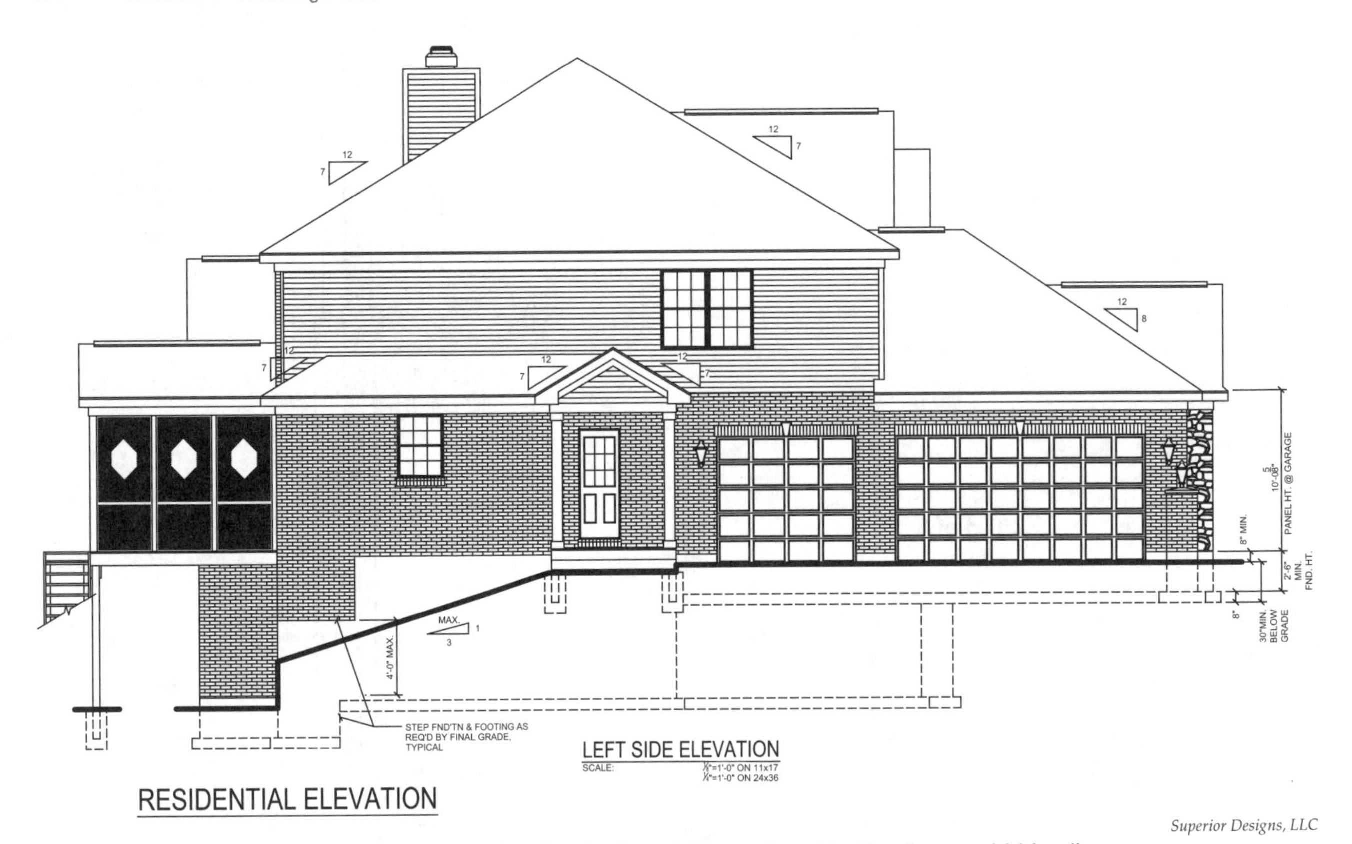

Superior Designs, LLC

Figure 10-5. Elevation view of a residence showing the foundation wall and footing lines as hidden lines.

Auger Cast Piles

An ***auger cast pile*** foundation system is drilled with a 12″–14″ steel auger to the appropriate design depth. The system uses a rock substrate for support or it uses the earth's friction to resist the building loads. While the drilling shaft is being raised out of the hole, concrete is pumped down the shaft to fill up the cavity. When you view an auger cast pile system on a drawing, you will see that there are usually several auger cast piles clustered together. They are topped with a pile cap to make the cluster work as a total load-carrying system, **Figure 10-9**.

Caissons

Caissons are also drilled, and are usually from 18″–72″ in diameter, **Figure 10-10**. Caissons are drilled to design depth. The drill is then removed and the bottom of the caisson tested for soil load capacity. After the hole passes inspection, a steel reinforcing cage is installed, and the hole is filled with concrete.

Steel Piles

A ***steel pile*** is a long H-shaped (H-pile) or round (pipe pile) steel member that is hammer-driven into the earth. The pile is driven to a suitable support stratum or driven to friction resistance of the soil (this type of pile is referred to as a ***friction pile***). A friction pile works on the principle of frictional resistance on the sides of the pile from the soil into which it has been driven. When you look at this type of foundation system on a drawing, you will usually see several steel piles clustered together and topped with a pile cap to make the cluster work as a total load-carrying system. See **Figure 10-11**.

Foundation Elevations

A critical part of starting a building properly is constructing the foundation at the designed elevation. An ***elevation*** is a measurement referenced to a known point. Elevations for a foundation are marked on the foundation plan view. Depending on the type of foundation, the elevations will be given for various parts of the system. For wall footings, the top of the footing is given an elevation marked, for example, as TOF = 91′-4″ or T/FTG = 91′-4″. Refer to **Figure 10-2A** and **Figure 10-3**. Sometimes, the bottom of the footing will also be marked as an elevation. For auger cast piles, caissons, and steel piles, the top of the pile cap (upon which the structure is going to be built) is the elevation given. All the information is defined in details and schedules. Each project will have its own special elevation designation system. Become familiar with the plans early in the preconstruction phase to work out any coordination problems. Referring to **Figure 10-2A**, note that the top of footing elevation is given as 95′-6″. If the first floor elevation is typically 100′-0″, this means the top of the footing is 4′-6″ below grade.

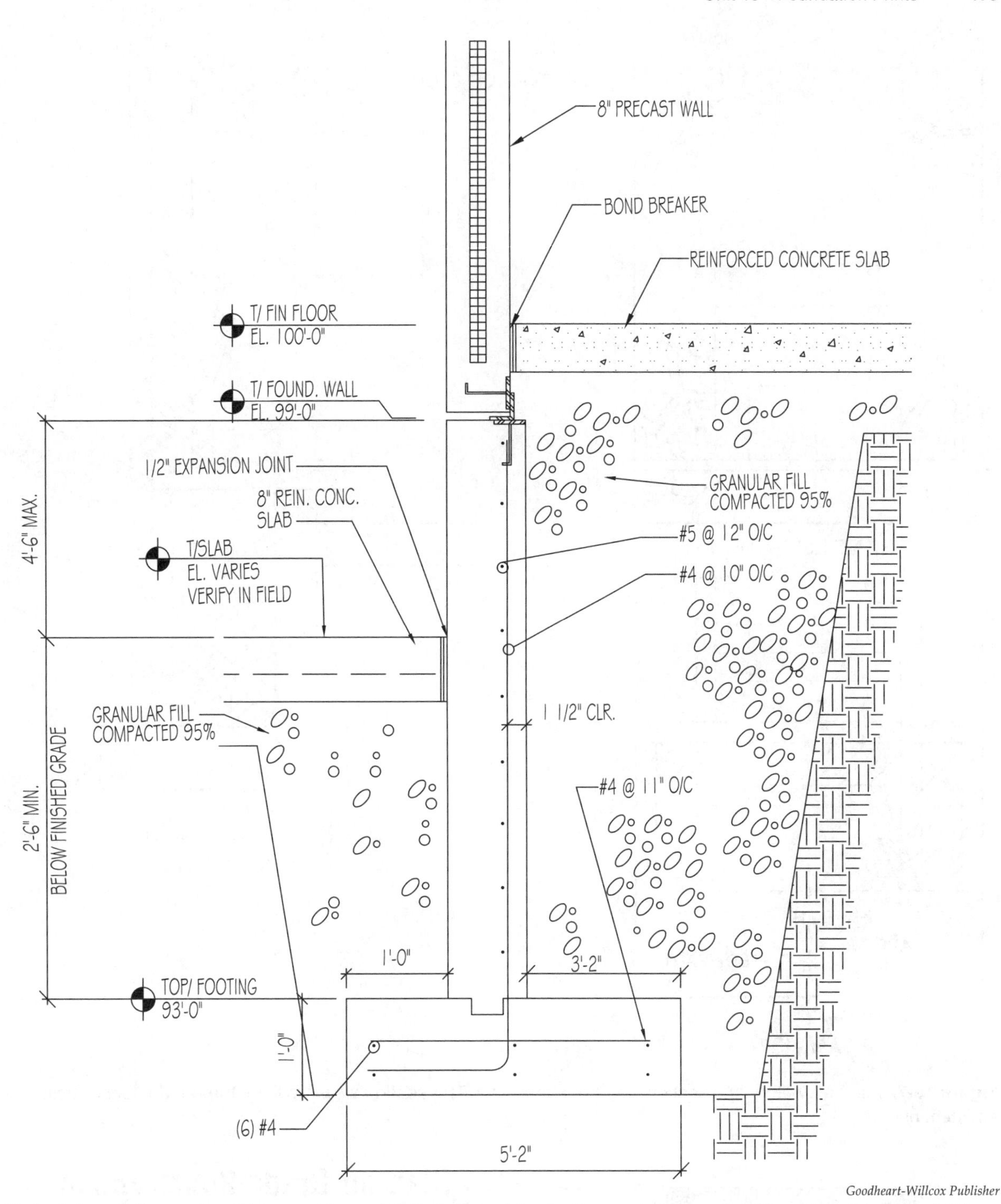

Goodheart-Willcox Publisher

Figure 10-6. A foundation wall section provides a detailed view of the footing and wall construction. Note the wider footing, which indicates that this wall is being used as a retaining wall.

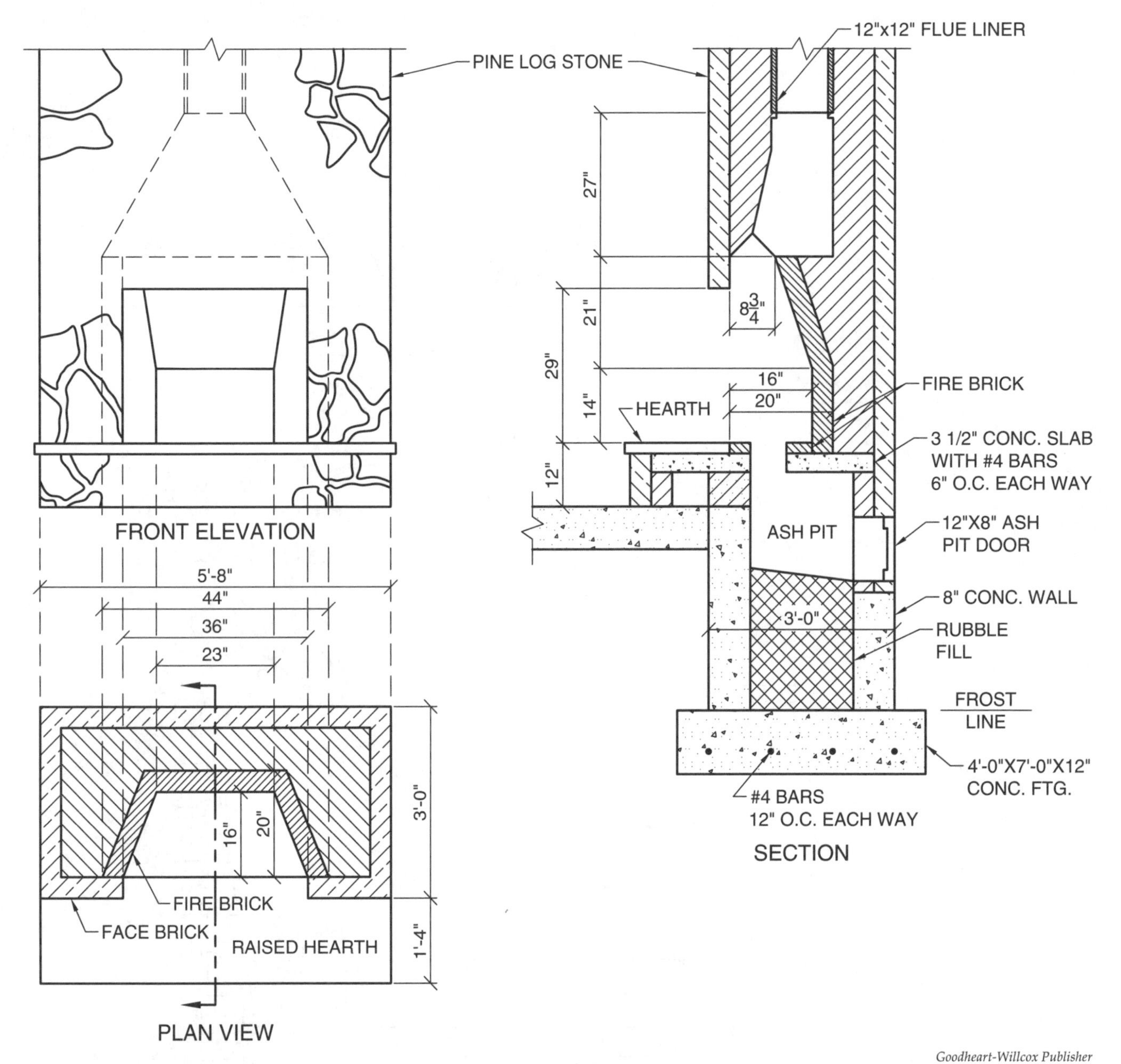

Goodheart-Willcox Publisher

Figure 10-7. Plan, elevation, and section views of a fireplace. The section view provides the needed construction details and dimensions.

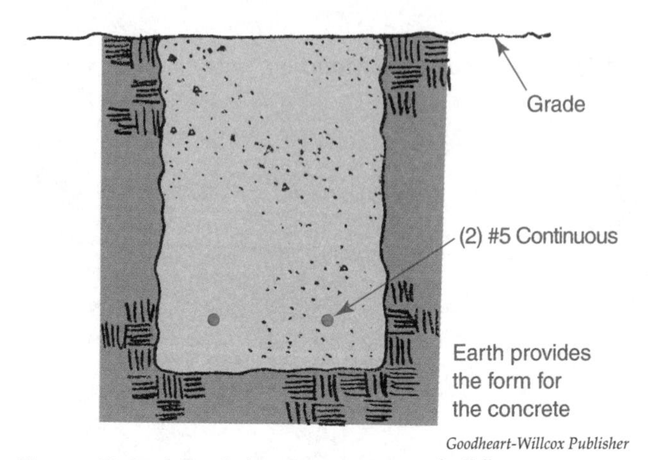

Goodheart-Willcox Publisher

Figure 10-8. A grade beam serves as a building support.

Slab-on-Grade Construction

A concrete slab poured at ground level is called a ***slab-on-grade***. Concrete slabs are used as both basement floors and main floors. Basement floors are poured after the footings and foundation walls are in and sometimes, in residential construction, before the rough framing starts.

Floating slab construction is a type of slab-on-grade construction that uses a ***monolithic slab*** (where the slab and footings are cast as one continuous unit). See **Figure 10-12**. The footings in this type of construction are sometimes called *turned-down footings*. Another method of producing a slab floor is to first pour the foundation walls to floor height. Then, the area within the walls is

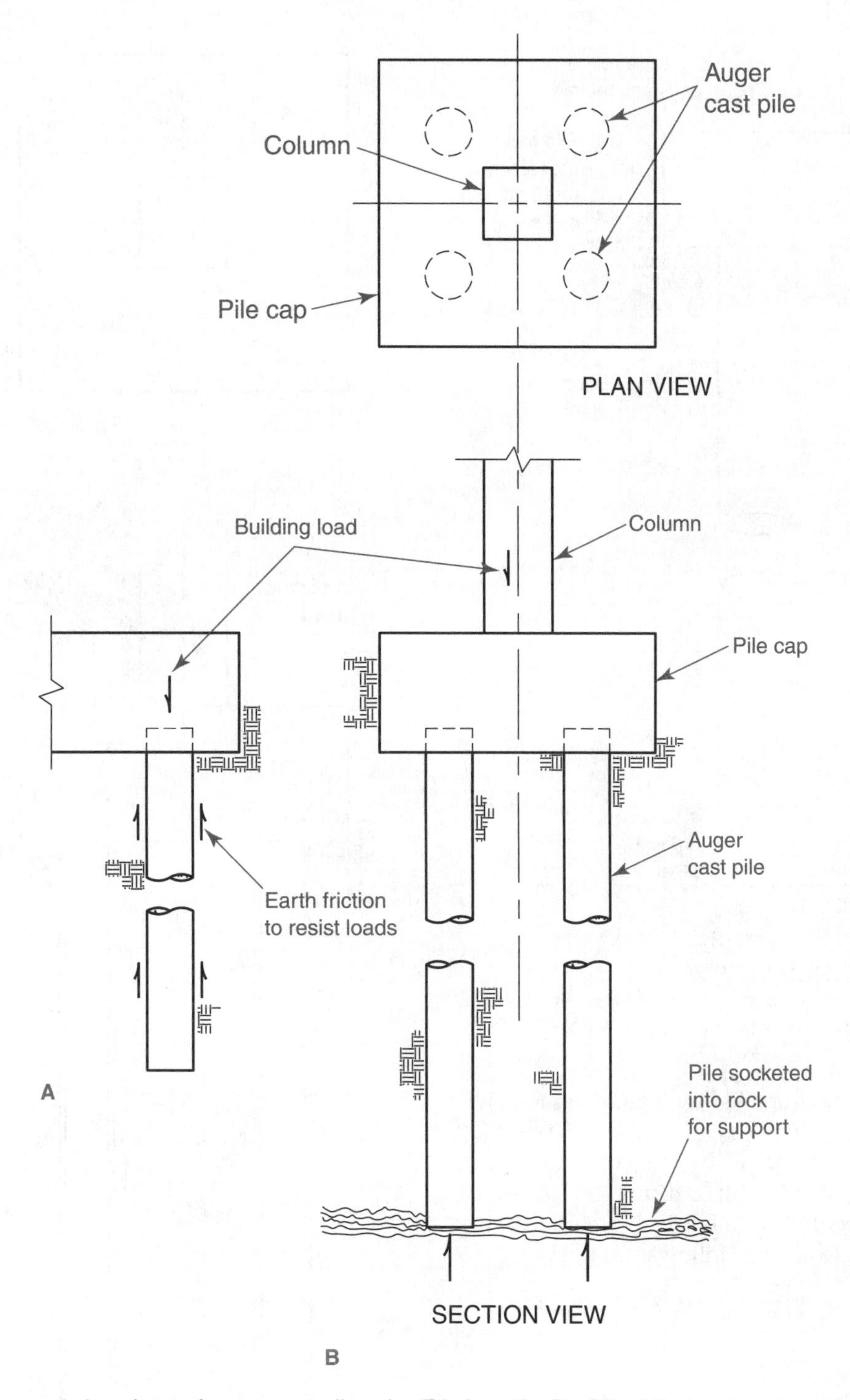

Goodheart-Willcox Publisher

Figure 10-9. Section and plan views of auger cast piles. A—Friction pile. B—Pile drilled and socketed into rock.

filled with soil and gravel. Finally, the floor is poured within the walls, separated by an expansion joint, **Figure 10-13**.

Loadbearing walls over slab floors require a thickened slab, **Figure 10-14**. On plan views, these areas are indicated by hidden lines indicating the deeper portion of concrete and a note.

Slab Reinforcement

A slab-on-grade is not considered a structural slab, but can have reinforcement to assist in load-carrying and resistance to cracking due to the slab being subjected to drying shrinkage. Steel reinforcing rods ("rebar") or welded wire fabric (WWF) are cast in the concrete. When the concrete hardens, the steel helps resist the

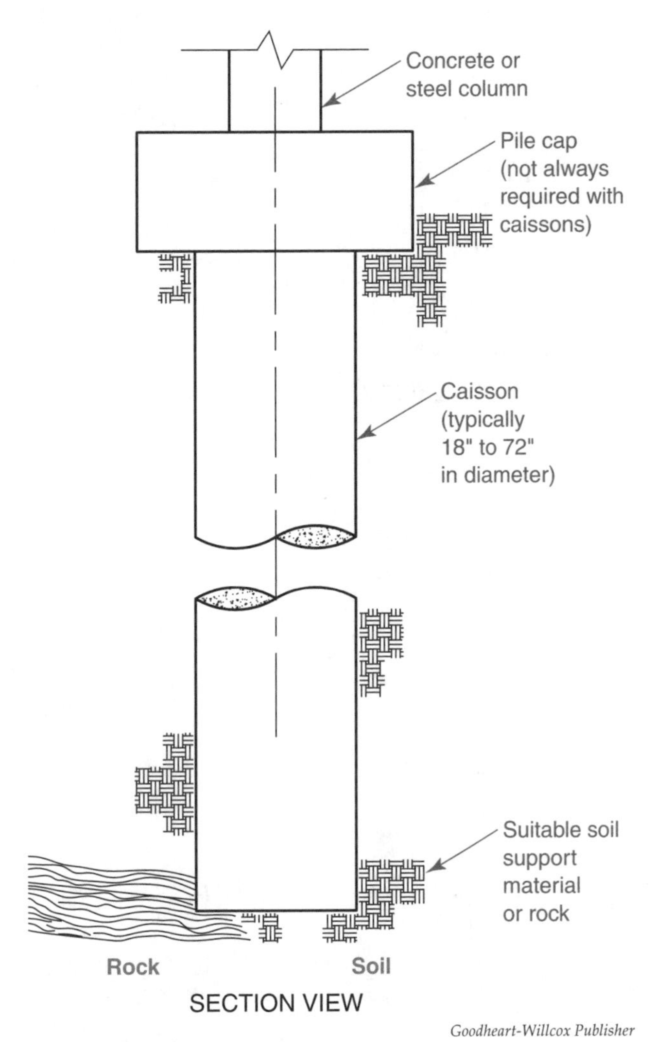

Figure 10-10. Drilled concrete caisson.

shrinkage. If the slab cracks, the reinforcement will keep the crack tight. Steel rebar is also used when the concrete slab is expected to be subjected to tension due to the settling of a dirt fill or heavy load. A typical note specifying welded wire fabric or steel rebar in a concrete floor would read as follows:

5″ THK SLAB w/6×6–W10×W10 WWF *(for mesh)*

or

#4 @ 18″ O.C. EW OVER ABC GRAVEL SUBBASE *(for steel rebar)*

Expansion Joints

Expansion joints are placed between a slab and a wall, and also around columns, **Figure 10-15**. These joints prevent the slab from cracking due to expansion and contraction. They are used with slabs-on-grade that do not settle evenly with the footings and foundation walls. Other expansion joints may be used in long slabs. These usually call for premolded joints and have metal coverings.

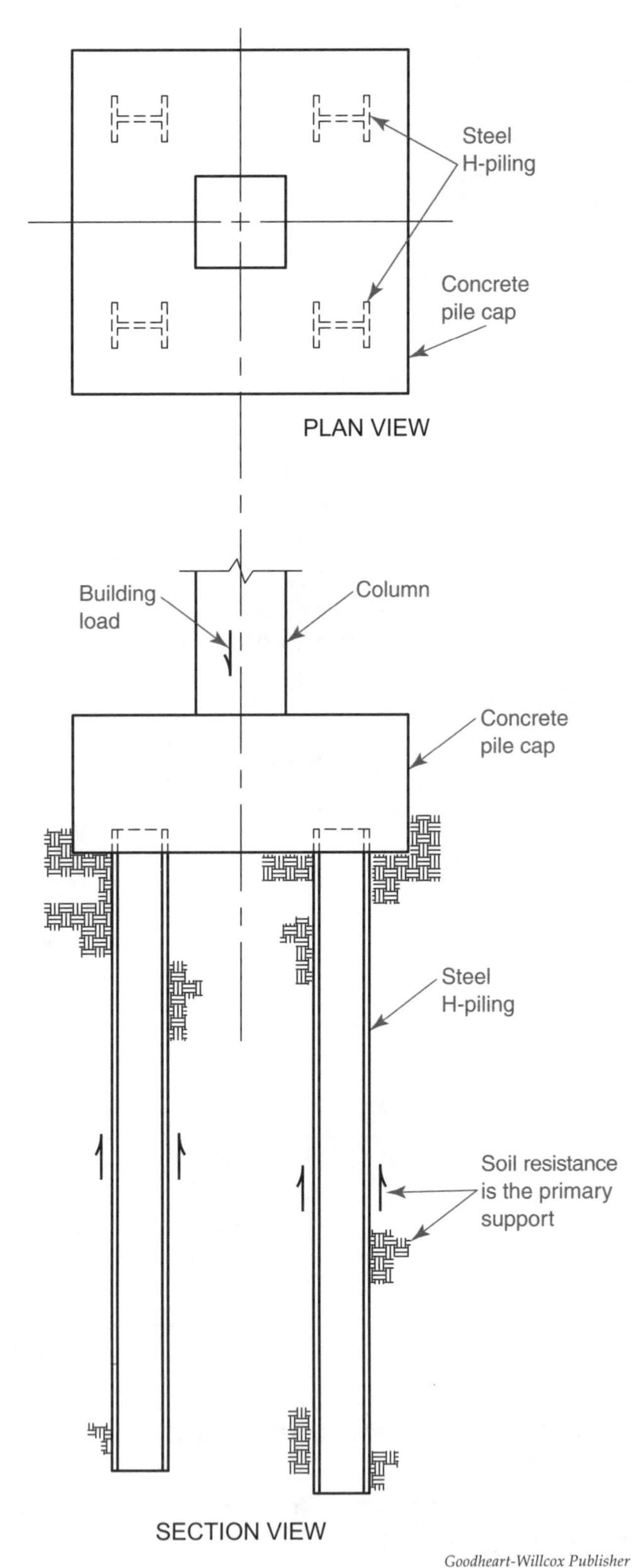

Figure 10-11. Building foundation system using steel H-piling.

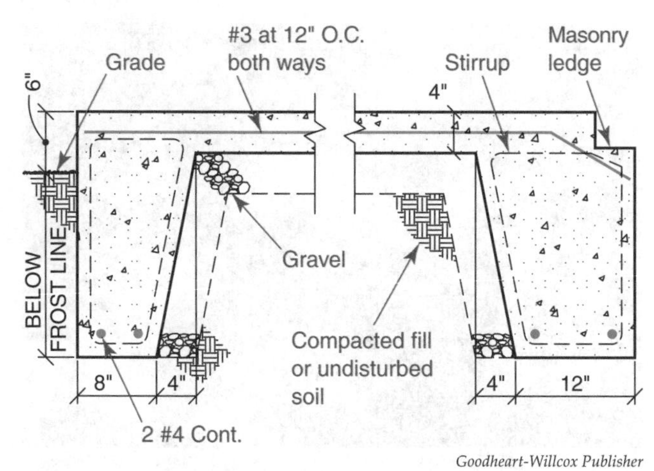

Goodheart-Willcox Publisher

Figure 10-12. A monolithic slab foundation.

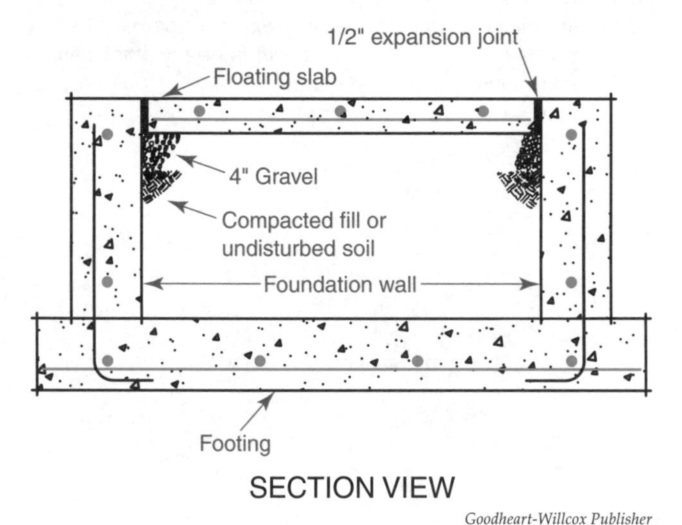

Goodheart-Willcox Publisher

Figure 10-13. A slab floor poured within the foundation walls.

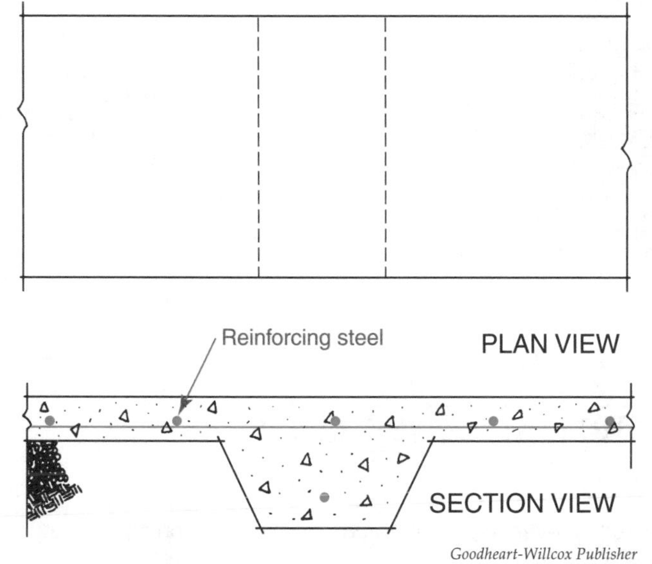

Goodheart-Willcox Publisher

Figure 10-14. The slab is thickened to create a beam below loadbearing walls.

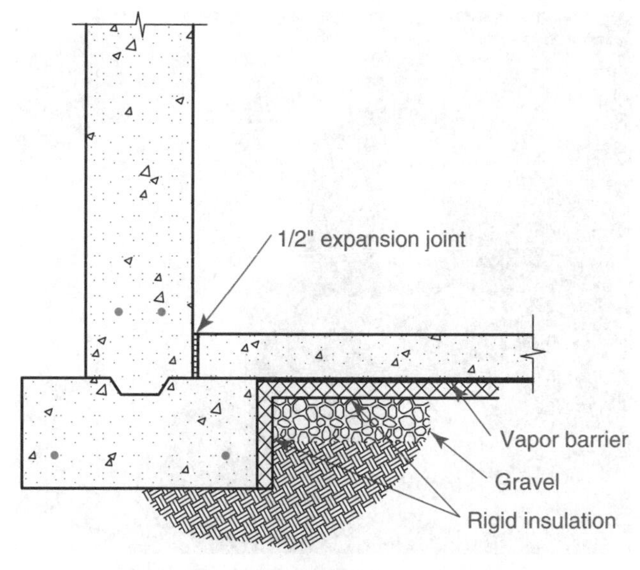

Goodheart-Willcox Publisher

Figure 10-15. An expansion joint made of a flexible material is placed between the concrete slab and wall.

Some architects indicate on a drawing where it is permissible to make a joint when the concrete placement is terminated at the end of the day. If no indication is given, the project engineer or architect must approve the location.

Foundation Waterproofing

Waterproofing of foundation walls is needed in areas where soil and climatic conditions demand protection from underground water. Residential waterproofing can be done with a variety of materials, but usually consists of mopping the outside of the foundation wall with tar or asphalt, **Figure 10-16**. Sometimes, a polyethylene sheet is applied over the tar. Drawings for a foundation to be waterproofed will have a heavy black line on the exterior wall with a note indicating location. Also, the building specifications may specify the exact material and process to be used.

A layer of crushed rock or gravel is laid below the floor area. This layer is then covered with a heavy plastic vapor barrier to keep the dampness in the ground from transferring to the slab and into the building. See **Figure 10-17**.

Other features are used in concrete construction to protect the building from water seepage. To inhibit water transmission at construction joints, a special sealing device called a waterstop is used. A ***waterstop*** is a linear section of polyvinyl chloride (PVC) located in the concrete to serve as a moisture barrier. In **Figure 10-18**, a waterstop is located at the intersection of the wall and footing in the keyway.

Goodheart-Willcox Publisher

Figure 10-16. Tar being applied to a residential foundation wall for waterproofing.

1125089601/Shutterstock.com

Figure 10-17. A heavy plastic vapor barrier is placed over a layer of gravel prior to pouring a concrete slab. This residential foundation is an example of slab-on-grade construction using both interior and exterior beams. The gravel subbase has been covered with a heavy plastic film to protect against ground moisture.

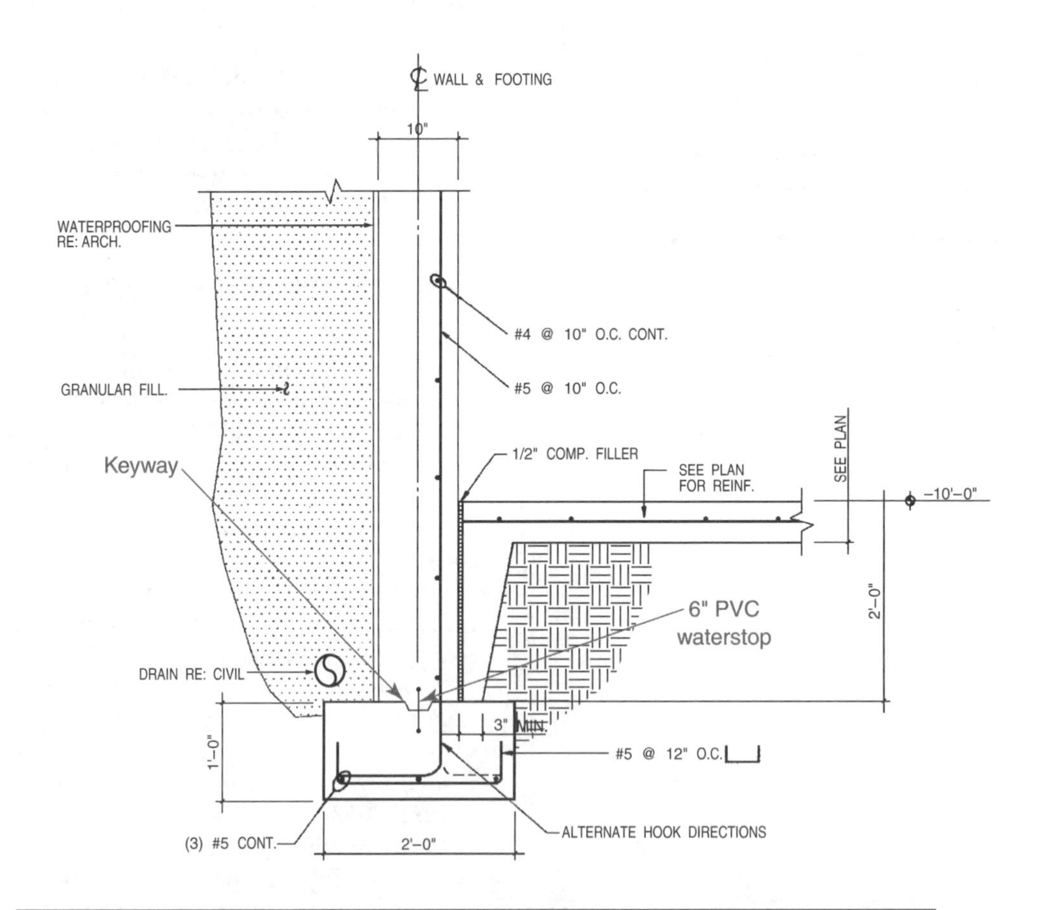

Goodheart-Willcox Publisher

Figure 10-18. Waterstops are used at concrete construction joints to prevent water seepage. This wall section shows a waterstop used at the keyway where the footing and foundation wall meet.

Test Your Knowledge

Name ________________________________

Write your answers in the spaces provided.

____ 1. When a footing must be placed above ground that has been disturbed, it should be ____.
A. twice as thick as usual
B. twice as wide as usual
C. reinforced with steel reinforcing bars
D. no different than usual
E. None of the above.

____ 2. On a residential foundation plan, footings are shown as ____.
A. continuous lines
B. hidden lines
C. phantom lines
D. dotted lines
E. Footings are not shown on foundation plans.

____ 3. ____ rods are used to reinforce footings, slabs, and foundation walls.
A. Iron
B. Steel
C. Aluminum
D. Tungsten
E. All of these are used.

____ 4. A(n) ____ is placed between a slab and a wall.
A. air gap
B. roll of insulation
C. reinforcing bar
D. expansion joint
E. All of these are commonly placed between a slab and a wall.

____ 5. Welded wire fabric is used to ____.
A. reinforce steel beams
B. protect basement windows
C. reinforce concrete slabs
D. protect welders from heat
E. None of the above.

____ 6. A ____ is a slotted joint in concrete used to connect portions cast at different times.
A. keyway
B. caisson
C. waterstop
D. pile
E. None of the above.

____ 7. *True or False?* Footings are only needed below the perimeter walls of a building.

____ 8. *True or False?* The frost line is the depth at which the ground is frozen when the temperature is 25°F (–4°C).

____ 9. *True or False?* Normally, a column footing is larger than a footing below a wall.

____ 10. *True or False?* Expansion joists placed around a slab expand and prevent the slab from moving.

____ 11. *True or False?* The footings used in a monolithic slab are called turned-up footings.

12. What are the five types of foundation systems described in this unit?
A. ________________________________
B. ________________________________
C. ________________________________
D. ________________________________
E. ________________________________

____ 13. What is the primary way in which a steel friction pile supports a load?
A. Bottom support
B. Pile cap support
C. Soil resistance along the side of the pile
D. Steel lugs
E. None of the above.

14. What does the term TOF indicate?

____

15. What is the purpose of the reinforcement in a slab-on-grade?

____

Activity 10-1

Concrete Foundation for a Residential Building Project

Name ________________________________

Refer to Sheets 1, 2, and 4 from the Sullivan residential building plans in the Large Prints supplement to answer the following questions.

1. Refer to the foundation plan to answer the following questions.

   A. On which sheet is the foundation plan? ________________________________

   B. What is the scale of the plan? ________________________________

   C. What areas are unexcavated? ________________________________

2. What is the specification for the garage slab?

   ____

   ____

3. What are the size and reinforcement requirements for the column footings in the basement?

   ____

   ____

4. What is the size of the foundation wall footing and what are the reinforcement requirements?

   ____

   ____

5. What is the size of the foundation wall and what are the reinforcement requirements?

   ____

   ____

6. What is the minimum depth below grade for the footings?

   ____

7. Where are the grade beam details found and what details are given?

   ____

Activity 10-2

Concrete Foundation for a Residential Building Project

Name ______________________________

Refer to Sheets 1, 2, 3, and 5 from the Marseille residential building plans in the Large Prints supplement to answer the following questions.

1. Give the specified psi rating for the following items:

   A. Garage slab ______________________________

   B. Interior slabs ______________________________

   C. Walls and footings ______________________________

2. What is the slab thickness and reinforcing for the following items?

   A. Garage slab ______________________________

   B. Basement ______________________________

   C. Terrace ______________________________

3. What size column footings are specified for the interior columns and what is the required thickness and reinforcing?

   ____

   ____

   ____

4. Referring to the grade beam details, how is the foundation reinforcing to be installed?

   ____

5. Where are the shear wall details located? What are they referring to?

   ____

   ____

6. What is the thickness of the typical foundation wall? What reinforcing is specified?

   ____

   ____

7. What is the floor pitch of the garage slab?

   ____

8. What are the outside dimensions of the foundation walls?

   ____

9. What is the total dimension from the inside of the right-side foundation wall to the center column?

____

10. What is the minimum depth from grade to the bottom of the foundation wall?

____

Activity 10-3

Structural Plans for a Commercial Building Project

Name ___________________________________

Refer to Sheets S1.1, S1.2, S3.1, and S5.1 from the Delhi Flower and Garden Centers commercial building plans in the Large Prints supplement to answer the following questions.

1. What is the size of the slab in the hardlines area and what are the reinforcement requirements?

____

2. What is the top of slab elevation for the slab in the hardlines area?

____

3. What is the typical top of footing elevation unless noted otherwise?

____

4. Give the size of the following column footings and the required reinforcing:

A. F1 ____

B. F3 ____

C. F4 ____

5. What is required for the greenhouse sales floor?

____

____

6. What type of column is footing F5 supporting?

____

7. What kind of angle is indicated on the dock leveler pit?

____

8. What is the contractor supposed to do in relation to the dock leveler pit?

____

____

9. What are the floor flatness and levelness specifications for the concrete slab-on-grade?

____

____

10. What is the specified psi rating for concrete for exterior flat work?

____

11. Where are the retaining walls located?

____

12. Refer to Section 12A/12B on Sheet S3.1 to answer the following questions.

    A. What is the width of the footing for a 6′-0″ high wall?

    ____

    B. What size rebar is required for the horizontal bars in the wall for this section?

    ____

    C. Referring to the "O" bars, what is required for a 10′-0″ high wall?

    ____

13. What section is referenced for the footing under the doorway between the Display Atrium Sales and Outside Sales areas?

____

14. What sizes are specified for the base plate and anchor bolts for column C01?

____

15. Referring to Section 1 on Sheet S3.1, what do the angled lines indicate?

____

Goodheart-Willcox Publisher

A variety of building materials are used in above-grade structural construction. This worker is setting anchor bolts in a freshly poured concrete wall. Metal forms are used for the concrete formwork.

UNIT 11
Structural Prints

TECHNICAL TERMS

beams
bolsters
cast-in-place concrete
chairs
column schedule
columns
I-beam
one-way slab system
open-web steel joists
posttensioned concrete
precast concrete
prestressed concrete
pretensioned concrete
reinforced concrete
splicing
stirrups
structural steel
ties
two-way slab system
waffle slab
wide-flange beam

LEARNING OBJECTIVES

After completing this unit, you will be able to:

- Identify different types of structural construction systems.
- Recognize various types of concrete floor systems.
- Understand concrete reinforcing notation.
- Recognize reinforcing steel on prints.
- Read beam and column schedules.
- Identify various structural steel members.

Large commercial and industrial buildings are typically constructed around a framework of reinforced concrete or structural steel, **Figure 11-1**. This type of construction is also used for large residences such as high-rise condominiums and light commercial buildings.

To make print reading easier, this unit explains how above-grade structural systems are denoted and constructed. Most typical systems are constructed from cast-in-place concrete and structural steel. Concrete and steel were introduced in Unit 7.

Dan Dorfmueller

Figure 11-1. Steel and concrete members make up the structural framework of this building.

GREEN BUILDING

Reusing Waste

Evaluate waste for other uses on or off the job site before disposal. Assess "wasted" gravel, concrete material, lumber, and steel. Determine whether these can be used elsewhere in the project or on another job site.

Above-Grade Structural Construction

Above-grade structural systems are divided into two major groups: concrete construction and steel construction. There are other above-grade systems, such as wood framing construction (typically used in residential construction). This unit focuses on concrete construction and steel construction. Concrete can be either cast in place (cast on site) or cast off site (precast). Both methods have advantages and disadvantages. Structural steel is fabricated off site and assembled at the job site.

Concrete Construction and Types

Cast-in-place concrete is cast at the construction job site. Formwork is built, steel reinforcing bars ("rebar") are set inside the formwork, and then freshly mixed concrete is placed into the forms. The concrete hardens, the forms are removed, and the concrete stays in place. This is called *cast-in-place reinforced concrete.* Since most concrete placements occur outdoors, the pour can be negatively impacted by adverse weather. Formwork dimensions and the setting of reinforcement steel are not as precise as in the controlled precast environment (described later in this unit). Most large concrete construction projects must be built from cast-in-place concrete, due to the extreme weight of concrete material and transportation limitations. Concrete is a moldable material that is formed by the shape of the forms used for the building structure design. Concrete exposed to view with the intent to create a finished space that defines a mood or atmosphere is called *architectural concrete.* Working with architectural concrete requires more concern with design and appearance, form construction, and the placement of the concrete in comparison to normal structural concrete.

Reinforced Concrete

Were it used for only slabs, drives, and walks, concrete would be a valuable building material. However, its applications become even more valuable when combined with steel reinforcement rods ("rebar") in structural applications. The combination of concrete's high compressive strength and steel rebar's high tensile strength makes ***reinforced concrete*** a unique, durable, and widely used building material.

Concrete Floor Systems

Concrete floor systems are constructed from reinforced concrete and are typically referred to as *slab systems.* Slab systems are classified as one-way slabs and two-way slabs. There are different designs based on each system.

One-Way Slab Systems

A ***one-way slab system*** is constructed with the main reinforcing steel spanning in one direction. In this system, reinforced concrete members resting on columns run parallel (one way) and support the floor slab. One-way slab systems may use either girders and beams or beams and joists in the design. A one-way slab system with concrete girders and beams is shown in **Figure 11-2.** A building using one-way slab construction with concrete beams and pan joists is shown in **Figure 11-3.**

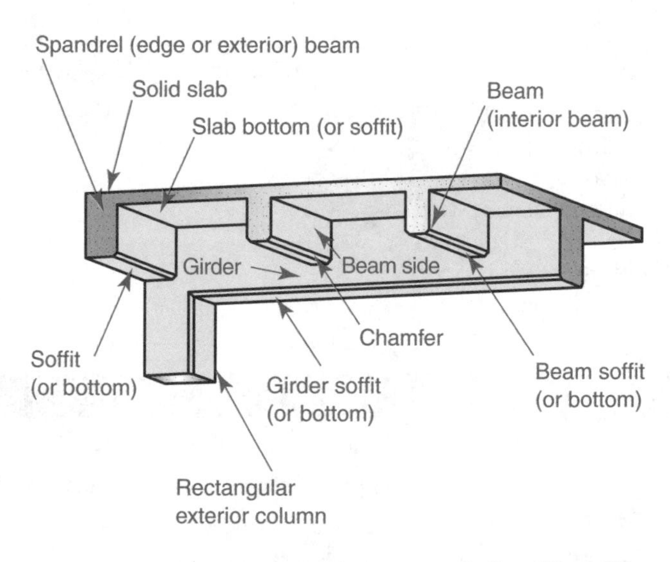

Goodheart-Willcox Publisher

Figure 11-2. This one-way slab system is formed by using forms in a monolithic (continuous) pour and is tied directly to the column supports.

Dan Dorfmueller

Figure 11-3. A parking garage constructed using a one-way slab system with beams and pan joists.

Two-Way Slab Systems

A ***two-way slab system*** is constructed with the main reinforcing steel spanning in two directions. Depending on the specific type of system, support members such as girders, beams, and joists may or may not be used. The *flat plate slab system* is a basic two-way slab system using a flat slab of uniform thickness. See **Figure 11-4A**. In this system, the slab rests directly on columns and loads are distributed evenly to the columns. The flat plate slab system is used to carry lighter loads in structures such as condominiums, office buildings, and hotels. For heavier construction that carries greater floor loads, the *flat slab system* is used. In this system, drop panels are used underneath the slab to provide additional support at the column locations. See **Figure 11-4B**.

A ***waffle slab*** is supported by joists running in two directions. In this system, steel forming "pans" or "domes" are used to form the support structure. See **Figure 11-5**. The resulting joists have a ribbed shape and span from column to column. A building with waffle slab construction is shown in **Figure 11-6**.

Steel Reinforcement

Reinforcing steel is used in all types of concrete structural systems. Floors, beams, and columns are the most common elements where steel reinforcement is used. Steel reinforcement comes in either plain or deformed bars. Deformed bars provide a better bond between

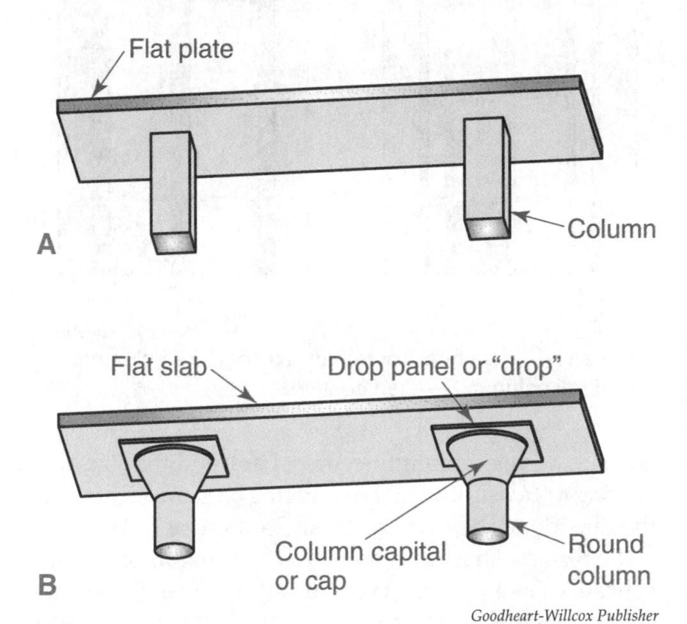

Goodheart-Willcox Publisher

Figure 11-4. Variations on the two-way slab system. A—Flat plate slab on columns. B—Flat slab with drop panels and capitals above the columns.

Dan Dorfmueller

Figure 11-6. The waffle slab constructed for this building provides structural support and serves as a finished ceiling.

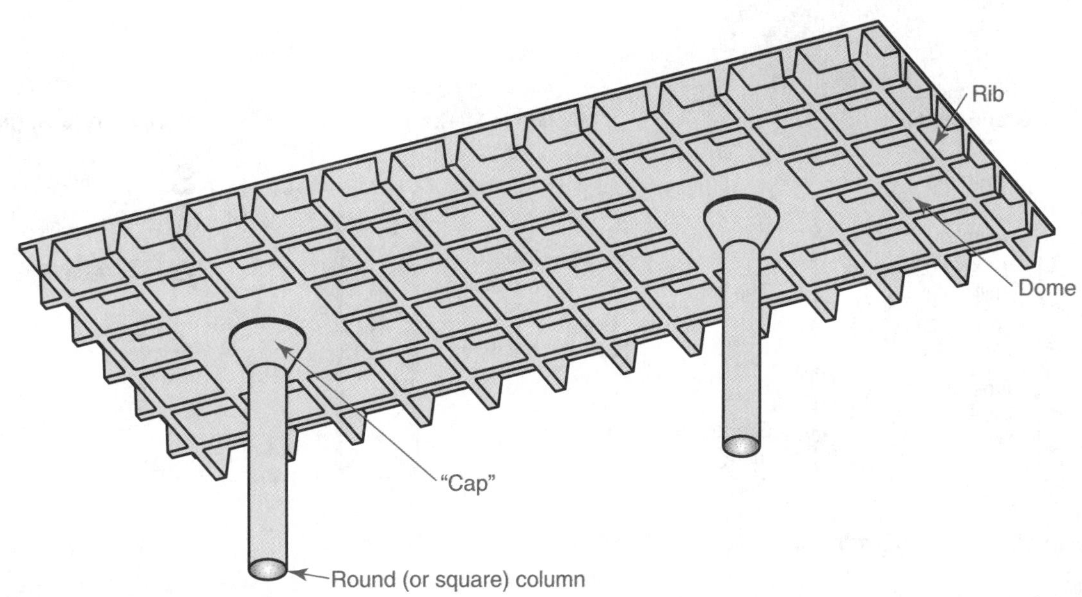

Goodheart-Willcox Publisher

Figure 11-5. Steel "pans" or "domes" are used to form the support structure for a waffle slab.

the concrete and the reinforcing steel, improving the load-carrying capacity of the structural element. The placement of the steel, which is determined by the designer or engineer, provides maximum resistance to forces of compression, tension, and shear. Reinforcing steel to be placed in concrete is designated with a note on the drawing giving the number of bars, bar size, spacing, and placement:

(2) #5 × 19′-0″ @ 2′-6″ O.C.

PLACE ON 1 1/2″ BAR CHAIRS

The size of the bar is indicated by a number representing eighths of an inch. That is, a #5 bar is one that has an actual diameter of 5/8″. Identification marks rolled into the bars identify the producing mill, bar size, and type of steel. For Grade 60, a grade mark indicating yield strength is included (Grade 40 and Grade 50 do not show a grade mark). See **Figure 11-7**. Reinforcing steel comes in 11 different sizes: #3, #4, #5, #6, #7, #8, #9, #10, #11, #14, and #18. Remember, the number represents eighths of an inch (to determine the bar diameter, multiply the bar number by 1/8″). Reinforcing steel was introduced in Unit 7.

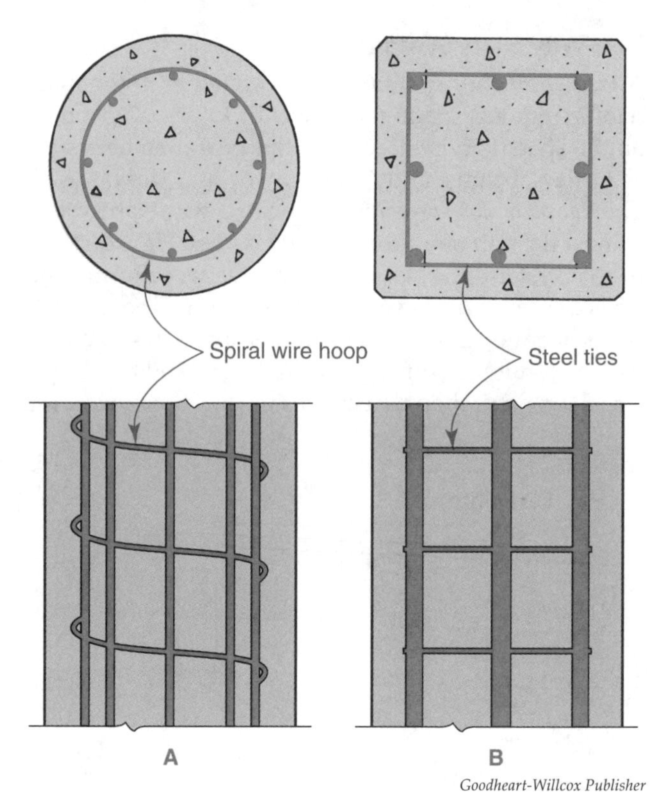

Figure 11-8. Two types of reinforced concrete columns. A—Spiral column. B—Tied column.

Columns and Beams

Columns are vertical support members designed to safely carry the anticipated final load of the building structure and its contents. Concrete columns are reinforced with steel bars or structural shapes, **Figure 11-8**.

Columns are normally specified on the drawing. Otherwise, they are identified by a grid system of numbers and letters on the plan view and specified on a ***column schedule***, **Figure 11-9**. Identification should include the location, the size of the column, the number and size of vertical reinforcement bars, and the size and spacing of ***ties*** (horizontal reinforcing bars placed around the outside of the vertical bars). See **Figure 11-10**.

Beams are structural members running horizontally beneath a floor system. They are supported at their ends by other beams, girders, or columns, **Figure 11-11**. Some beams are rectangular in cross-section. Others are thinner at the bottom, where they are reinforced with steel to provide tension strength, and wider at the top to provide compression strength. See **Figure 11-12A**.

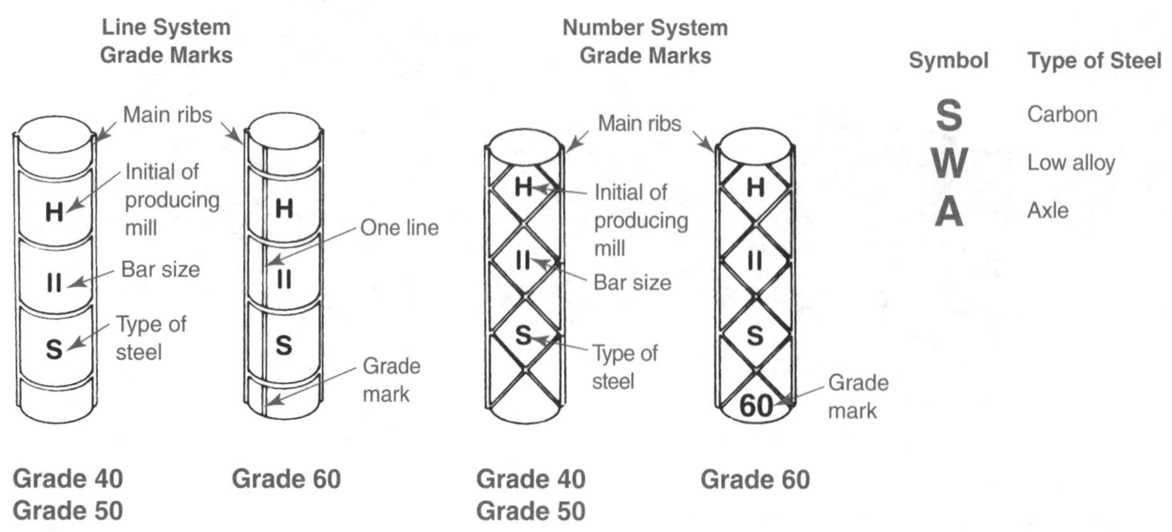

Figure 11-7. Identification marks for steel reinforcing bars.

COLUMN SCHEDULE

MARK	SIZE	B/ BASE PLATE ELEVATION	BASE PLATE SIZE	ANCHOR BOLTS
C01	HSS5x5x1/4	99'-5 1/2"	3/4"x12"x12"	(4)-3/4"ø
C02	HSS5x5x1/4	98'-1 1/2"	3/4"x12"x12"	(4)-3/4"ø
C03	HSS6x6x1/4	99'-5 1/2"	3/4"x12"x12"	(4)-3/4"ø

9" ANCHOR BOLT EMB.

EQ. EQ.

1 1/2" TYP.

EQ. EQ.

1 1/2" TYP.

HSS COLUMN BASE PLATE

N.T.S.

Delhi Flower and Garden Center
Arch/Image 2 Architects
Pinnacle Engineering Services, Inc.

Figure 11-9. The column schedule contains the specifications for all of the columns in the building.

Goodheart-Willcox Publisher

Figure 11-10. Ties are horizontal reinforcing bars used to hold vertical reinforcement bars in position when the concrete is cast. The size and spacing of ties is included when specifying information for columns.

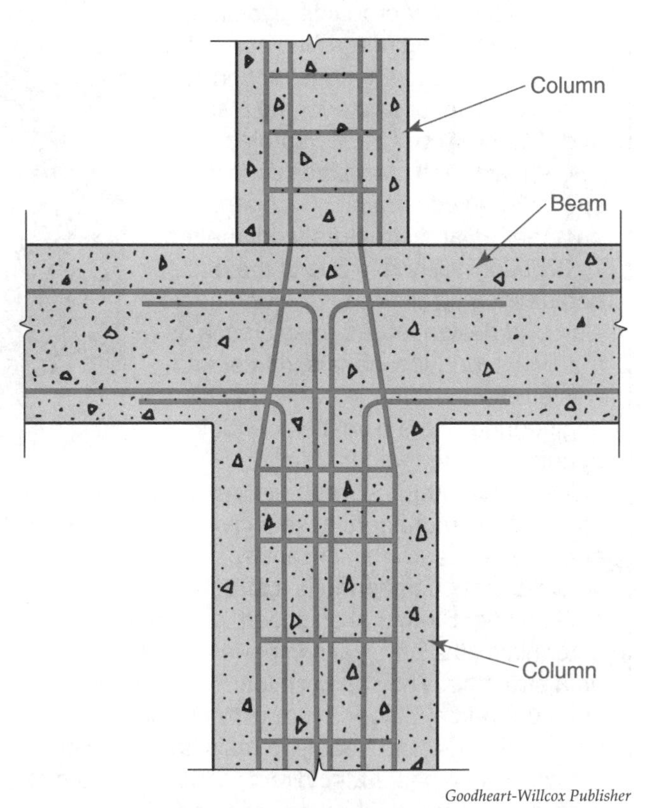

Goodheart-Willcox Publisher

Figure 11-11. Beams are structurally tied into other structural members. Shown is steel reinforcement tying a beam to a column.

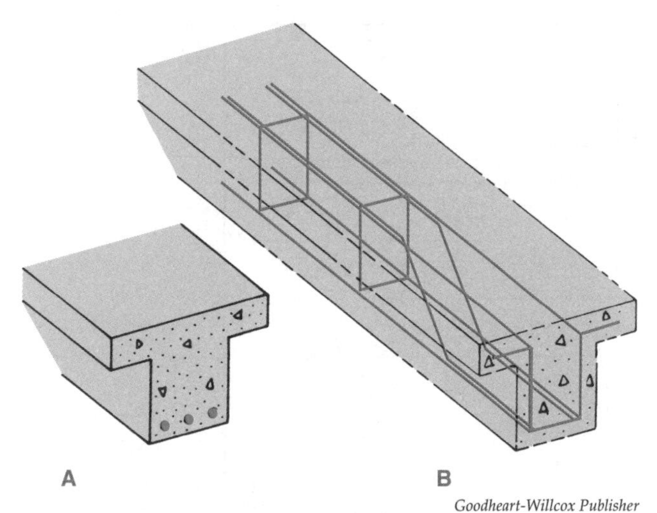

Goodheart-Willcox Publisher

Figure 11-12. Cross-sections of reinforced "T" beams. A—The beam is wider at the top, where compressive strength is needed. Tensile strength is needed in the bottom of the beam, so steel is added. B—Stirrups hold the reinforcing bars in place and make the beam stronger.

Beams are given added strength by adding ***stirrups*** (U-shaped steel bars) that hold the horizontal bars in place and increase the resistance to shear stresses. See **Figure 11-12B**. Devices called ***bolsters*** and ***chairs*** are used to support reinforcing bars at the desired level and clearance from the formwork, **Figure 11-13**. A chart showing the standard types and sizes of bar supports and their symbols is included in the Reference Section of this textbook.

Beams are normally specified on the drawing, **Figure 11-14**. If there are many beams, they may be identified on the drawing and specified in a beam schedule. Identification should include the location, beam size, the number and size of reinforcing bars at the bottom and top of the beam, and the number, size, and type of stirrups. See **Figure 11-15**.

As is the case with columns, beams and other structural components are typically referenced on the plan view using a grid system of numbers and letters. In **Figure 11-14**, note the use of grid lines to identify beam lines and columns. Using this system, columns are

CAREERS IN CONSTRUCTION

Engineer

An engineer is a trained professional who designs new products, systems, or structures. Engineers who specialize in work relating to the construction industry are responsible for the planning and design of construction projects. Two engineering disciplines that specialize in construction work are civil engineering and structural engineering.

Civil engineering is one of the oldest engineering disciplines. Civil engineers plan and design large projects such as roadways, bridges, canals, dams, and wastewater systems. A career in civil engineering offers a vast amount of opportunities in the construction industry. Although not required, many construction managers have a civil engineering degree and license.

bibiphoto/Shutterstock.com

Civil engineering is concerned with the planning and design of large construction projects.

Structural engineering is a subdiscipline of civil engineering. Structural engineers are responsible for determining the physical components that are to hold up a structure. They analyze loads and forces on a structure and engineer the structure's foundation, walls, columns, and beams to resist these forces. Structural engineers also direct CADD operators to develop the structural construction documents used in a building project.

Becoming an engineer requires earning a bachelor's or master's degree, gaining professional experience, and passing a state licensing examination. Engineers who work in the construction industry are often employed by a large firm or a city government.

Whitacre Engineering

Figure 11-13. Bolsters and chairs hold the reinforcing steel at the correct height in the forms. These are fastened to the forms and wired to the reinforcing bars.

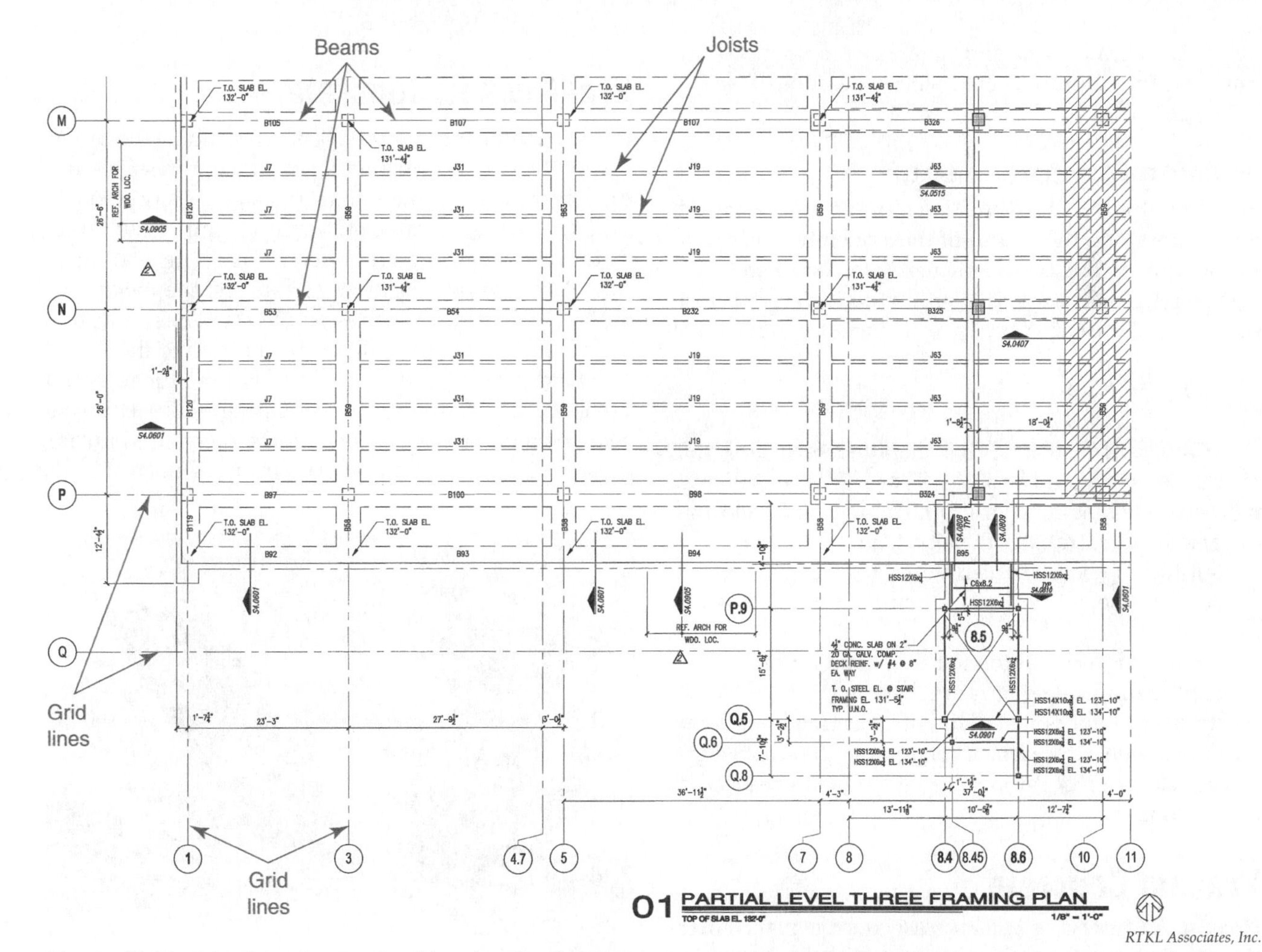

RTKL Associates, Inc.

Figure 11-14. A portion of a structural framing drawing for a commercial building. Note that beams are represented with centerlines. In addition, note the use of grid lines to identify beams and columns.

BEAM SCHEDULE								
BEAM MARK	BEAM SIZE W X D (IN.)	REINFORCING (NUMBER-SIZE-TYPE)				STIRRUPS		NOTES
					SIZE	TYPE	SPA.EA.END	
B1	24.00 X 24.63	TOP BARS	4-5-B	4-5-F		#3	A	@8"	
		BOT BARS	4-9-S						
B2	24.00 X 24.63	TOP BARS	4-5-J			#3	A	@8"	
		BOT BARS	4-9-S						
B3	28.25 X 24.63	TOP BARS	4-6-B	4-7-F	4-7-E	#3	A	@8"	
		BOT BARS	4-7-S						
B4	28.25 X 24.63	TOP BARS	4-6-F			#3	A	@8"	
		BOT BARS	4-8-S						
B5	28.25 X 24.63	TOP BARS	4-5-J			#3	A	@10"	
		BOT BARS	4-7-S						
B6	39.25 X 24.63	TOP BARS	4-5-B	4-7-F	2-7-E	#3	D	9@6" R@10"	
		BOT BARS	4-9-S						
B7	39.25 X 24.63	TOP BARS	4-8-F			#3	D	@10"	
		BOT BARS	4-9-S						

Goodheart-Willcox Publisher

Figure 11-15. A beam schedule providing member dimensions and reinforcement information.

identified with a letter and number combination. For example, the column located where grid lines P and 1 intersect is identified as column P1.

Splicing Reinforcement

Splicing of reinforcing bars in columns and beams is necessary when the length of these members exceeds the length of the bars. Splicing is accomplished by overlapping the bars by a length of a given number of diameters of the bar. A typical splicing note would read:

30 DIA MINIMUM LAP

If No. 4 bars (4/8″ or 1/2″ diameter) were specified, the overlap length would be 30 × 1/2″, or 15″. Other splicing methods, such as mechanical fastening and welding, are also used to transfer the load from bar to bar.

Reinforcing bars are also added to strengthen wall corners. These corner bars are spliced onto the wall reinforcing bars.

Piping and electrical conduit are placed in slabs at the same time reinforcing bars are installed, **Figure 11-16**. Electrical conduit is installed with electrical boxes on the ceiling side of the slab, or conduit may be stubbed up at a wall location or through a grid system to be located in the floor.

Precast Concrete

Precast concrete is cast at a precast concrete plant. After the member hardens, it is transported to the construction site, where it is lifted into place with a crane and

Goodheart-Willcox Publisher

Figure 11-16. Plumbing pipes and electrical conduit extend out of the poured concrete slab. They are put in place at the same time as the reinforcing bars before the concrete is placed. They are typically covered with subbase material prior to concrete placement.

installed. Because the casting is done in a controlled environment, higher quality is attained. Beams, columns, and panels can all be precast. However, precast components must be small enough to allow for transportation. **Figure 11-17** shows a precast wall panel being hoisted into place at the building site.

Prestressed Concrete

When a load is placed on a beam, it bends—the upper half of the beam is pushed together (compression) and the lower half is pulled apart (tension). As previously stated, concrete is much stronger in compression than in tension. ***Prestressed concrete*** uses high-strength steel cables to pull together (compress) the lower half of the beam before a load is placed on it. Basically, this is the same as pushing up on the bottom of the beam. The beam can then support an additional load equal to the load "pushing up." Prestressing makes concrete beams much stronger, and allows them to span far greater distances than normal reinforced beams.

Jack Klasey

Figure 11-17. Precast wall panels are brought by truck to the building site. A crane is used to place the panels.

Concrete is prestressed by casting it around steel cables or rods that are placed under tension (pulled) to as much as 250,000 psi (pounds per square inch). Prestressed concrete members can be designed to use half the concrete and one-fourth the steel required for conventional reinforced concrete members of equivalent strength. Concrete bridge beams and beams in parking decks are normally prestressed.

There are two types of prestressed concrete members—pretensioned and posttensioned:

- ***Pretensioned concrete*** members are normally precast at a precast concrete plant and transported to the construction site. The high-strength prestressing cables are first pulled to produce tension. While the cables are held, the concrete beam is cast. Once the beam hardens, the ends of the cables are cut. The cables within the concrete attempt to contract, causing the concrete to go into compression. See **Figure 11-18**.
- ***Posttensioned concrete*** members have the cables contained in tubes so that they do not bond with the concrete. The tube, with the cable already inside it, is cast in the beam. A plate is attached to the cable at one end of the beam. At the other end, a jack is used to pull the cable. When the correct tension is reached, a plate is secured to the pulling end. The tensioned cable pulls on the end plates, compressing the concrete. See **Figure 11-19**. Posttensioned members can be either precast or cast-in-place.

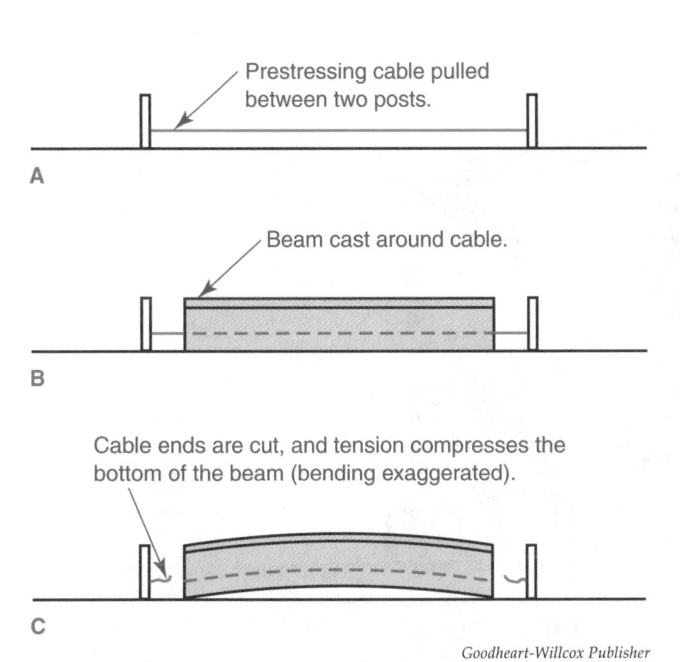

Goodheart-Willcox Publisher

Figure 11-18. Constructing a pretensioned beam. A—The cable is put under tension. B—Concrete is cast around the cable and allowed to harden. C—The ends of the cable are cut. The cable causes the beam to compress.

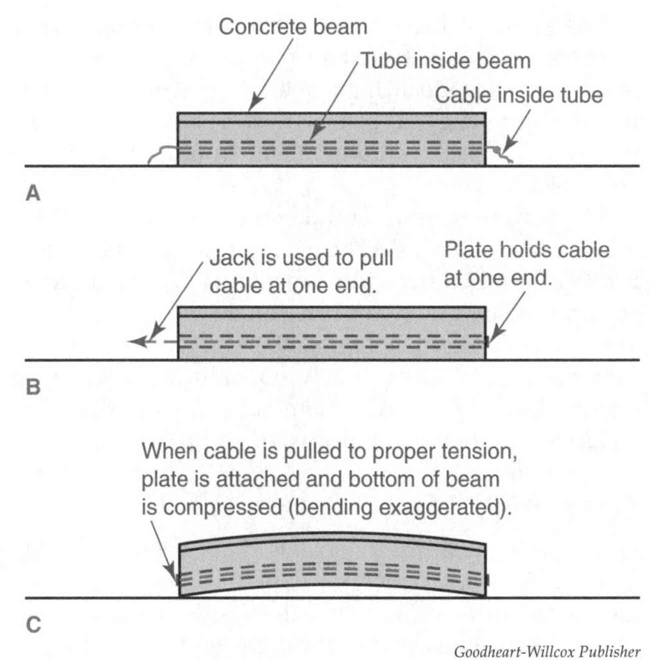

Goodheart-Willcox Publisher

Figure 11-19. Constructing a posttensioned beam. A—A tube containing the posttensioning cable is cast in the beam. B—One end of the cable is pulled to create tension. C—The cable is cut and held at the ends with plates.

Structural Steel

Structural steel construction refers to a system of building in which the framework consists of steel shapes. The most widely used shapes are the ***wide-flange beam***, designated by a (W), and the ***I-beam***, designated by a (S). See **Figure 11-20**. Other structural steel shapes include pipes, tubes, channels, angles, and tees.

Goodheart-Willcox Publisher

Figure 11-20. Structural steel columns and beams form the supporting frame of this office building. The beams are fastened to steel plates on the columns with bolts. Connection requirements are specified on the construction prints for the project.

The framework in heavy construction is designed so floor loads are transferred to columns of steel, so there is no need for loadbearing walls. Often, structural steel members are encased in concrete or some other fireproofing material to protect them from failing in the event of a fire.

Members used in structural steel construction include girders, beams, spandrel beams, columns, and trusses. See **Figure 11-21**. Individual members may be dimensioned on the drawing or listed in a schedule (if there are a sizable number of members). A typical beam dimension is W12×40, which indicates the following: a wide-flange beam (W), a nominal depth of 12″, and a weight of 40 pounds per linear foot.

Open-Web Steel Joists

Open-web steel joists are widely used in light commercial construction for floor and roof structures. Open-web joists use less steel than wide-flange beams and support loads more efficiently, so they cost less. The Steel Joist Institute has standardized the specifications and labeling of these joists.

Steel joists are grouped into three major series. K-series joists, used for typical light commercial construction, are available in depths of 10″ to 30″ and spans up to 60′. LH-series (longspan) joists have been standardized in depths from 18″ to 48″ and spans to 96′. DLH-series (deep longspan) joists have been standardized in depths from 52″ to 120″ and spans up to 240′.

Open-web joist designations consist of three parts. First, the *depth* of the joist (in inches) is given. Then the *type* of joist is given. Finally, a number corresponding to the size of top and bottom chords is listed. For example, a typical designation for a K-series joist would be 12K3, which means a K-series joist 12″ deep. The 3 refers to the relative size of chords. The usual range of allowable span for this size joist is 12′ to 24′, depending on the load. The heavier the load, the shorter the span. A 60DLH15 joist is a deep longspan high-strength joist with a depth of 60″. From a load table, the typical range of span for this size joist is 70′ to 120′ depending on the load required.

Steel joists are manufactured as both top-bearing and bottom-bearing. Top-bearing joists are supported at the ends of the top chord, which is the common method of support. Bottom-bearing joists are supported at the ends of the bottom chord. Longspan and deep longspan joists come with parallel chords or pitched chords for roof drainage. Different designs for steel joists are shown in **Figure 11-22**.

Metal Decking

In structural steel buildings, the floors are almost always concrete supported by metal decking, **Figure 11-23**. The metal decking spans the structural steel members and serves as permanent formwork. Metal form decking supports concrete floors until the concrete hardens and the slab is able to carry its own weight. Composite decking also serves as permanent formwork and adds tensile reinforcement to the concrete (the concrete bonds to deformations on the decking). Sometimes, shear studs are added to improve the bond between the two members.

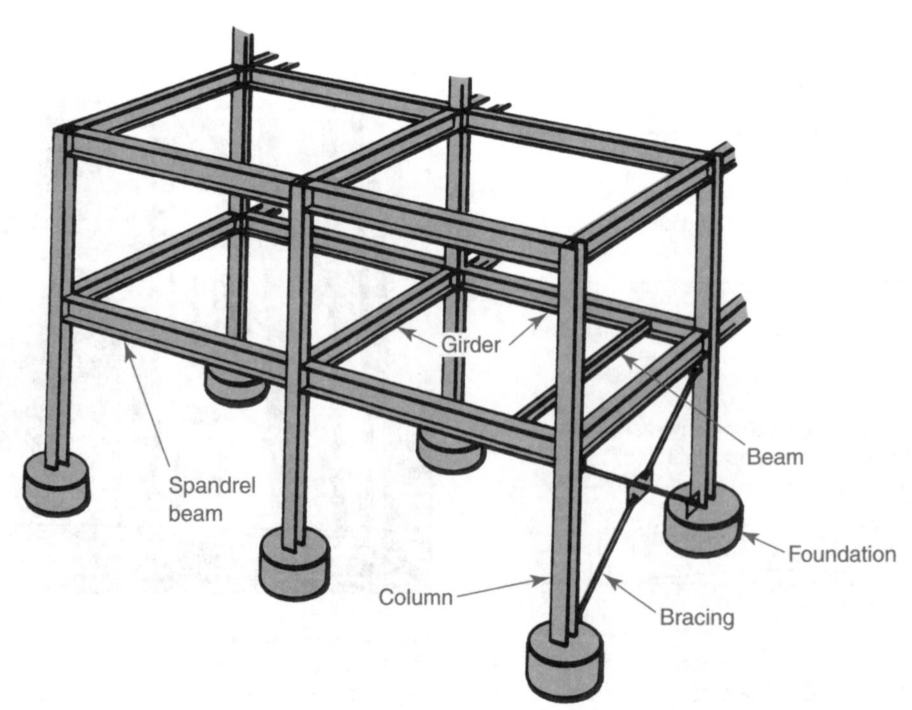

Goodheart-Willcox Publisher

Figure 11-21. Structural steel members.

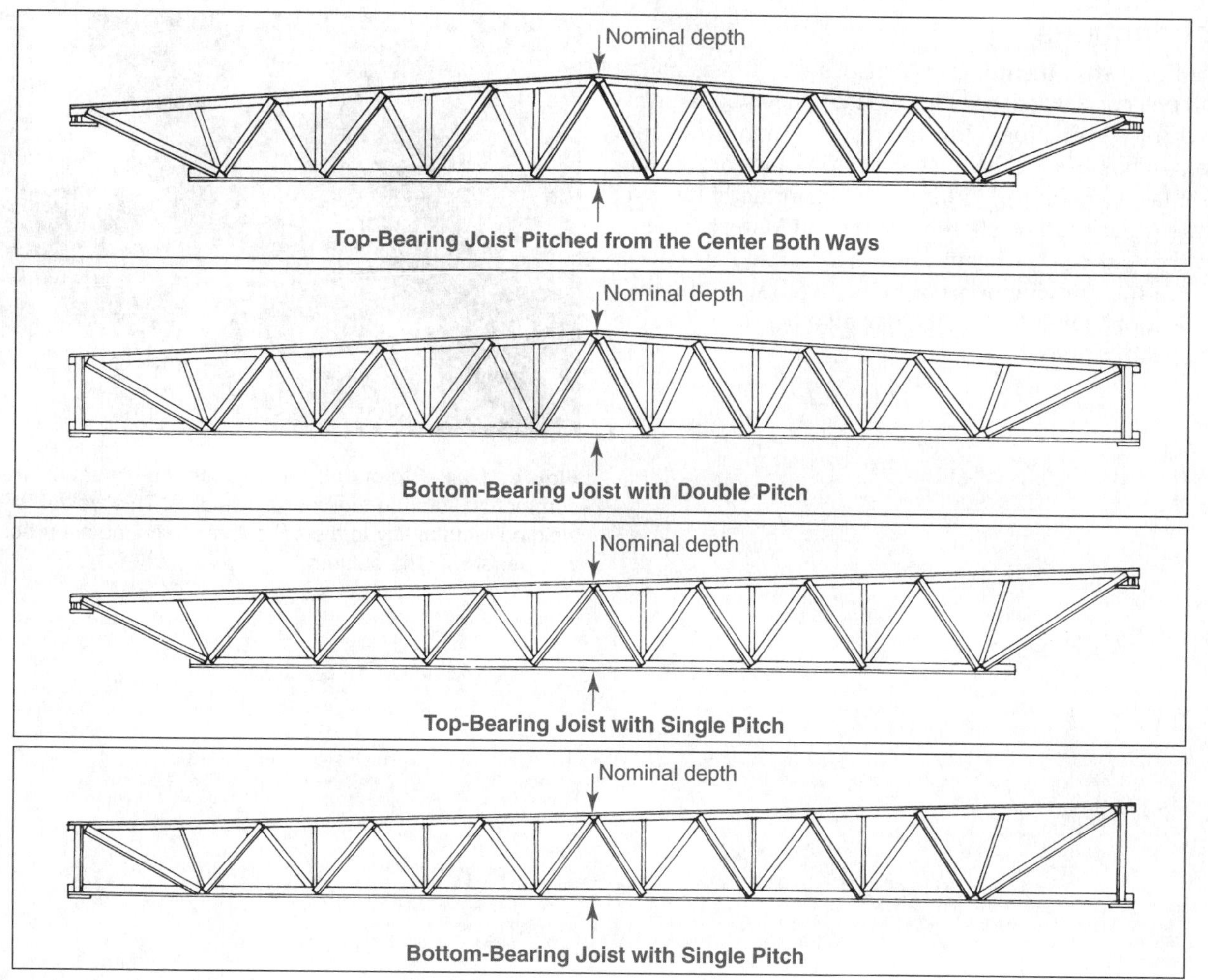

Armco Steel Corporation

Figure 11-22. Open-web steel joists.

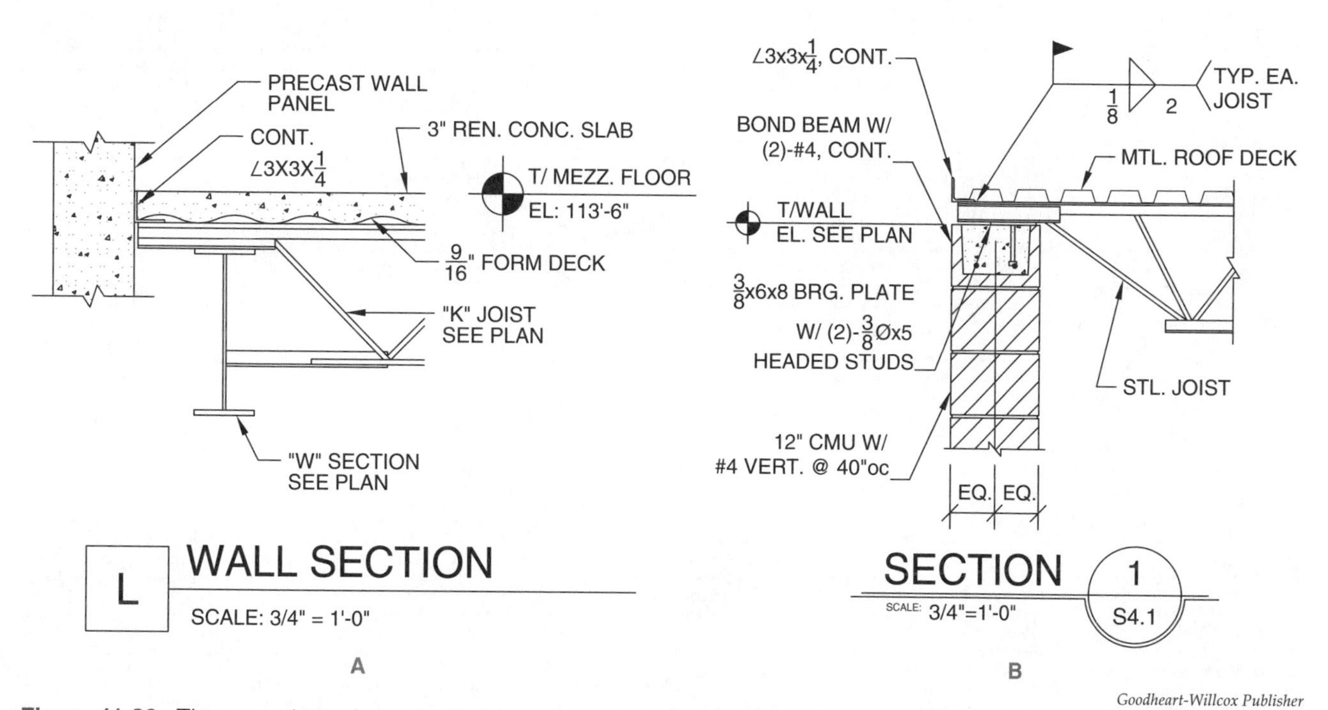

Goodheart-Willcox Publisher

Figure 11-23. These sections show steel joists and metal decking being used in two different situations. A—The joist is being used as a floor support and as support for the form deck that will receive the 3″ concrete slab. B—The joist supports the metal deck that serves as final support for the roofing system.

Connections

Structural steel members and shapes may be fastened with bolts or welded. See **Figure 11-24**. Rivets were once used for connections, but the development of high-strength bolts ended their use. Test procedures are set for these fasteners by ASTM International, formerly known as the American Society for Testing and Materials. The type of bolts or welding electrodes to be used are specified in the drawing notes or in the detail showing the connection. These materials may also be identified in the specifications.

Goodheart-Willcox Publisher

Figure 11-24. Structural steel members and shapes are commonly fastened together by welding. This welder is welding a metal key to a steel column to connect a precast wall panel with the column.

Test Your Knowledge

Name ______________________

Write your answers in the spaces provided.

_____ 1. In order to be an effective structural material, concrete must be combined with _____.

A. plastic rings

B. a wax coating

C. reinforcing steel

D. weathered lumber

E. All of these products are needed to make concrete useful.

_____ 2. A waffle slab is a type of _____.

A. steel floor system

B. metal decking

C. framing procedure

D. concrete floor system

E. None of the above.

_____ 3. The diameter of a #5 reinforcing bar is _____.

A. 5″

B. 5/8″

C. 5/16″

D. 5 mm

E. None of the above.

_____ 4. If the minimum lap for reinforcing bars is 40 diameters, how long would the splice be for a #4 bar?

A. 16″

B. 20″

C. 40″

D. 160″

E. None of the above.

_____ 5. Which of the following is *not* a horizontal support member?

A. Spandrel beam

B. Girder

C. Joist

D. Column

E. All of the above.

_____ 6. *True or False?* After slabs are cast, holes are cut where needed for pipes and electrical conduit.

_____ 7. *True or False?* The purpose of a one-way slab system is to only allow pedestrian traffic in one direction.

_____ 8. *True or False?* Reinforcing steel is added to concrete to improve tensile (pulling) strength.

_____ 9. *True or False?* Steel beams can be protected from fire by encasing them in concrete.

_____ 10. *True or False?* Precast concrete is cast in a precast concrete plant and transported to the job site for erection.

_____ 11. *True or False?* Metal decking on a steel building is commonly used as formwork for the concrete to be cast.

_____ 12. What does the 12 designate in "12K3"?

A. 12 pounds/lineal foot

B. 12″ deep joist

C. 12′ maximum span for this size joist

D. 12″ O.C. joist spacing

Activity 11-1

Structural Drawings for a Commercial Building Project

Name ______________________________

Refer to Sheets S1.1, S1.2, S2.1, S2.2, S3.1, S4.1, and S5.1 from the Delhi Flower and Garden Centers commercial building plans in the Large Prints supplement to answer the following questions.

1. What is the total dimension from grid line F to the typical columns set on the F5 footings?

____

2. What is the main roof beam specified in the hardlines area?

____

3. Give the column mark and size specification for the column supporting the beam referred to in Question 2.

____

4. What types of supporting members (beams, trusses, etc.) are supporting the entrance canopy?

____

____

5. What does the mark (symbol) L1 indicate? Give the size specification and describe how members are set in the masonry.

____

____

6. What is specified for the typical metal roof decking in the hardlines and display atrium sales areas?

____

7. On which sheets are the plan and detail drawings for the mezzanine framing?

____

8. What size slab is specified for the mezzanine and what are the reinforcement requirements?

____

____

9. What are the main horizontal framing members for the trellis work above the outdoor sales area?

____

10. Referring to Section 11 on Sheet S3.1, what is the vertical reinforcing specified? What type of column (post) is being supported?

____

____

11. Referring to Section 2 on Sheet S5.1, how is the metal deck supported?

____

____

12. What is used to connect the framing members for the trellis work above the outdoor sales area to the columns (posts)?

____

13. What detail is referenced on the east wall for the steel joist?

____

14. On the detail referenced in Question 13, how is the steel joist supported?

____

____

15. What size steel angle is required for the HVAC curb over the opening above the checkout area? What is the load limit on the support?

____

APA-The Engineered Wood Association

Engineered wood products, such as the glue-laminated beams used in the roof system of this building, have sustainable benefits in construction and help conserve wood. They are designed to be stronger than solid sawn members and can be manufactured from smaller trees that are not suited for sawn lumber.

UNIT 12 Residential Framing Prints

TECHNICAL TERMS

balloon framing
bird's mouth
blocking
bridging
collar beam
common rafters
cripple jacks
cripple studs
fire stop
header
hip jacks
hip rafters
joist
knee walls
ledger
loadbearing partitions
on center (O.C.)
post-and-beam framing
platform framing
purlin
rafter
ribbon
ridge board
rise
riser
rough sill
run
sill plate
slope
sole plate
span
stringer
top plate
tread
valley jacks
valley rafters
winders

LEARNING OBJECTIVES

After completing this unit, you will be able to:

- List the differences between heavy framing and light framing.
- Recognize the construction of various floor, wall, and roof framing systems.
- Read framing drawings.
- Explain the differences between platform, balloon, and post-and-beam framing.
- Understand stair details and terms.
- Recognize metal framing systems.

The framing (structural) systems commonly used for residential construction are similar to the large steel and concrete framing systems used in commercial construction. Beams and joists are used to support floors. Columns are used in situations where a wall would be inappropriate, such as open areas found in basements or open areas between large gathering areas.

However, there are several differences between heavy commercial frames and light residential frames:

- Wood members are more commonly used, rather than structural steel or concrete. See **Figure 12-1**. Wood frames are lightweight and easier to construct.
- Many more members are used in wood-framed residential structures, because the wood is relatively weak as compared to concrete or steel.
- Walls are also constructed as a frame, with wood studs spaced every 12″, 16″, or 24″ on center.
- The weight from a floor is supported by the entire length of the wall, rather than by columns.
- Residential framing plans may include additional information, such as the roofing material, sheathing, and finish details.

Light-gage steel is sometimes used in residential frames instead of wood. However, the framing systems do not change significantly when steel members are used. When using light-gage metal framing, the system should be designed by an engineer for structural integrity.

Residential foundation systems are similar to heavier structural systems used in commercial construction, except that residential foundations are much simpler in most situations. Usually, a residential foundation consists of concrete footings and loadbearing walls, and sometimes simple retaining walls. Residential foundation systems are discussed in Unit 10.

Dan Dorfmueller

Figure 12-1. Wood is the most widely used framing material in residential construction, due to its flexibility and ease of use.

Wood Framing

Wood is the most widely used residential construction material. Its availability and affordability lead to its popularity and the methods of wood framing are widely known. Wood has proved to be a durable, dependable material for houses. The following sections describe the common wood framing systems used in residential construction and the manner in which each of these systems is shown on prints.

Floor Framing

The basic component terminology related to floor framing is described as follows. Typical floor framing and wall framing members used in residential construction are shown in **Figure 12-2**.

- **Sill plate**. The ***sill plate*** is a board attached to the top surface of the foundation wall. Anchor bolts cast in the concrete are used for the connection. A 2×4 or 2×6 member is often used for the sill plate.
- **Header**. The ***header***, also called a *rim joist*, is nailed to the top of the sill plate at its exterior edge. The header is positioned with its longer cross-sectional dimension vertical. The header is the same size as the joists that attach to it.
- **Joists**. ***Joists*** are horizontal floor-supporting beam members. The ends of the joists rest on the sill plate and are nailed to the header. Joists commonly span from the sill plate to an interior wall or a beam. See **Figure 12-3**. Joists are normally spaced 12″–16″ apart. Common lumber sizes used for joists are 2×8, 2×10, and 2×12. The joists shown in **Figure 12-3** are solid wood joists. Engineered wood joists are also available and are used for spanning longer distances. Wood I-joists,

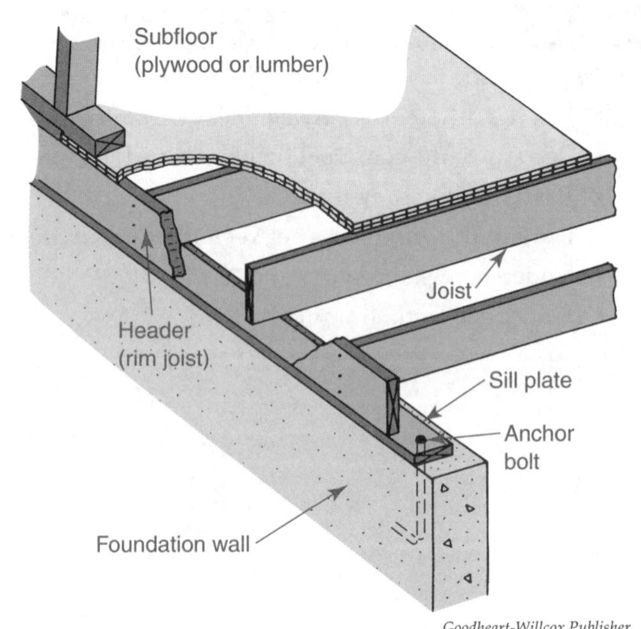

Goodheart-Willcox Publisher

Figure 12-2. Basic components of floor framing construction.

Dan Dorfmueller

Figure 12-3. Wood floor joists framed over an interior loadbearing partition and foundation walls.

Figure 12-4, are engineered wood members commonly used in floor construction. Wood I-joists have flanges made from solid sawn lumber or laminated veneer lumber (LVL) and a web made from oriented strand board (OSB) or plywood.

- **Subfloor**. The joists and header are covered with subflooring. A sheet material, such as plywood or chip board, is normally used. The subfloor is fastened to the joists using nails or screws. Often, construction adhesive is used to attach the subfloor to the joists before installing fasteners to strengthen the assembly. The finished floor, such as carpet, wood, or ceramic tile, will cover the subfloor.

Additional floor framing members are shown in **Figure 12-5**:

- **Double header**. When an opening that disrupts the framing pattern is needed, a double header is installed perpendicular to the joists. The same size member that is used for the joists is used.

APA-The Engineered Wood Association

Figure 12-4. Wood I-joists are lightweight, high-strength framing members used for long spans. Common depths range from 9 1/2″ to 16″ and flange widths vary from 1 1/2″ to 3 1/2″. Member lengths up to 66′ are available.

GREEN BUILDING

Engineered Wood Products

The development of modern engineered wood products has done much to support the movement toward green construction. According to the Engineered Wood Association, I-joists use 50% less wood than solid sawn 2× joists. LVL, frequently used for girders and rim boards, also conserves wood by providing the required strength characteristics while using less wood. Laminated strand lumber (LSL) further conserves resources because it can be manufactured from small-diameter, misshapen trees that would otherwise be unusable. Because engineered wood products can be purchased in any length, the waste at the construction site is greatly reduced.

- **Double trimmer**. A double trimmer consists of two joists nailed together next to an opening.
- **Tail joist**. This is a joist interrupted by an opening. Tail joists normally run between the double header and the sill plate.
- **Ledger**. A ***ledger*** is a small piece of lumber, such as a 2×2, nailed to the side of the double header, at its bottom edge. This piece serves as a ledge on which each tail joist rests. A notch must be cut into the joist to keep the top of the joist even.
- **Bridging**. ***Bridging*** consists of small members connected between the sides of adjacent joists to provide bracing. Bridging provides lateral stability for the joists and helps to transmit the load between the joists. Many types of bridging are used, including joist-sized members, crossed 2×4s, and crossed sheet metal bars.

The floor framing system is often shown on the floor framing plan. The sizes of the members are given and marks or arrowheads are used to indicate the direction of the span. A note is typically used to specify dimensions for joists:

2×12 JOISTS 16″ O.C.

This designation means that a 2×12 member is used for each joist and the joists are spaced 16″ from one another ***on center (O.C.)***. A measurement specified with the abbreviation *O.C.* or *o/c* refers to the distance between the centers of the adjoining building components.

Joists on a plan represent the joists above the level shown. For example, joists shown on a foundation plan would be located *above* the basement and below the first floor. See **Figure 12-6**.

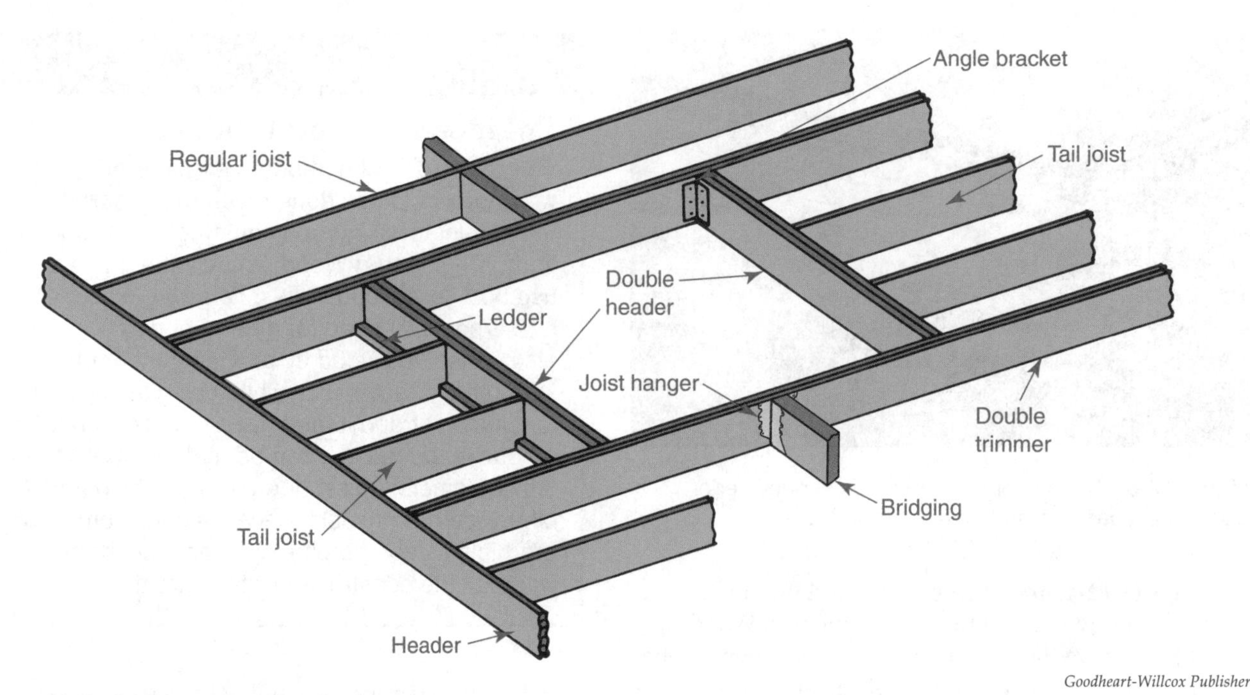

Goodheart-Willcox Publisher

Figure 12-5. Additional floor framing members.

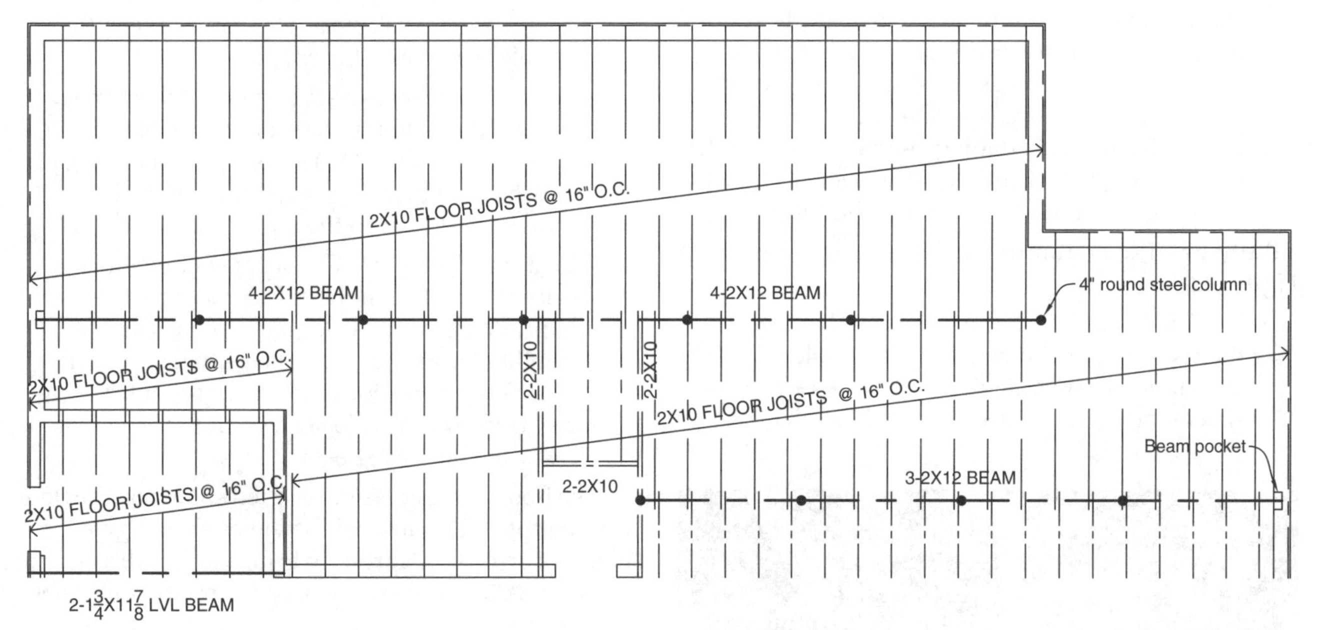

Goodheart-Willcox Publisher

Figure 12-6. A partial residential floor framing plan showing locations and dimensions for joists.

Dimensioning Floor Framing

Normally, dimensions for exterior walls are given to the outside of the stud wall for frame and brick veneer buildings. See **Figure 12-7**. A note may be added to the drawing to read:

NOTE: EXTERIOR DIMENSIONS ARE TO OUTSIDE EDGE OF STUDS; INTERIOR DIMENSIONS ARE ALSO TO EDGE OF STUDS WITHOUT DRYWALL OR FINISH MATERIAL. SPECIAL ATTENTION MUST BE GIVEN BETWEEN THE OUTSIDE EDGE OF THE CONCRETE AND THE EDGE OF THE WOOD FRAMING SO THAT ALL EXTERIOR FINISHES COORDINATE PROPERLY.

As noted in Unit 5, some architects follow a different practice. They start the dimensions for single-story frame buildings at the surface of the wall sheathing (which should align with the foundation wall), rather than the outside of the studs.

Drawings should be checked carefully to verify the dimensioning practice used. Usually, interior walls of frame construction are dimensioned to their edges, but

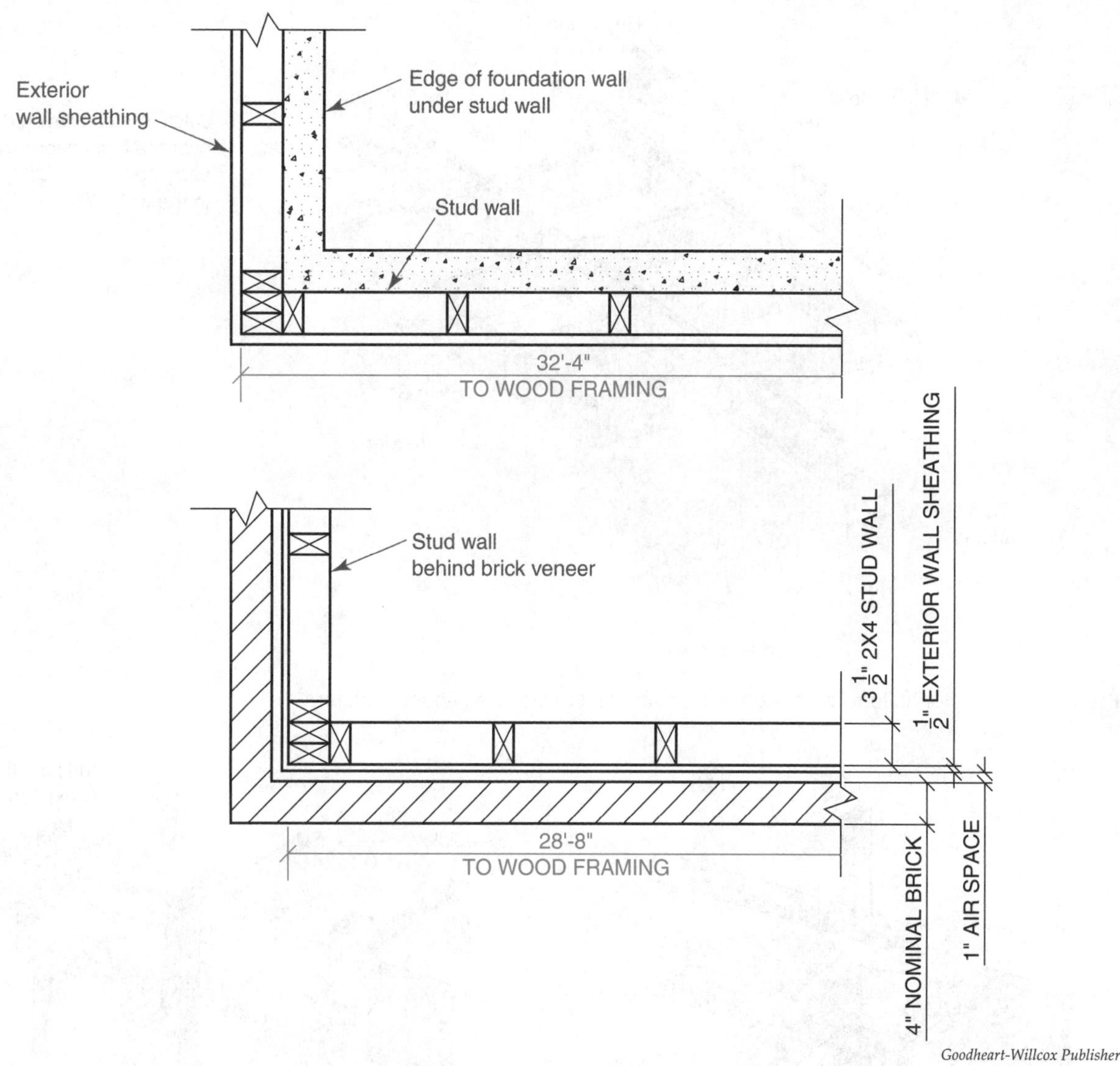

Goodheart-Willcox Publisher

Figure 12-7. The dimensions shown in color are to the face of the stud wall.

sometimes they are dimensioned to their centerlines. Masonry interior walls are dimensioned to their faces, with the wall thickness also dimensioned.

When studying the floor plans of buildings with two or more stories, note the adjoining stairs, chimneys, and loadbearing partition walls.

In certain building styles, the floor is cantilevered out over the foundation wall. In these cases, special framing details are shown. See **Figure 12-8**. Houses that have second stories smaller than the first are called one-and-a-half story houses. These houses usually involve ***knee walls*** (short walls joined by a sloping ceiling) and dormers.

Split-level houses have floor plans in which the levels are separated by a half-flight of stairs. Many variations are called for in framing of this type of structure, so the plans should be studied carefully.

Wall Framing

There are three basic types of light frame construction: platform, balloon, and post-and-beam. The construction worker should be familiar with the three types and be able to distinguish between them on drawings.

Platform framing, also known as *western framing*, is the most widely used type. It gets its name from its appearance. The first floor is built on top of the foundation, so it resembles a platform when the subflooring is complete.

The first-floor wall sections are raised and a second-floor platform is built on top of these walls. Then, the second-floor wall sections are raised and another platform for the second-story ceiling is constructed. See **Figure 12-9**. Each floor is a separate unit built on the structure below.

Balloon framing is not used to any large extent today. In this type of framing, the studs extend unbroken from the first floor sill plate to the top plate of the highest floor. Second-floor joists rest on a member called a ***ribbon***, which is set into the studs, **Figure 12-10**. Balloon framing has some advantages: it reduces lumber shrinkage problems in masonry veneer and stucco structures; it also simplifies running ducts and electrical conduit from floor to floor.

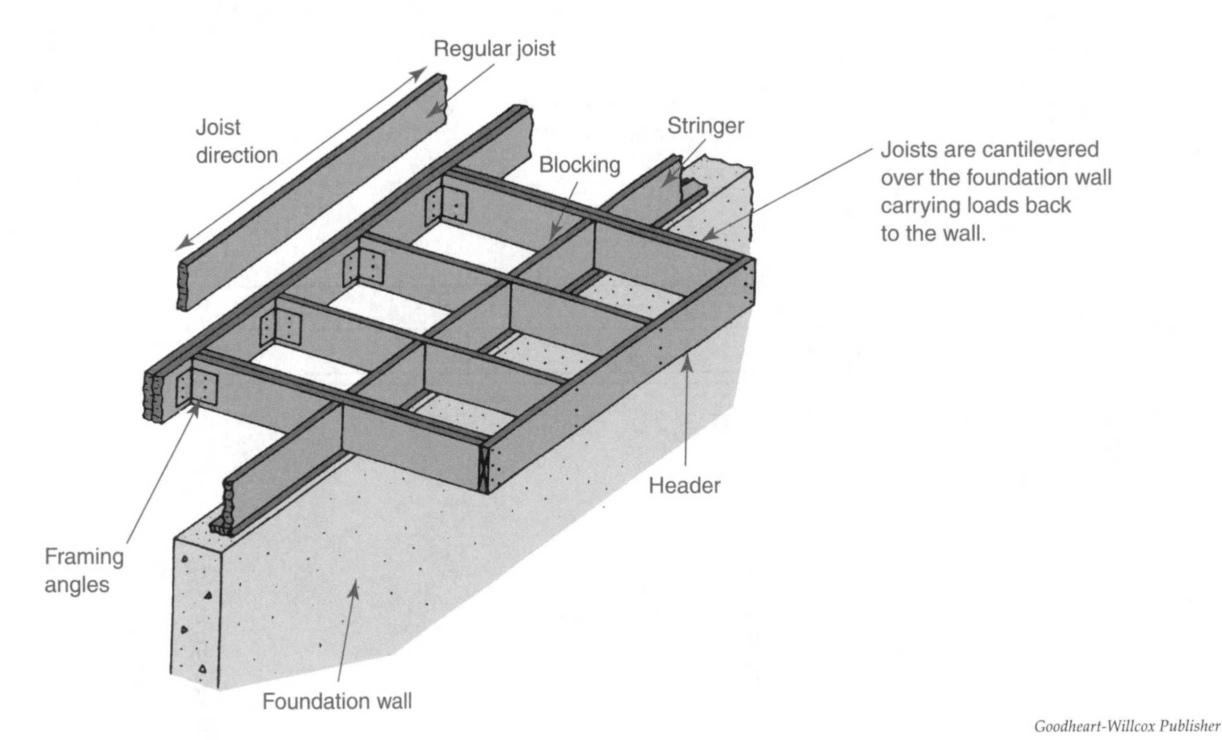

Goodheart-Willcox Publisher

Figure 12-8. Pictorial detail showing the framing of a cantilevered flooring system.

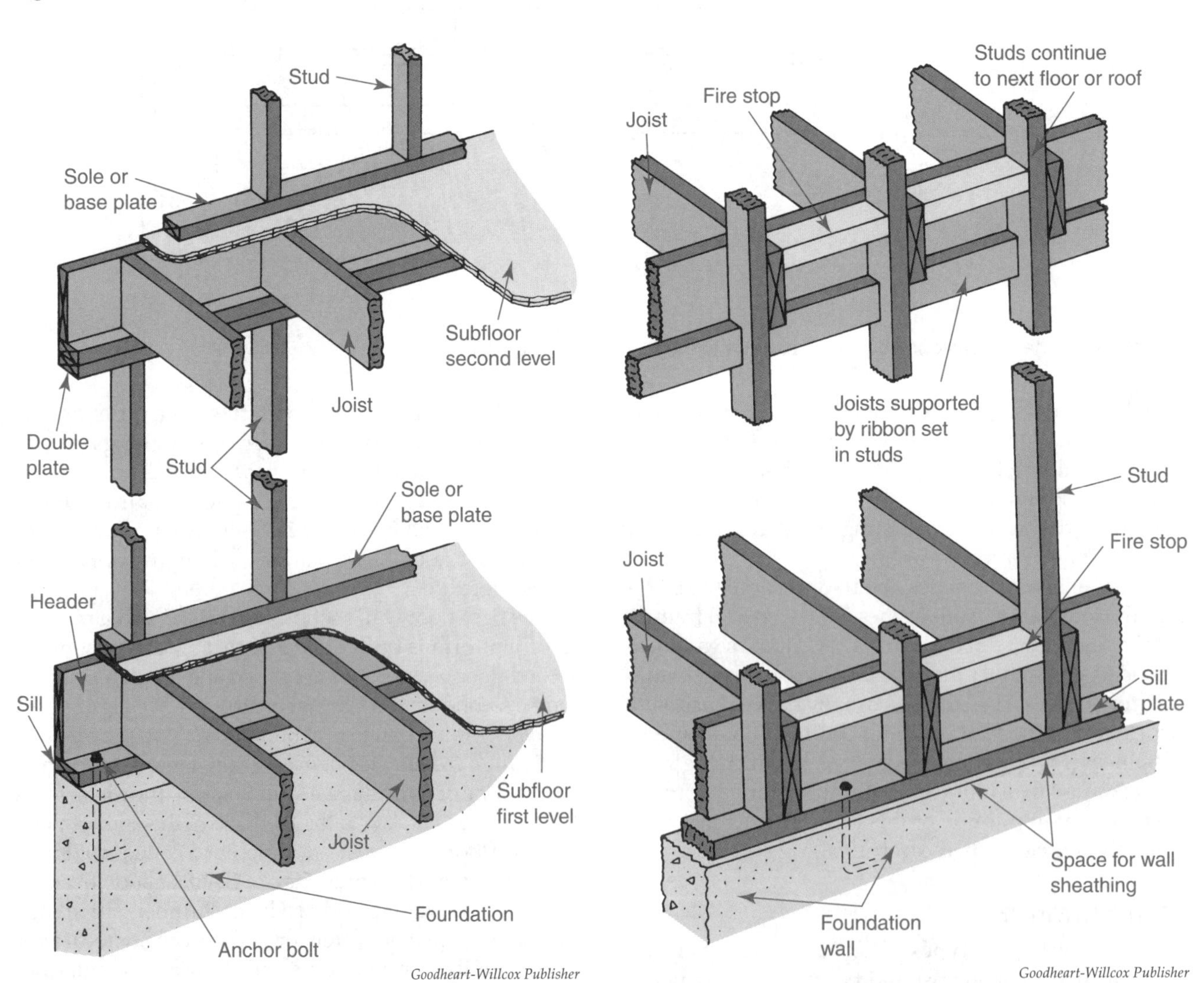

Goodheart-Willcox Publisher

Figure 12-9. An example and details of platform framing.

Goodheart-Willcox Publisher

Figure 12-10. An example and details of balloon framing.

The main disadvantage of balloon framing is the tendency of the walls to act as flues in spreading fires from floor to floor unless blocking is added between studs. A block member used in this manner is called a ***fire stop***. Balloon framing is also more difficult to manage in assembling a wall section.

Post-and-beam framing consists of heavy timber material for vertical posts in wall sections and horizontal beams supporting floor and roof sections. The floor and roof sections are typically made from plank material 2″ thick. See **Figure 12-11**. The structural members are placed at wider intervals than in other methods of framing. This type of framing lends itself to interesting architectural effects and extensive use of glass and exposed wood sections.

Various components of wall framing are illustrated in **Figure 12-12**:

- **Sole plate**. This serves as a base for the wall frame. The ***sole plate*** is the same size member as the studs (normally 2×4 or 2×6) and is nailed to the subfloor.
- **Studs**. Studs are the vertical members in the wall frame, running from the sole plate to the top plate. Studs are normally 2×4 or 2×6 members.
- **Header**. When some studs must be left out to make room for a window or door, a header is used to distribute the weight of the building around the opening. Headers can be constructed in many ways: one of the most common is to turn two 2×4 members sideways and insert a 1/2″ spacer (to make the assembly 3 1/2″ thick, the same width as a 2×4). The header is then nailed in place. Some headers are made up of larger 2×6 or 2×8 members so they can span larger openings. A header schedule may be provided to indicate the 2× member size required to span each opening.

Artazum/Shutterstock.com

Figure 12-11. A residence making use of the post-and-beam system of framing.

CAREERS IN CONSTRUCTION

Carpenter

As one of the oldest trades in construction, carpentry is still needed in almost every facet of the construction process. Carpenters are highly skilled workers who perform framing and finish work in residential and commercial structures.

Carpenters who construct the framing of a building are referred to as *rough framing carpenters*. Carpenters who perform finish carpentry work are called *finish carpenters*. The rough framing carpenter assembles the floor, wall, and roof framing. In addition, the rough framing carpenter builds the formwork used for concrete construction. The finish carpenter installs interior and exterior trim, builds or sets kitchen cabinets, sets windows and doors, installs finished hardware on doors, and so on. A finish carpenter must be much more precise in installation work than a rough framing carpenter.

Carpenters primarily work with wood components and light steel framing products. Carpenters must be able to work from prints and make accurate measurements when cutting and joining members. Carpenters must also be able to use a variety of hand and power tools safely and accurately.

Monkey Business Images/Shutterstock.com

Carpenters are highly skilled workers. They must be able to measure accurately and use a variety of hand and power tools.

Most carpenters learn the trade on the job or through training in a trade school or apprenticeship program. Carpentry is a rewarding career that offers many benefits. With experience, a carpenter can advance into a position as a foreman, superintendent, and even project manager.

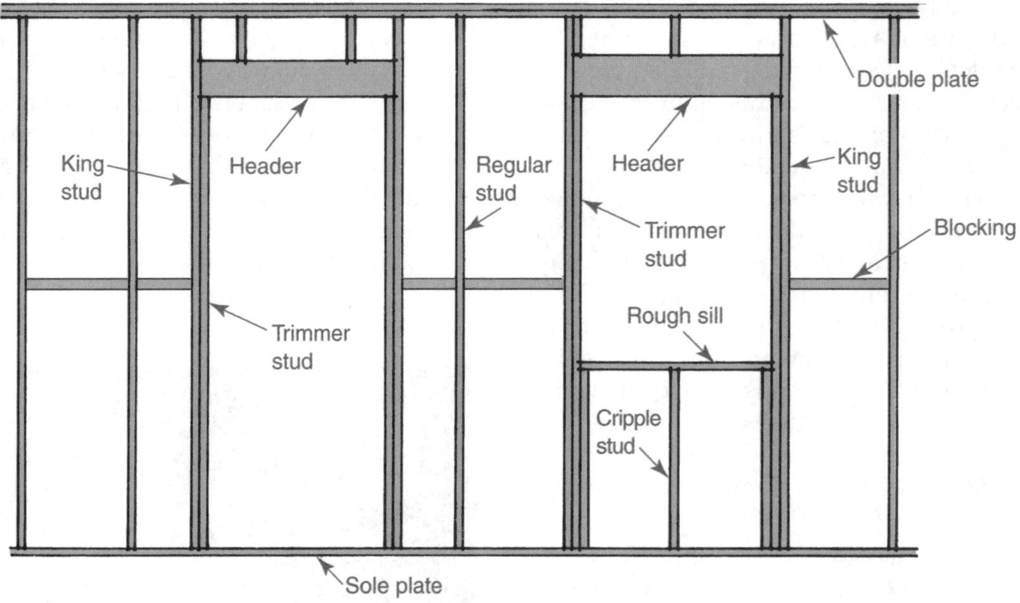

Goodheart-Willcox Publisher

Figure 12-12. Components of stud wall framing.

- **King stud**. A king stud is a stud located on either side of a header. It is a full-height stud that runs from the sole plate to the top plate.
- **Trimmer stud**. Next to the king studs and below the header, trimmer studs are placed. A trimmer stud extends from the sole plate to the bottom of the header. It is attached to both the king stud and to the header.
- **Rough sill**. A ***rough sill*** is positioned to support a window.
- **Cripple stud**. Short studs, called ***cripple studs***, extend between the top plate and header or the sole plate and rough sill. They are similar to trimmer studs, but are not paired with an adjacent stud.
- **Blocking**. ***Blocking*** is used to provide structural support and to prevent the spread of fire from floor-to-floor through stud spaces.
- **Top plate**. The ***top plate*** (shown in **Figure 12-12** as a *double plate*) rests above the studs. The next level of joists or rafters is supported by the top plate.

GREEN BUILDING

Insulated Headers

Headers are often sized to fill the space above the rough opening to the wall plate. Without any added insulation, this creates a tremendous thermal bridge. To cut heat loss and build greener headers, size headers properly and add a layer of rigid foam insulation between a built-up header's layers or to the exterior face.

Interior walls that carry the ceiling or floor load from above are called ***loadbearing partitions***. Usually, they are located over a beam or bearing wall, **Figure 12-13**.

Schedules

Door and window schedules give the number and size of all doors and windows in the building. See **Figure 12-14**. Units listed in the schedule are referenced to the plan view with a letter or number. Some architects provide the rough opening size in the schedules to speed

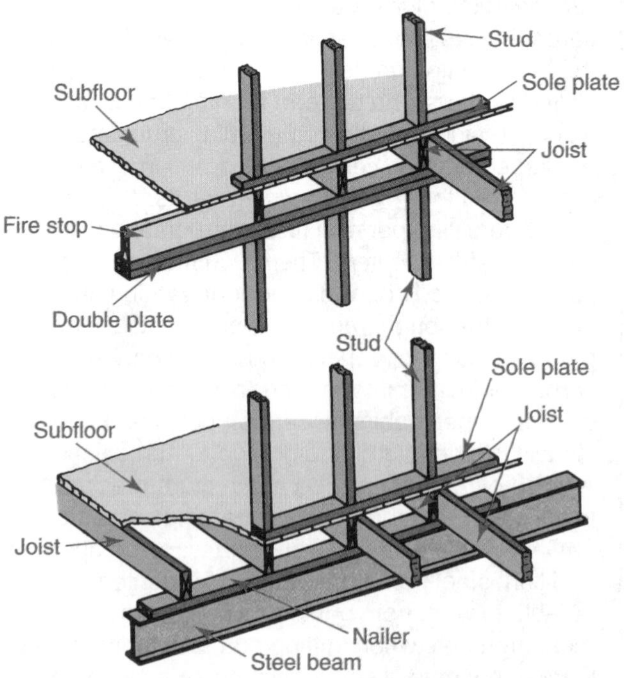

Goodheart-Willcox Publisher

Figure 12-13. Loadbearing partitions are used to transfer loads from floors above to the bearing structure below.

Door Schedule					
Mark	**Type**	**Size**	**Material**	**Frame**	**Remarks**
A	1	3'-0" × 7'-0" × 1¾"	Hollow Metal	Hollow Metal	Closer and Threshold
B	1	3'-0" × 7'-0" × 1¾"	Hollow Metal	Hollow Metal	Closer
C	2	2'-8" × 7'-0" × 1¾"	Hollow Metal	Hollow Metal	Closer and Kick Plate
D	2	3'-0" × 7'-0" × 1¾"	Hollow Metal	Hollow Metal	Closer
E	2	3'-0" × 7'-0" × 1¾"	Hollow Metal	Hollow Metal	Closer

Smith & Neubek & Associates

Figure 12-14. A typical door schedule.

construction and ensure a correct fit. If rough opening sizes for doors and windows are not provided, the construction worker must calculate them.

Sectional Views

Sectional views of walls are drawn to a larger scale and included on the drawings to clarify construction details. The section location is identified on the plan view with a section cutting line and reference symbol.

Full sections are cut through the width or length of a building, **Figure 12-15**. They are prepared for buildings with more complex frames, such as split-level houses or those with unusual interiors. These sectional views show features such as floors, walls, and ceilings as sections. Features beyond the cutting plane are shown as they appear in the interior of a building.

Roof Framing

Construction workers should be familiar with the different types of roof styles and how they are framed. Sketches of typical roof styles found in residential construction are shown in **Figure 12-16**. The style of the roof is most easily identified in elevation drawings.

Some architects do not supply a roof framing drawing for the more common roofs, such as gables or flat roofs, but rely on the elevation and detail drawings to guide the construction workers and the truss fabricator. When the roof is more complicated, or when the architect desires to specify the manner of construction, a roof framing plan is prepared, **Figure 12-17**. Roof framing can become very complicated and requires construction by skilled carpenters.

Figure 12-18 illustrates some common terms used in roof framing:

- **Rafter**. A ***rafter*** is one of a series of angled beam members that support the roof. A rafter is normally a 2×6, 2×8, or 2×10 member.
- **Ridge board**. A ***ridge board*** is the horizontal member at the peak of the roof. The upper end of each rafter is connected to the ridge board.
- **Collar beam**. This horizontal member ties the rafters together. A ***collar beam*** makes the roof frame more stable.
- **Rise**. The ***rise*** is the vertical distance between the top plate and the ridge board.
- **Run**. The ***run*** is the horizontal distance from the wall supporting the bottom of the rafter to the ridge board.
- **Slope**. The ***slope*** of the roof is the relationship between the rise and run. The slope describes the vertical rise in inches per 12" of horizontal run. For example, a roof slope of 4:12 indicates that the increase in vertical height (rise) is 4" for every 12" of horizontal distance (run). On section and elevation drawings showing the roof of the structure, the roof slope is indicated with a slope symbol. See **Figure 12-19**.
- **Span**. The distance between the walls supporting the rafters is the ***span***. It is twice the length of the run.
- **Bird's mouth**. For the rafter to fit flush on the top plate, this cut must be made. The ***bird's mouth*** consists of two cuts: the *seat cut* and the *plumb cut*.

Figure 12-20 shows various types of rafters:

- **Common rafters**. ***Common rafters*** run at right angles from the wall plate to the ridge.
- **Hip rafters**. Rafters that extend from an outside corner of the building to the ridge board, usually at a 45° angle, are ***hip rafters***.
- **Valley rafters**. ***Valley rafters*** extend from an inside corner of a building to the ridge board, usually at a 45° angle.

GREEN BUILDING

Green Building Details

In high-wind or seismic zones, some green building techniques do not provide enough stability. Know regional and local building codes to account for environmental changes. Submit house plans with green building technique details noted to a building official for review and approval.

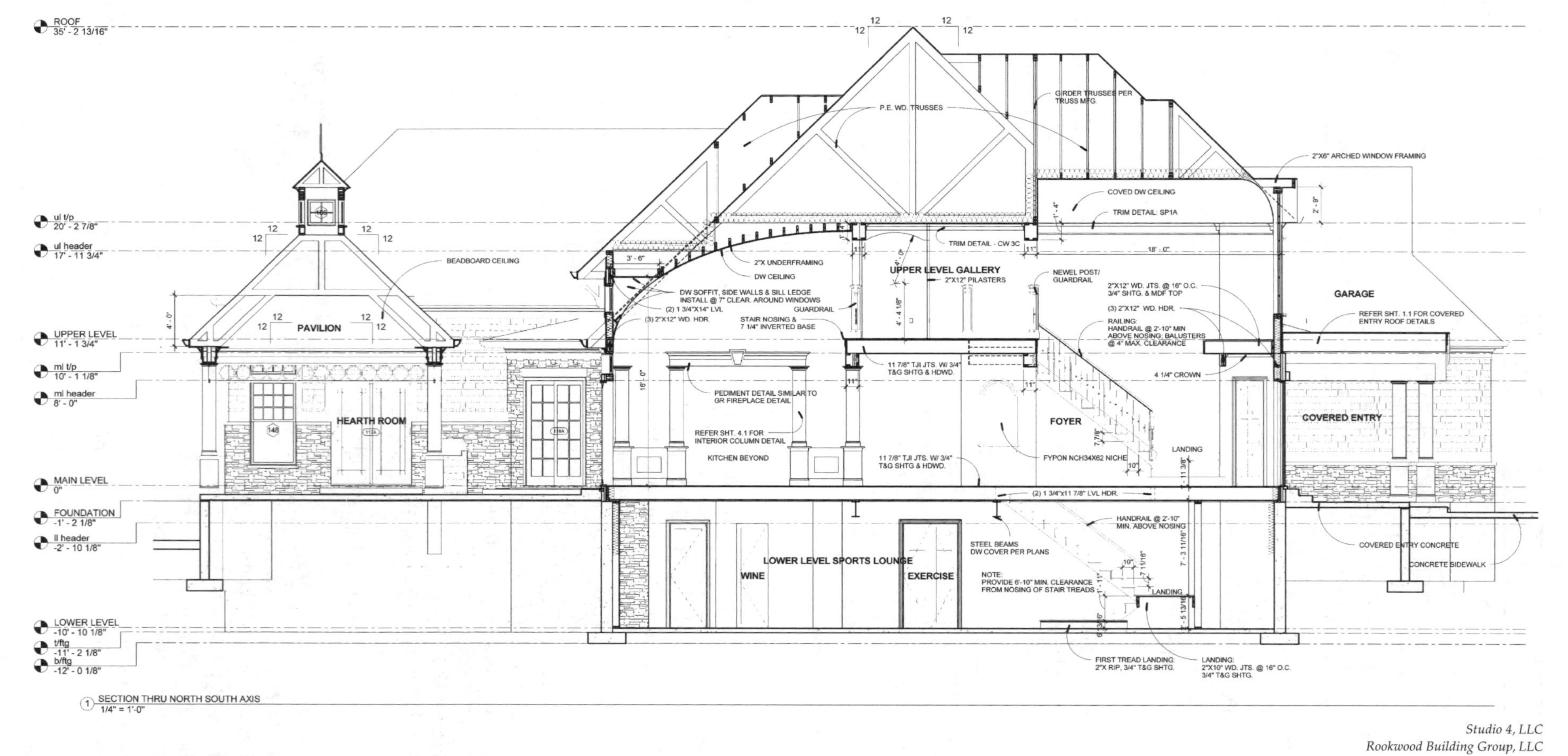

Studio 4, LLC
Rookwood Building Group, LLC

Figure 12-15. A full section shows the entire building.

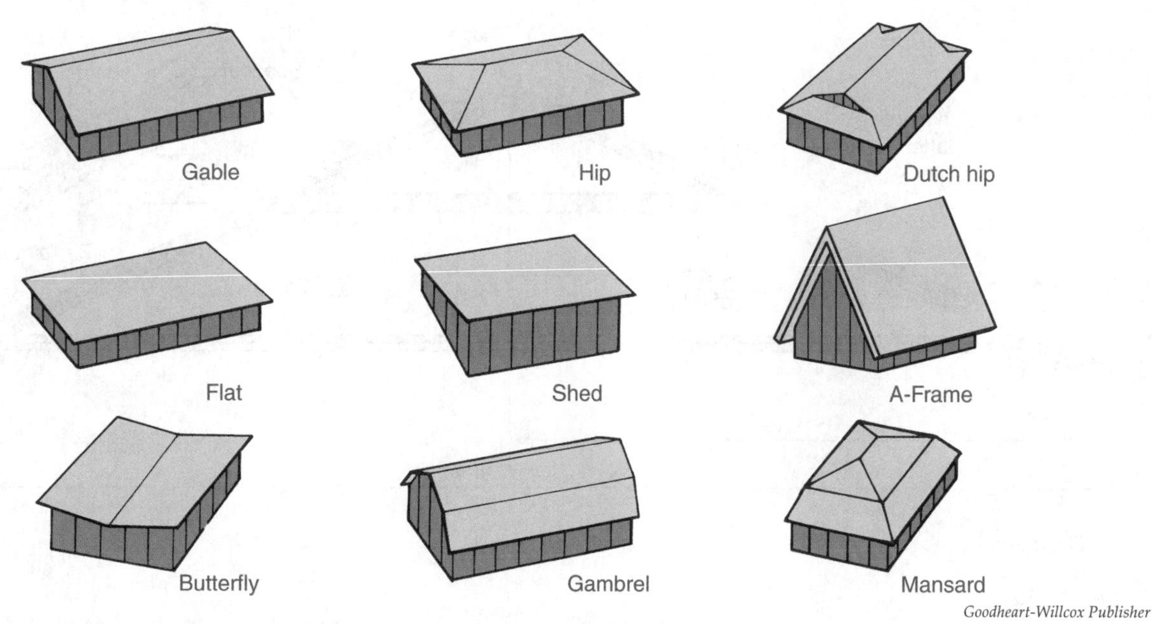

Goodheart-Willcox Publisher

Figure 12-16. Common roof styles and shapes used in residential construction.

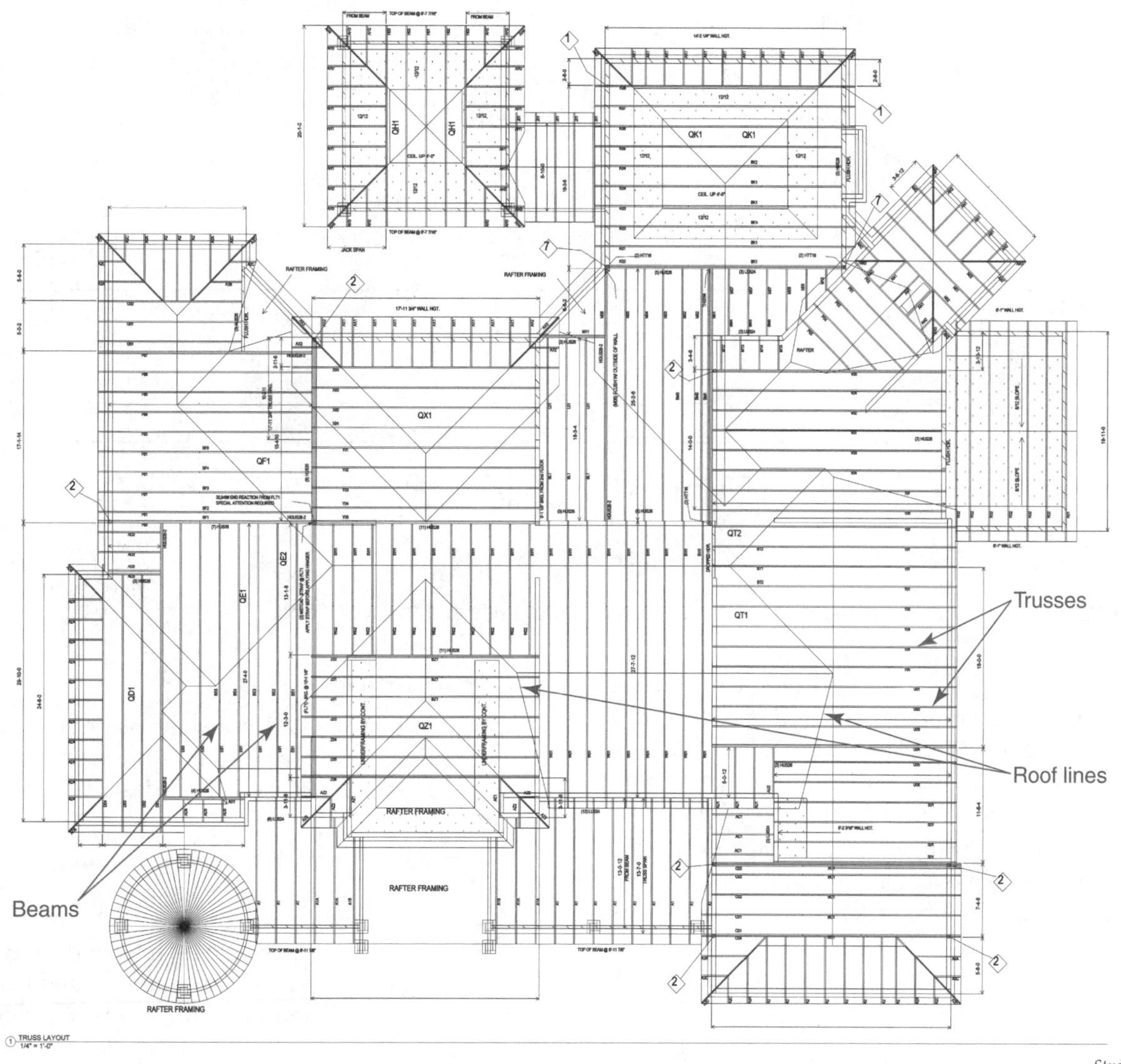

Studio 4, LLC
Rookwood Building Group, LLC

Figure 12-17. A typical roof framing plan for a residence.

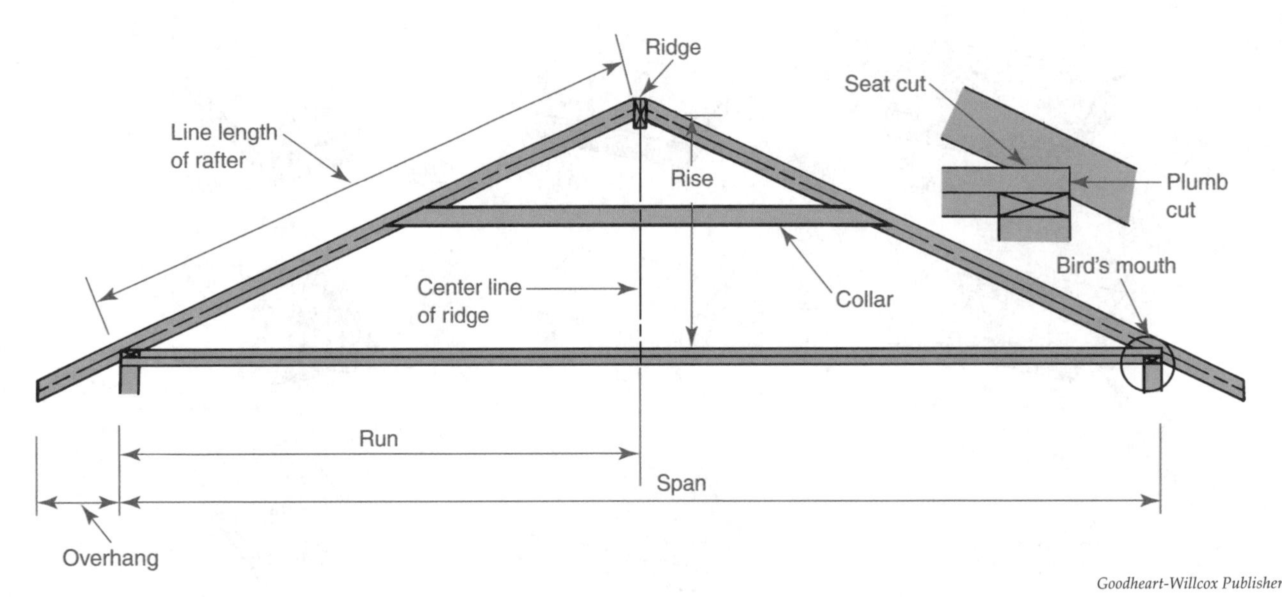

Figure 12-18. Common roof and truss framing terms.

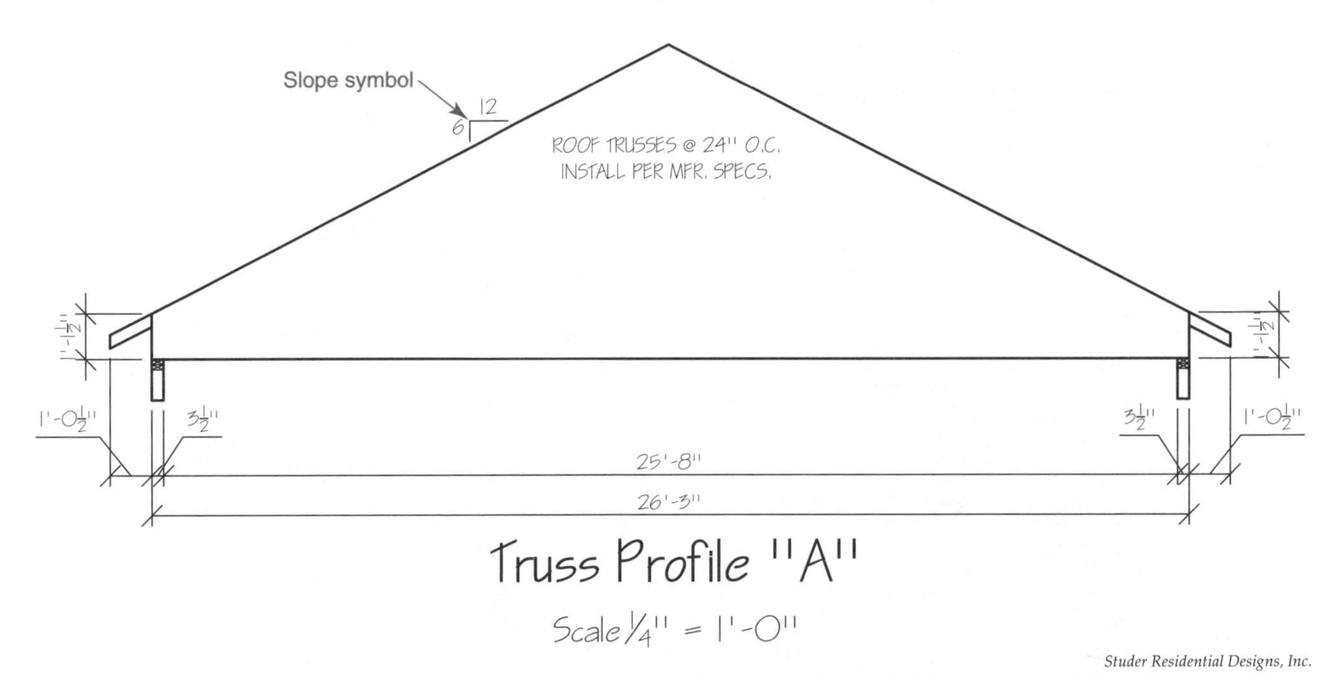

Figure 12-19. A detail drawing showing the dimensions of a roof truss and the roof slope.

- **Jack rafters**. Jack rafters have shorter spans and extend to either hip or valley rafters. ***Hip jacks*** extend from the top plate to a hip rafter. ***Valley jacks*** extend from a valley rafter to the ridge. ***Cripple jacks*** run between valley and hip rafters.

Another roof framing member is the ***purlin***, a horizontal member laid over truss rafters or under a series of rafters to support long rafters.

Stair Framing

To correctly install a staircase, some planning is needed. An elevation view of the stair is normally provided. There are several terms used to describe a stair. See **Figure 12-21**.

- **Stringer**. The angled member running between the lower and upper floors that supports the stairs is the ***stringer***. A 2×8, 2×10, or 2×12 can be used for a stringer, depending on how wide and long the staircase is. The stringers are cut and installed while the wall and floor framing is completed.

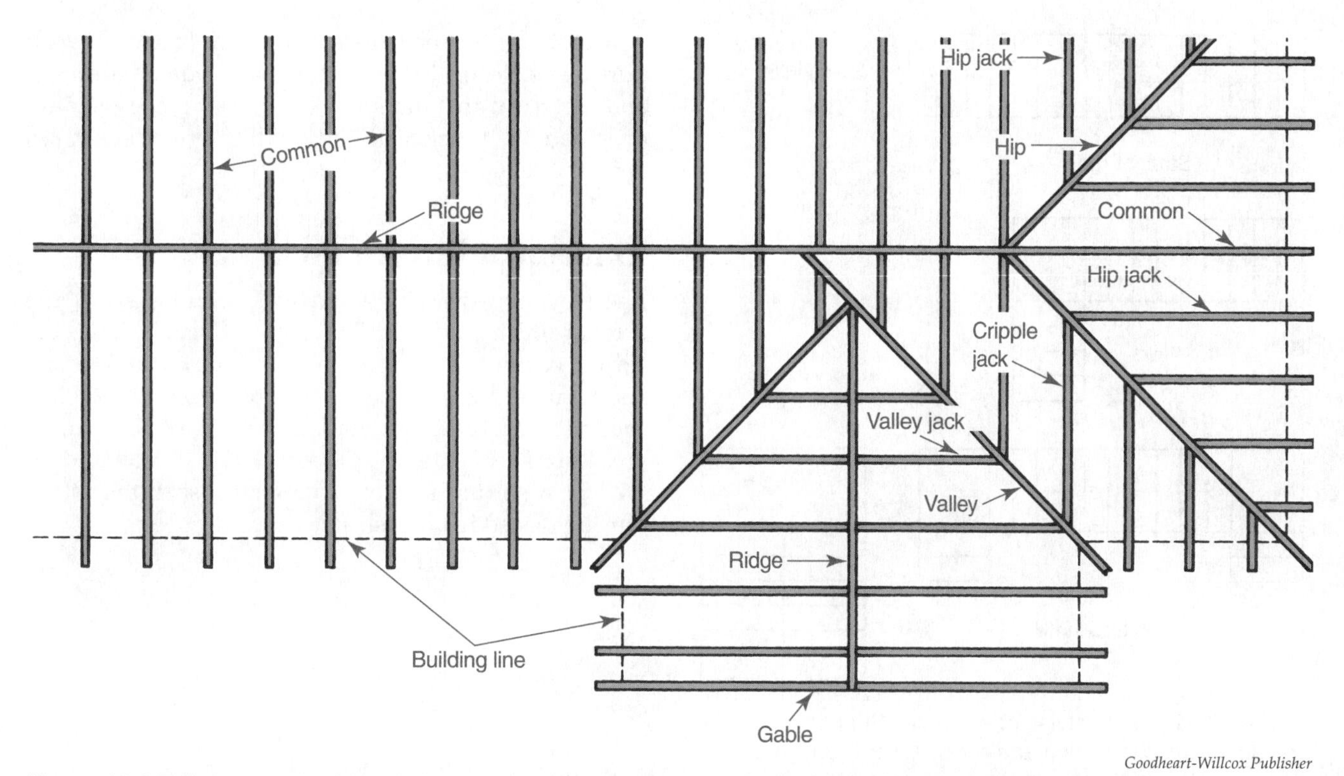

Goodheart-Willcox Publisher

Figure 12-20. Terms used to identify various kinds of rafters.

- **Tread**. The ***tread*** is the horizontal member that forms the "step." The distance between either the front or the back of adjacent stairs is the tread width. The number of treads in the stairway and the width of each tread are given on the drawing.
- **Riser**. The ***riser*** is the vertical member that provides the change in elevation between two adjacent stairs. The number of risers and the height of each are given on the drawing. There is always one more riser than the number of treads.

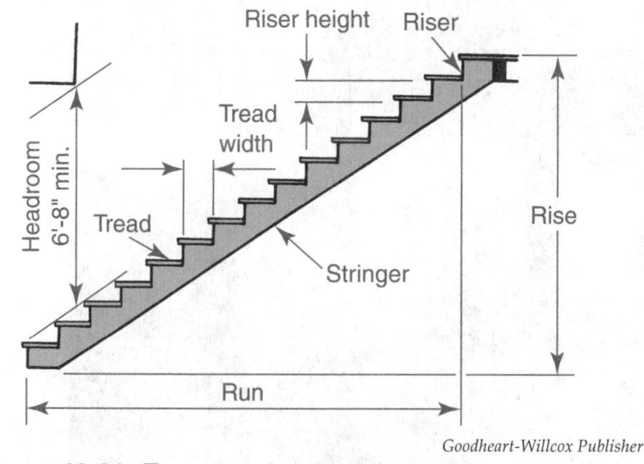

Goodheart-Willcox Publisher

Figure 12-21. Terms used to describe a stair.

- **Run**. The total horizontal length of the stairway is the ***run***. The run should be the same as the number of treads multiplied by the tread width.
- **Rise**. If the riser height is multiplied by the number of risers, the vertical distance between floors is found. This distance is the ***rise***.

There are many stair arrangements, three of which are shown in **Figure 12-22**. A modification of the long "L" style features winders (pie-shaped treads) instead of a landing. Using ***winders*** conserves space. Also, there are circular winding stairs (called *spiral stairs*) that gradually change direction as they ascend or descend.

Rough treads are installed for use during construction. These are replaced when finish work is performed.

Metal Framing

Builders are using metal framing more often in both residential and commercial construction. Components are available for metal framing to meet all construction requirements, **Figure 12-23**. In addition to joists, wall studs, tracks, and blocking, many accessories are manufactured to provide for any need in metal framing, **Figure 12-24**. Buildings constructed with metal stud framing are designed by structural engineers.

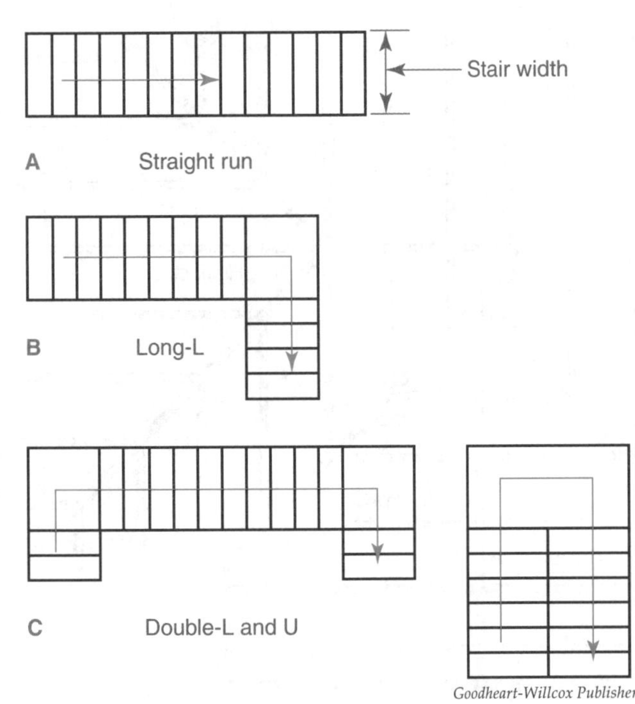

Goodheart-Willcox Publisher

Figure 12-22. Three styles of stairs. A—Straight run stairs are the simplest to design and construct. B—Long-L stairs are used when a change of direction is needed or if the rise is large enough to require a landing. C—Either the double-L or U is used when the stair must turn 180°.

Metal studs and joists are manufactured with punched holes in the web to accommodate plumbing and electrical installations. The holes in the web are provided 12″ from each end, with intermediate holes at intervals of 24″.

Exterior/Interior Finish

The elevation drawings and wall sections show the exterior finish materials. The exterior finish includes the siding, cornice, roofing, exterior windows, and doors. See **Figure 12-25**. Also, dimensions of ceiling heights, roof pitch, and cornice details are given. Window units are designated with an "O" for a fixed section and an "X" for a sliding section. Interior finish information may be placed in a finish schedule.

Goodheart-Willcox Publisher

Figure 12-23. A variety of metal framing components is used in this commercial building project. Although residential use continues to grow, most metal framing today is used in commercial buildings.

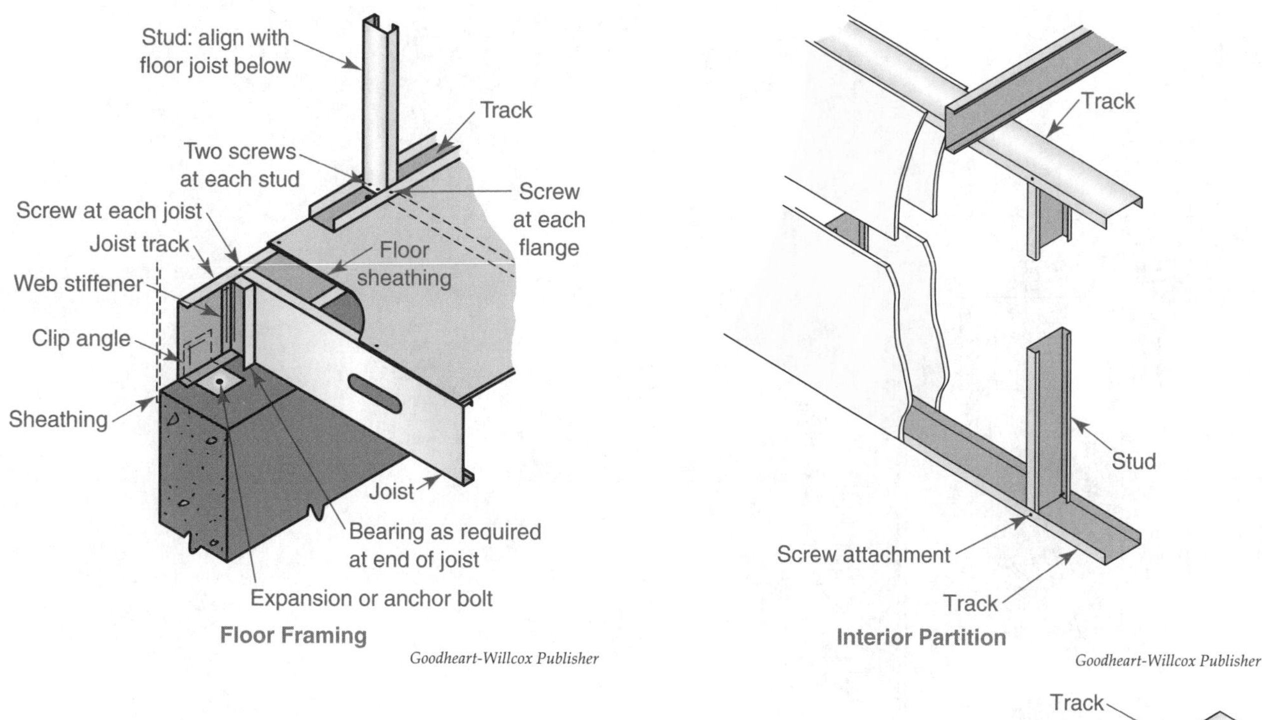

Floor Framing

Goodheart-Willcox Publisher

Interior Partition

Goodheart-Willcox Publisher

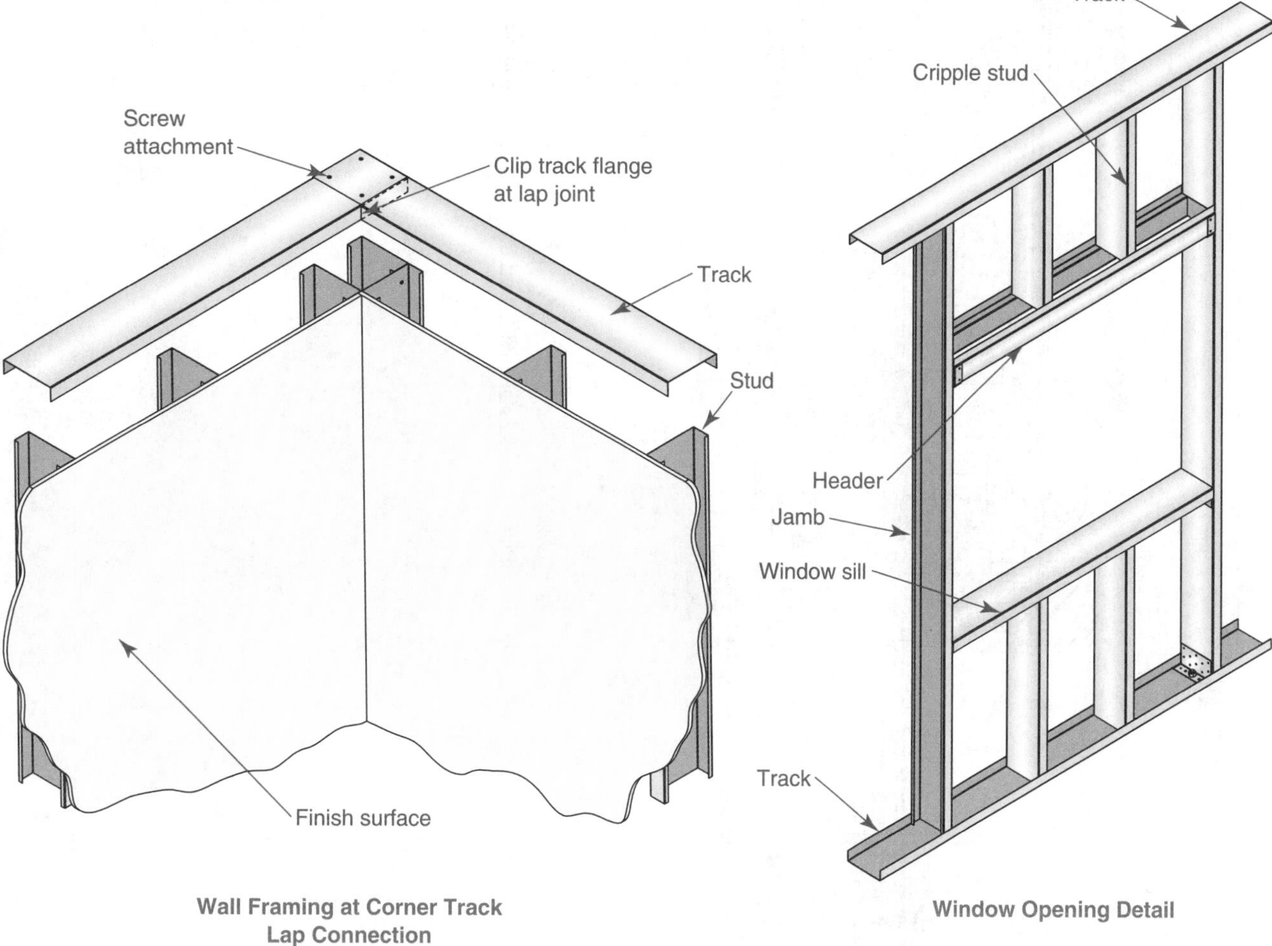

Wall Framing at Corner Track Lap Connection

SCAFCO Corporation

Window Opening Detail

SCAFCO Corporation

Figure 12-24. Details for residential construction using metal framing.

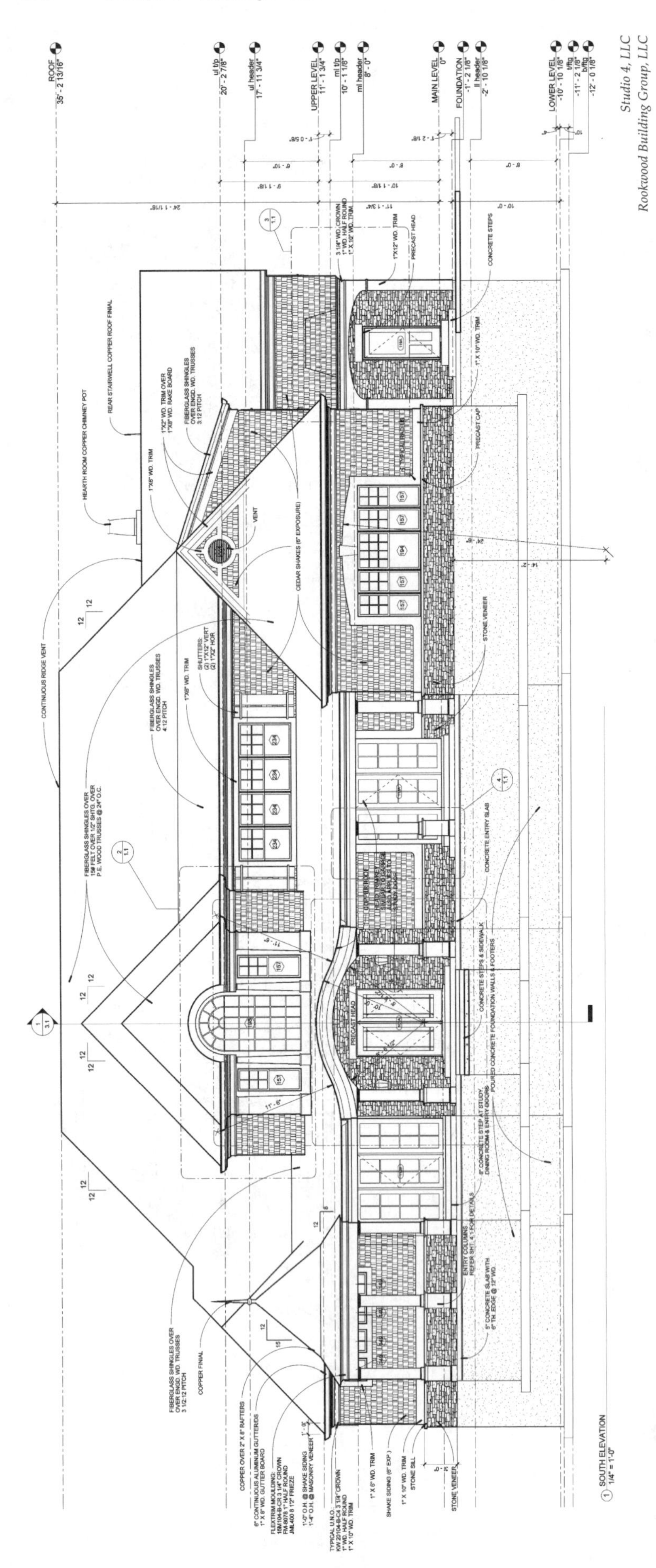

Studio 4, LLC
Rookwood Building Group, LLC

Figure 12-25. The exterior siding and other exterior finish materials are specified on elevation drawings.

Test Your Knowledge

Name ________________________________

Write your answers in the spaces provided.

_____ 1. Which of the following is *not* a reason for the use of wood in residential construction?
A. It is readily available.
B. It protects well against fire.
C. It is relatively affordable.
D. Dependable construction procedures have been established.
E. All of these are reasons for the popularity of wood.

_____ 2. Which member rests directly on top of the foundation wall?
A. Header
B. Joist
C. Ledger
D. Sill plate
E. None of the above.

_____ 3. Which member is located above a window?
A. Rough sill
B. Trimmer stud
C. Header
D. Ledger
E. None of the above.

_____ 4. Which member connects rafters to make the roof frame more stable?
A. Collar beam
B. Joist
C. Hip rafter
D. Valley rafter
E. None of the above.

_____ 5. A *winder* is a _____.
A. member to which joists are attached
B. flexible pipe
C. stair handrail
D. stair tread
E. None of the above.

_____ 6. *True or False?* A door schedule specifies the order in which doors are to be installed.

_____ 7. *True or False?* A loadbearing partition is an interior wall that supports a load from above.

_____ 8. *True or False?* Sectional views are normally drawn at a larger scale than plan views.

_____ 9. *True or False?* Exterior wall dimensions are normally shown to the outside of a stud wall.

_____ 10. *True or False?* In platform framing, wall studs run continuously from the sill plate to the roof.

11. Name the following parts of stair construction.

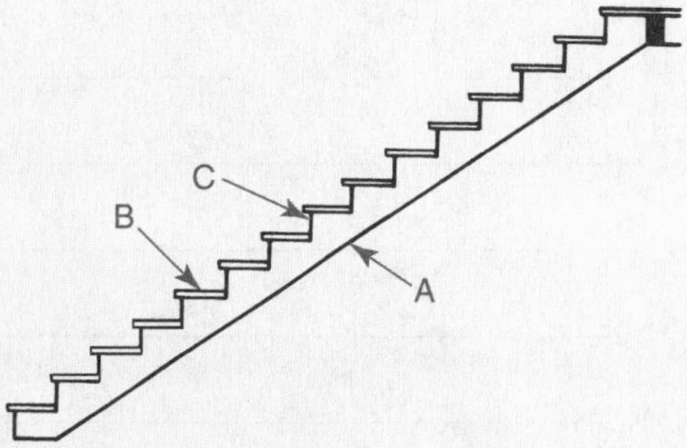

A. ________________________________

B. ________________________________

C. ________________________________

12. Name the following parts of the roof framing plan.

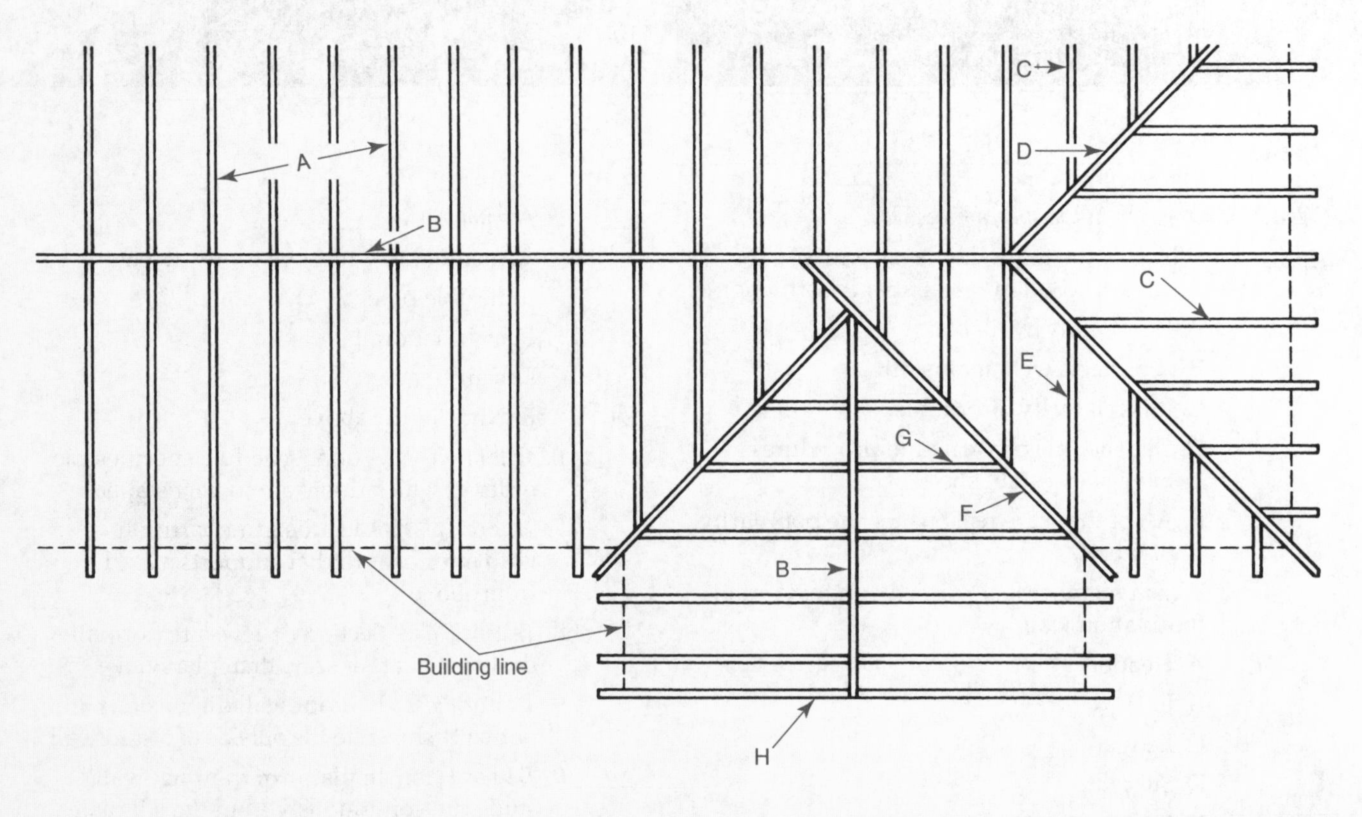

A. ______________________________

B. ______________________________

C. ______________________________

D. ______________________________

E. ______________________________

F. ______________________________

G. ______________________________

H. ______________________________

13. Name the following parts of the given framing system.

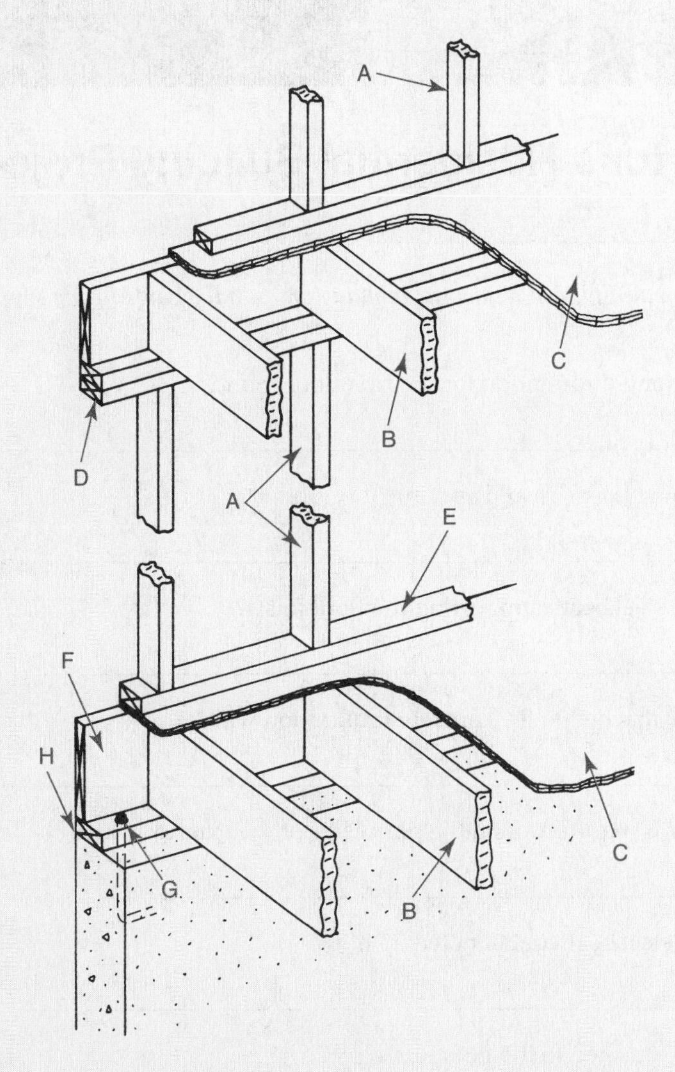

A. ______

B. ______

C. ______

D. ______

E. ______

F. ______

G. ______

H. ______

14. Referring to Question 13, what type of framing does the illustration represent?

______

Activity 12-1

Framing Plans for a Residential Building Project

Name ________________________________

Refer to Sheets 2, 4, and 5 from the Sullivan residential building plans in the Large Prints supplement to answer the following questions.

1. Where is the floor framing information for the first floor found?

____

2. What is the floor framing joist size and spacing?

____

3. What is the size of the steel beam supporting the floor joists?

____

4. How is the steel beam attached to the concrete foundation wall?

____

5. What kind of blocking is required along the perimeter of the exterior wall?

____

6. On which sheet is the steel beam detail noted?

____

7. How is the steel beam attached to the floor joists?

____

8. What is the roof slope over the following areas?

   A. Garage ____

   B. Main house ____

   C. Back porch ____

9. What is the overhang at the roof soffits?

____

10. What is the typical header requirement over openings on loadbearing partitions?

____

Activity 12-2

Framing Plans for a Residential Building Project

Name ____________________________

Refer to Sheets 1, 2, and 3 from the Marseille residential building plans in the Large Prints supplement to answer the following questions.

1. What is the drawing scale for the following?

   A. Foundation Plan ____________________________

   B. First Floor Plan ____________________________

2. What is the size of the beam supporting the floor joists for the first floor?

   ____

3. What is the specification for the column resting on the 30″×30″ footing and supporting the beam?

   ____

4. What is the size and spacing of the floor joists for the first floor?

   ____

5. How is the steel column connected to the steel plate anchored to the 30″×30″ footing?

   ____

6. What are the specified wood types for the following framing members?

   A. 2×4 and 2×6 ____________________________

   B. 2×8, 2×10, and 2×12 ____________________________

7. What is the specified repetitive bending stress for 2×10 floor joists?

   ____

8. What is the symbol for rough wood framing? Sketch the symbol freehand.

9. What is the live load for the first floor?

10. What is to be used for the header above the garage door?

11. What type of floor framing is required for the second-floor area?

12. Where on the first floor is the location of the loadbearing partition that carries the second-floor framing?

13. What type of drywall is required in the garage?

14. What are the ceiling heights in the following areas?

    A. Great Room

    B. Dining

    C. Master Bedroom

15. What type of window is required to the right side of the front door?

Activity 12-3

Framing Details and Elevation Drawings for a Residential Building Project

Name ______________________________

Refer to Sheets 1, 3, 5, and 6 from the Marseille residential building plans in the Large Prints supplement to answer the following questions.

1. What is the drawing scale for the following?

   A. Roof Plan ______________________________

   B. Shear Wall Detail at Garage ______________________________

2. What type of subfloor is specified over the floor joists and how is it attached?

   ____

3. What are the insulation requirements for the following areas of construction?

   A. 2×4 exterior walls ______________________________

   B. Roof ______________________________

   C. Basement ______________________________

4. Refer to the roof truss detail for the roof truss over the garage.

   A. What profile is referenced? ______________________________

   B. What spacing is used for roof trusses? ______________________________

   C. What is the specified slope? ______________________________

5. What is typically used to tie down the house framing to the foundation?

   ____

   ____

6. Referring to the stairs, what are the floor-to-floor heights and the number of risers for the following?

   A. Basement to first floor ______________________________

   B. First to second floor ______________________________

7. What is specified for the weather-resistant membrane?

____

____

8. What is specified for the curved part of the roof over the upper-level middle windows?

____

UNIT 13
Plumbing Prints

TECHNICAL TERMS

cleanouts
copper piping
control valves
distribution pipes
drain/waste/vent (DWV) system
finish plumbing
galvanized steel pipe
isometric drawing
main
plastic pipe
rough-in plumbing
sweated fittings
waste stack

LEARNING OBJECTIVES

After completing this unit, you will be able to:

- Identify various piping systems.
- Recognize plumbing fixture symbols on prints.
- Explain the three stages of plumbing installation.
- Read a piping diagram.

In most residences and light commercial buildings, plumbing consists of the water distribution system, sewage disposal system, and piping needed for heating and cooling systems. The water distribution system consists of the piping used to supply water to different service locations in the building. The sewage disposal system provides for the removal of water and waste from the building. In some buildings, other piping systems are installed as part of a heating and cooling system. This unit is designed to familiarize you with the way in which these systems are detailed on construction prints.

Sometimes, piping diagrams are unnecessary. Symbols on the plan drawings locate plumbing fixtures such as sinks, water closets, floor drains, and exterior hose bibbs. See **Figure 13-1**. The plumber installs the system in accordance with the specifications and local building codes. Larger construction jobs have complete plumbing drawings showing the supply and sewage piping.

Specifications

The plumbing specifications should be read carefully before the job is bid and started. These specifications detail the work included in the plumbing contract—piping materials, plumbing fixtures, and tests to be performed on the piping systems.

Plumbing is installed in three stages and must be carefully coordinated with all trades and their work assignments:

- **Stage 1.** Provisions for the service entrance of the water supply and sewer drain to the building structure are made prior to the pouring of the foundation. See **Figure 13-2**.
- **Stage 2.** The next stage is the ***rough-in plumbing***, which includes installing water supply pipes and sewage drain pipes, **Figure 13-3**. The rough-in work is performed before the slab is poured in slab-on-grade construction and before wall-covering materials are placed on the wall framing.

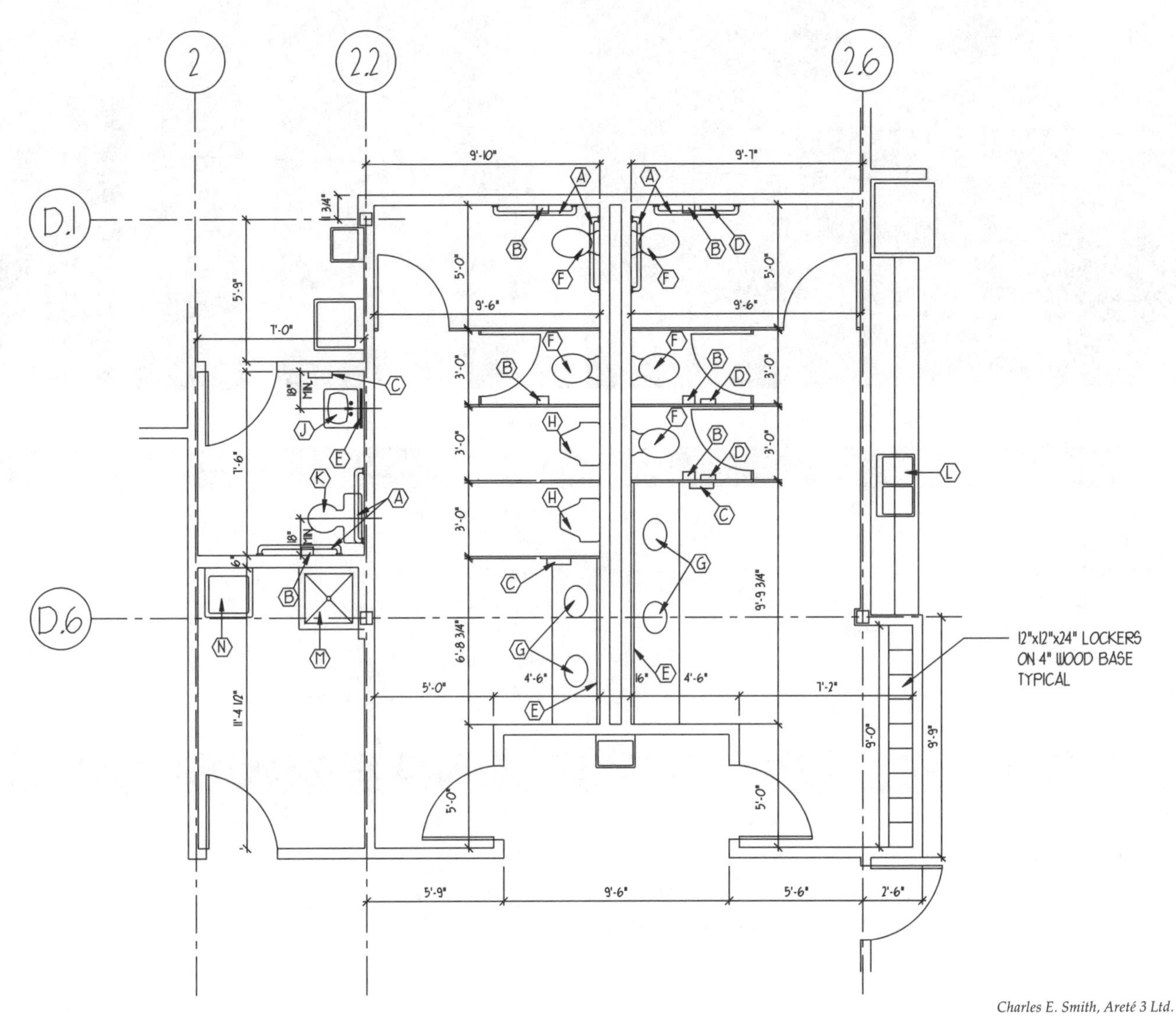

Charles E. Smith, Areté 3 Ltd.

Figure 13-1. This floor plan shows the location of plumbing fixtures in restrooms for an office building.

Goodheart-Willcox Publisher

Figure 13-2. In the first stage of the plumbing installation, underground drains are installed prior to placing the concrete floor slab.

Goodheart-Willcox Publisher

Figure 13-3. The rough-in plumbing is installed before concrete is placed for floor slabs. This installation is being done in a commercial building.

- **Stage 3.** The final stage of the plumber's work is the ***finish plumbing***. This includes the installation and connection of plumbing fixtures (such as toilets, sinks, dishwashers, and drinking fountains) after the walls and floors are finished.

Water Distribution System

The water distribution system includes the main supply line to the building from the municipal water meter, individual well, or other source of water supply. All pipes that take the water from the ***main*** to the various service outlets (water heaters, sinks, water closets, hose bibbs, etc.) are called ***distribution pipes***. The distribution system also includes all of the ***control valves***.

As previously discussed, symbols are used to represent plumbing fixtures on drawings. Symbols are also used to represent other plumbing equipment, such as different types of valves and fittings. Common plumbing symbols are shown in **Figure 13-4**. Unless the piping layouts are unusual, they generally are not shown on residential drawings. The piping layouts are included on commercial drawings, however. Piping layouts for commercial buildings can be shown on plan and elevation views, but often an ***isometric drawing*** of the system is provided, **Figure 13-5**. An isometric plumbing drawing is a three-dimensional (3D) representation of the piping, fittings, and fixtures in a piping system. Lines are drawn at 30° angles in order to better describe the flow and distribution of water and sewage. The detail provided on an isometric plumbing drawing, combined with the information on the floor plans, helps the estimator and plumber better understand the project.

Distribution Piping Materials

Piping materials used for water distribution include copper, galvanized steel, and plastic. For residential installations, copper and plastic are the most common.

Copper piping comes in heavy, medium, and light wall thickness. Heavy and medium pipes are suitable for underground and interior plumbing systems. Light pipes are suitable for interior plumbing applications. Copper is available in hard and soft tempers. The hard tempers are more suitable for straight runs of exposed pipe, where appearance is a factor. The hard-temper pipe can be bent to a limited radius with proper tools. Soft-temper copper pipe is easily bent by hand or with a tube bender.

Copper piping should not be embedded in concrete slabs, masonry walls, or footings. When it is necessary for the pipe to go through a slab or wall, a plastic sleeve or a larger pipe should be placed between the copper water pipe and concrete. This will permit movement due to expansion and contraction of the copper.

GREEN BUILDING

Tankless Water Heaters

One alternative to the traditional water heater with a large storage tank is one or more "tankless" water heaters. These units, also called *on-demand* water heaters, produce instant hot water and require that only a cold water line be piped to the unit. The water is heated by an electric heating element or a gas burner, depending on the model.

Although tankless water heaters are more expensive to install than ordinary water heaters, they last longer, and they require less energy to use. The biggest advantage of using a tankless water heater is that energy is not wasted keeping a large storage tank of water hot at all times. Water is heated only as it is needed. However, the amount of hot water that can be produced by a tankless water heater is limited. Typical flow rates range from 2 to 5 gallons per minute. In larger homes or homes that use a lot of hot water, more than one unit may be needed.

Copper has many advantages and is widely used in water distribution systems. Copper pipe should never come into contact with other metals, such as steel pipe or steel reinforcing. Rapid corrosion caused by electrolytic action can be induced. Joints in copper pipe are either sweated or threaded. ***Sweated fittings*** are soldered using lead-free solder.

Galvanized steel pipe has great strength and dimensional stability. The galvanized coating protects the pipe against rusting. There are three grades (wall thicknesses) for galvanized steel pipe: standard weight (S), extra strong (XS), and double extra strong (XXS). The standard weight pipe is used for most residential and light commercial water distribution systems.

Plastic pipe is used extensively. Three of the most common types are polyvinyl chloride (PVC), chlorinated polyvinyl chloride (CPVC), and cross-linked polyethylene (PEX). PEX is flexible, lightweight tubing that is connected using special fittings. PVC and CPVC pipes and fittings are available in threaded form, but more commonly, connections are made using solvent-welded joints. Plastic pipe is typically rated according to its wall thickness and its ability to carry hot water. PVC can be used for cold water supply piping, but it is not approved for hot water supply piping. CPVC can be used for both cold and hot water supply piping. The most common grades of PVC and CPVC pipe are Schedule 40 and Schedule 80 pipe. The schedule number indicates the wall thickness (the higher the number, the greater the thickness).

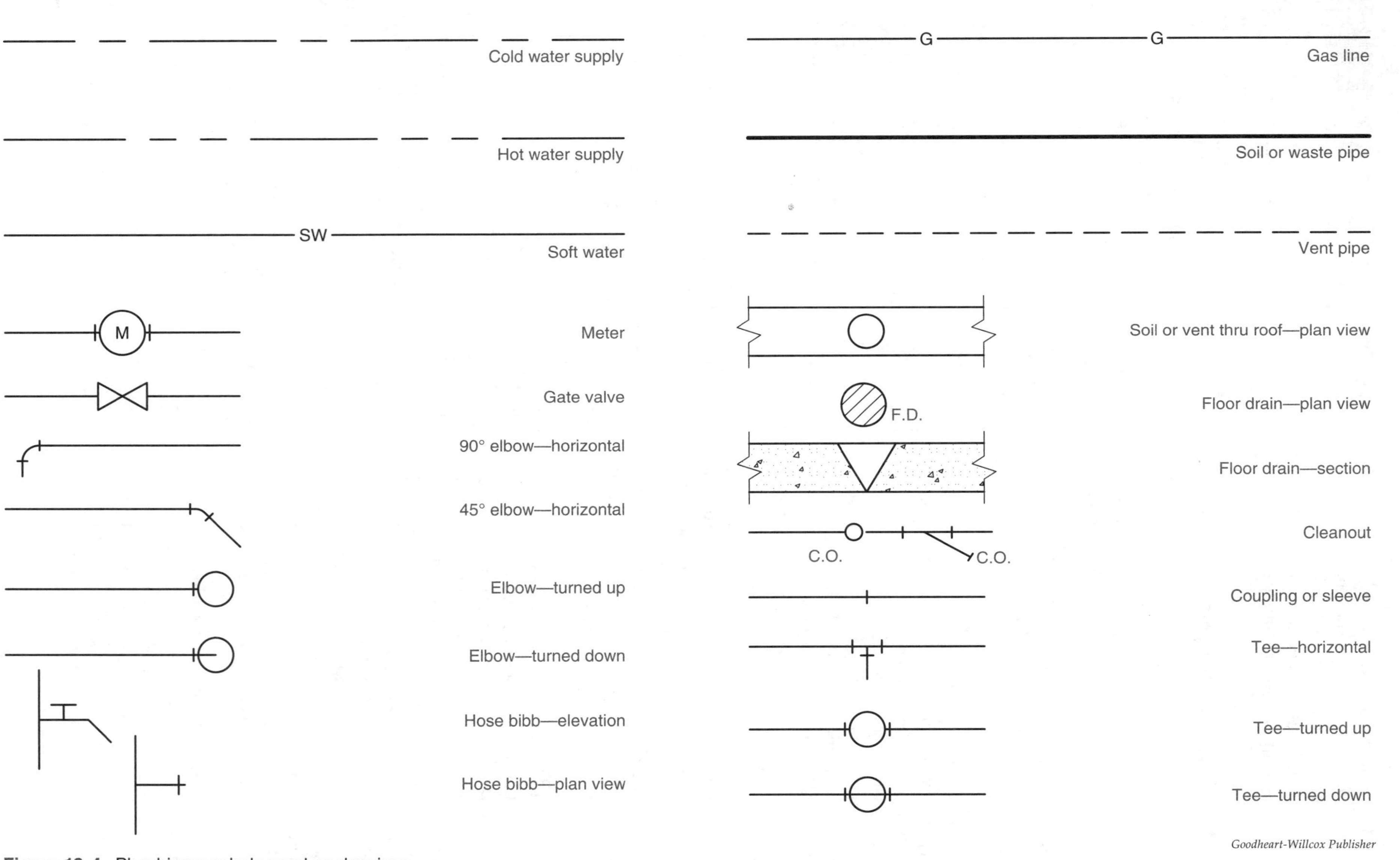

Figure 13-4. Plumbing symbols used on drawings.

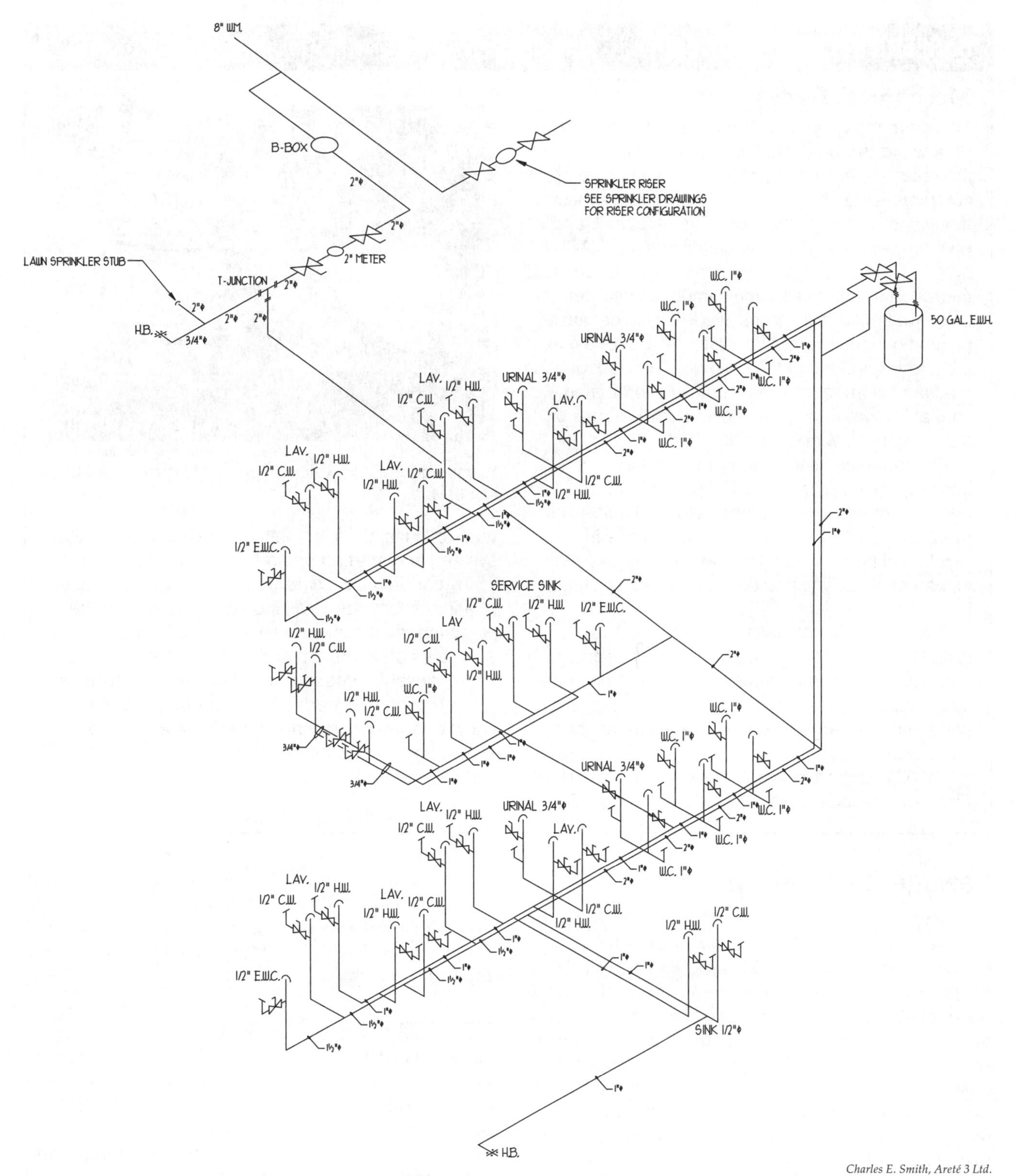

Charles E. Smith, Areté 3 Ltd.

Figure 13-5. An isometric drawing of a water distribution system for a commercial building. The isometric style makes visualization easier.

CAREERS IN CONSTRUCTION

Mechanical Trades

There are many opportunities in construction for tradeworkers who design, install, and maintain the mechanical systems of a residential home or commercial building. The mechanical systems of a building consist of the plumbing, electrical, and heating, ventilating, and air-conditioning (HVAC) systems. These systems are typically installed once the rough framing has been completed and before finish wall materials are installed. This is because plumbing pipe, wiring, and ductwork are routed in the interior framing of the building. The work of each mechanical trade must be carefully coordinated so that interference is prevented between runs of plumbing pipe, wiring, and ductwork.

Plumbers install the supply and wastewater lumbing systems used in a building. They also install plumbing for any other gases or liquids that need to be distributed, such as natural gas in a residential home or commercial building, oxygen in a medical facility, and compressed air or steam in an industrial plant.

Electricians install electrical systems. They install the wiring for lights, switches, and receptacles in a residential home. In more complex structures, electricians are responsible for installing systems used to power large mechanical equipment and machinery.

HVAC technicians install climate control systems. They are responsible for fabricating and installing ductwork, installing the unit used in the air distribution system, and starting up and fine-tuning the system.

Lisa F. Young/Shutterstock.com

Plumbers install the water supply and waste piping used in a building.

Preparation for a career in the mechanical trades requires specialized training. Many workers in the plumbing, electrical, and HVAC trades complete 3–5 years of on-the-job and classroom training in an apprenticeship program. Installers and technicians in the mechanical trades must know how to read prints and work safely and must have a thorough knowledge of building codes.

Sewage Disposal System

The sewage disposal system is also known as the ***drain/waste/vent (DWV) system***. It includes a large diameter vertical soil (waste) stack, a vent, and a trap for each fixture. The ***waste stack*** carries the waste materials to the building drain, which connects to the building sewer line outside the building. The building sewer line carries the building waste to the public sewer or septic tank. At the base of each waste stack, fittings called ***cleanouts*** (CO) are installed so the drain lines may be cleaned out with a plumber's rod or snake. The installation of DWV systems is carefully controlled by plumbing codes to prevent contamination of the water supply and to keep sewer gases from entering the building.

As previously discussed, piping drawings are provided for most commercial construction projects. A plan view diagram of a water distribution system and a DWV system for a commercial building is shown in **Figure 13-6**. Note the larger size of pipe specified for drain pipes and sanitary lines. Walls that contain vertical waste stacks serving water closets must be built thicker, using at least 6″ studs, to accommodate a 4″ soil pipe and its joints.

Sewage Piping Materials

For water distribution systems, piping materials that can pollute the water cannot be used. However, this is not a concern with wastewater. Sewage disposal systems can be made from many different kinds of pipe materials.

Cast iron pipe has good qualities of strength and resistance to corrosion. Copper and plastic pipe are used extensively because of ease of installation. Special plastics are used for drainage systems where chemicals are disposed. Plastic pipe is used extensively for sewer piping. The most common types used are acrylonitrile butadiene styrene (ABS) and PVC.

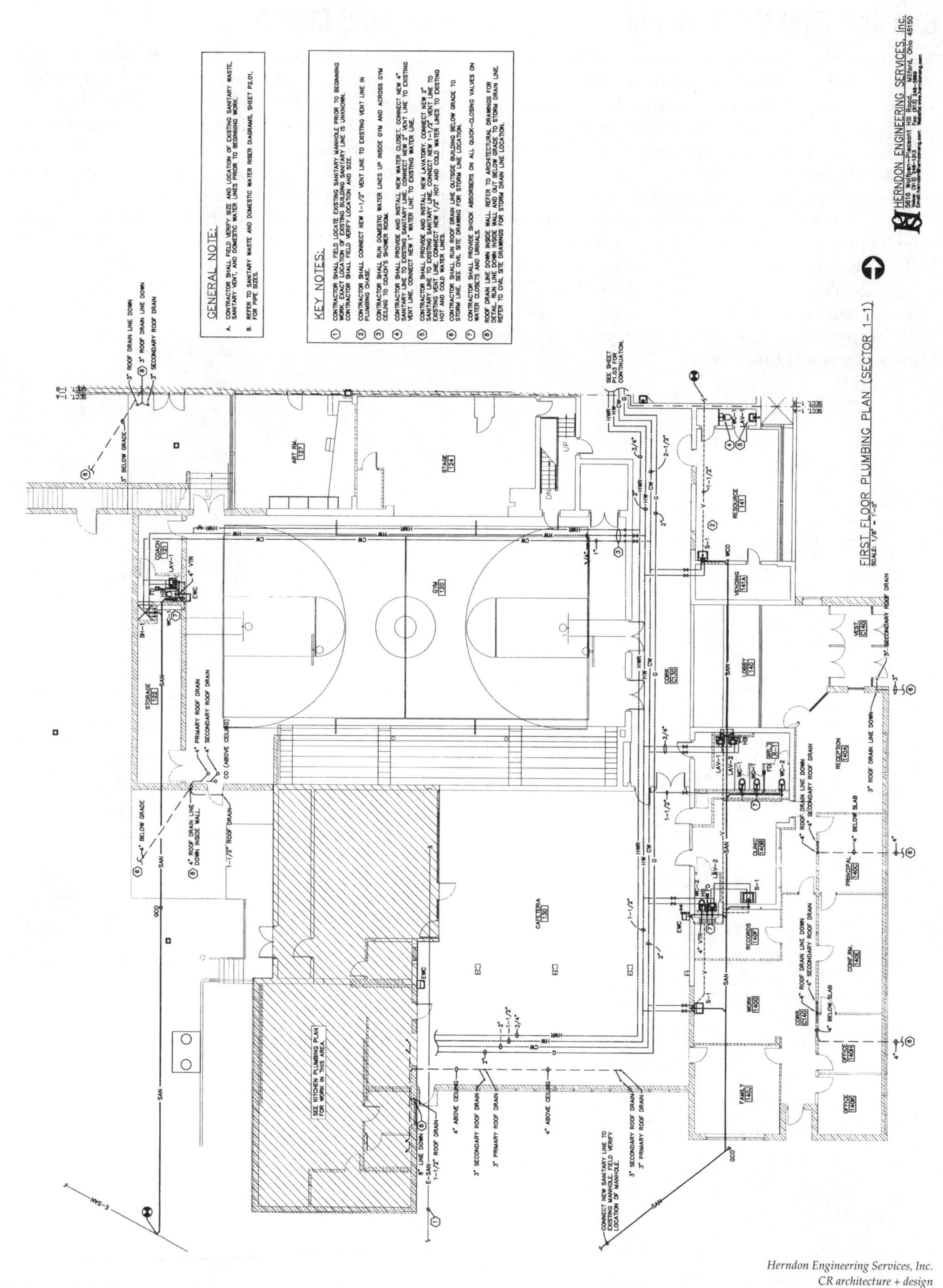

Herndon Engineering Services, Inc.
CR architecture + design

Figure 13-6. Plan view diagram of a water distribution system and drain/waste/vent (DWV) system for a commercial structure.

Gas and Fuel Oil Systems

Sometimes, the piping for a gas or oil heating system is included in the plumbing contract. Other than the location of the particular service desired, piping for heating purposes usually is not shown on the drawing. Specifications for the construction job will detail the size and kind of piping to use.

The types of piping most commonly used for gas systems are black iron pipe, galvanized steel pipe, and yellow brass pipe. Copper tubing is banned by most building codes because it corrodes when exposed to some gases. Black iron pipe is often required by building codes for piping combustible gases, such as natural gas. This pipe is not galvanized, but typically painted black and made of carbon steel.

Plumbing Codes

Model codes—such as the Uniform Plumbing Code and the International Plumbing Code—and the local building code control all aspects of plumbing work. Building code provisions include the kinds and sizes of pipe used, locations of traps and cleanouts, plumbing fixture requirements, venting provisions, and connections to water supply and sewer lines. Plumbing codes also specify the leak testing to be conducted on water supply lines and waste lines.

Test Your Knowledge

Name ________________________

Write your answers in the spaces provided.

____ 1. *True or False?* The first stage of plumbing installation is attaching fixtures to walls.

____ 2. Installation of water supply pipes and sewage drain pipes before the slab is poured is considered part of the ____.
- A. finish plumbing stage
- B. distribution plumbing stage
- C. main plumbing stage
- D. rough-in plumbing stage

____ 3. *True or False?* Copper pipes should *not* come in contact with concrete.

____ 4. *True or False?* Cleanouts must be included in the water supply system.

5. Match the following plan symbols with the correct nomenclature.

A. ——⋈——

B. ⊘

C. ⊢+

D. —— - - —— - - ——

____ Hose bibb

____ Gate valve

____ Floor drain

____ Hot water supply

____ 6. *True or False?* The DWV system is used to supply water to different service locations in the building.

____ 7. *True or False?* Drawings that show the path of pipes through the building are required for all structures.

____ 8. What are two common types of plastic pipe for plumbing installations?
- A. ABC
- B. ABS
- C. PVC
- D. OXY
- E. PCL

____ 9. At the base of each waste stack, there is a fitting used to provide access to clogged lines. What is this fitting called?
- A. Plumber's access fitting
- B. Floor drain
- C. Vented access trap
- D. Cleanout

____ 10. What type of pipe is *not* commonly used for gas piping?
- A. Black iron
- B. Galvanized steel
- C. Copper tubing
- D. Yellow brass

Activity 13-1

Plumbing Plans for a Residential Building Project

Name ______________________________

Refer to Sheets MP2, MP3, and MP4 from the Marseille residential building plans in the Large Prints supplement to answer the following questions.

1. What size line is going out to the sewer?

   ____

2. What symbol is used to identify vents?

   ____

3. What size is the floor drain in the basement? Where is it located?

   ____

4. On which sheet is the isometric drawing of the drainage system shown?

   ____

5. What size drain lines are specified for the following fixtures?

   A. Master Bath tub ______________________________

   B. Toilet in bathroom next to laundry room ______________________________

   C. Laundry room clothes washer ______________________________

6. What typical size vent is going through the roof?

   ____

7. On what side of the house is the water meter located?

   ____

8. How many cleanouts are there?

   ____

Activity 13-2

Plumbing Plans for a Commercial Building Project

Name ________________________________

Refer to Sheets P1.1 and P2.1 from the Delhi Flower and Garden Centers commercial building plans in the Large Prints supplement to answer the following questions.

1. In which room is the water meter located?

   ____

2. Give the pipe size of the supply line for the following items:

   A. Hot water heater EWH-1 ________________________________

   B. Non-potable water line from meter ________________________________

   C. Main domestic water line from meter ________________________________

   D. HB-1 at main entry ________________________________

3. What is the size of the hot water heater for the men's and women's restrooms?

   ____

4. What is the contractor to do prior to submitting a bid?

   ____

5. On which sheet is the sanitary isometric drawing shown?

   ____

6. Give the pipe size of the following waste system items:

   A. Floor drains ________________________________

   B. Cleanouts ________________________________

   C. Urinals ________________________________

7. What is specified for the following plumbing fixtures?

   A. Water closet WC-1 ________________________________

   B. Hose bibb ________________________________

   C. Water cooler ________________________________

8. What is the size of the main gas line from the meter?

9. What is the size of the downspouts at the greenhouse sales area?

10. What is the slope of the storm line (ST) in the greenhouse sales area?

UNIT 14
HVAC Prints

TECHNICAL TERMS

ducts
electric radiant heating system
evaporative cooling systems
forced-air system
heating, ventilating, and air-conditioning (HVAC) systems
hydronic heating system
remote cooling systems
series-loop system
unit cooling systems

LEARNING OBJECTIVES

After completing this unit, you will be able to:

- Identify the purpose of HVAC systems.
- Explain different types of heating and cooling systems.
- Identify HVAC symbols on prints.

Heating, ventilating, and air-conditioning (HVAC) systems produce the movement of air within a building to condition the space for its intended use. This air may be heated or cooled, then moved to another location to change the air conditions. The HVAC system makes a space more comfortable for the building occupants. In an industrial setting, the HVAC system may serve a more specialized purpose, such as providing refrigeration for a freezer or air sterilization for a hospital operating room.

Air treatment involves controlling the temperature, the humidity (moisture in the air), and air cleanliness. During colder periods of the year, the air in a building is heated and moisture is added for comfort. In warmer periods, the air is cooled and moisture is removed.

To accomplish the desired conditioning of air in a building, a heating system and a cooling system are needed. Sometimes, a system for cleaning and treating the air for desired humidity is also included. In this unit, you will learn about heating and cooling systems and how these systems are shown on drawings.

HVAC Plans

For residential and light commercial buildings, HVAC plans are prepared by the HVAC subcontractor. Generally, these plans are drawn on the floor plan of the structure for the approval of the architect and/or owners. For larger commercial structures, the heating and cooling plans are prepared by a mechanical engineer under the direction of the architect. These plans are also usually drawn over the floor plans for reference purposes. Typically, some items that would normally be shown on the floor plans are either not shown or "grayed out" so that the HVAC system and equipment stand out.

Symbols used on HVAC plans are shown in **Figure 14-1**. A symbol legend is included on the HVAC drawing to identify the symbols used.

Heating Systems

There are three basic types of heating systems used in new construction: forced-air, hydronic (hot water), and electric radiant heating. Forced-air systems are popular and common in residential and light commercial construction.

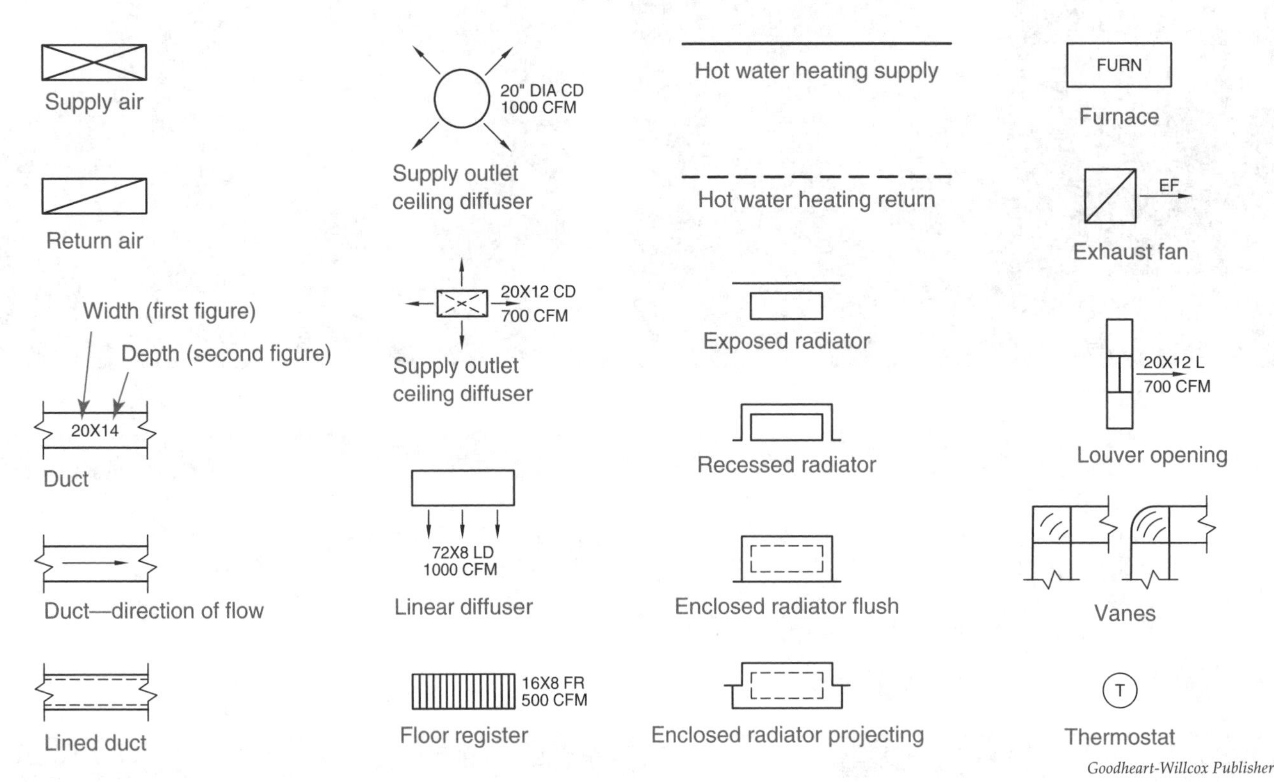

Figure 14-1. Symbols for HVAC equipment.

In a ***forced-air system***, the heated air from the furnace or heat pump chamber is transferred by means of a motor-driven fan through a series of ***ducts*** (rectangular or round pipes) to registers or diffusers in the various rooms. See **Figure 14-2**. Cool air is gathered through registers near the floor, called cold air returns, and returned to the heating unit through ducts and a filtering system to be reheated and recirculated.

In residences or light commercial structures, the forced-air heating system is considered to be a *closed-circuit system* (one with no provision for outside air to be added). In large commercial structures, provision is made for the addition of fresh air from the outside and no air is returned from kitchens and restrooms.

Sources of heat for forced-air systems are natural gas, liquefied petroleum gas (LPG), oil, coal, and electricity. Solar energy (in active solar systems) is also used as a fuel source in forced-air systems.

An HVAC plan showing the layout of ducting for a forced-air system is shown in **Figure 14-3**. Note that the size of ducts is specified. In addition, HVAC plans often specify the airflow rate at a given point. The flow rate is specified in cubic feet per minute (cfm). Note also in **Figure 14-3** that the ducts reduce in size toward the end of each run. This allows the system to maintain even pressure as the air is distributed to the supply vents.

A ***hydronic heating system*** begins by heating water in a boiler. Then, the hot water is circulated by a pump and piping system to convectors in the spaces to be heated. Hydronic systems for residences normally use the ***series-loop system*** to carry the heated water to the convectors, **Figure 14-4**. For larger areas, a one-pipe system or two-pipe system is specified. The one-pipe and two-pipe systems provide more uniform heat for a larger area than does the series-loop system.

An ***electric radiant heating system*** usually consists of wires embedded in the ceilings, walls, or floors and baseboards of the building. The wires are spaced in a grid pattern, with approximately 1 1/2″ between wires.

Goodheart-Willcox Publisher

Figure 14-2. Ducts are used to distribute air in forced-air systems. Shown is a ceiling layout of ducting in a commercial building.

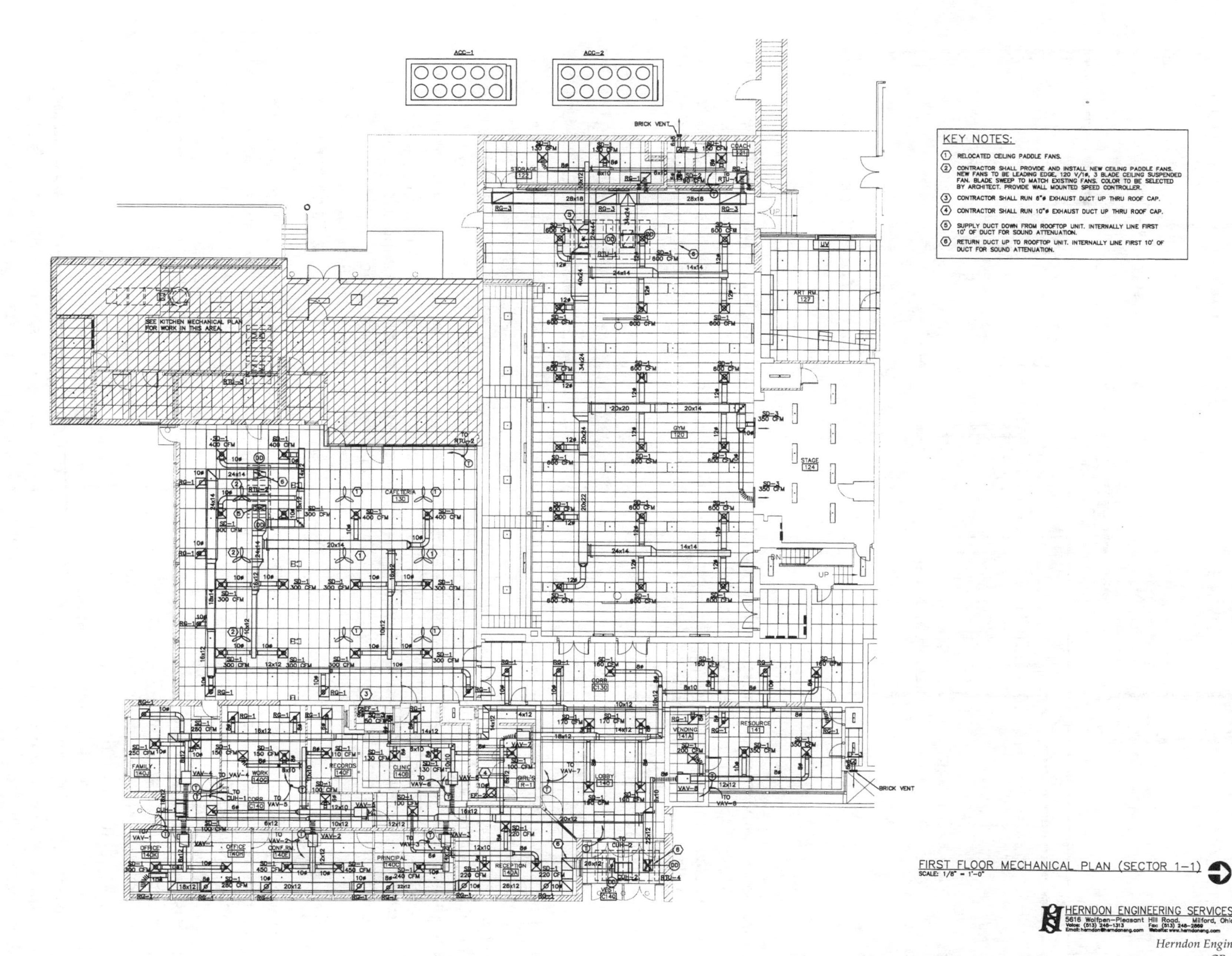

Herndon Engineering Services, Inc.
CR architecture + design

Figure 14-3. An HVAC plan for a forced-air system in a commercial building with sizes of ducts noted.

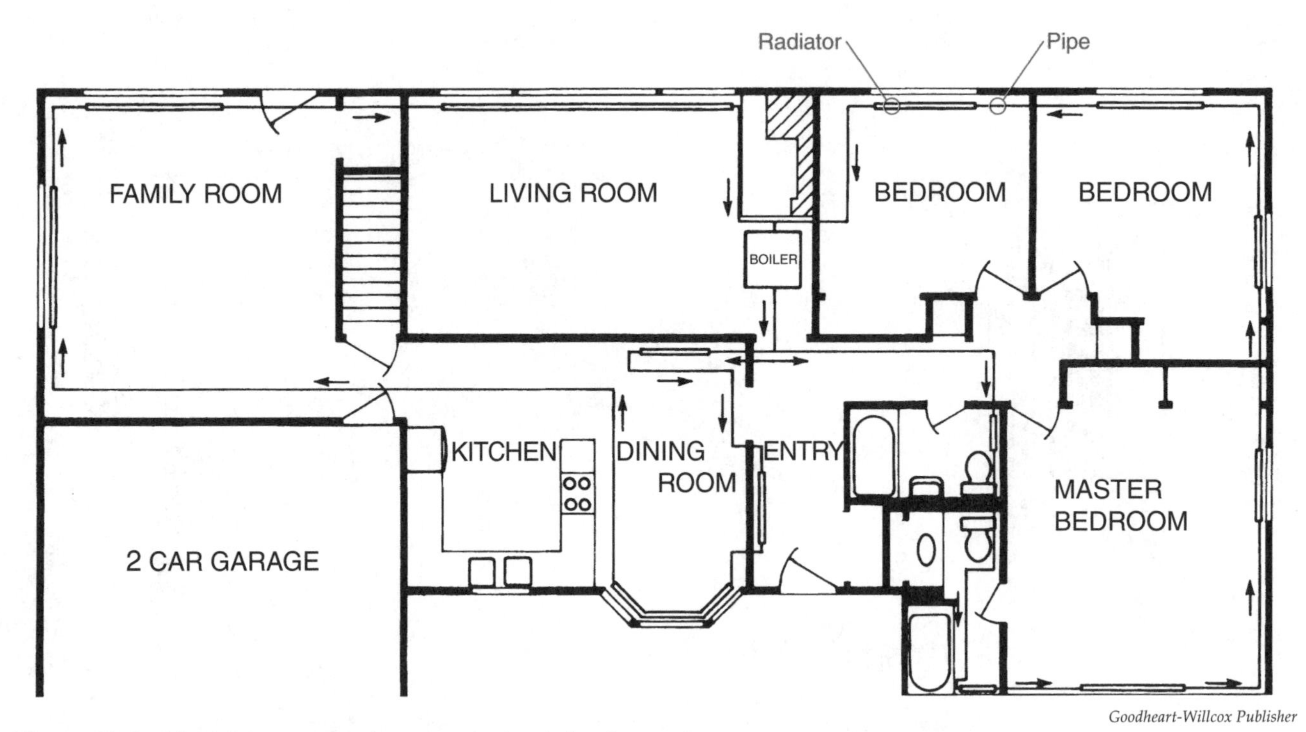

Goodheart-Willcox Publisher

Figure 14-4. A layout for a series-loop type hydronic heating system.

They are stapled to the gypsum lath of the ceiling or wall and covered with the finish covering material. Floor cables are heavier. These are placed in a grid pattern 1 1/2″–3″ below the concrete surface. Radiant heat is given off by the floor, ceiling, or wall materials (such as concrete or plaster) that are warmed by resistance induced in the wires embedded in them.

Electric radiant heating panels are also available for ceiling and wall installation. These panels can be covered with a finish covering material, plastered, or mounted directly to the ceiling or wall.

When heating system drawings are provided, they are usually superimposed over the floor plans. They can also be provided as isometric diagrams. In either case, the appropriate notes are included. When drawings are not provided, the amount of heat required for each space is noted on the floor plan.

Cooling Systems

Cooling systems can be grouped as unit systems (window or wall-mounted units) and remote systems (where refrigeration equipment is located away from the area to be conditioned). A third type of system, the evaporative system, is used only in special circumstances.

Unit cooling systems, which are installed in a window or space provided in an exterior wall, are used to cool a room. Very little construction is involved in their installation. Often, a unit cooling system is purchased for an older home that was not equipped with a built-in system and the unit is retrofit into an existing space. Units can be purchased to heat as well as cool the air.

Remote cooling systems have the condensing unit in a remote space away from the area to be cooled. The evaporator is in the main duct, where a fan forces air past the cooling coils and circulates the air to the

GREEN BUILDING

Smart Climate Control Devices

Many responsible homeowners set their thermostats to a reasonable level before they leave on vacation. This is a “green” practice that can help conserve energy while no one is in the home. Programmable thermostats take this a step further by allowing people to program the climate control system to make the house comfortable only when they are at home, such as on nights and weekends. At other times, the thermostat is programmed to conserve energy.

Some homes have “smart” thermostats that can be controlled by voice, telephone, computer, and electronic devices via the Internet. Suppose a homeowner lives in Michigan and leaves for a two-week vacation in October. The owner leaves the air conditioner on and set to 82°, and the furnace is off. Three days before the owner is scheduled to return, a major snowstorm occurs. Temperatures drop into the 20s. With a smart thermostat, the owner can contact the climate control system and change the settings so that the pipes do not freeze and water damage does not occur.

rooms to be cooled. A variation of this cooling system uses a remote condensing unit and evaporator, with cooled water or brine circulated to the heat exchangers in each room.

Evaporative cooling systems are most effective in dry climates where the relative humidity is low (preferably 20% or less). They will work, but less effectively, at higher levels of humidity. The evaporative system functions by moving air rapidly over a pad of loose fibers that is kept moist by a water spray mist. The air is cooled as it passes through the pad, and is then carried through a duct system to the rooms. This system of cooling raises the humidity in the space being cooled.

The supply duct layout for an evaporative cooling system is similar to that of a forced-air system. However, the evaporative system has larger ducts, and no return ducts are necessary, since outside air is used.

Air Filters

Most heating and cooling systems provide a means of filtering the air that flows through the system. Air filters usually have an adhesive coating that collects lint and dust particles. Air filters may be either disposable or washable.

Most systems have a designated space for inserting the filter. An electrostatic filter is usually a separate unit added to the system. It is noted on the heating and cooling plan and detailed in the specifications.

Goodheart-Willcox Publisher

HVAC system installations must be closely coordinated with the work of other trades. The ductwork in this commercial building is suspended below the structural steel framing members and above the fire sprinkler piping.

Test Your Knowledge

Name ________________________________

Write your answers in the spaces provided.

____ 1. *True or False?* An air-conditioning system changes the temperature of the air, but does nothing to clean the air.

2. Match the following plan symbols with the correct nomenclature.

A. 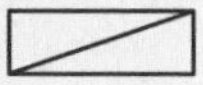

B. 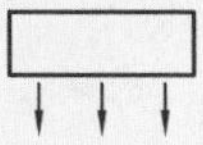

C. 

D. 

____ Supply air

____ Thermostat

____ Return air

____ Linear diffuser

____ 3. *True or False?* A forced-air heating system transports air through ducts.

____ 4. *True or False?* A unit cooling system is the most difficult type of cooling system to install.

____ 5. Which of the following is *not* a heating system?

A. Hydronic system
B. Remote system
C. Forced-air system
D. Electric radiant system

____ 6. *True or False?* To improve comfort during warm weather, moisture should be added to the air.

____ 7. *True or False?* In a forced-air system, heat is always provided by burning natural gas.

____ 8. In an evaporative cooling system, the air is cooled by ____.

A. electrostatic evaporative condensation
B. a water spray mist over fiber pads
C. condensing units
D. blocks of ice

____ 9. In which type of climate is an evaporative cooling system most effective?

A. Hot and humid
B. Dry climate where relative humidity is below 20%
C. Cold
D. Rainy

____ 10. HVAC plans are typically shown on the ____ plans.

A. reflected ceiling
B. floor
C. structural
D. site

Activity 14-1

HVAC Plans for a Residential Building Project

Name ______________________________

Refer to Sheets MP2, MP3, and MP4 from the Marseille residential building plans in the Large Prints supplement to answer the following questions.

1. What is the design TD (temperature differential) for the following?

   A. Summer ______________________________

   B. Winter ______________________________

2. What size A/C unit is specified for the house?

   ____

3. On which sheet are the supply ducts for the basement shown?

   ____

4. What sizes are the main trunk lines that supply the basement?

   ____

5. Where are the return vents located on the first floor and what size are they?

   ____

   ____

   ____

   ____

6. Referring to the first floor, where is the supply air duct up to the second floor located?

   ____

7. What size are the supply lines off the main trunk on the second floor?

   ____

8. What size is the return main trunk on the second floor?

   ____

9. What size are the ceiling distribution vents on the second floor?

   ____

10. For which city is the HVAC system of the house designed?

    ____

Activity 14-2

HVAC Plans for a Commercial Building Project

Name ______________________________

Refer to Sheets M1.1 and M1.2 from the Delhi Flower and Garden Centers commercial building plans in the Large Prints supplement to answer the following questions.

1. What is the scale of the HVAC floor plan?

____

2. Which RTU units supply the hardlines and checkout areas? What is the cfm capacity of each?

____

3. What is the size of the main duct supplying air to the lines in the checkout area?

____

4. What unit supplies air conditioning to the men's and women's restrooms?

____

5. What model is specified for the condensing unit that goes with the unit referred to in Question 4?

____

6. What type of insulation is to be used on all ductwork?

____

7. What type of thermostat is to be used?

____

8. How many thermostats are specified and where are they located?

____

____

9. How is the loading/shipping area heated? What is the cfm output for each unit?

____

____

10. Where are the FR-1 air devices located?

____

____

11. Answer the following questions related to item EF-2.

    A. Where is this item located? ____________________

    B. What is the cfm output? ____________________

    C. What manufacturer is specified? ____________________

12. What is specified to be done with the ductwork in the count room?

    ____

UNIT 15
Electrical Prints

TECHNICAL TERMS

ampere
arc fault circuit interrupter (AFCI)
branch circuits
circuit
circuit breaker
conductor
conduit
distribution panel
equipment schedule
ground
ground fault circuit interrupter (GFCI)
one-line diagram
lighting schedule
National Electrical Code (NEC)
panel schedule
raceway
receptacle
relay
service entrance
voltage
watt
wiring diagram

LEARNING OBJECTIVES

After completing this unit, you will be able to:

- Identify electrical symbols.
- Explain various electrical terms.
- Recognize different types of electrical drawings.

The electrical prints for residential and light commercial buildings usually consist of the outlets and switches shown on the floor plan. Wire routing is left to the electrical contractor and provisions established by the electrical codes. Large industrial and commercial buildings involve complex electrical systems. For these, detailed electrical diagrams are prepared by electrical engineers and make up their own set of drawings.

Electricians must be able to read prints in order to plan where circuits will run in the basement, crawl spaces, and attics. They also need to know the direction of floor and ceiling joists. To help you read and understand electrical prints, this unit will explain electrical terms, symbols, diagrams, codes, and simple circuit calculations.

Electrical Terminology

The following are some common terms used in electrical construction:

- **Ampere.** The unit of measurement of current (electricity flowing through a conductor). ***Ampere*** is abbreviated as *amp.*
- **Voltage.** The electromotive force that causes current to flow through a conductor (wire) is called ***voltage***.
- **Watt.** The ***watt*** is the unit of measurement of electrical power. Current (amps) × voltage = watts. Most appliances are rated in watts.
- **Circuit.** A ***circuit*** is a path consisting of two or more conductors (wires) carrying electricity from the source (distribution panel) to an electrical device and back.

- **Circuit breaker.** A ***circuit breaker*** is a switching device that automatically opens a circuit when it becomes overloaded.
- **Conductor.** A ***conductor*** is a wire or other material used to carry electricity.
- **Conduit.** A channel or pipe in which conductors run. Required by code where conductors need protection.
- **Convenience receptacle.** An outlet where current is taken from a circuit to serve electrical devices such as lamps, clocks, and toasters. Usually referred to as merely ***receptacle***.
- **Service entrance.** The service equipment including the conductors from the utility pole, the service head, and the mast that bring the electrical current to the distribution panel.
- **Distribution panel.** The insulated panel or box, sometimes called a *breaker panel*, that receives the current from the source and distributes it through branch circuits to various points throughout the building. The ***distribution panel*** often contains the main disconnect switch and always contains fuses or circuit breakers protecting each circuit.
- **Ground.** A ***ground*** is a wire connecting the electrical circuit or device to the earth to minimize injuries from shock and possible damage from lightning.

Ground Fault Circuit Interrupters

A ***ground fault circuit interrupter (GFCI)*** should be installed in areas where moisture may be present or where the user of an electrically powered tool or appliance could come in contact with a grounded metal surface. The use of a GFCI is defined in the ***National Electrical Code (NEC)***. Examples of situations where GFCI devices must be used include the following:

- Receptacles in bathrooms
- Receptacles serving kitchen countertops
- Receptacles in laundry areas
- Receptacles in crawl spaces at or below grade
- Receptacles in unfinished basements
- All receptacles in garages
- All outdoor receptacles

A GFCI will open the circuit if a current leakage or fault (to ground) occurs in excess of 0.006 amperes. These interruptions occur when there is a difference detected in the amount of current entering and current leaving the circuit. The GFCI automatically senses the fault and immediately turns off the power.

GREEN BUILDING

Energy Star Program

Have you ever noticed the Energy Star label that appears on products such as home appliances and electronic devices? The Energy Star program was developed by the Environmental Protection Agency (EPA) and the Department of Energy to rate products and guarantee that they are energy-efficient. Using products with the Energy Star label saves owners money and helps reduce greenhouse gas emissions. The owner can be confident that these products have superior energy efficiency and will help protect the environment.

Arc Fault Circuit Interrupters

An ***arc fault circuit interrupter (AFCI)*** is a device that detects arcing conditions and opens the circuit if an arc fault occurs. Arcing currents can produce elevated levels of heat exceeding temperatures of 10,000°F and have the potential to ignite nearby materials, posing a fire hazard. Arcing currents can be caused by worn or damaged wiring, worn insulation, loose wiring connections, or damaged electrical or appliance cords. AFCI devices monitor arcing conditions and open a circuit when abnormal arcing is detected. The use of AFCI devices is specified in the National Electrical Code. AFCI protection is required for 120-volt, 15-amp and 20-amp branch circuits supplying power to outlets and devices in dwelling units. Examples of rooms requiring protection are kitchens, family rooms, dining rooms, living rooms, bedrooms, closets, hallways, laundry areas, and similar rooms and areas.

Electrical Symbols

Some of the more common symbols used on electrical drawings are shown in **Figure 15-1**. A legend listing symbols is usually shown on the electrical plan.

Any standard symbol can be used to designate some variation of standard equipment by the addition of lowercase subscript lettering to the symbol. This would be identified in the legend and, if necessary, further described in the specifications.

Electrical Prints

There are a number of types of drawings used to detail the electrical construction portion of a project. These include drawings such as electrical site plans, **Figure 15-2**, floor lighting plans, reflected ceiling plans, wiring diagrams, one-line diagrams, electrical power plans, etc. The size and complexity of the project will determine the number and types of prints required to define the scope of work.

Lighting Outlets

- Ceiling outlet
- Drop cord
- Fan outlet
- Junction box
- Lamp holder with pull switch
- Exit light outlet
- Outlet controlled by low voltage switching when relay is installed in outlet box
- Surface mounted fluorescent fixture
- Recessed light fixture
- Recessed square light fixture

Signaling System Outlets Residential Occupancies

- Push button
- Buzzer
- Bell
- Telephone
- Intercom
- Electric door opener
- Chime
- Television outlet
- Thermostat

Receptacle Outlets

- Duplex receptacle outlet
- Duplex receptacle ground fault circuit interrupter
- Weatherproof receptacle outlet
- Triplex receptacle outlet
- Quadruplex receptacle outlet
- Duplex receptacle outlet—split wired
- Single special-purpose receptacle outlet
- Range outlet
- Special purpose connection
- Clock hanger receptacle
- Floor single receptacle outlet
- Under floor duct and junction box for triple, double, or single duct system as indicated by number of parallel lines

Panels, Circuits, and Miscellaneous

- Ground
- Lighting panel
- Power panel
- Wiring, concealed in ceiling or wall
- Wiring, concealed in floor
- Conduit run to panel board
- *Indicates number of conductors
- Externally operated disconnect switch

Switch Outlets

- Single pole switch
- Double pole switch
- Three-way switch
- Four-way switch
- Key operated switch
- Switch and pilot light
- Weatherproof circuit breaker
- Weatherproof switch
- Switch for low voltage switching system
- Time switch
- Ceiling pull switch
- Switch and single receptacle
- Switch and double receptacle
- Circuit breaker
- Remote control switch
- Fused switch
- Master switch for low voltage switching system
- Automatic door switch

*Indicates number of conductors (in this case, 4). Any circuit without cross hatches indicates two-conductor circuit. Some electrical engineers show number of hot conductors with full marks; neutral conductors with half marks (= 3 hot conductors, 1 neutral).

Goodheart-Willcox Publisher

Figure 15-1. Common symbols used on electrical plan drawings and diagrams.

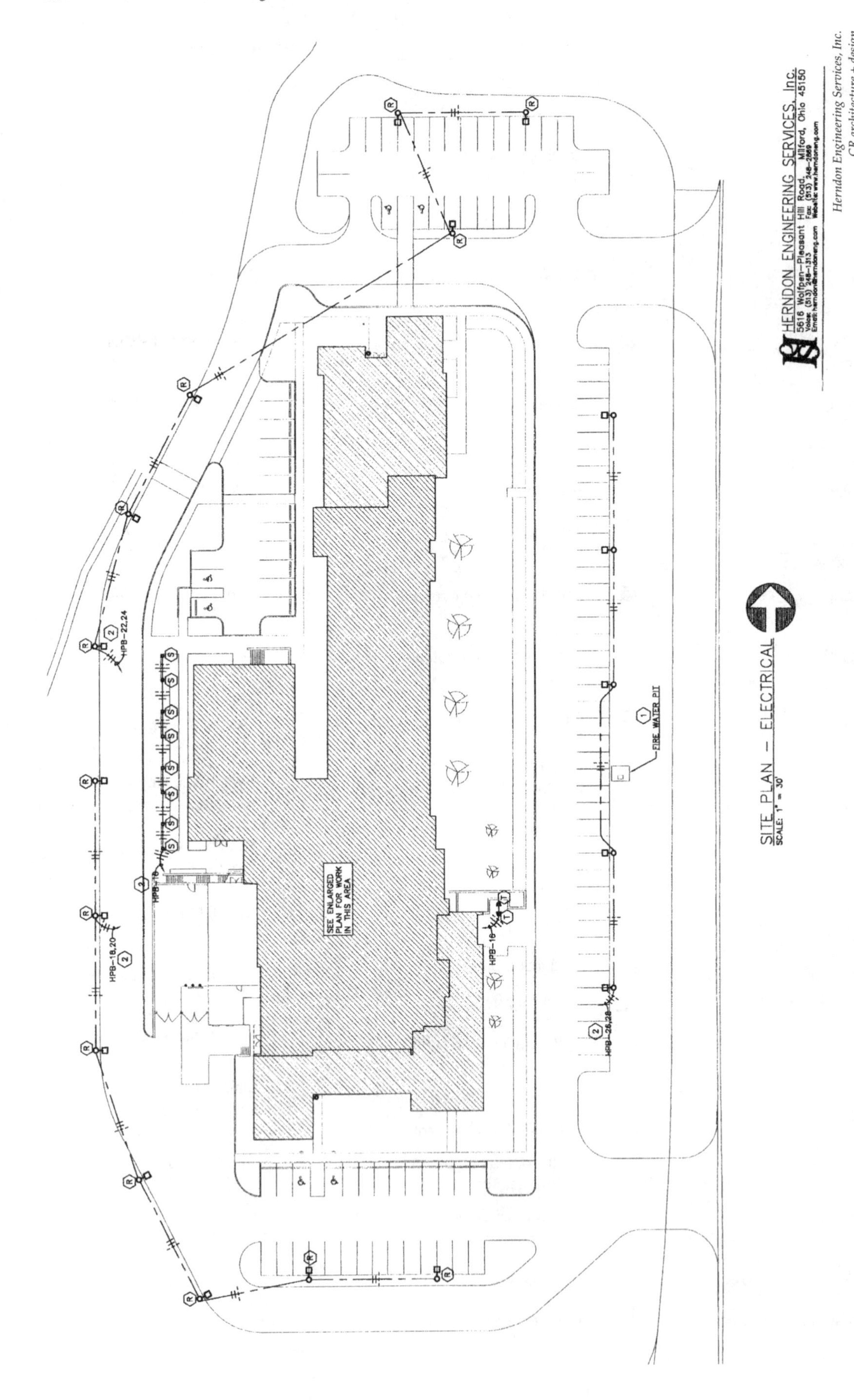

Figure 15-2. An electrical site plan for a commercial building. Shown are locations of exterior light fixtures to be installed on the building site.

Herndon Engineering Services, Inc.
CR architecture + design

Electrical Plans

An electrical plan shows the locations of receptacles, switches, lights, and all powered devices. See **Figure 15-3**. Straight or curved lines indicate which outlets and switches are connected. However, the actual path of the wire is not necessarily where the lines are drawn.

The electrical plan may also show conduit runs, **Figure 15-4**. The lines on the drawing representing the conduit are used to show the starting and ending points of the conduit run. Again, the line does not show the exact location where the conduit should be located. Conduit location is normally determined in the field by the electricians.

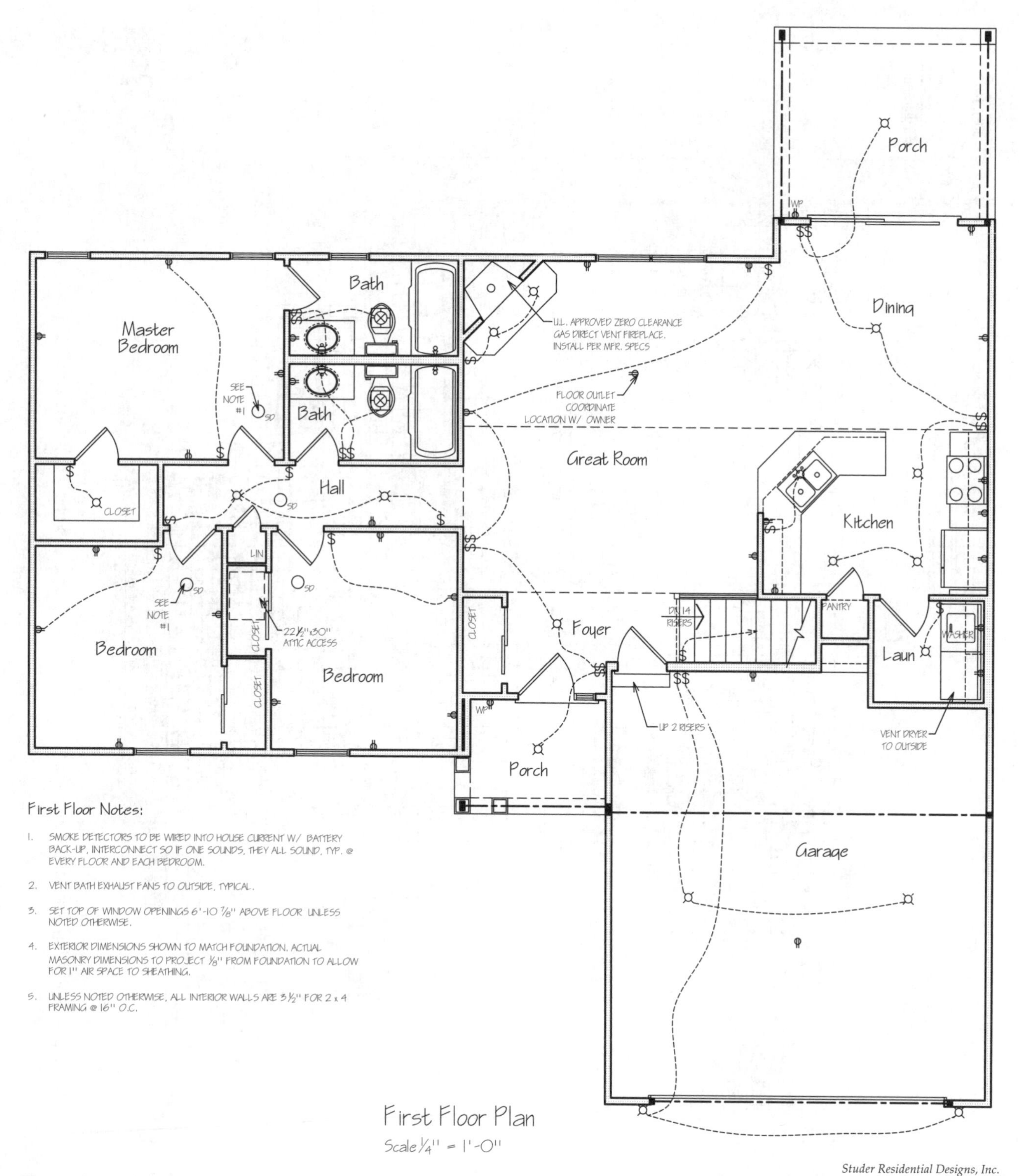
Porch
Dining
Master Bedroom
Bath
Bath
U.L. APPROVED ZERO CLEARANCE GAS DIRECT VENT FIREPLACE. INSTALL PER MFR. SPECS
FLOOR OUTLET COORDINATE LOCATION W/ OWNER
SEE NOTE #1
Great Room
Hall
CLOSET
Kitchen
LIN
SEE NOTE #1
22½"x30" ATTIC ACCESS
CLOSET
CLOSET
CLOSET
Bedroom
Bedroom
Foyer
PANTRY
WASHER
Laun
UP 2 RISERS
VENT DRYER TO OUTSIDE
Porch
Garage
First Floor Notes:
1. SMOKE DETECTORS TO BE WIRED INTO HOUSE CURRENT W/ BATTERY BACK-UP, INTERCONNECT SO IF ONE SOUNDS, THEY ALL SOUND, TYP. @ EVERY FLOOR AND EACH BEDROOM.
2. VENT BATH EXHAUST FANS TO OUTSIDE, TYPICAL.
3. SET TOP OF WINDOW OPENINGS 6'-10 7/8" ABOVE FLOOR UNLESS NOTED OTHERWISE.
4. EXTERIOR DIMENSIONS SHOWN TO MATCH FOUNDATION. ACTUAL MASONRY DIMENSIONS TO PROJECT 1/8" FROM FOUNDATION TO ALLOW FOR 1" AIR SPACE TO SHEATHING.
5. UNLESS NOTED OTHERWISE, ALL INTERIOR WALLS ARE 3½" FOR 2 x 4 FRAMING @ 16" O.C.
First Floor Plan
Scale ¼" = 1'-0"

Studer Residential Designs, Inc.

Figure 15-3. A residential electrical plan.

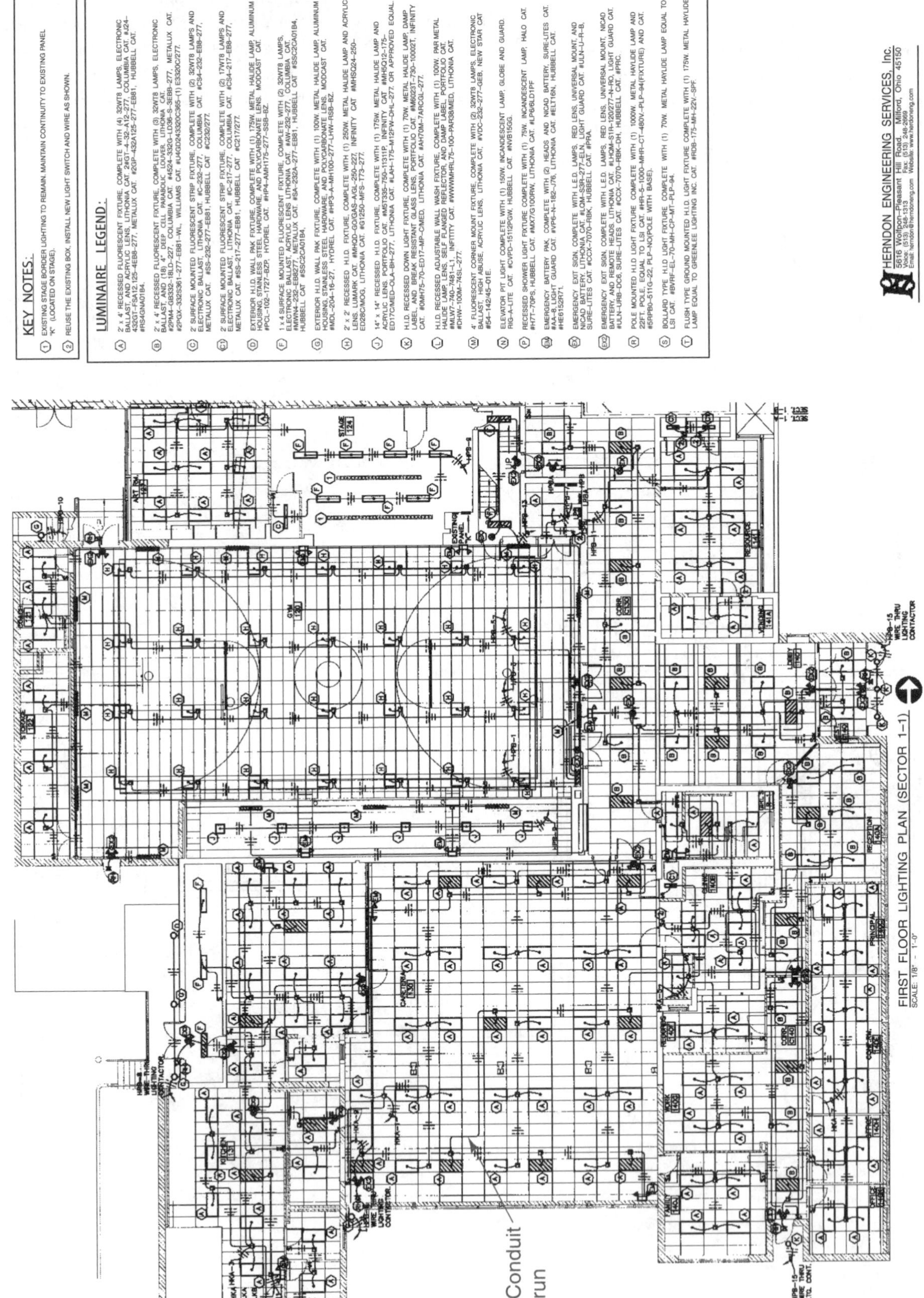

Figure 15-4. A lighting plan for a commercial building showing locations of conduit runs and light fixtures.

For larger construction jobs, one electrical plan may be prepared for outlets, another for lighting, and still another for the service entrance. These plans, together with the set of specifications, detail the electrical work to be done and the materials and fixtures to be used. To prevent interference between electrical conduit and other installed work in the field, such as plumbing lines, conduit runs must be carefully coordinated with the work of other trades.

Wiring Diagrams

A ***wiring diagram***, **Figure 15-5**, is used when wiring details cannot be shown clearly on the plan. Wiring diagrams correspond to a specific piece of equipment. The types of wire running between the equipment and its power source, sensors, gauges, and other related equipment are shown.

Schedules

There are many types of schedules used with electrical drawings:

- **Panel schedule.** All of the information associated with a distribution panel (also called a *circuit breaker box, breaker panel, lighting panel,* or *power panel*) is included in the ***panel schedule***, **Figure 15-6**. The voltage entering the box, the numbers and sizes of the breakers, and a brief description of the devices protected by the breakers are included.

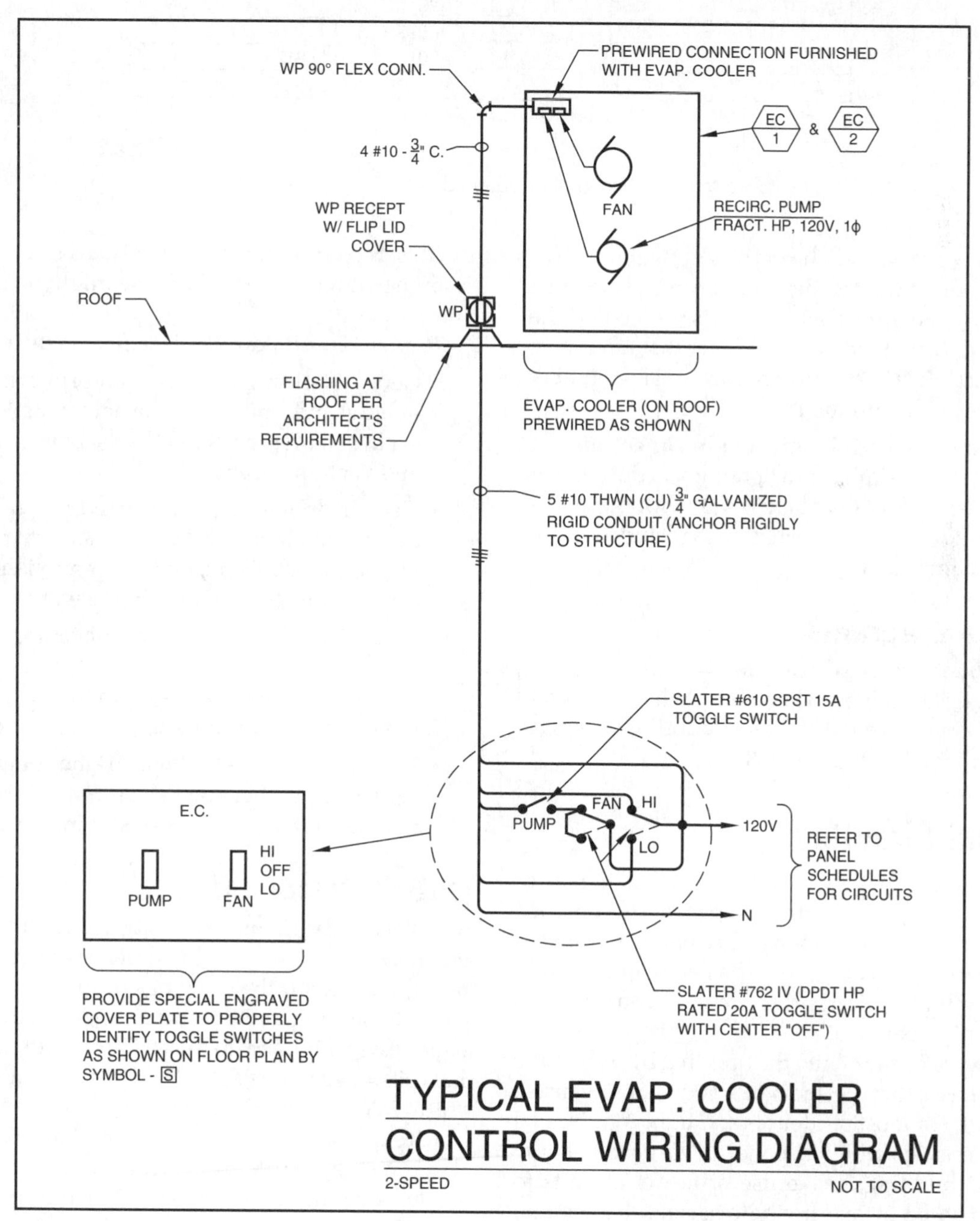

Goodheart-Willcox Publisher

Figure 15-5. A wiring diagram is used to clearly illustrate the wiring for complex equipment.

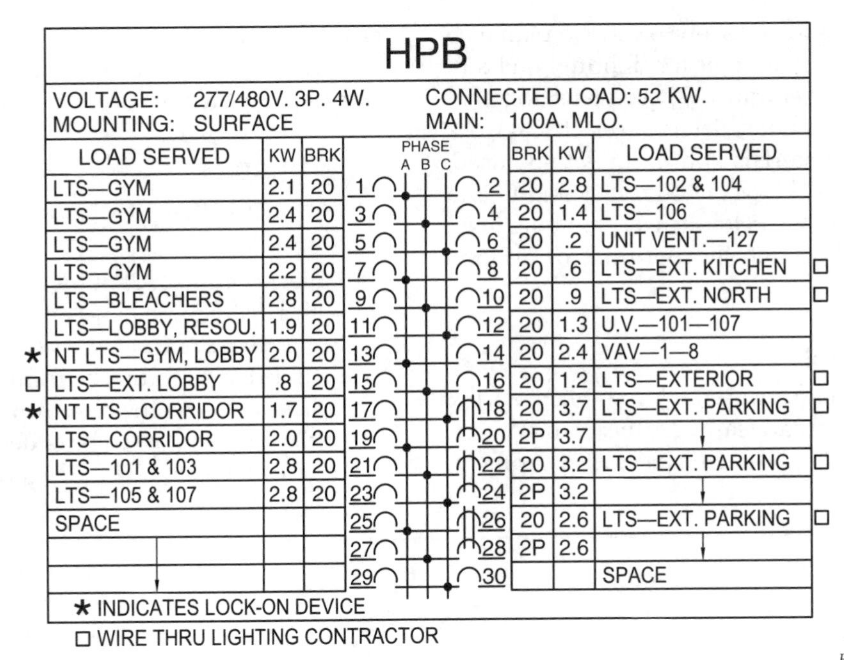

HPB

VOLTAGE: 277/480V. 3P. 4W. CONNECTED LOAD: 52 KW.
MOUNTING: SURFACE MAIN: 100A. MLO.

	LOAD SERVED	KW	BRK	CKT	CKT	BRK	KW	LOAD SERVED	
	LTS—GYM	2.1	20	1	2	20	2.8	LTS—102 & 104	
	LTS—GYM	2.4	20	3	4	20	1.4	LTS—106	
	LTS—GYM	2.4	20	5	6	20	.2	UNIT VENT.—127	
	LTS—GYM	2.2	20	7	8	20	.6	LTS—EXT. KITCHEN	□
	LTS—BLEACHERS	2.8	20	9	10	20	.9	LTS—EXT. NORTH	□
	LTS—LOBBY, RESOU.	1.9	20	11	12	20	1.3	U.V.—101—107	
★	NT LTS—GYM, LOBBY	2.0	20	13	14	20	2.4	VAV—1—8	
□	LTS—EXT. LOBBY	.8	20	15	16	20	1.2	LTS—EXTERIOR	□
★	NT LTS—CORRIDOR	1.7	20	17	18	20	3.7	LTS—EXT. PARKING	□
	LTS—CORRIDOR	2.0	20	19	20	2P	3.7		
	LTS—101 & 103	2.8	20	21	22	20	3.2	LTS—EXT. PARKING	□
	LTS—105 & 107	2.8	20	23	24	2P	3.2		
	SPACE			25	26	20	2.6	LTS—EXT. PARKING	□
				27	28	2P	2.6		
				29	30			SPACE	

★ INDICATES LOCK-ON DEVICE
□ WIRE THRU LIGHTING CONTRACTOR

Herndon Engineering Services, Inc.
CR architecture + design

Figure 15-6. A panel schedule shows the details of a distribution panel.

- **Lighting schedule.** The permanently mounted light fixtures used in the project can be listed in a ***lighting schedule.*** Each fixture is marked on the drawing with an identifying letter that references the schedule. The brand of fixture, catalog number, and power requirements are listed.
- **Equipment schedule.** An ***equipment schedule,*** **Figure 15-7,** is similar to a lighting schedule. It lists equipment rather than light fixtures. More detailed wiring and power information is included in the equipment schedule.

One-Line Diagrams

One-line diagrams (also referred to as *riser diagrams*) are schematic drawings. They show which pieces of equipment are connected electrically and what is used to connect them. See **Figure 15-8**.

Electrical Circuits

A ***circuit*** is the path of electricity from a source (distribution panel) through the components (receptacles, lights, etc.) and back to the source. Circuits are numbered on the drawing and connected by a heavy line, ending in an arrow that indicates the circuit is connected to the distribution panel.

Electricity is brought into the building by way of the ***service entrance*** through the meter and on to the distribution panel. For most residences and light commercial buildings, one distribution panel is sufficient. Large commercial buildings make use of feeder circuits to further distribute the electricity to several distribution panels. Larger conductor sizes should be used in feeder circuits. Feeder circuits help avoid excessive voltage drop in branch circuits that otherwise might be in excess of 100′ in length.

Branch circuits can be classified as the following:

- General lighting circuits used primarily for lighting and small, portable devices and appliances such as radios, clocks, TV sets, home computers, and vacuum cleaners.
- General appliance circuits used for outlets such as those along the kitchen counter serving toasters, waffle irons, mixers, and other appliances. These circuits are also used for home workshops.
- Individual circuits used for major appliances that require large amounts of electricity (such as ranges, washers, dryers, and refrigerators) or motor-driven equipment.
- Dedicated circuits for mainframe computers and servers and other special equipment that cannot tolerate voltage fluctuations or interruptions.

Circuit Calculations

Calculations for circuits should be made by the electrical engineer or contractor, if the number of circuits has not been indicated on the drawings or in the specifications. Circuits serving light fixtures and small appliances usually are planned to serve 2400 watts (20 amps × 120 volts = 2400 watts). These circuits normally are wired with 12 AWG (*American Wire Gage*) wire.

Circuits serving individual appliances would be wired with a size of wire and circuit breaker or fuse to safely carry the current required by the appliance. For example, an individual circuit for a 240-volt cooktop using

MECHANICAL EQUIPMENT SCHEDULE						
MK	HP	KW	LOAD/EA.	VOLTS-Ø	DISC. SW./LPN-RK FUSES	FEEDERS/CONDUIT
IU -- INDOOR UNITS						
1–4, 6, 8	$\frac{3}{4}$	8.0	41.7A	230 - 1	60/2P W/ (2) 60A	2 #6 & 1 #8 (BOND) THWN (CU)-1"C
5, 9	$\frac{1}{3}$	5.0	25.4A	230 - 1	60/2P W/ (2) 35A	2 #8 & 1 #8 (BOND) THWN (CU)-$\frac{3}{4}$"C
7	$\frac{1}{5}$	3.0	15.2A	230 - 1	30/2P W/ (2) 20A	2 #10 & 1 #10 (BOND) THWN (CU)-$\frac{1}{2}$"C
10, 11	-	1.0	8.3A	120 - 1	MANUAL MOTOR STARTER	2 #12 & 1 #12 (BOND) THWN (CU)-$\frac{1}{2}$"C
12	$\frac{1}{2}$	5.0	27.0A	230 - 1	60/2P W/ (2) 35A	2 #8 & 1 #8 (BOND) THWN (CU)-$\frac{3}{4}$"C
OU -- OUTDOOR UNITS						
1, 2, 4, 8	-	-	27.9A	230 - 1	WP, 60/2 W/ (2) 40A	2 #8 & 1 #8 (BOND) THWN (CU)-$\frac{3}{4}$"C
3, 6	-	-	36.3A	230 - 1	WP, 60/2 W/ (2) 50A	2 #6 & 1 #8 (BOND) THWN (CU)-1"C
5, 9	-	-	15.1A	230 - 1	WP, 30/2 W/ (2) 20A	2 #10 & 1 #10 (BOND) THWN (CU)-$\frac{1}{2}$"C
7	-	-	10.3A	230 - 1	WP, 30/2 W/ (2) 15A	2 #12 & 1 #12 (BOND) THWN (CU)-$\frac{1}{2}$"C
10, 11	-	-	10.0A	120 - 1	WP, 30/2 W/ (1) 15A	2 #12 & 1 #12 (BOND) THWN (CU)-$\frac{1}{2}$"C
12	-	-	25.4A	230 - 1	WP, 30/2 W/ (2) 35A	2 #8 & 1 #8 (BOND) THWN (CU)-$\frac{3}{4}$"C
EC -- EVAP. COOLERS						
1, 2	$\frac{1}{2}$	-	9.8A	120 - 1	SEE CONTROL DIAGRAM THIS SHEET	

The Fleischer Estancia
Christensen, Cassidy, Billington and Candelaria, Inc., Architects

Figure 15-7. An equipment schedule includes the power requirements for the machinery.

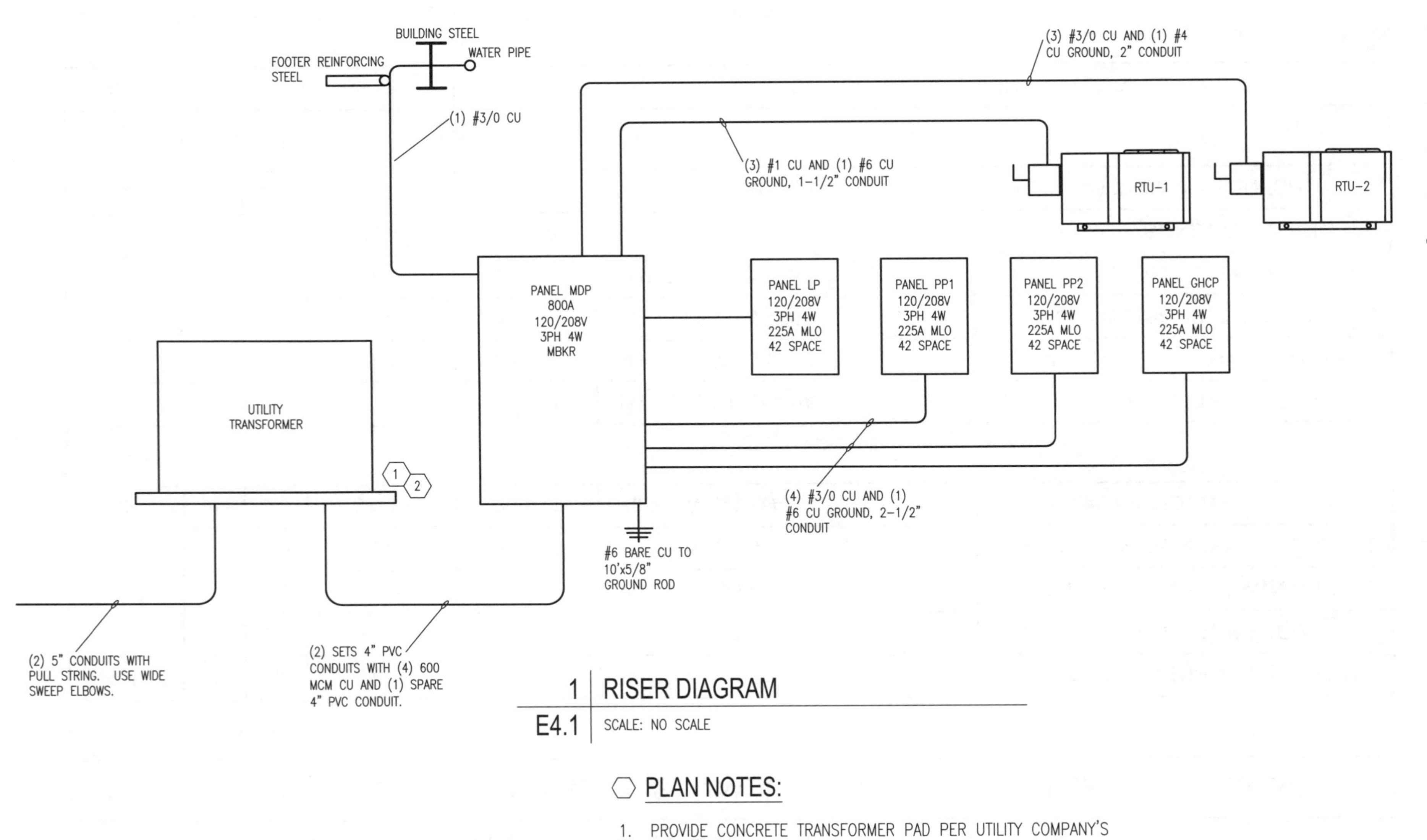

Delhi Flower and Garden Center
PE-Services

Figure 15-8. Example of a one-line diagram for a commercial building project.

8000 watts would require a 40-amp circuit (8000 watts ÷ 240 volts = 33.33 amps). This circuit would be wired using 8 AWG wire.

Low-Voltage Systems

The use of a low-voltage (24-volt) wiring system makes it possible to control the switching of any light or outlet in a building from any location in the building. The lights or outlets are wired using standard wire sizes of 12 AWG or 14 AWG wire. The switches are operated through a low-voltage system using low-voltage wiring. The switch (located anywhere in the building) in the low-voltage system activates a ***relay*** (electrically operated switch) at the outlet, which turns the device on or off. This system provides a method of remote control and allows for flexibility. Low-voltage (12 or 24-volt) wiring for doorbells or chimes can be wired with bell wire.

Electrical Codes

It is the architect's responsibility, with the assistance of consulting engineers, to design a building to meet existing codes. This includes the provisions of the NEC and any state or local codes. In the case of residential or light commercial buildings where no detailed electrical diagram or specifications are provided, the electrical subcontractor is responsible for meeting the provisions of the codes.

Types and Sizes of Conductors

Conductors (wires) in building projects are typically specified as copper. Aluminum is also used, most typically for service conductors bringing power to a home or building and for feeder conductors supplying power to distribution panels. Copper is typically used for branch circuit wiring. Sizes of wire are designated by gage numbers, based on the diameter of the wire. Some typical wire sizes are shown in **Figure 15-9**. Note that as the size number decreases, the wire size increases. Wire sizes larger than 1 AWG are referred to as 1/0 (1 "aught"), 2/0 (2 "aught"), etc. Most residential wiring calls for 12 AWG copper wire. On some electrical prints, this designation may appear as #12 AWG or No. 12 AWG. Other common wire sizes used in residential wiring are 10 AWG and 14 AWG (the smallest conductor permitted in branch circuits by the National Electrical Code is 14 AWG). Wire sizes are increased with longer runs.

Wire sizes up to 4/0 AWG are specified by diameter. Larger wire sizes are specified in circular mils, which is a measure of the cross-sectional area of the wire. One mil is equal to one-thousandth of an inch (.001"). When a wire diameter is specified in mils, the square of the diameter is equal to mils squared, or circular mils. In this system, for simplification, wire sizes are most commonly expressed in kcmil (thousand circular mils, where k = 1000). For example, a 250 kcmil wire refers

GREEN BUILDING

Alternative Energy Sources

Alternative energy sources such as solar panels and wind turbines provide additional options to meet the energy requirements of a building. These sources reduce energy costs and serve as alternatives to using electricity from the utility-operated "power grid." Alternative energy systems typically allow the owner to recover the initial cost of installation within a period of several years. Wind-powered electricity has grown in use in recent years. Some small-scale wind turbine systems for residential and commercial use are connected to the utility so that the building uses power supplied by the utility when the system is not generating enough electricity. When the system produces more electricity than needed by the building, excess electricity is sent to the electrical grid. This enables the owner to sell electricity back to the local utility company.

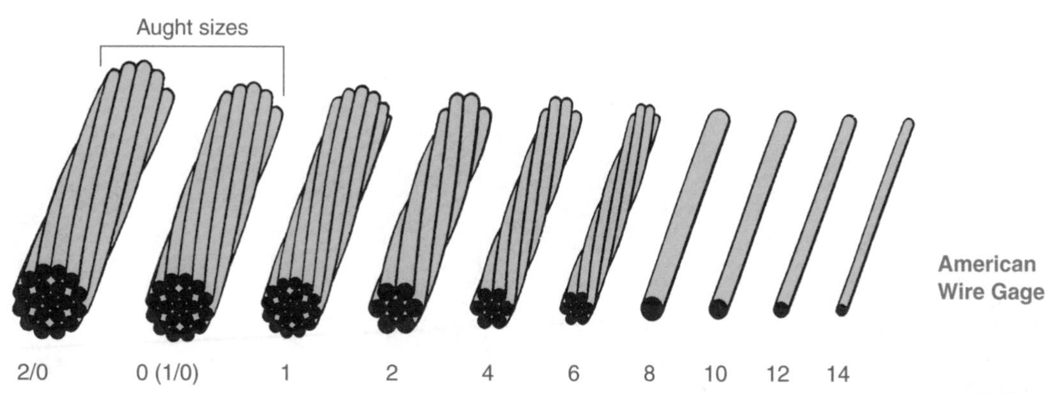

Figure 15-9. Relative sizes of electrical conductors (wires).

to a measurement of 250,000 circular mils. On some electrical prints, the kcmil designation may be specified as *MCM* (where *M* = 1000).

The National Electrical Code uses letters to designate the type of conductor insulation. This governs the use of electrical conductors. Some of the more common designations are listed in **Figure 15-10**. For more specific applications, refer to the latest edition of the National Electrical Code.

Types of Conductor Insulations	
Letter Designation	Type of Insulation
THHN	Flame-retardant, heat-resistant thermoplastic
THHW	Flame-retardant, moisture- and heat-resistant thermoplastic
THWN	Flame-retardant, moisture- and heat-resistant thermoplastic
THW	Flame-retardant, moisture- and heat-resistant thermoplastic
UF	Moisture resistant

Goodheart-Willcox Publisher

Figure 15-10. Common types of wire insulations.

Conduit

Conduit is enclosed tubing used to carry electrical conductors. Conduit and other types of enclosed channels used to hold conductors are referred to as ***raceways***. Common types of conduit include rigid metal conduit (RMC), electrical metallic tubing (EMT), and flexible metal conduit (FMC). Conduit is also available in nonmetallic materials.

Most rigid metal conduit is galvanized or enameled steel with threaded joints. Rigid metal conduit comes in 10′ lengths. The inside dimensions of rigid metal conduit are nominal sizes (1″ is actually 1.049″).

Electrical metallic tubing, also referred to as *thinwall*, has thinner walls than rigid metal conduit and can be bent to shape. It is usually made of galvanized steel, although it is available in bronze for use in corrosive atmospheres. It is available in sizes up to 4″. Electrical metallic tubing is connected with compression or setscrew fittings.

Flexible metallic conduit can be bent by hand and is more easily routed in construction. This type of conduit is also called *Greenfield*.

Conductors can also be run through metal or plastic surface raceways with snap-on covers. Raceways are also made to be placed in concrete floors that have surface-covered junction boxes at regular intervals so that electrical connections can be made where needed.

Test Your Knowledge

Name ____________________________________

Write your answers in the spaces provided.

____ 1. With regard to electrical construction, *NEC* stands for ____.

A. National Electric Company

B. Nominal Electrical Conductor

C. Negative Electric Circuit

D. National Electrical Code

E. None of the above.

____ 2. A ____ is used to connect a distribution panel to a light.

A. receptacle

B. conductor

C. relay

D. GFCI

E. None of the above.

____ 3. A ground fault circuit interrupter should be installed in a(n) ____.

A. bathroom

B. closet

C. overhead light

D. lighting panel

E. None of the above.

____ 4. For branch circuit wiring in residential construction, ____ wire is normally used.

A. aluminum

B. steel

C. copper

D. brass

E. None of the above.

____ 5. *True or False?* A lighting schedule lists equipment rather than light fixtures.

____ 6. *True or False?* A panel schedule includes a list of all the circuits running from the box.

____ 7. *True or False?* Electrical plans show the precise locations where conduit is to be run.

____ 8. *True or False?* An ampere is a measure of electrical power.

____ 9. *True or False?* A wire carrying 12 volts is considered a low-voltage wire.

10. Match the following plan symbols with the correct nomenclature.

A.

B.

C. $\$_3$

D.

E.

____ Telephone outlet

____ Duplex receptacle outlet

____ Ceiling outlet

____ Fluorescent fixture

____ Three-way switch

Activity 15-1

Electrical Plans for a Residential Building Project

Name ______________________________

Refer to Sheets 2, 3, and 4 from the Marseille residential building plans in the Large Prints supplement to answer the following questions.

1. What amp service is provided to the main breaker panel? Where is the main breaker panel located?

____

2. Where are the smoke detectors located on each level?

A. Basement ______________________________

B. First floor ______________________________

C. Second floor ______________________________

3. What model is specified for the smoke detectors?

____

4. What kind of exhaust fan is specified for the bathrooms?

____

5. What do the following symbols indicate?

A. ⦶ ______________________________

B. $ ______________________________

C. ✲ ______________________________

D. ⊗ ______________________________

6. What type of light fixture is specified for the front terrace entry lights?

____

7. Where are the front terrace entry lights switched? What else is also lit by this switch?

____

____

8. How many lights are in the following rooms?

   A. Kitchen ______________________________

   B. Garage ______________________________

   C. Master Bath ______________________________

9. What type of light fixture is specified for the dining room? How is it switched?

   ____

   ____

10. What type of fixture is specified for the garage car door lights? How many fixtures of this type are there?

    ____

    ____

11. How many ceiling fans are there in the house? What are their specifications?

    ____

    ____

12. What is specified for the electrical service for the clothes dryer?

    ____

Activity 15-2

Electrical Plans for a Commercial Building Project

Name ______________________________

Refer to Sheets E1.1, E1.2, E2.1, and E4.1 from the Delhi Flower and Garden Centers commercial building plans in the Large Prints supplement to answer the following questions.

1. Which circuits serve the following?

   A. Water heater EWH-2 ______________________________

   B. Unit heater 3 ______________________________

   C. Receptacles in count room ______________________________

   D. RTU-2 ______________________________

   E. Cash registers ______________________________

2. Where are the electrical breaker panels located?

   ____

3. What circuit serves the sliding doors at the main entrance?

   ____

4. What is to be done below the restroom sinks?

   ____

5. What light fixtures are specified at the following locations?

   A. Checkout ______________________________

   B. Greenhouse sales ______________________________

   C. Trellis at outside sales ______________________________

   D. Loading/shipping ______________________________

6. Identify the symbol for each of the following items. Sketch the symbol freehand.

   A. In-floor quad receptacle ______________________________

   B. Weatherproof switch ______________________________

   C. Dimmer switch ______________________________

   D. Wall mounted emergency lighting fixture—battery type ______________________________

7. What size wire and conduit is supplied to the RTU-2 unit?

____

8. How is the exterior light pole grounded?

____

9. What model is specified for ceiling fans and who is responsible for installation?

____

10. Which panel and circuit is designated for the ceiling fans in the storage room?

____

Goodheart-Willcox Publisher

Welding is commonly used in the construction industry to fasten structural steel framing members. In order to make welds correctly, a welder must have a high level of skill and the ability to interpret information provided on prints.

UNIT 16
Welding Prints

TECHNICAL TERMS

American Welding Society (AWS)
arrow side
contour symbol
field weld symbol
groove angle
groove weld
melt-through symbol
other side
reference line
spot welds
tail
weld all-around symbol
weld dimensions
weld symbol
welding symbol

LEARNING OBJECTIVES

After completing this unit, you will be able to:

- Identify different portions of a welding symbol.
- Explain weld details based on the welding symbol on a print.

Welding is one of the principal means of fastening members in structural steel work, **Figure 16-1**. The ***American Welding Society (AWS)*** has developed standard procedures for using symbols to indicate the location, size, strength, geometry, and details of a weld. The welding symbols studied in this unit will assist you in reading and interpreting drawings involving welding processes. Welding requires knowledge of specific welding processes and careful attention to safety and should be performed by certified personnel.

Goodheart-Willcox Publisher

Figure 16-1. A welder at work.

Welding Symbols

It is important to distinguish between the *weld symbol* and the *welding symbol*. The ***weld symbol*** indicates the specific type of weld. The ***welding symbol*** consists of the weld symbol and any additional information needed to completely specify the required weld. A welding symbol includes the following elements:

- **Reference line.** The ***reference line*** is the horizontal line portion of a welding symbol, **Figure 16-2**. It has an arrow at one end and a tail at the other. On some prints, the reference line may be vertical.
- **Arrow.** An arrow is used to connect the welding symbol reference line to one side of the joint to be welded. This is considered the ***arrow side*** of the joint. The side opposite the arrow is termed the ***other side*** of the joint.

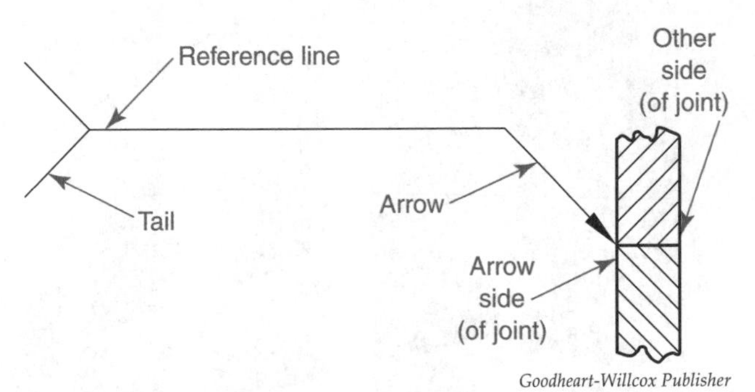

Goodheart-Willcox Publisher

Figure 16-2. Basic welding symbol.

- **Tail.** Notes are placed within the ***tail*** to designate the welding specification, process, or other reference. See **Figure 16-3**.
- **Weld symbol.** The basic weld symbols for various types of welds are shown in **Figure 16-4**. If the symbol is below the reference line, the weld is made on the arrow side of the joint. If the symbol is above the reference line, the weld is made on the other side of the joint. If the symbol is present on both sides of the reference line, the weld is made on both sides of the joint. See **Figure 16-5**. A more comprehensive list of weld symbols and their applications is given in the Reference Section of this textbook.

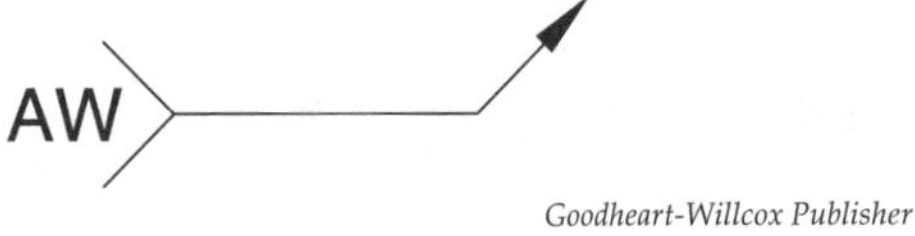

Goodheart-Willcox Publisher

Figure 16-3. A welding symbol with a tail section designating arc welding.

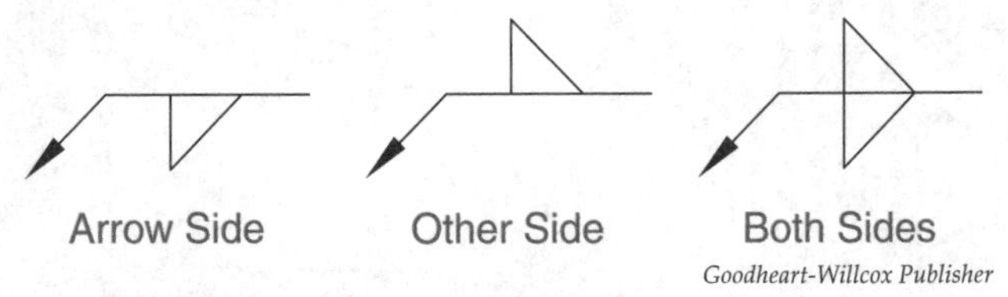

Goodheart-Willcox Publisher

Figure 16-5. Location of welds.

- **Weld dimensions.** ***Weld dimensions*** are drawn on the same side of the reference line as the weld symbol, **Figure 16-6A**. When the dimensions are covered by a general note, the welding symbol need not be dimensioned, **Figure 16-6B**. When there are welds on both sides of the joint and both welds have the same dimensions, one or both can be dimensioned, **Figure 16-6C**. The pitch of staggered intermittent welds is shown to the right of the length of the weld, **Figure 16-6D**.

Supplementary symbols, **Figure 16-7**, are used on welding symbols to provide additional information about the weld to be made. Supplementary symbols used with the welding symbol include the following:

- **Contour symbol.** A ***contour symbol*** next to the weld symbol indicates welds that are to be flat-, convex-, or concave-faced. See **Figure 16-8**. When welds are to be finished by machining or some other process, a finish symbol is included with the contour symbol. The finish symbol indicates the finishing method to be used (C = chipping, G = grinding, H = hammering, M = machining, P = planishing, R = rolling, U = unspecified). Refer to **Figure 16-8D**.

Groove							
Square	Scarf	V	Bevel	U	J	Flare-V	Flare-Bevel

Fillet	Plug	Slot	Stud	Spot or Projection	Seam	Back or Backing	Surfacing	Edge
	Ø Ø							

Goodheart-Willcox Publisher

Figure 16-4. Basic weld symbols.

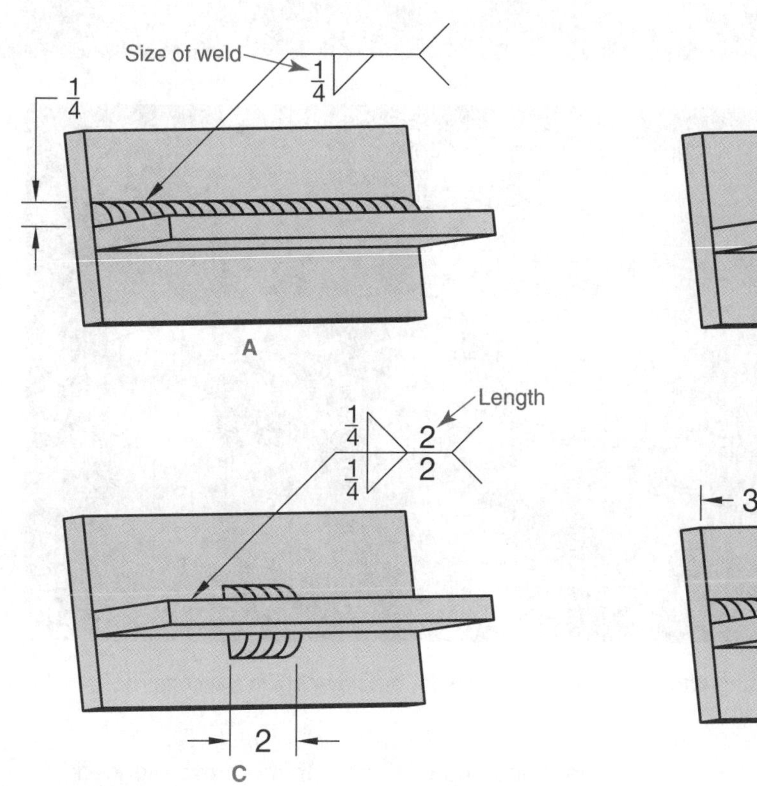

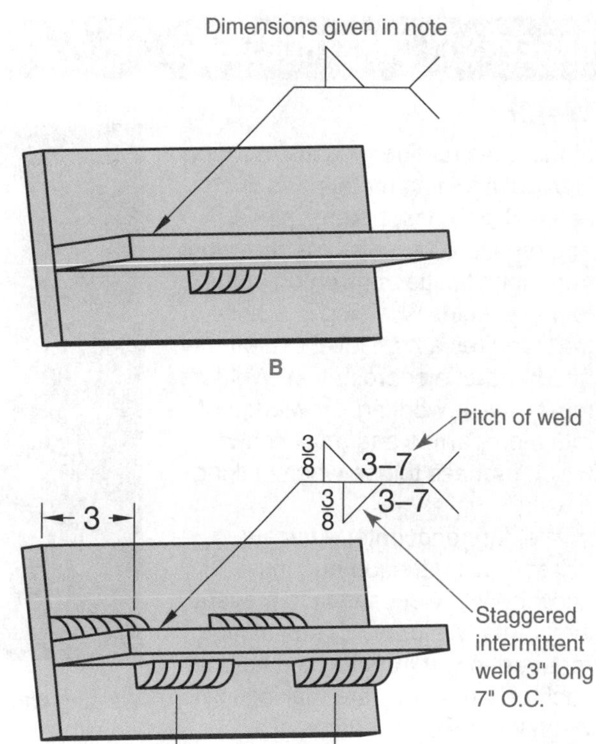

Goodheart-Willcox Publisher

Figure 16-6. Examples of weld dimensions.

- **Weld all-around symbol.** The *weld all-around symbol,* **Figure 16-9**, indicates that the weld extends completely around a joint. The symbol consists of a small circle surrounding the intersection of the reference line and the arrow.
- **Field weld symbol.** The *field weld symbol,* **Figure 16-10**, consists of a small line and triangle originating at the intersection of the reference line and arrow. This symbol identifies welds to be made at the construction site, rather than in the assembly shop.
- **Melt-through symbol.** A *melt-through symbol* indicates where 100% joint or member penetration is required in welds made from one side. See **Figure 16-11**.

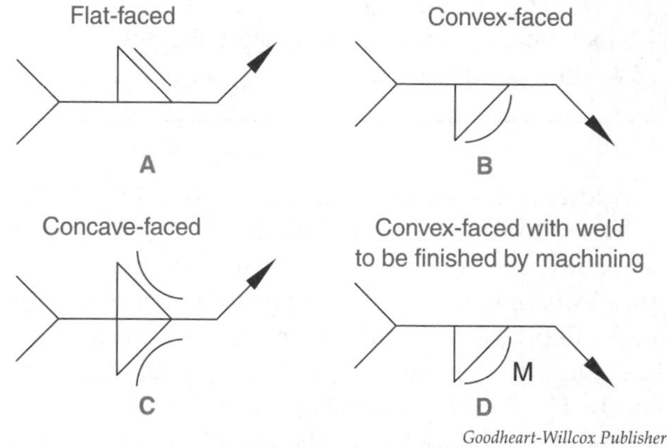

Goodheart-Willcox Publisher

Figure 16-8. Contour symbols show whether a weld is to be flat-, convex-, or concave-faced. A finish symbol (shown in D) is used to indicate a finishing process for the weld.

Weld All-Around	Field Weld	Melt-Through	Consumable Insert (Square)	Backing or Spacer (Rectangle)	Contour		
					Flush or Flat	Convex	Concave
				Backing Spacer			

Goodheart-Willcox Publisher

Figure 16-7. Standard supplementary symbols.

CAREERS IN CONSTRUCTION

Welder

In the construction industry, welding is used in joining metal parts such as steel columns, beams, pipes, and pressure vessels. Welders work from specifications given on prints to make welds. Welding is a skill that requires a high level of dexterity and hand-eye coordination. Welders must have a working knowledge of metal properties and must know which process to use when making a weld.

Welders encounter a variety of hazards on the job and must know how to work safely to prevent accidents. Welders are sometimes required to work at elevated heights, such as when they are making welds on the steel beams of a multilevel structure. Often, welders are required to work in unusual positions and must be able to stand or bend for long periods of time. Welders must always wear appropriate clothing and eye protection.

Training to become a welder usually begins in a technical school or community college program. Welders acquire skills by gaining on-the-job experience and working with more experienced welders. Welders in construction are employed by steel fabricators, structural steel installation subcontractors, and construction companies. Welders are generally in demand in a variety of industries because of the need to fabricate, maintain, and repair welded products.

Kokliang/Shutterstock.com

Welding offers many opportunities for employment in the construction industry.

Additional specifications are used with welding symbols to designate requirements for a specific weld. For example, a welding symbol indicating a ***groove weld*** may include specifications for the groove angle, depth of preparation for the joint, groove weld size (depth of penetration), and root opening. See **Figure 16-12**. The ***groove angle*** is shown on the same side of the reference line as the weld symbol. The weld size (depth) of a groove weld is shown to the left of the weld symbol. The depth of preparation for the groove joint is shown to the left of the weld size. The root opening of a groove weld is shown inside the weld symbol.

Spot welds are specified by their diameter, strength in pounds, pitch (center-to-center spacing), and number of welds. See **Figure 16-13**.

Welds are verified for quality by conducting inspections and tests. The most common inspection method is visual inspection. Other inspection methods include magnetic particle inspection, liquid penetrant inspection, and X-ray inspection.

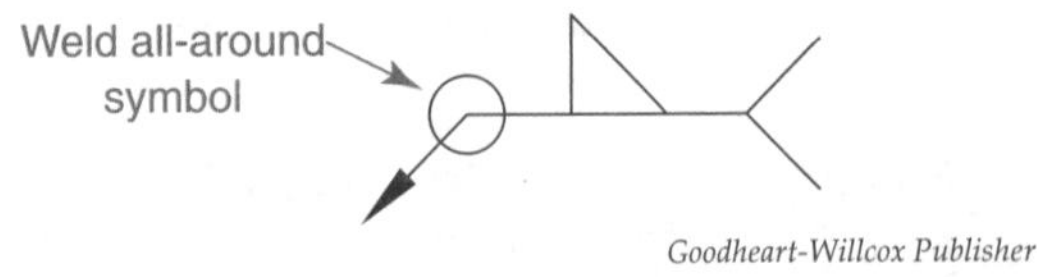

Goodheart-Willcox Publisher

Figure 16-9. A welding symbol with the weld all-around symbol.

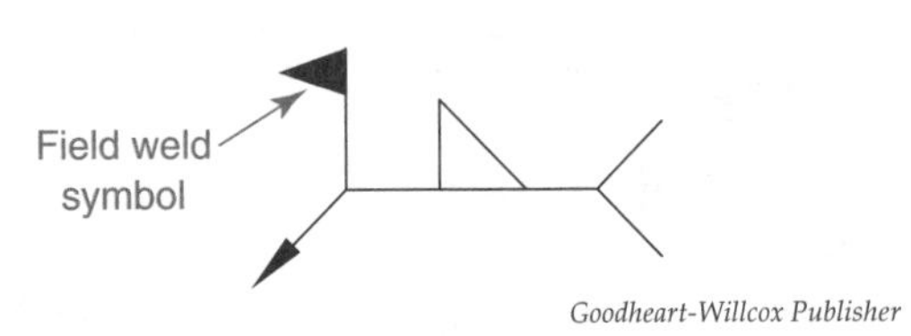

Goodheart-Willcox Publisher

Figure 16-10. A welding symbol with the field weld symbol.

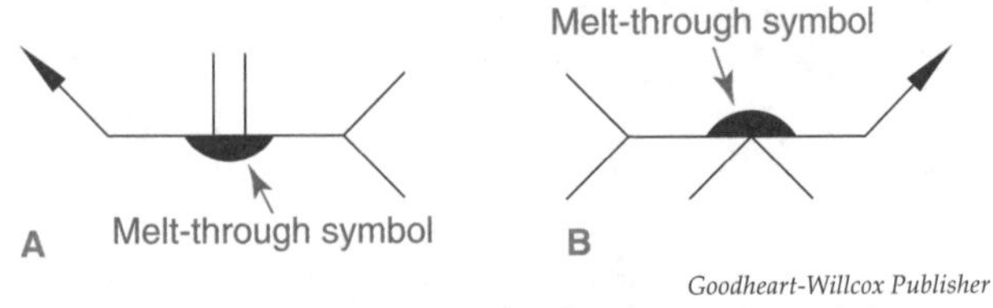

Goodheart-Willcox Publisher

Figure 16-11. Welding symbols with the melt-through symbol. A—Square groove. B—V-groove.

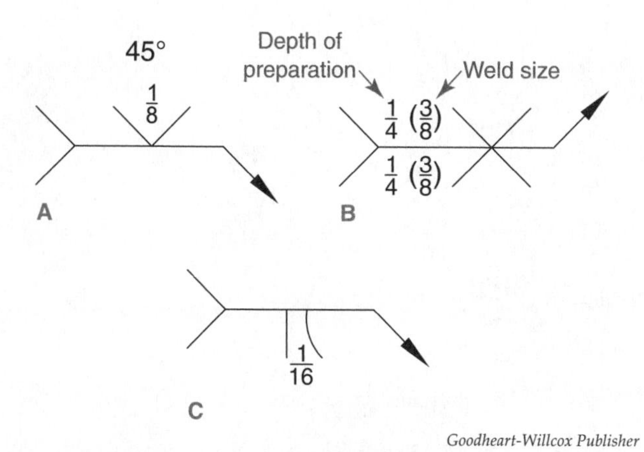

Goodheart-Willcox Publisher

Figure 16-12. Specifications for groove welds. A—The groove angle is shown on the same side of the reference line as the weld symbol. B—The weld size (depth of the weld) is shown to the left of the weld symbol in parentheses. The depth of preparation for the groove joint is shown to the left of the weld size. C—The root opening is shown inside the weld symbol.

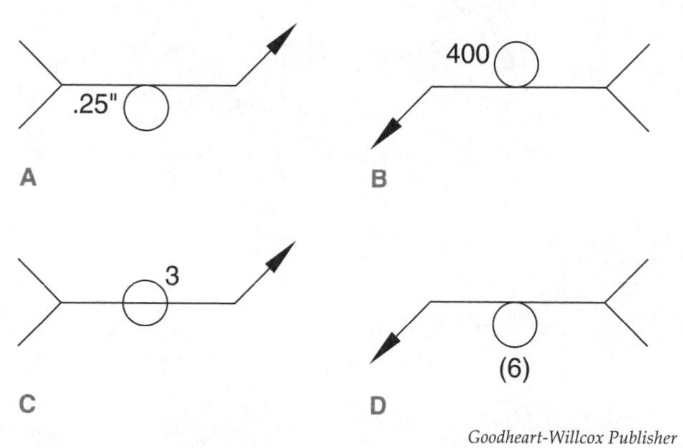

Goodheart-Willcox Publisher

Figure 16-13. Specifications for spot welds. A—Diameter. B—Strength in pounds. C—Pitch. D—Number of welds.

fotoinfot/Shutterstock.com

The ability to accurately read dimensions from a print is important on the job site. An error in reading dimensions can be costly in terms of wasted material and time.

Test Your Knowledge

Name ________________________________

Write your answers in the spaces provided.

_____ 1. *True or False?* The American Welding Society develops welding standards.

_____ 2. *True or False?* On the welding symbol, the arrow touches the tail.

_____ 3. *True or False?* A weld symbol below the reference line signifies a weld on the arrow side of the joint.

_____ 4. *True or False?* If a weld is to be made in the field (at the job site), a note must be included within the tail.

_____ 5. *True or False?* The size of the weld is indicated inside the weld symbol.

6. Interpret the welds specified by the following welding symbols:

A.  ________________________________

B.  ________________________________

C. 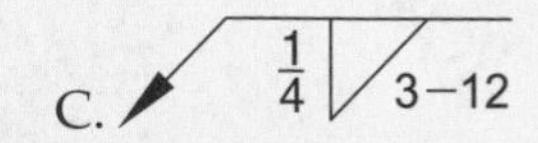

____

D. 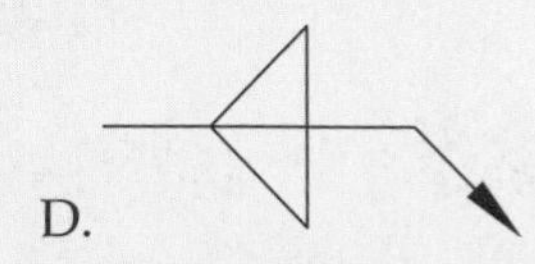 ________________________________

7. In the space provided, draw a symbol used to indicate a fillet weld that is to be made on the arrow side and finished by machining.

8. In the space provided, draw a symbol used to indicate a 1/4″ spot weld that is to be made on the arrow side at six locations.

9. In the space provided, draw a symbol used to indicate a V-groove melt-through weld to be made on the arrow side.

10. Briefly explain the difference between a weld symbol and a welding symbol.

____

____

____

____

Activity 16-1

Commercial Welding Prints

Name ________________________________

Refer to Sheets S4.1 and S5.1 from the Delhi Flower and Garden Centers commercial building plans in the Large Prints supplement to answer the following questions.

1. Referring to Section 1 on Sheet S4.1, describe how the 3″×3″×1/4″ continuous angle is attached to the steel joist.

   ____

   ____

2. Referring to Section 8 on Sheet S4.1, what type of weld is specified to connect the peak bent plate to the HSS8×6 beam?

   ____

   ____

3. Referring to Section 9 on Sheet S4.1, what is the specification to attach the W16 beam to the HSS8×6 purlin?

   ____

   ____

4. Referring to Section 13 on Sheet S4.1, how is the HSS4×4 purlin welded to the 1/4″×6″ continuous plate?

   ____

   ____

5. What type of welding rods are specified for field welds?

   ____

   ____

SECTION 5
Estimating

Unit 17 Estimating Construction Costs

Goodheart-Willcox Publisher

UNIT 17 Estimating Construction Costs

TECHNICAL TERMS

addendum
approximate method
budget parameter estimating
detailed method
labor cost
overhead
production rates
profit
takeoffs

LEARNING OBJECTIVES

After completing this unit, you will be able to:

- Explain how building costs, overhead costs, and profit are calculated.
- Recognize the proper estimating method for a given situation.
- Use the approximate and detailed methods of estimating.

Estimating the cost of a construction project is one of the most important operations in a contracting business. If the estimator is careless and misses items that are included in the scope of work or contract, or does not compute costs accurately, the business could suffer severe losses. If some items are included more than once (by carelessness or by overlapping subcontractor bids), or if production rates are too low on self-performed work activities, the contractor may bid too high and lose the job to the competition. The unending challenge to the estimator is to balance between bidding too low on a job to make a profit and bidding too high by adding extra money for unexpected costs and losing the job. The estimator must carefully evaluate a number of factors in order to bid a job competitively while maintaining profitability.

Estimating can be a relatively simple procedure, if a systematic approach is used. Reasonable care must be given to the details of preparing the estimate. In this unit, you will have an opportunity to study the general procedure used in estimating building costs.

In construction projects, an *estimate* is simply what the word means—it is an estimate of the projected cost of the project. Most jobs rarely come out exactly as estimated. The job will most likely have both overruns and underruns. The goal is to have more underruns than overruns.

Building Costs

Almost all building costs are made up of labor, material, and equipment. Often, subcontractors are used on a job and they too will break down their estimates into labor, material, and equipment. These costs must be calculated very carefully in order to develop an accurate estimate.

Labor

Labor refers to the people (construction workers, also called tradeworkers) required to put the construction materials in place. When developing the ***labor cost***, you must determine the labor rates for each trade, such as carpenter, electrician, or plumber. You also must determine ***production rates***, which establish how much material can be installed per hour or day. For example, to install 10 base cabinets and 8 wall cabinets in a kitchen, you must consider the following: It takes two carpenters one hour to set a base cabinet and 30 minutes to set a wall cabinet. If the carpenter wage rate is $15 per hour, the cost would be:

- 10 base cabinets × $30 (two carpenter-hours at $15 each), or $300
- 8 wall cabinets × $15 (two carpenter half-hours at $7.50 each), or $120

Adding $300 to $120 yields $420, or the labor cost to install the kitchen cabinets. Labor can become very complex as the construction components get broken down into many small detailed activities.

The actual labor cost used for estimating typically will be higher than the hourly rate paid to the worker. This higher labor cost reflects other factors that impact the final figure. Among these factors are the taxes and benefits (such as health insurance, vacation pay, retirement contributions, etc.) that the employer pays on behalf of the employee. Factors such as insurance will vary considerably by location and the type of work involved. Becoming familiar with the rates in your local area or the area in which the project is located is extremely important in developing accurate labor rates for the estimate.

Labor cost is also broken down based on the experience of the tradeworker. A new worker or apprentice will receive less pay per hour than a highly experienced or journeyman worker. Most work crews are made up of a composite of experienced individuals. To estimate labor costs for crews, the hourly rate of each individual is multiplied by the number of hours of work. Then, the individual totals are added to arrive at a final figure. The number of hours allotted for each task is based on how much production each crew can accomplish per hour or day. **Figure 17-1A** shows labor rates for each crew involved in decorative concrete work. The labor rates include hourly wages and fringe benefits. **Figure 17-1B** shows a sample crew makeup for stamped concrete work. The work is crewed out per task by designating the number of individuals needed. The total labor cost is calculated based on the labor rates and number of hours per individual for each task involved. This type of labor development is highly specific to each company and the ability of the crews to produce the work.

Material

Material costs are broken down into two categories. The first category consists of the materials that are actually installed in the building project. The second category consists of the materials that it takes to build the structure, such as the formwork for cast-in-place concrete or the paintbrushes used for painting the building (these materials will be removed or discarded at the completion of the project).

Depending on the project and scope of work involved, construction bids can also be included in the process of estimating material costs. Construction bids are typically prepared by material suppliers or subcontractors for materials such as structural steel, reinforcing steel, windows, and interior doors. In some cases, certain materials are furnished and installed. For example,

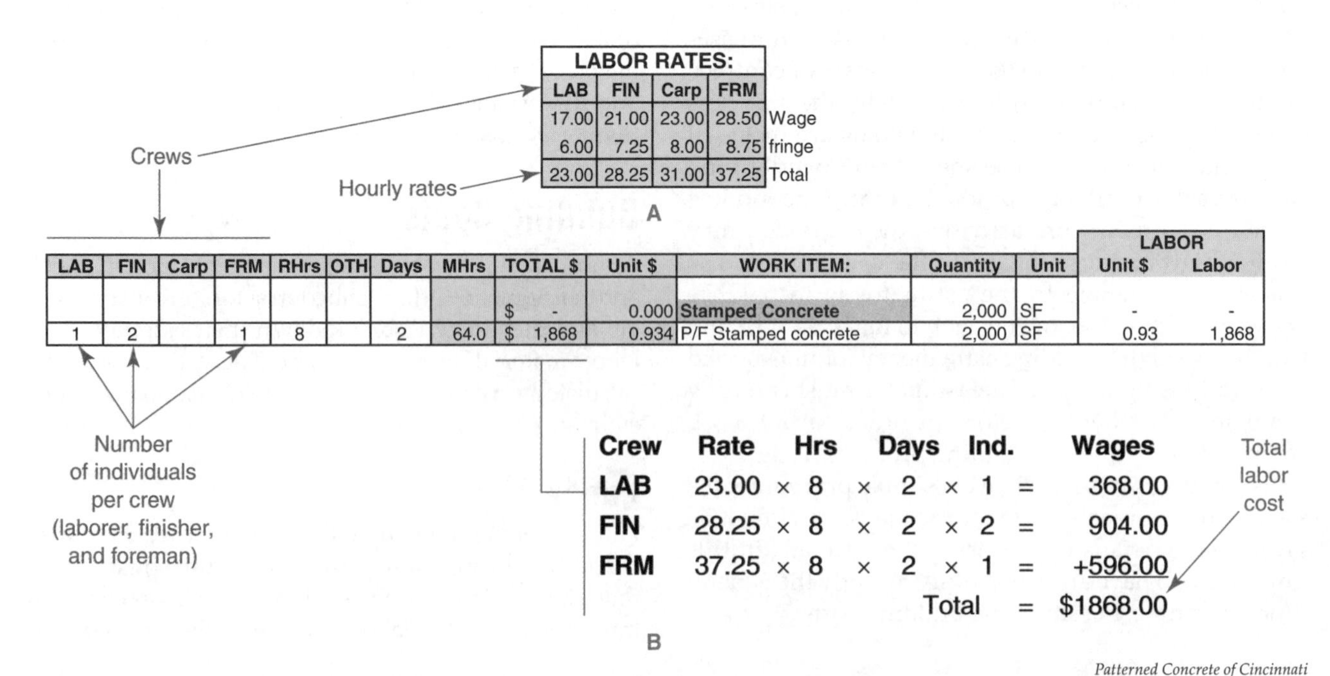

Patterned Concrete of Cincinnati

Figure 17-1. A breakdown of the total labor cost for stamped concrete work in a construction project. A—A table showing labor rates for each crew involved in the work. B—The number of individuals required for each crew and hours of work are specified in the table. Wages are calculated for each crew member and added to arrive at the total cost.

CAREERS IN CONSTRUCTION

Estimator

The estimator plays an important role on the construction team. All of the stakeholders involved in a construction project employ an estimator to analyze costs. The general contractor employs an estimator to put together the cost of the entire building process. The subcontractor employs an estimator to determine the cost of providing the subcontractor's trade on the job. The material supplier employs an estimator to provide firm bids on material goods for the project.

An estimator examines the plans for the entire job or as they relate to an area of interest and prepares takeoffs of all labor and material costs. Usually, the chief estimator assigns a project to a lead estimator. Besides being responsible for the labor and material costs, the estimator must also develop a schedule, general conditions, and contingencies for the project. Several team members finalize the estimate on a major project by checking and cross-checking each other's work for errors and scope exclusions.

While some estimators enter the field by earning a college degree in engineering, construction management, or architecture, estimating is a skill learned by experience more than anything else. Most estimators have field experience in constructing the kinds of structures they estimate. The experience they have gained is invaluable in providing an accurate cost of construction.

There are many training courses for estimating taught by trade associations, technical schools, and individual trainers. The American Society of Professional Estimators (ASPE) provides education and a certification program for the estimating profession.

Cineberg/Shutterstock.com

Estimators are responsible for calculating all costs relating to a construction project.

a roofing subcontractor may supply and install the materials for a roofing system, the mason will furnish and lay the brick, a flooring contractor will supply and install the ceramic tile, wood, and other products used for flooring, and so on. It is the responsibility of the estimator to develop the materials list, calculate material costs, and calculate the related labor costs.

Equipment

Equipment for a project includes the items that it takes to build the building, such as cranes, air compressors, high lifts, electric generators, and so on. Equipment is broken down into large and small equipment. If the job is large enough, the company may decide to purchase small hand and power tools, such as drills and saws, and expense them to the job. For larger equipment, the company may invest in items that can be used on current jobs and in future projects, such as cranes, bulldozers, and generators. The company can then develop a daily, weekly, or monthly usage and maintenance rate for the equipment and calculate an accurate cost estimate for a given job. Or, the company can simply rent the equipment from an outside rental company and prepare an estimate based on rental fees. Determining the types of equipment needed and the amount of usage on a given project is critical to preparing an accurate estimate.

Overhead and Profit

Overhead is the cost of doing business for a company. Overhead costs include expenditures for office space, company vehicles, phone use, office utilities, office supplies, and so on. Overhead costs also include salaries for office employees, such as estimators and general management personnel. A company's overhead cost divided by annual revenue is the percentage cost of overhead. This percentage represents the amount of money the company needs to bring in to break even for the year. The percentage cost of overhead can vary throughout the course of a year and needs to be monitored very closely. However, depending on the scope of a job, a company may elect to use a fixed overhead cost as a base for calculating job expenses. For example, if a company has $320,000 in fixed overhead costs and earns $2 million in annual sales, the percentage cost of overhead is 16% ($320,000 ÷ $2,000,000 × 100% = 16%). Based on annual sales of $2 million, this company would have to mark all jobs up a minimum of 16% to break even for the year. Note that if sales or overhead costs increase or decrease during the year, this percentage is affected and calculations must be adjusted accordingly.

Profit is the money a company intends to make above and beyond construction and overhead costs. This is usually calculated based on the risk of a specific type of job and the work performed. A self-performed job by a company is more risky and requires a higher mark-up for profit than a job in which a subcontractor is hired to supply material and perform a defined scope of work. Each company must develop its own profit/risk schedule and allocate the appropriate amount of profit to the job estimate.

Estimating Methods

There are several ways to estimate the costs for a construction project. Two broadly classified methods are the approximate method and the detailed method. Each has a place in construction.

Approximate Method

The ***approximate method,*** also known as ***budget parameter estimating,*** is used by owners, architects, or contractors to determine a *rough estimate* of cost. This estimate is made by breaking down the building into major components in either square feet or cubic feet, cubic yards, or other units and then multiplying the units by dollars per unit cost (based on similar projects or experience).

For example, a customer's finished residential foundation required 150 cubic yards of concrete for a 4000 square foot home. The foundation cost was $25,050. That breaks down into a cubic yard price of $167, and a square foot price of $6.26. A prospective customer has decided to build a new home of similar style with an area of 3570 square feet. Based on your experience with the previous home, you multiply the square footage by $6.26 for a budget of $22,348. It may be safer to round up to $6.50 until the drawings are produced and a more accurate estimate can be prepared. This kind of approximation can include more or less detail, depending on the completion of the construction documents and the information provided.

The accuracy of the estimate is affected by many variables. A more accurate estimate will result if the following are true:

- The cost per unit was arrived at on a sound basis of experience.
- Costs have not increased since the unit cost was established (or an appropriate increase is included).
- The construction project to be built is nearly identical to the one with which it is being compared.

The approximate method of estimating is often used in helping an owner decide whether or not to build and establish the financing for the project. This method is usually used early in the project planning stages when drawings and specifications are incomplete.

Detailed Method

The ***detailed method*** is the only accurate basis of cost estimation when the project is unique in structure, size, site, geographic area, or any other aspect. The detailed method involves:

- A careful study of the completed prints (contract documents).
- A detailed understanding of the specifications.
- Creating ***takeoffs*** (lists) of all labor costs, material costs, subcontractor costs, equipment costs, project management and supervision costs, energy costs, overhead, and profit.
- Developing a schedule listing all construction activities for the duration of the project.

Steps in Preparing a Detailed Estimate

The following steps will assist you in developing a systematic approach to detailed estimates:

1. Study the construction prints.
2. Study the construction specifications.
3. Study the addenda on the project. An ***addendum*** is a change to the contract documents made by the architect or design team prior to the construction bid.
4. Create labor and material takeoffs (these usually follow the headings of the specifications).
5. Price labor and materials and calculate total costs.
6. Create a subcontractor list based on the listed specification items. For example, list items supplied by the landscaping contractor, drywall contractor, flooring contractor, and plumbing contractor. This list can become very extensive and should be properly managed during the bid development stage.
7. Add other costs, including those covered under the General Conditions in the specifications. Use the following as a guide.
   A. Fees for permits.
   B. Job site trailers and supervision.
   C. Utilities and fuel—equipment fuel, energy for tools, and electric, gas, telephone, water, and sewer hookups.
   D. Insurance protection for workers and materials.
   E. Overhead for the business or office operation.
   F. Profit for the job.

Forms for Preparing a Detailed Estimate

Most contractors develop a set of forms or sheets that suit their particular needs in preparing an estimate. **Figure 17-2** shows a typical estimating form detailing the job tasks related to decorative concrete work. The work items include stamping concrete, exposed aggregate work, and staining concrete. For a major building project, a complete estimate would consist of many pages with multiple categories defining the entire scope of the job. The estimating form in **Figure 17-2** is a simplified version using sample entries for typical items. This form is for reference only and is not representative of a comprehensive estimating form for a competitive bid. This type of form is typically generated by creating an electronic spreadsheet document using software such as Microsoft Excel. A form created in this manner can have data entries that are linked to formulas based on current labor rates and material costs. Total costs are calculated automatically and are updated when different labor or material costs are entered.

Study the organization of items and data in **Figure 17-2**. Note the three columns breaking down labor costs, material costs, and equipment and subcontractor costs. On the left side of the sheet, note the labor rates and note that most labor activities are specified in 8-hour full-day increments. This is due to the fact that most people like to work full days. A construction company will generally try to accommodate this scheduling for its employees.

The rows near the bottom of the sheet show calculations for taxes and insurance, overhead and profit, and the total cost of the project. A specified percentage is used to calculate taxes and insurance for company employees. This percentage accounts for payroll taxes, unemployment taxes, and worker's compensation insurance for each employee and would vary from state to state and contractor to contractor. The overhead and profit calculations for the project are also based on a specified percentage. The same overhead cost percentage is used for labor, materials, and equipment and subcontractors. Different percentages are used for profit mark-ups. The different mark-ups are for self-performed labor by the company, materials, and equipment and subcontractors. The mark-up for self-performed work is higher than the mark-ups for materials, equipment, and subcontractors. The risk in the job is the labor, not the material, equipment, and subcontractors. This is how a self-performing contractor would estimate a job.

The total bid for the job is listed as $47,116. Note that the cost per square foot is also given. This is based on the total square footage of the project and provides a rough estimate of the cost of construction.

Checking a Detailed Estimate

After completing a detailed estimate, the estimator should carefully check through the prints and specifications for all items in the contract. Then, if all things have been considered, the estimate should be checked by the approximate method, preferably by another person. The cost can be roughly compared to a similar local project that has already been completed. This manner of checking will help avoid any serious loss due to omission of an item or an error in calculations.

Calculations Used in Construction Estimating

To prepare an estimate accurately, the estimator must be able to read the prints and specifications for the building project and calculate material quantities and costs. This requires the ability to work from the dimensions given on the prints and make basic math calculations. For example, dimensions on prints are usually given in inches and feet. Often, these dimensions are used to make area calculations (in square units) and volume calculations (in cubic units). As an example, the amount of concrete required for a concrete foundation wall or footing is determined by calculating the cross-sectional area in square feet and multiplying it by the height of the wall or footing, providing the volume in cubic feet. The volume calculation is then converted to cubic yards (concrete is sold by the cubic yard). This information is used by the estimator in calculating the cost of the concrete. The estimator must be certain to make these calculations carefully in order to prepare an accurate estimate of costs. For more information about basic math calculations used in construction, refer to Unit 2.

When making volume calculations for a concrete wall foundation or footing, special attention must be given to corner intersections on the plan drawing. For example, **Figure 17-3** shows two examples requiring a calculation of the amount of concrete needed for an 8″ thick foundation footing. In **Figure 17-3A**, this is a simple matter of computing the square footage of the surface area and multiplying it by the thickness of the footing:

$$\text{Area} = 50'\text{-}0'' \times 2'\text{-}0'' = 100 \text{ ft}^2$$
$$\text{Volume} = 100 \text{ ft}^2 \times 8'' \ (0.667') = 66.7 \text{ ft}^3$$

There are 27 cubic feet in a cubic yard, so the amount in cubic feet is divided by 27.

$$\text{Volume} = 66.7 \text{ ft}^3 \div 27 = 2.47 \text{ cubic yards (cy)}$$

To allow for enough material, the computed volume is rounded up to 3 cy.

The concrete footing shown in **Figure 17-3B** has two segments and a 90° corner intersection where there is overlapping material. In this case, the overall dimensions are not used directly to calculate the area. The overlap at the corner between the two segments must be deducted from the calculation. One way to do this is to break down the footing into two rectangular areas. See **Figure 17-4**. In the example shown, the two rectangular

Job Name: *Estimating example*
Bid Date: 9-Jun
Contractor:

LABOR RATES:				
LAB	FIN	Carp	FRM	
17.00	21.00	23.00	28.50	Wage
6.00	7.25	8.00	8.75	fringe
23.00	28.25	31.00	37.25	Total

E S T I M A T E

BID:
$ 47,116

LAB	FIN	Carp	FRM	RHrs	OTH	Days	MHrs	TOTAL $	Unit $	WORK ITEM:	Quantity	Unit	LABOR Unit $	LABOR Labor	MATERIAL Unit $	MATERIAL Material	EQUIP / SUB Unit $	EQUIP / SUB Eqp/Sub	SUB TOTALS
								$ -	0.000	**Stamped Concrete**	2,000	SF	-	-		-		-	94 hrs
							-	$ -	0.000	Fine Grade	2,000	SF	-	-		-		-	
							-	$ -	0.000	Gravel subase x 3" thk	44.4	TNS	-	-	17	756		-	
1				3		1	3.0	$ 69	0.035	Protection of surfaces	2,000	SF	0.03	69	0.03	60		-	
1		1		3		1	6.0	$ 162	0.070	Mesh 42#	2300	SF	0.07	162	0.3	690		-	
1		1		2		1	4.0	$ 108	0.432	Form 4 - 5" edges	250	LF	0.43	108	0.65	163		-	
							-	$ -	-	pool edge form		LF	-	-	3.75	-		-	
1				1		1	1.0	$ 23	0.460	1/2" expansion x 4"	50	LF	0.46	23	0.6	30		-	
1	2		1	8		2	64.0	$ 1,868	0.934	P/F Stamped concrete	2,000	SF	0.93	1,868		-		-	
							-	$ -	0.000	4000psi Concrete x 4"	32.1	CY	-	-	96	3,083		-	
							-	$ -	0.000	Litho Grp B x 60#/CSF	25.0	BGS	-	-	38	950		-	
							-	$ -	0.000	Colored Concrete (Group 1)	34.1	CY	-	-		-		-	
1	1			8		1	16.0	$ 410	0.205	Clean & Seal	2,000	SF	0.21	410		-		-	
							-	$ -	0.000	Decorative Sealer Material	20.0	GLS	-	-	30	600		-	
							-	$ -	0.000	Release agent	3.3	PLS	-	-	65	212		-	
							-	$ -	-	Clean out trucks		EA	-	-	60	-		-	
							-	$ -	0.000	Stair Risers	24	LF	-	-	12	288		-	
							-	$ -	0.000	Demolition of existing patio	450	SF	-	-		-	3.00	1,350	
							-	$ -	-				-	-				-	
							-	$ -		**Cost w/T&I & Sales tax**			**3,247**		**7,309**		**1,350**		**$ 11,906**
							-	$ -	0.000	**Exposed agg.**	700	SF	-	-		-		-	48 hrs
							-	$ -	0.000	Protection of surfaces	700	SF	-	-	0.03	21		-	
							-	$ -	0.000	#3 @ 24" o/c e/w x 20'	39	Pcs	-	-	3.5	136		-	
							-	$ -	0.000	Form 4 - 5" edges	88	LF	-	-	0.65	57		-	
							-	$ -	0.000	1/2" expansion x 4"	88	LF	-	-	0.55	48		-	
1	2	1	1	8		1	40.0	$ 1,182	1.689	P/F concrete	700	SF	1.69	1,182		-		-	
							-	$ -	0.000	4000psi Concrete	12	CY	-	-	96	1,141		-	
							-	$ -	0.000	(Group 1)	12	Bags	-	-	50	594		-	
1				8		1	8.0	$ 184	26.286	pressure wash & Seal	7	GLS	26.29	184	21	147		-	
							-	$ -	0.000	Top Coat retarder	7	GLS	-	-	40	280		-	
							-	$ -	0.000	Exposed agg mtrl	1	TN	-	-	200	200		-	
							-	$ -	-				-	-		-		-	
							-	$ -	-				-	-		-		-	
							-	$ -	-				-	-		-		-	
							-	$ -		**Cost w/T&I & Sales tax**			**1,680**	**-**	**2,809**		**0**		**$ 4,489**
							-	$ -	0.000	**Stain Floors**	3,000	SF	-	-		-		-	112 hrs
							-	$ -	0.000	Clean Floor w/ rotary brush	3,000		-	-		-		-	
1				8		1	8.0	$ 184	0.061	Protection of area	3,000		0.06	184	0.03	90		-	
1	2		1	8		2	64.0	$ 1,868	0.623	Apply Chemstain (1 coat)	3,000		0.62	1,868		-		-	
1			1	8		1	16.0	$ 482	0.161	Rotary Clean/vacum floor	3,000		0.16	482		-		-	
							-	$ -	0.000	Stain Mtl	12		-	-	65	780		-	
2				8		1	16.0	$ 368	0.123	Clean & Seal	3,000		0.12	368		-		-	
							-	$ -	0.000	Sealer Material	30		-	-	21	630		-	
1				8		1	8.0	$ 184	0.061	Apply wax (2 coats)	3,000		0.06	184	0.05	150		-	
							-	$ -	-				-	-		-		-	
							-	$ -	0.000	Brushes	2		-	-	100	200		-	
							-	$ -	0.000	Commercial vacumns	2		-	-	100	200		-	
							-	$ -	0.000	Rotary Machines	2		-	-	100	200		-	
							-	$ -	-				-	-		-		-	
							-	$ -	-				-	-		-		-	
							-	$ -	-				-	-		-		-	
							-	$ -		**Cost w/T&I & Sales tax**			**3,796**	**-**	**2,408**		**0**		**$ 6,203**
							-	$ -	0.000	**General Conditions**	16.0	DYS	-	-		-		-	
							-	$ -	0.000	Safety	16	DYS		-	50	800		-	
							-	$ -	0.000	Superintendent	16	DYS	225.00	3,600		-		-	
							-	$ -	0.000	FormanTruck	16	DYS	-	-		-	300.00	4,800	
							-	$ -	0.000	Misc Small Tools	16	DYS	-	-		-	50.00	800	
							-	$ -	-				-	-		-		-	
							-	$ -	-	Winter Concrete Charges		CY	-	-		-		-	
							-	$ -	-	Blankets		DYS	-	-		-		-	
							-	$ -	-	Travel		DYS	-	-		-		-	
1	1			8		1	16.0	$ 410	410.000	Samples	1	LS	410.00	410	300	300	100.00	100	
							-	$ -	-				-	-		-		-	
							-	$ -	0.000	pressure washer	1	DYS	-	-		-	25.00	25	
							-	$ -	0.000	Bobcat	1	DYS	-	-		-	175.00	175	
							-	$ -	0.000	Power Buggy	2	DYS	-			-	150.00	300	
							-	$ -	-	**Cost w/T&I & Sales tax**			**4,932**	**-**	**1,177**		**6,200**		**$ 12,309**
1	1	0	0	8	0	16	270	$ 7,502	-				-	-					
										Sub-Total - 1				11,102		12,805		7,550	
	Total MHrs						270			Tax & Ins	23.0%	%		2,553					
										Sales Tax on Material	7.0%	%				896			
										Sub-Total - 2 **Cost**	$ 34,907			13,655		13,702		7,550	
										Overhead	16.0%	%		2,185		2,192		1,208	
										Profit		%	24.00%	3,802	15.00%	2,384	5.00%	438	
										Job Total Revenue	**47,116**	**100%**		19,642		18,278		9,196	

Job Total Revenue	**47,116**	**100%**
Labor	19,642	42%
Material	18,278	39%
Equipment/Sub	9,196	20%
Total Cost:	47,116	100%
Overhead:	5,585	12%
Profit:	6,624	14%
	5,700	SF
Selling price per $/SF	**$ 8.27**	SF

Total Gross Profit	
$ 12,209	25.9%

Tax and insurance calculations

Overhead and profit calculations

Total bid

Cost per square foot

Patterned Concrete of Cincinnati

Figure 17-2. A sample estimating form for decorative concrete work. The data shown is for reference only and is not representative of a complete construction bid. Note that the profit mark-up for labor is higher than the profit mark-ups for materials, equipment, and subcontractors. Self-performed labor involves greater risk compared to materials and equipment.

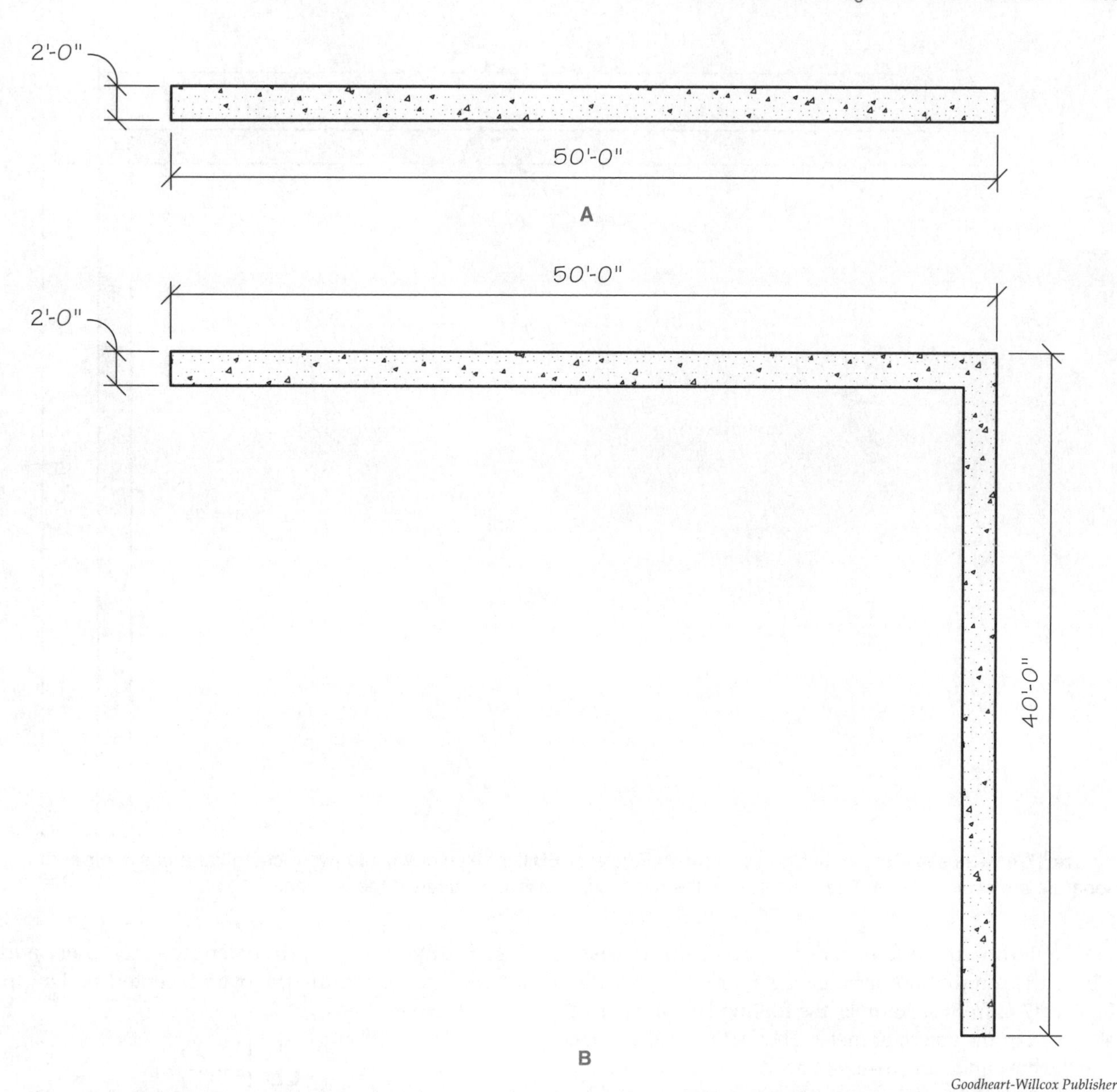

Goodheart-Willcox Publisher

Figure 17-3. Calculating the required volume of concrete for a concrete foundation footing. A—A simple calculation is made for a single footing without a corner intersection. B—This footing has a 90° corner intersection with overlapping material. The overall dimensions are not used directly to calculate the area.

areas are calculated and added together. Then, the volume is calculated:

$$\text{Area 1} = 50'\text{-}0'' \times 2'\text{-}0'' = 100 \text{ ft}^2$$

$$\text{Area 2} = 38'\text{-}0'' \times 2'\text{-}0'' = 76 \text{ ft}^2$$

$$\text{Area 1} + \text{Area 2} = 100 \text{ ft}^2 + 76 \text{ ft}^2 = 176 \text{ ft}^2$$

$$\text{Volume} = 176 \text{ ft}^2 \times 8'' \ (0.667') = 117.40 \text{ ft}^3$$

$$\text{Volume} = 117.40 \text{ ft}^3 \div 27 = 4.35 \text{ cy}$$

$$\text{Rounded volume} = 5 \text{ cy}$$

Figure 17-5 shows another method used to calculate the required volume of concrete for the footing. This method uses the centerline dimensions of the footing instead of the overall dimensions to make the area calculations. Study the centerline dimensions shown in **Figure 17-5A**. These are calculated by deducting half the footing width (1′-0″) at the corner end. Subtracting this amount from the overall dimension eliminates the overlap at the corner. **Figure 17-5B** illustrates the resulting centerline areas and shows how areas in the corner are represented in the calculation. The area and volume calculations are as follows:

$$\text{Centerline Area 1} = (50'\text{-}0'' - 1'\text{-}0'') \times 2'\text{-}0'' = 98 \text{ ft}^2$$

$$\text{Centerline Area 2} = (40'\text{-}0'' - 1'\text{-}0'') \times 2'\text{-}0'' = 78 \text{ ft}^2$$

$$\text{Centerline Area 1} + \text{Centerline Area 2} = 98 \text{ ft}^2 + 78 \text{ ft}^2 = 176 \text{ ft}^2$$

$$\text{Volume} = 176 \text{ ft}^2 \times 8'' \ (0.667') = 117.40 \text{ ft}^3$$

$$\text{Volume} = 117.40 \text{ ft}^3 \div 27 = 4.35 \text{ cy}$$

$$\text{Rounded volume} = 5 \text{ cy}$$

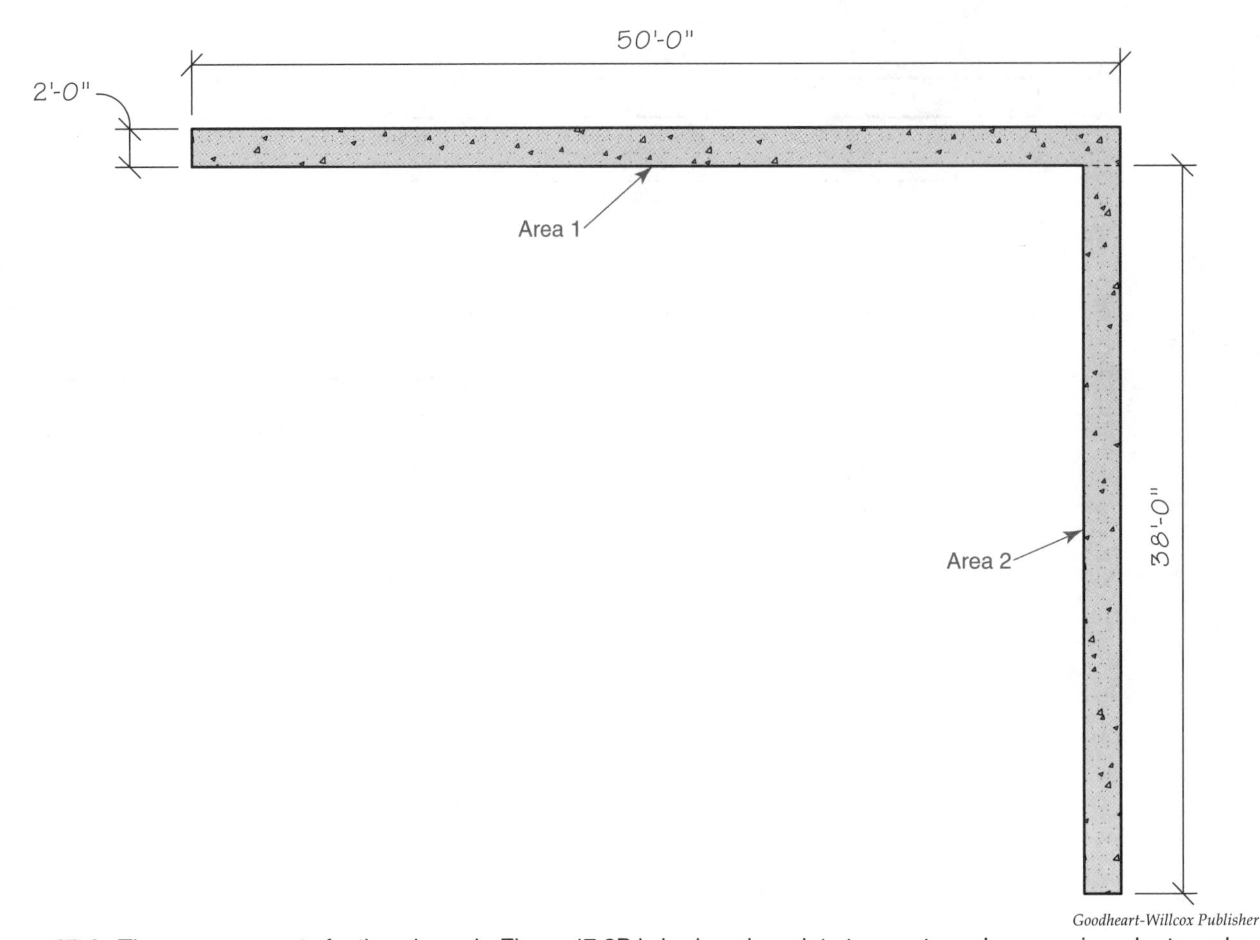

Figure 17-4. The same concrete footing shown in Figure 17-3B is broken down into two rectangular areas in order to make separate area calculations. Then, the sum of the two areas is used to calculate the volume.

The centerline method becomes more useful in cases where the foundation perimeter is more complex. See **Figure 17-6**. In this example, the footing layout has two offsets from the outer perimeter. The footing is 2′-0″ wide and 8″ thick and can be broken down into six segments. The centerline length of each segment is calculated by deducting the corner overlap from the overall dimension at each end. Then, the individual centerline lengths are added together to calculate the overall centerline length in lineal footage. This measurement is multiplied by the footing width to calculate the total area. Note in **Figure 17-6** that each centerline is labeled and noted with a calculated measurement. For each centerline, half the footing width (1′-0″) is deducted from the overall dimension at each corner end. The total amount deducted is 2′-0″. For example, the length of centerline CL_1 is calculated as follows:

$$\begin{aligned} CL_1 &= 50'\text{-}0'' \\ &\underline{-\ 2'\text{-}0''} \\ &= 48'\text{-}0'' \end{aligned}$$

The individual centerline lengths are then added together to determine the overall centerline length in lineal footage:

$$\begin{aligned} &CL_1\ (48'\text{-}0'') \\ &CL_2\ (58'\text{-}0'') \\ &CL_3\ (9'\text{-}0'') \\ &CL_4\ (8'\text{-}0'') \\ &CL_5\ (39'\text{-}0'') \\ +&\underline{CL_6\ (50'\text{-}0'')} \\ =&\ 212 \text{ lf} \end{aligned}$$

The area is calculated by multiplying the measurement in lineal footage by the footing width. Finally, the volume is calculated and the computed volume is converted to cubic yards:

$$\text{Area} = 212 \text{ lf} \times 2'\text{-}0'' = 424 \text{ ft}^2$$

$$\text{Volume} = 424 \text{ ft}^2 \times 8'' \ (0.667') = 282.81 \text{ ft}^3$$

$$\text{Volume} = 282.81 \text{ ft}^3 \div 27 = 10.47 \text{ cy}$$

$$\text{Rounded volume} = 11 \text{ cy}$$

Notice that the centerline method requires fewer calculations in comparison to breaking down the perimeter layout into rectangular areas. This method can be used in any layout with inside or outside corner intersections provided that all of the corners meet at 90°.

Estimating Software

There are many software programs for construction estimating available on the market today. These programs provide tools for creating takeoffs, schedules, and reports from imported drawing data. These tools help simplify the estimating process and automate tasks traditionally

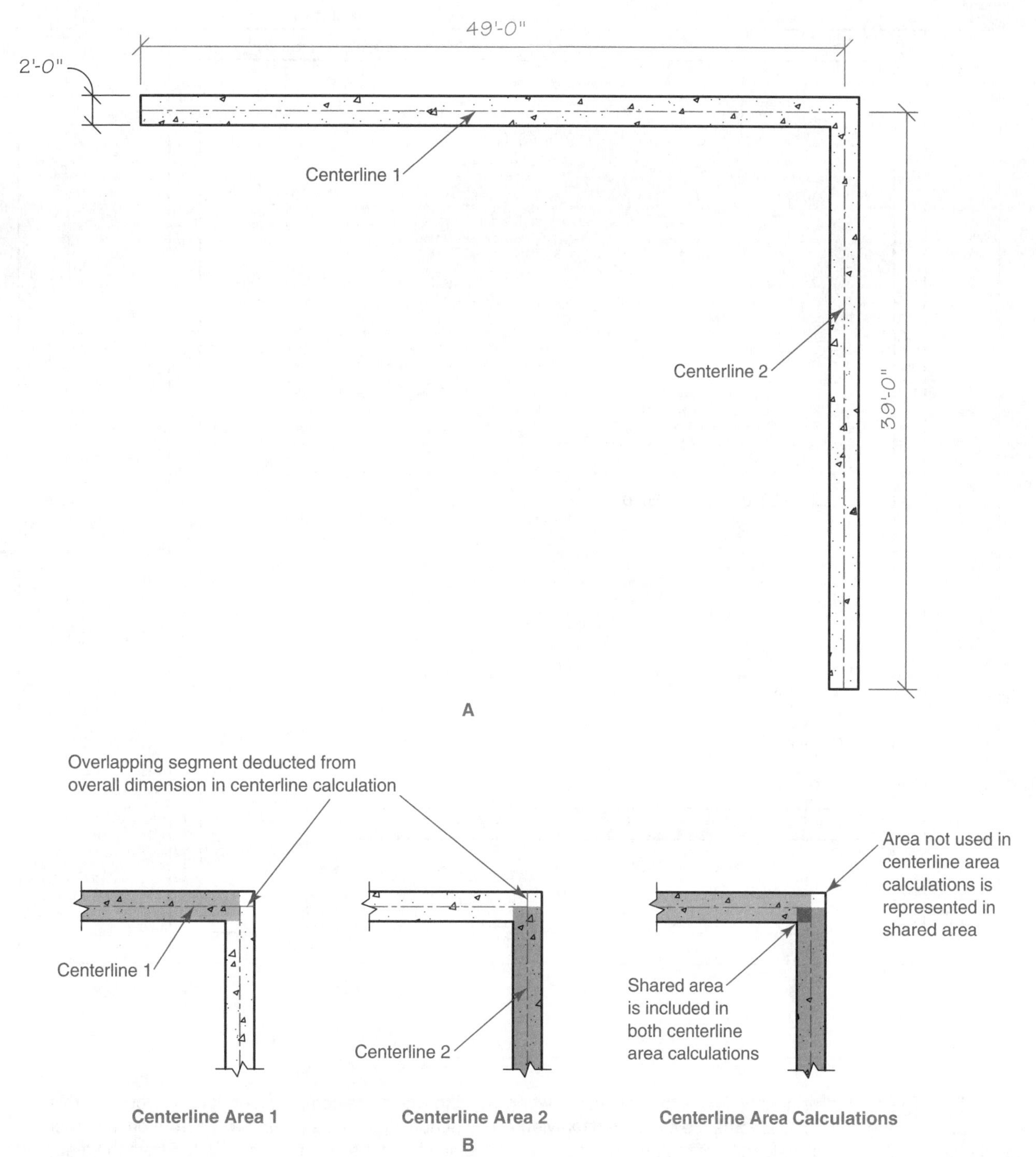

Goodheart-Willcox Publisher

Figure 17-5. Using the centerline method to calculate the required volume of concrete for a concrete foundation footing. A—The footing centerline dimensions are used instead of the overall dimensions to make area calculations. This eliminates the overlap at the corner. B—The highlighted areas indicate the centerline areas calculated from the centerline lengths. Note the segments removed from the calculations. Also, note that the shared area shown in the illustration on the right is included in both centerline area calculations and accounts for the unused area in the upper-right corner.

completed by hand. Choosing and setting up the appropriate estimating software for you and your company requires a great deal of investigation and effort. However, once the software is properly set up, it will greatly assist in completing repetitive estimating tasks. To find out more information about the software programs available, conduct an Internet search for "construction estimating software."

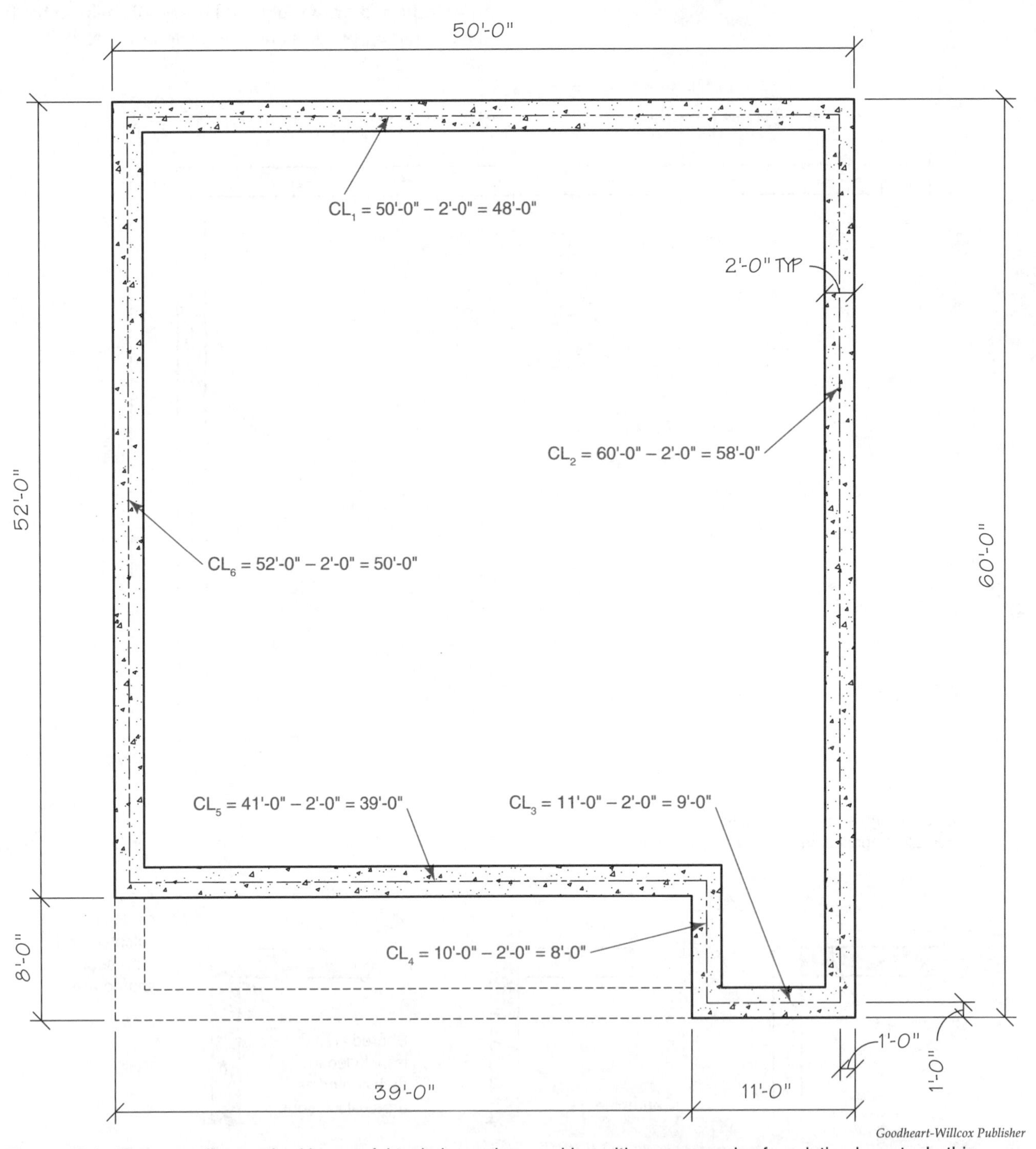

Goodheart-Willcox Publisher

Figure 17-6. The centerline method is a useful technique when working with more complex foundation layouts. In this example, the footing centerline lengths are calculated individually by deducting half the footing width from each corner end. The sum of the centerline lengths is then used in the area calculation and the volume is calculated from the area. The dashed outline in the drawing represents the extents of the outer perimeter and appears for reference only.

Test Your Knowledge

Name ______________________________

Write your answers in the spaces provided.

_____ 1. *True or False?* When calculating labor rates, fringe benefits such as vacation pay are excluded.

_____ 2. *True or False?* A company's overhead costs include salaries paid by the company.

_____ 3. *True or False?* An apprentice is usually paid a higher rate on the job than a journeyman worker.

_____ 4. *True or False?* The detailed method of estimating is also known as budget parameter estimating.

_____ 5. A(n) _____ is a change to the contract documents made by the architect prior to the construction bid.

A. parameter

B. schedule

C. takeoff

D. addendum

_____ 6. *True or False?* Estimating the cost of a job based on the cost per cubic yard is an example of using the approximate method.

_____ 7. When developing an estimate, which of the following items is typically marked up with the highest percentage rate?

A. Equipment

B. Materials

C. Safety items

D. Labor

_____ 8. A project has 225 cubic yards of concrete in the footings, walls, and slabs. The total cost for this work is $35,325. What is the cubic yard unit cost?

A. $165

B. $157

C. 15.70

D. None of the above.

_____ 9. A crew of workers is made up of two laborers each paid $15 per hour, one carpenter paid $21 per hour, and a foreman paid $25 per hour. What is the composite per-hour rate?

A. $19 per hour

B. $22 per hour

C. $76 per hour

D. $680 per day

_____ 10. A landscaping business is bidding a job that requires 3 trees at $150 each, 5 large bushes at $45 each, 2 dozen roses at $19 each rose, and 3 dozen assorted flowers at $30 each dozen. Based on these prices and the labor rates and production rates shown below, what is the cost of doing this job? No taxes, mark-ups, or overhead costs are to be included. Use the space below for calculating your answer.

A. $657

B. $1221

C. $1878

D. None of the above.

Crew Member	Labor Rate
Laborer	$12/hour
Foreman	$18/hour

Item	Production Rate
Trees	3 crew hours each
Large Bushes	0.5 crew hour each
Roses	4 crew hours each dozen
Assorted Flowers	1 laborer, 2 hours each dozen

Activity 17-1

Estimating Construction Costs

Name ______________________________

Refer to the set of prints for the Marseille residential building in the Large Prints supplement to answer the following questions. Use the detailed method of estimating. List all costs including the allowances indicated in the material list provided.

Compute all labor and materials required to complete the construction of the house. Include mark-up and administrative costs.

Note: The purpose of this exercise is to become familiar with the estimating process. Prices for materials and subcontractor costs will vary greatly from region to region and will be impacted by local codes and practices.

The following outline will help you organize your estimate.

Outline for Preparing a Detailed Estimate

1. Prepare a cost estimate using the form *Estimate Takeoff and Cost Sheet* shown in **Figure 17-7**.
2. Consider each of the following items, and enter your estimates on the cost estimate form.

   A. Plans and specs
   B. Building permits and fees
   C. Utilities
   D. Excavation and fill
   E. Concrete footings, foundations, floors
   F. Unit masonry
   G. Steel beams and columns
   H. Framing lumber
   I. Finish lumber
   J. Hardware
   K. Roofing
   L. Windows and screens
   M. Doors and screens
   N. Electrical wiring
   O. Electrical fixtures
   P. Sheet metal ducts and gutters
   Q. Air conditioning
   R. Plumbing
   S. Insulation
   T. Gypsum board
   U. Cabinets
   V. Countertops
   W. Tile
   X. Floor sanding
   Y. Painting and decorating
   Z. Glazing and mirrors
   AA. Floor coverings
   AB. Window cleaning
   AC. Built-ins and accessories
   AD. Range-oven
   AE. Vent hood
   AF. Dishwasher
   AG. Garbage disposal
   AH. Tub-shower enclosures
   AI. Fireplace mantel
   AJ. Cleanup

   In addition to these items, you may want to consider business overhead, finance charges, and profit on the job.
3. The list of materials that follows includes most items except electrical, plumbing, and air conditioning materials, which are included as allowances. Use this list as a basis for your estimate, adding items other than materials listed under Step 2. Remember, it is your job to provide a complete and detailed estimate.
4. When you have completed your detailed estimate of costs, check your estimate by using the approximate method. Your instructor will assist you in arriving at an approximate cost per unit for your area.

Estimate Takeoff and Cost Sheet

Project ____________________

Item no.	Item identification	Location on job	Cost			Total cost
			Labor	Material	Equipment	

Goodheart-Willcox Publisher

Figure 17-7. Estimating form for Activity 17-1.

MATERIALS LIST

Foundation

Footers

18 cy	Concrete (no waste)
63 pcs.	1/2″×20′ #4 rebar

House Walls

41 cy	Concrete (no waste)
52 pcs.	1/2″×20′ #4 rebar
38 pcs.	1/2″×12″ J bolt
3 ea.	2817 basement sash, steel buck frame
3 ea.	36″×30″ window well
12 pcs.	3/8″×2 1/2″ wedge anchor

Garage, Porch Wing Walls

10 cy	Concrete (no waste)
15 pcs.	1/2″×20′ #4 rebar
13 pcs.	1/2″×12″ J bolt

House, Garage, Porch Slabs

40 cy	Concrete
40 tons	Gravel
58 pcs.	1/2″×20′ #4 rebar
3 rolls	10′×100′ poly vapor barrier
220 lin. ft.	4″ footer drain pipe

Steel

14 lin. ft.	W8×15 I-beam
48 lin. ft.	W8×21 I-beam
3 pcs.	3″ steel pipe column, #A-53, 7′-10″ pour

First Floor Subfloor

6 rolls	3 1/2″ sill sealer
14 pcs.	2×4×16′ PT plate
1 pc.	2×6×14′ SPF beam plate
3 pcs.	2×10×16′ SYP beam plate
3 pcs.	2×10×8′ SYP joist
10 pcs.	2×10×10′ SYP joist
49 pcs.	2×10×14′ SYP joist
31 pcs.	2×10×16′ SYP band board
2 pcs.	Double 2×10 hanger
5 pcs.	Single 2×10 hanger
90 pcs.	Pr. 1×3 wood bridging, 2×10 at 16″ O.C.
48 pcs.	4×8×3/4″ T&G plywood

16 tubes	Adhesive
1 50 lb. box	16d sinker nails
1 50 lb. box	8d sinker nails
2 lbs.	Hanger nails

First Floor Exterior Walls

6 pcs.	2×6×14′ SPF plate
10 pcs.	2×6×16′ SPF plate
12 pcs.	2×4×14′ SPF plate
19 pcs.	2×4×16′ SPF plate
9 pcs.	2×6×104 5/8″ SPF stud
52 pcs.	2×6×12′ SPF stud
12 pcs.	2×6×18′ SPF stud
158 pcs.	2×4×104 5/8″ SPF stud
2 pcs.	1 3/4″×9 1/4″×10′ LVL
4 pcs.	2×8×8′ SYP lintel at porch, flush beam
1 pc.	2×12×8′ SYP header
1 pc.	2×12×10′ SYP header
2 pcs.	2×12×12′ SYP header
4 pcs.	2×12×14′ SYP header
58 pcs.	4×8×1/2″ CDX plywood sheathing
4 pcs.	2×6×12′ SPF fire blocking
6 pcs.	2×4×12′ SPF fire blocking
176 lin. ft.	18″ flash and seal
2 rolls	10′×150′ Tyvek
1 roll	1 7/8″ Tyvek tape
1 box (2000 ct.)	1 1/4″ cap nails at Tyvek

Garage Walls

2 rolls	3 1/2″ sill sealer
3 pcs.	2×4×16′ PT plate
5 pcs.	2×4×14′ SPF plate
4 pcs.	2×4×16′ SPF plate
68 pcs.	2×4×10′ SPF stud
1 pc.	2×12×8′ SYP header
1 pc.	2×12×14′ SYP header
2 pcs.	1 3/4″×16″×18′ LVL
2 pcs.	Simpson HTT22 tension tie
1 pc.	Simpson HTT16 tension tie
3 pcs.	5/8″×12″ wedge anchor
21 pcs.	4×8×1/2″ CDX plywood sheathing
3 pcs.	2×4×12′ SPF fire blocking
46 lin. ft.	18″ flash and seal
1 roll	10′×150′ Tyvek
1 roll	1 7/8″ Tyvek tape
1 box (2000 ct.)	1 1/4″ cap nails at Tyvek

First Floor Interior Walls

2 pcs.	2×6×14′ SPF plate
3 pcs.	2×6×16′ SPF plate
11 pcs.	2×4×14′ SPF plate
18 pcs.	2×4×16′ SPF plate
10 pcs.	2×6×104 5/8″ SPF stud
18 pcs.	2×4×12′ SPF stud
138 pcs.	2×4×104 5/8″ SPF stud
1 pc.	2×12×14′ SYP header

Second Floor Subfloor

2 pcs.	2×10×8′ SYP joist
2 pcs.	2×10×10′ SYP joist
13 pcs.	2×10×14′ SYP joist
14 pcs.	2×10×18′ SYP joist
2 pcs.	2×10×16′ SYP band board
2 pcs.	Double 2×10 hanger
5 pcs.	Single 2×10 hanger
17 pcs.	Pr. 1×3 wood bridging, 2×10 at 16″ O.C.
14 pcs.	Pr. 1×3 wood bridging, 2×10 at 12″ O.C.
19 pcs.	4×8×3/4″ T&G plywood
6 tubes	Adhesive
2 lbs.	Hanger nails

Second Floor Exterior Walls

3 pcs.	2×6×12′ SPF plate
7 pcs.	2×4×14′ SPF plate
10 pcs.	2×4×16′ SPF plate
11 pcs.	2×6×92 5/8″ SPF stud
80 pcs.	2×4×92 5/8″ SPF stud
2 pcs.	2×12×8′ SYP header
1 pc.	2×12×14′ SYP header
22 pcs.	4×8×1/2″ CDX plywood sheathing
1 roll	9′×150′ Tyvek
1 roll	1 7/8″ Tyvek tape

Second Floor Interior Walls

7 pcs.	2×4×14′ SPF plate
11 pcs.	2×4×16′ SPF plate
76 pcs.	2×4×92 5/8″ SPF stud

Trusses

1 lump sum (LS)	Trusses per plan, per manufacturer layout, gables, commons, girders, vaults, stepped, raised heel, hips, hangers

Roof System

14 pcs.	1×4×14′ #2 WP truss stay
34 pcs.	2×4×14′ SPF truss blocking
2 pcs.	2×10×20′ SYP over frame
3 pcs.	2×10×18′ SYP over frame
3 pcs.	2×10×16′ SYP over frame
2 pcs.	2×10×12′ SYP over frame
5 pcs.	2×10×10′ SYP over frame
16 pcs.	4×8×1/2″ CDX plywood sheathing at truss radius ends
4 pcs.	4×8×1/2″ CDX plywood gable sheathing
138 pcs.	4×8×1/2″ CDX plywood roof sheathing
11 pcs.	2×4×8′ SPF dormer framing
4 pcs.	4×8×1/2″ CDX plywood dormer framing
3 pcs.	4×8×3/4″ CDX plywood dormer truss material
17 pcs.	2×6×16′ SPF gutter, rake backer
500 pcs.	1/2″ ply clips
1 50 lb. box	16d sinker nails
1 50 lb. box	8d sinker nails
5 lbs.	Hanger nails

Roofing

10 rolls	15 lb. felt
39 2/3 squares	25-year shingles
100 sq. ft.	Copper standing seam
3 pcs.	W valley
32 pcs.	Drip edge
11 pcs.	48″ shingles over ridge vent
2 bundles (100 ea.)	5″×7″ step flashing
4 tubes	Flashing sealant
130 lbs.	1 1/4″ roofing nails
4 lbs.	3″ roofing nails at ridge vent
1 lump sum (LS)	Copper standing seam fasteners, per manufacturer
2 boxes (2000 ct. ea.)	1 1/4″ cap nails at felt

Overhead Door Blocking

2 pcs.	2×6×10′ SPF
1 pc.	2×6×16′ SPF

Landings, Stairs, Walls at Basement

2 pcs.	2×4×12′ PT plate
4 pcs.	2×4×12′ SPF plate
22 pcs.	2×4×92 5/8″ SPF stud
11 pcs.	2×4×8′ PT furring
2 pcs.	2×10×10′ SYP joist
6 pcs.	2×10×14′ SYP joist
2 pcs.	4×8×3/4″ T&G plywood

1 split set	Carpet grade stairs, 37″ wide; 1 at 6 risers, 1 at 8 risers
1 split set	Oak treads, poplar stringers and risers, 37″ wide; 1 at 8 risers, 15 total open ends
1 set	2 risers PT at garage, 42″ wide

Tray Ceiling Framing at Great Room

4 pcs.	2×4×16′ SPF
6 pcs.	2×4×8′ SPF

Exterior Doors

1 ea.	3′-0″ entry w/2 at 14″ sidelites, 24″×72″ custom arch transom, grills 6 9/16″
1 ea.	FWH 31611, lockset, grills 6 9/16″
1 ea.	2′-8″ 9 lite, panel bottom 4 9/16″
1 ea.	2′-8″ fire rated, self close 4 9/16″, door unit jambs as noted, deadbolt bore, weatherstrip
1 ea.	16′×8′ steel overhead door
1 roll	7″ flex wrap
1 roll	9″ flex wrap
3 bundles	Shims
3 lbs.	16d galvanized finish nails
2 tubes	Sealant

Windows

1 ea.	TW210510-2 at master bedroom 4 9/16″
1 ea.	Cxw16-3 with A31-3 at great room 6 9/16″
1 ea.	CX14 at kitchen 4 9/16″
1 ea.	TW20310 at bath
1 ea.	TW24310 at laundry 6 9/16″
1 ea.	CXW16 with AFCP301 at garage 4 9/16″
1 ea.	CXW16 at garage 4 9/16″
1 ea.	CW255 with ATCW21 at dining 6 9/16″
1 ea.	A31 at lower landing 6 9/16″
1 ea.	CXW16 at master closet 4 9/16″
1 ea.	C23 at master bath, tempered 4 9/16″
1 ea.	TW34410 at bedroom 4 9/16″
2 ea.	AFC13 at dormers
1 ea.	CXW16 with AFCP302 at upper landing, tempered 6 9/16″
1 ea.	CXW15 with AFCP301 at bedroom 4 9/16″
1 ea.	TW2032 at bath 6 9/16″, window units manufactured by Andersen, jambs as noted, screens, grills, sash hardware
60 lin. ft.	18″ EPDM
2 rolls	9″ flex wrap
3 rolls	7″ flex wrap
3 rolls	Straight flash tape
10 tubes	Sealant

Masonry

32 pcs.	4″×8″×16″ H.C. block
9 pcs.	4″×8″×8″ H.C. block
18,400 pcs.	Standard size brick
720 sq. ft.	Wire lath
720 sq. ft.	Cultured stone
1900 pcs.	Brick ties
96 pcs.	Weep hole covers
166 bags	Cement mortar
23 tons	Mortar sand
35 lin. ft.	3 1/2″×3 1/2″×3/8″ steel angle
22 lin. ft.	5″×3 1/2″×3/8″ steel angle
18 lin. ft.	8″×4″×1/2″ steel angle

Rough Exterior Trim

5 pcs.	2×4×8′ PT plate, at column post
5 pcs.	2×4×8′ SPF plate, at column post
6 pcs.	2×4×12′ SPF stud, at column post
4 pcs.	4×8×1/2″ CDX plywood, at column post
10 pcs.	2×4×10′ SPF blocking
200 lin. ft.	8″ drop ladder

Exterior Trim

1 pc.	Round top dormer louver, manufactured by Beach Metal
2 pcs.	48″ flower box by Flower Framers
30 lin. ft.	1×4 primed redwood
270 lin. ft.	1×6 primed redwood
300 lin. ft.	1×8 primed redwood
20 lin. ft.	1×12 primed redwood
40 lin. ft.	5/4×6 primed redwood
10 lin. ft.	1 5/8″ primed barge mold
40 lin. ft.	2 1/4″ primed barge mold
100 lin. ft.	8″ redwood bevel siding
8 pcs.	4×8×3/8″ primed AC plywood
176 lin. ft.	2″ soffit vent
15 lbs.	7d siding nails
5 lbs.	4d galvanized box nails
40 sq. ft.	Bead board at entry
19 lin. ft.	Wrought iron rail

Insulation

2850 sq. ft.	R-15 faced
1050 sq. ft.	R-19 faced
20 sq. ft.	R-38 batt
1980 sq. ft.	R-38 blown-in type

520 sq. ft.	R-6 rigid foam, Dow board
96 pcs.	Insulation baffle

Drywall

9 pcs.	4×8×1/2″ M.R. board
179 pcs.	4×12×1/2″ Reg. board
22 pcs.	4×12×1/2″ Type X board
6 pcs.	8′ corner bead
24 pcs.	10′ corner bead
15 rolls	Joint tape
10 5-gal. pails	Joint compound
6 5-gal. pails	Topping compound
40 lbs.	1 1/4″ drywall screws
10 lbs.	1 3/8″ drywall nails, ring shank

Laundry Room Cabinet, Tops

1 ea.	48 base, sink
1 ea.	Square edge top 48 1/4″, 1 cutout, 1 end cap

Bath Cabinets, Tops

3 ea.	39″×21″ vanity base
1 ea.	54″×21″ vanity base
1 ea.	40″×22″ single bowl top
1 ea.	55″×22″ single bowl top
1 ea.	79″×22″ double bowl top

Kitchen Cabinets, Tops

2 ea.	2742 wall
1 ea.	3030 wall, microwave
3 ea.	3042 wall
1 ea.	3924 wall
1 ea.	3642 wall
2 ea.	27 base
2 ea.	30 base
1 ea.	15 base, drawer
1 ea.	24 base
1 ea.	42 base, corner sink
1 ea.	48 base
1 ea.	3″ filler
2 ea.	Custom corner base
1 ea.	36″×96″ panel
1 ea.	Square edge built up top 76 1/2″, 4 angle clips, all edge finish
1 ea.	Square edge top 60 1/4″, 1 end cap
1 ea.	Square edge top, L shape, with angle 108 1/4″, front angle at 24″, right leg 69 1/4″, 2 end caps, 1 cutout
1 ea.	Square edge top 27 1/4″, 2 end caps

Interior Trim

Pre-hung units, 4 9/16″ jamb, use ball catches @ doubles

1 pc.	1′-8″
1 pc.	1′-10″
6 pcs.	2′-4″
4 pcs.	2′-6″
1 pc.	2′-8″
2 pcs.	Double 1′-4″
1 pc.	Double 2′-6″
8 pcs.	4 9/16″ jamb side
108 pcs.	7′-0″ casing
336 lin. ft.	Casing
720 lin. ft.	Base
192 lin. ft.	Base shoe
25 lin. ft.	Wire linen shelf
20 lin. ft.	Wire pantry shelf
44 lin. ft.	Wire closet shelf
16 lin. ft.	2 1/4″ wall rail
18 lin. ft.	2 1/4″ top rail
1 pc.	4″×4″×48″ newel post
34 pcs.	1 3/8″×1 3/8″×34″ tapered baluster
5 pcs.	Handrail brackets
1 pc.	72″ poplar mantel, boxed raised panel legs
24 lin. ft.	2″ mantel mold
23 ea.	Hinge type door stops
1 ea.	Entry lock
2 ea.	Key lock
2 ea.	Deadbolt
5 ea.	Privacy
8 ea.	Passage
6 ea.	Dummy
6 bundles	Shims
25 lbs.	8d finish nails
15 lbs.	6d finish nails
2 lbs.	4d finish nails

Fireplace

1 pc.	36″ zero clearance, top vent gas unit
1 pc.	Firestop
1 pc.	Flue support
25 lin. ft.	Flue
1 pc.	Flashing kit
1 pc.	Storm collar
1 pc.	Top cap
1 pc.	Gas log set
1 pc.	Sealed glass kit
1 pc.	Fan, blower kit
1 pc.	Remote starter switch

Sealants

8 tubes	Exterior caulk
3 tubes	Tub and shower
2 pt.	Wood filler

Allowances to Be Included in the Estimate

Item	Allowance
Floor coverings	$11,000
Bath tile	$2500
Fireplace tile or marble	$3500
Lighting fixtures	$4000
Exterior paint or stain	$3000
Interior paint or stain	$9000
Foundation waterproofing	$1800
Gutters, downspouts, sub-surface drain pipe	$3000
HVAC	$6800
Plumbing	$9200
Electrical	$11,300
Driveway, sidewalk, patio	$6000
Lot cost	$90,000
Construction interest	$5000
Lot survey	$500
Building permits	$4000
Sewer line to house	$2000
Gas line to house	$700
Landscaping	$2200
Appliances	$7000
Contingency	$3000

SECTION 6
Advanced Print Reading Projects

This section of the text contains four comprehensive projects. Each project comprises several sets of questions based on a set of drawings from the *Large Prints* supplement. Each set of questions focuses on a different aspect of the building; one set is based on the foundation, one set is based on the site plan, and so on.

Project A and Project B involve residential drawings. These projects will test your familiarity with the symbols and conventions used in house construction. Both projects are based on standard home designs.

Project C and Project D involve commercial drawings. Project C involves drawings for a combined office and warehouse facility. Project D is based on a set of drawings for a parking and office building. These projects are more extensive than the residential construction projects.

As you work through the projects, refer back to the appropriate units when you are unsure of an answer. Your instructor may assign additional activities using the project drawings.

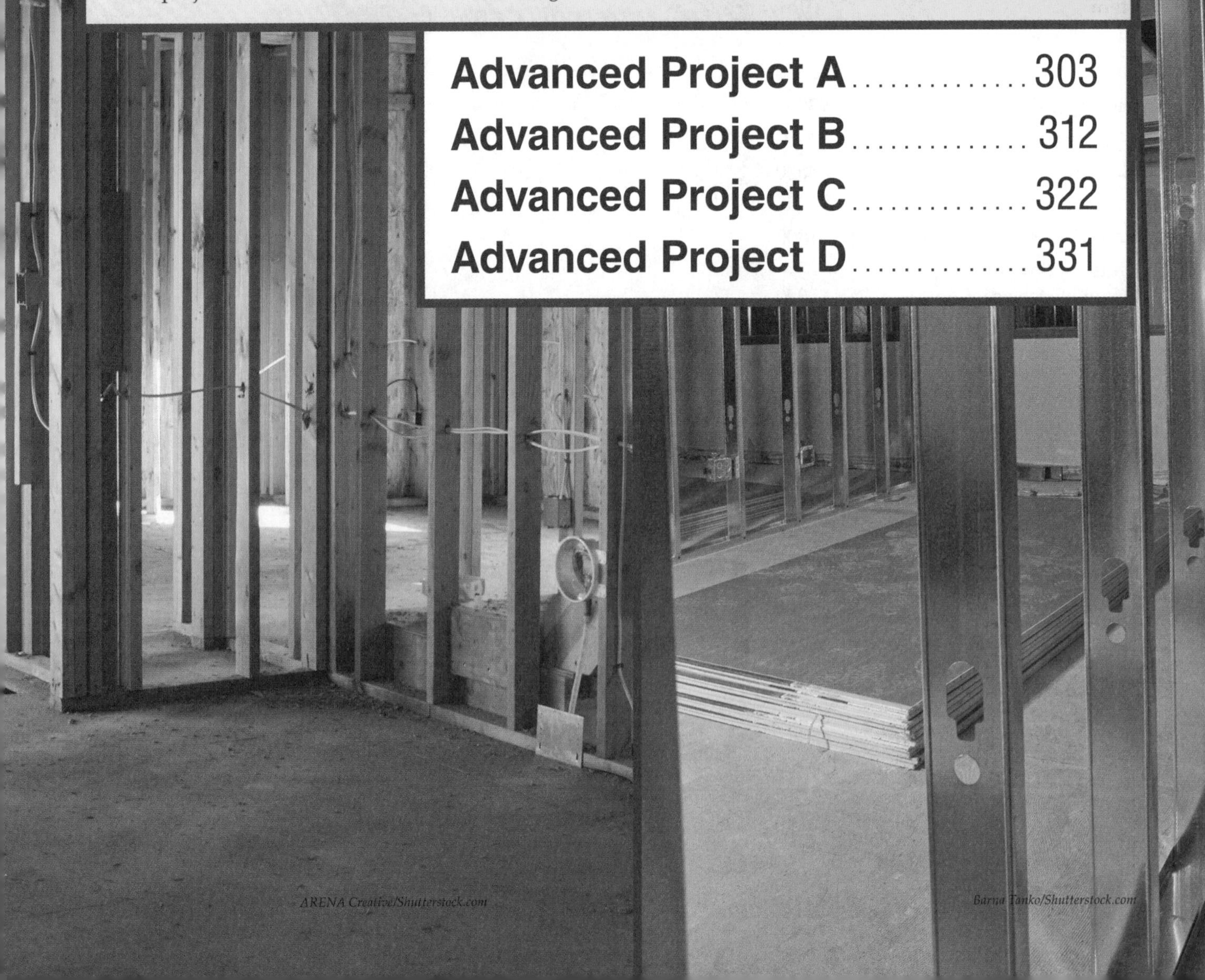

ARENA Creative/Shutterstock.com

Barna Tanko/Shutterstock.com

Advanced Project A

Residence Mercedes Pointe

This project is based on drawings for the two-story brick and stone veneer residence shown in **Figure A-1**. *This large, four-bedroom, 3 1/2 bath home includes a 3-car garage, a study, and a screened porch and deck. Refer to Prints RMP-1 through RMP-8 in the Large Prints supplement when working on the following activities. An index is provided below. Note: The prints provided do not represent a complete set of drawings for the building.*

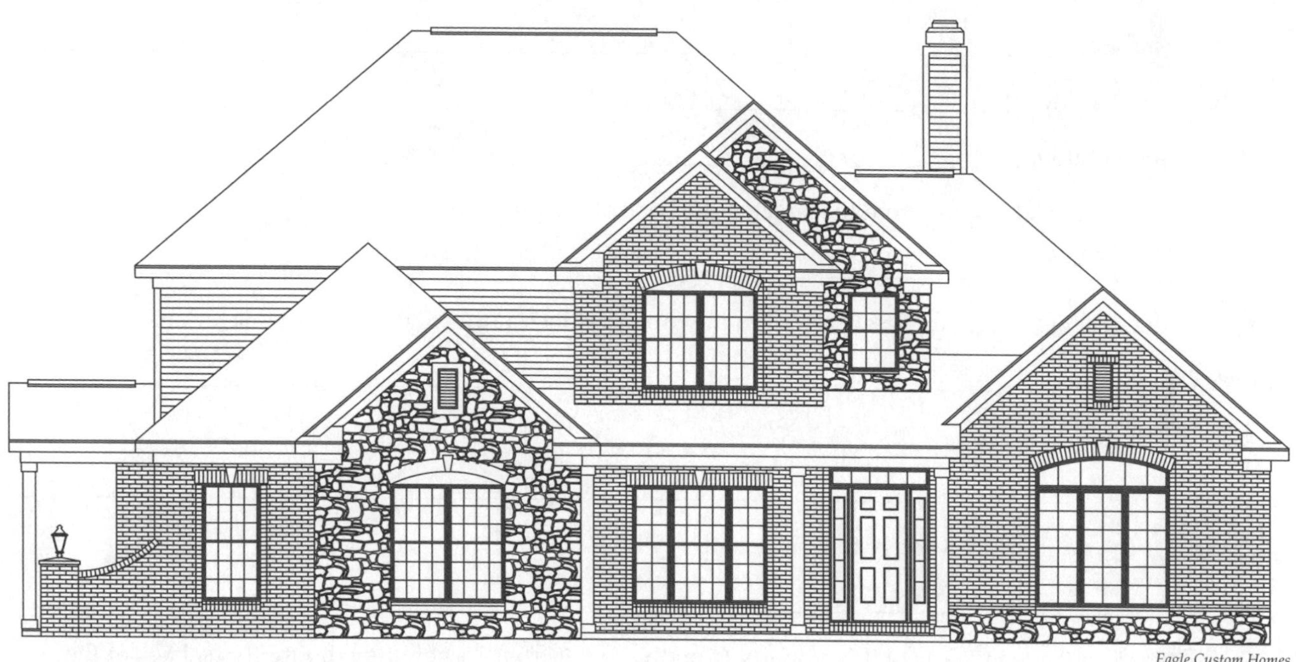

Eagle Custom Homes
McGill Smith Punshon, Inc.
Superior Designs, LLC

Figure A-1. Front elevation of the Residence Mercedes Pointe residence.

PRINT INDEX • RESIDENCE MERCEDES POINTE

PRINT LABEL	SHEET NUMBER	SHEET TITLE
RMP-1	1	FRONT & RIGHT ELEVATIONS
RMP-2	2	LEFT & REAR ELEVATIONS
RMP-3	3	FOUNDATION PLAN, SITE PLAN
RMP-4	4	FIRST FLOOR PLAN
RMP-5	5	SECOND FLOOR PLAN
RMP-6	7	FIRST FLOOR ELECTRICAL PLAN
RMP-7	9	WALL SECTIONS, STAIR SECTIONS
RMP-8	10	GENERAL NOTES, WALL SECTIONS

Activity A-1

Site Plan

Name ______________________________

1. What is the scale of the site plan?

____

2. What is the bearing and length of the property line along the north side of Lot 5?

____

3. What is the size in acres of Lot 5?

____

4. What are the elevations for the following?

A. First floor ______________________________

B. Top of footing ______________________________

C. Basement floor ______________________________

5. How is the ground sloping from the street to the back of the property?

____

6. Give the distance the house is set back from the property lines at the following corner locations.

A. Center front ______________________________

B. North side ______________________________

C. South side ______________________________

7. What is the radius dimension of the street property line?

____

8. What is the change in grade of the property from the center front corner of the house to the back of the house where the deck meets the back wall?

____

9. What is running along the south property line?

____

10. Where is the water utility line located?

____

Activity A-2

Foundation

Name ____________________

1. What is the scale of the foundation plan?

2. What is the size of the typical foundation wall footing? Where is this found?

3. Identify the following in relation to the full height foundation wall.

   A. Wall thickness ____________________

   B. Reinforcing ____________________

   C. Type of waterproofing ____________________

   D. Height of wall above footing ____________________

4. What size are the interior column footings? What reinforcing is specified?

5. What is the specification for the garage slab, including the psi and air entrainment requirements?

6. What is shown at locations where the structural steel beams extend into the concrete wall?

7. What is the difference between the slab specification in the garage and the slab specification in the basement?

8. What is the foundation wall height for the walls at the master bath, kitchen, and great room?

9. What is along the interior side of the walls referred to in Question 8?

10. For the walls referenced in Questions 8 and 9, what is the specification for anchor bolts holding down the sill plate?

Activity A-3

Floor Plans, Elevations, and Sections

Name ______________________________

1. What are the floor-to-floor heights for the following?

   A. Basement to first floor ______________________________

   B. First floor to second floor ______________________________

   C. Second floor to bottom of wood truss at dining room ______________________________

2. What are the overall dimensions for the following?

   A. Basement along the W8×28 beam ______________________________

   B. Garage ______________________________

   C. Kitchen outside wall to face of great room fireplace ______________________________

   D. Front of porch ______________________________

3. How many risers and treads are needed to go from the first floor to the second floor?

   ____

4. What flooring material is indicated for the following areas?

   A. Kitchen floor ______________________________

   B. Laundry ______________________________

   C. Master bedroom ______________________________

   D. Master bath ______________________________

5. What does the dashed line in the dining room indicate?

   ____

6. Identify the door sizes and types for the following.

   A. Garage door to house ______________________________

   B. Front door ______________________________

   C. Basement walk-out door ______________________________

7. Give the specification for the following windows.

   A. Typical basement window ______________________________

   B. Master bath window ______________________________

   C. Dining room window ______________________________

   D. Great room upper windows ______________________________

8. Identify the cabinet sizes for the following.

   A. Sink ______________________

   B. Island ______________________

   C. Oven ______________________

   D. Corner wall cabinet ______________________

9. What is the minimum clearance required around all aisles at cabinets, islands, and peninsulas?

   ____

10. What type of drywall is required for the garage walls and ceiling?

    ____

11. Where is the attic access panel located?

    ____

12. What are the "U" values at the following locations?

    A. 2×6 brick______________________

    B. Above garage______________________

    C. 2×4 frame______________________

13. What is the roof slope at the following locations?

    A. Above kitchen ______________________

    B. Above great room______________________

    C. Above study ______________________

14. What are the specifications for the brick and how is it attached to the framing?

    ____

15. What is between the drywall and studs?

    ____

16. What size is the fascia board?

    ____

17. What is required to be installed in the floor joist space at the exterior walls?

    ____

18. Where are the finish materials for the exterior of the house identified?

    ____

19. What material is specified for the wood decking?

    ____

20. What is the elevation change from the first floor to the deck?

    ____

Activity A-4

Floor and Roof Framing

Name ______________________________

1. Where is the floor framing for the first floor found?

____

2. Give the size of the steel beams at the following locations.

   A. Parallel to stairs toward the front of the house ______________________________

   B. Through the center of the house ______________________________

   C. Span between the steel beams near the "up" end of the stairs ______________________________

3. Give the typical size and spacing specification for floor joists.

____

4. Give the specification for joists perpendicular to the stair framing openings.

____

5. What type of wood is to be used for wood that contacts concrete?

____

6. What type of column is specified in wood-framed walls where a steel beam frames into the wall?

____

7. Give the size and spacing for the joists that frame the second floor.

____

8. What are the specifications for the headers used to frame the two garage door openings?

____

____

9. What is the specification for the framing members above the study, master bedroom, and master bath?

____

10. Give the specification for the header above the front porch.

____

11. What are the specifications for the typical flooring over the wood floor joists?

____

12. Give the size and spacing of studs used at the second-story space in the great room.

____

13. What is specified for the exterior covering attached to the studs?

____

14. What is specified for the sill plate at the great room wall?

____

15. What is specified for wood lintels for a 6′-0″ opening?

____

16. What is specified for steel lintels for a 16′-0″ opening?

____

17. Give the design loads for the following.

A. Floor load ____

B. Stair load ____

C. Snow load ____

D. Soil bearing capacity ____

18. How are the deck joists for the screened porch deck attached to the house?

____

____

Activity A-5

Electrical

Name ______________________________

1. Identify the electrical symbol for each of the following. Sketch the symbol freehand.

   A. 220V outlet

   B. Track light

   C. Smoke detector

   D. 3-way switch

2. How many 110V and 220V electrical outlets are in the kitchen?

   ____

3. What is the switch in the master bedroom controlling?

   ____

4. At the top of the stairs, there is a 3-way switch. Where is the other switch located?

   ____

5. What are the three switches at the front door controlling?

   ____

Activity A-6

Estimating

Name ____________________________

Estimate the total construction cost using the approximate "square footage" estimate. Then, select one aspect of the construction, such as foundation, framing, interior finishes, roofing and siding, etc. Make a detailed estimate of the cost for that portion of the project. Calculate material and labor costs separately. Use the estimate sheet below.

Estimate Takeoff and Cost Sheet

Project ____________________________

Item no.	Item identification	Location on job	Cost			Total cost
			Labor	Material	Equipment	

Advanced Project B

The North House Residence

This project is based on drawings for the two-story brick and stone veneer residence shown in **Figure B-1**. *This five-bedroom home includes four full bathrooms and two 1/2 bathrooms, a 3-car garage, a piano room, a study, and a concrete patio. Refer to Prints NH-1 through NH-13 in the Large Prints supplement when working on the following activities. An index is provided below. Note: The prints provided do not represent a complete set of drawings for the building.*

Norris & Dierkers Architects/Planners, Inc.

Figure B-1. Front elevation of the North House residence.

PRINT INDEX • THE NORTH HOUSE RESIDENCE

PRINT LABEL	SHEET NUMBER	SHEET TITLE
NH-1	A1	EXTERIOR ELEVATIONS
NH-2	A2	EXTERIOR ELEVATIONS
NH-3	A3	FOUNDATION/FOOTING/LOWER LEVEL PLAN
NH-4	A4	FIRST FLOOR PLAN
NH-5	A5	SECOND FLOOR PLAN, FOOTING PAD DETAILS
NH-6	A7	ROOF PLAN, TYPICAL RAKE DETAIL
NH-7	A8	BUILDING SECTIONS
NH-8	A9	WALL SECTIONS
NH-9	A10	LOWER LEVEL REFLECTED CEILING AND LIGHTING PLAN
NH-10	A11	FIRST FLOOR REFLECTED CEILING AND LIGHTING PLAN
NH-11	A12	SECOND FLOOR REFLECTED CEILING AND LIGHTING PLAN
NH-12	A13	GENERAL NOTES AND SPECIFICATIONS
NH-13	A14	GENERAL NOTES AND SPECIFICATIONS, WALL SECTION

Activity B-1

Foundation

Name ______________________________

1. What are the dimensions for the following?

   A. Overall dimension across the rear of the house ______________________________

   B. Overall dimension across the front of the garage ______________________________

   C. Rear right corner of the house to the centerline of the windows in the ping pong area______________

   ____

2. Refer to the foundation wall footings to answer the following questions.

   A. Where are the foundation wall sections found?______________________________

   B. What are the sizes of the footings? ______________________________

   C. What is the specified psi rating of the concrete? ______________________________

   D. What is the rebar specification?______________________________

3. Refer to the foundation wall sections to answer the following questions.

   A. What is the difference between the reinforcing shown in Section 1 and Section 2 on Sheet A9? ________

   ____

   B. Which section specifies foundation waterproofing?______________________________

   C. Of what size and type are the footing drains? ______________________________

4. Referring to the foundation wall sections, what is specified for the concrete, vapor barrier, and under-slab fill for the basement slab?

   ____

   ____

5. What is the specified grade for concrete reinforcing bars?

   ____

6. What are the psi and air entrainment specifications for the concrete for the garage slab?

   ____

7. Describe the discrepancy between the basement slab concrete psi rating indicated on plan drawings, sections, and notes and/or general notes and specifications. What should the contractor expect to do?

   ____

   ____

   ____

8. What is to be done at each garage door opening?

____

9. What is the meaning of the shaded area at the front porch?

____

10. What note refers to the column footings and what is specified?

____

11. How many column footings are there?

____

12. What is specified to support the stair framing?

____

Activity B-2

Floor Plans, Elevations, and Sections

Name ______________________________

1. What are the typical finish materials that are indicated on the elevations of the house?

____

____

2. Refer to the front porch columns to answer the following questions.

   A. How many columns are there? ______________________________

   B. What note refers to the columns? ______________________________

   C. What material is specified? ______________________________

3. What is the ceiling height at the covered front entry from the first floor level?

____

4. On the rear elevation drawing, the windows located toward the center of the house are "stepped" down. What is inside the house to require this layout?

____

5. What is the typical first floor window rough opening (R.O.) height?

____

6. What type of flashing is specified where the roof intersects the facade of the house?

____

7. What are the floor finishes in the following rooms?

   A. Powder room #2 ______________________________

   B. Exercise room ______________________________

   C. Kitchen ______________________________

   D. Study ______________________________

8. What is above the toilet in Bathroom #2?

____

9. Which note and which detail provide footing requirements for bearing stud walls?

____

10. What note refers to the builder's option projection screen and where is this item located?

____

11. How many furnaces are there on the lower level?

____

12. How is the singing platform constructed?

____

13. What note refers to insulation on the foundation wall and what is specified?

____

14. How high is the countertop in the bar area?

____

15. What note refers to the limits of the ceramic tile in the bar area and how are the limits designated?

____

16. What ceiling heights are specified in the following areas?

A. Storage/Mechanical #1 _______________

B. Media area _______________

C. Linen closet _______________

D. Above singing platform _______________

E. Bulkheads in media area _______________

F. Garage _______________

G. Great room _______________

H. Foyer _______________

I. Study _______________

J. Bedroom #3 _______________

K. Bulkheads and ceiling in great room _______________

17. What note refers to the dishwasher in the kitchen and where is this item located?

____

18. What type of doors are specified for the following openings?

A. Florida room to eating area _______________

B. Front door _______________

C. Hallway to garage _______________

D. Garage car door _______________

19. Where are the attic access panels located?

____

____

20. What is the minimum height of handrails and how is the height measurement taken?

____

21. What manufacturer of windows is specified and what type of glass is required?

____

22. What is the minimum percentage of aggregate glass area required for a habitable room?

____

23. What note refers to the panel between the whirlpool tub and shower area and what is specified?

____

24. Refer to the main stairs going "up" to answer the following.

    A. How many risers are there? _______________

    B. What is the riser height? _______________

    C. What is the total rise? _______________

25. What is the overall size of Bathroom #3, including the shower area?

____

26. What is different about the windows in Bedroom #2 and Bedroom #3 compared to the rest of the windows on the second floor?

____

27. On the second floor plan, what does the circle symbol represent in the upper-left corner of the upper great room?

____

28. How many lineal feet of clothes rod and shelf are required for Walk-in Closet #3? Give the answer to the nearest foot.

____

29. What is the overall area of Bedroom #3, including the walk-in closet and bathroom?

____

30. What is the typical roof overhang?

____

Activity B-3

Floor and Roof Framing

Name ______________________________

1. On which sheet is the first-floor framing shown?

____

2. What is the typical floor joist size and spacing specification for the first floor?

____

3. Give the specification for typical steel columns required for the steel beams.

____

4. What type of steel beam is specified at the following locations?

   A. Span across the media center ______________________________

   B. Span from the media center to the ping pong area ______________________________

   C. Span between the foundation walls in the exercise room ______________________________

5. What is the floor joist size and spacing specification for the typical joists that frame the second floor?

____

6. What is specified for the columns in the dining room, kitchen, and eating area?

____

7. What type and size is the beam spanning diagonally from column to column in the space between the kitchen and great room?

____

8. What size beam spans from the kitchen column to the piano room over the "down" portion of the stairs?

____

9. What is the typical spacing for the roof trusses?

____

10. What does the dashed outline indicate on the roof plan?

____

11. What is the typical pitch of the roofing system?

____

12. What is the total design load for the roof?

____

13. What pitch is specified for the back side of the garage roof?

____

14. What size steel angles are specified for openings 4′-0″ to 6′-0″?

____

Activity B-4

Electrical

Name ______________________________

1. What type of fixture is specified for the fixtures in the ping pong area? How are the fixtures switched?

____

____

2. What types of recessed can fixtures are specified for the exercise room and how many are shown?

____

3. What type of lighting fixture is specified above the bar area along the perimeter? How many are shown?

____

4. Where is the electrical panel located? What amp service is specified?

____

____

5. How many exterior sconces are there?

____

6. What is the difference between the two different types of paddle fan symbols?

____

7. Where is the lighting plan for the study shown?

____

8. What is specified for the chandeliers at the following locations?

   A. Piano room ______________________________

   B. Eating area ______________________________

   C. Study ______________________________

9. Where is fixture DCHF-2 switched?

____

10. Where are the lights in the hall from the garage to the kitchen area switched?

____

____

Activity B-5

Estimating

Name ___________________________

Estimate the total construction cost using the approximate "square footage" estimate. Then, select one aspect of the construction, such as foundation, framing, interior finishes, roofing and siding, etc. Make a detailed estimate of the cost for that portion of the project. Calculate material and labor costs separately. Use the estimate sheet below.

Estimate Takeoff and Cost Sheet

Project ___________________________

Item no.	Item identification	Location on job	Cost			Total cost
			Labor	Material	Equipment	

Advanced Project C

Office and Warehouse

This project is based on drawings for the office and warehouse shown in **Figure C-1**. *This building has 15,000 square feet of office space on two floors, plus over 50,000 square feet of warehouse space. The office provides working space for more than 70 employees and contains four conference rooms, a photo studio, and a classroom. Refer to Prints OW-1 through OW-40 in the Large Prints supplement when working on the following activities. An index is provided below. Note: The prints provided do not represent a complete set of drawings for the building.*

Charles E. Smith, Areté 3 Ltd.

Figure C-1. Rendering of a two-story office building with attached warehouse.

PRINT INDEX • OFFICE AND WAREHOUSE

PRINT LABEL	SHEET NUMBER	SHEET TITLE
OW-1	SP-1	ARCHITECTURAL SITE PLAN
OW-2	SP-2	ENLARGED SITE DETAILS
OW-3	A-1	FIRST FLOOR PLAN
OW-4	A-1.1	FIRST FLOOR ENLARGED PLAN
OW-5	A-2	MEZZANINE FLOOR PLAN
OW-6	A-2.1	MEZZANINE ENLARGED PLAN
OW-7	A-3	ELEVATIONS
OW-8	A-6	ENTRANCE ELEVATIONS AND DETAILS
OW-9	A-7	WALL SECTIONS
OW-10	A-8	WALL SECTIONS
OW-11	A-9	WALL SECTIONS
OW-12	A-10	ROOF PLAN
OW-13	A-12	STAIR PLANS & SECTIONS
OW-14	A-13	STAIR DETAILS
OW-15	A-14	VAULT & ELEVATOR SECTIONS
OW-16	A-15	ELEVATOR DETAILS
OW-17	A-16	DOOR SCHEDULE & DETAILS
OW-18	A-17	FIRST FLOOR REFLECTED CEILING PLAN
OW-19	A-18	MEZZANINE REFLECTED CEILING PLAN
OW-20	A-19	ENLARGED TOILET ROOM PLAN, TOILET ROOM ELEVATIONS, & TOILET ACCESSORIES
OW-21	A-20	MISC. ELEVATIONS, DOOR DETAILS, & PARTITION SECTIONS
OW-22	A-21	DOOR SCHEDULE & DETAILS
OW-23	S-1	NOTES
OW-24	S-2	FOUNDATION PLAN
OW-25	S-3	ENLARGED FOUNDATION PLANS
OW-26	S-4	FOUNDATION DETAILS
OW-27	S-5	FOUNDATION DETAILS
OW-28	S-6	MEZZANINE FRAMING PLAN
OW-29	S-7	MEZZANINE FRAMING DETAILS
OW-30	S-8	ROOF FRAMING PLAN
OW-31	S-9	ROOF FRAMING DETAILS
OW-32	S-11	CANOPY DETAILS
OW-33	S-12	ENLARGED STAIR FRAMING PLAN
OW-34	P-1	PLUMBING PLAN
OW-35	P-2	ENLARGED PLUMBING PLANS
OW-36	P-3	ENLARGED PLUMBING PLANS
OW-37	P-4	PLUMBING DIAGRAMS
OW-38	HVAC-1	ENLARGED SECOND FLOOR PLAN
OW-39	HVAC-2	ENLARGED FIRST FLOOR PLAN
OW-40	HVAC-3	FIRST FLOOR PLAN

Activity C-1

Site Plan

Name ______________________________

1. What is the scale of the site plan?

2. How wide are the parking stalls?

3. How thick are the concrete sidewalks?

4. The asphalt in the truck access areas consists of two layers of material. Describe these layers.

5. How does the asphalt in the dock areas differ from the asphalt in the parking lot?

6. What type of reinforcement is needed for the curbing at the entrance to the parking lot?

7. Along the north side of the building is a 5′-0″ square concrete stoop. Why is the stoop needed?

8. How wide is the sidewalk leading to the main (east) entrance of the office?

9. What slope is used for the curb ramp at the end of the sidewalk?

10. What is the setback distance for the front (east side) of the building?

Activity C-2

Architectural Drawings

Name ______________________________

1. What is the difference between the door to the vault on the first floor and the door to the vault on the mezzanine level?

2. What color are the pipe rails around the loading dock areas painted?

3. What type of lights are used in Room 200 (large conference room)?

4. What is the specification for the ceiling tile in the office?

5. In the stair located behind the reception area, how many risers are needed?

6. What is the narrowest space between racks in the warehouse?

7. What two sizes of overhead doors are used in the loading dock areas?

8. What is the elevation of the top of most of the precast wall panels?

9. What size are the lockers in the women's restrooms?

10. How thick are the restroom mirrors?

11. In the stairs next to the elevator, what are the dimensions of the landing?

12. How wide between masonry blocks is the opening for the elevator door?

13. How far below the first floor is the top of the slab at the bottom of the elevator pit?

14. What is the finish floor elevation in the warehouse?

15. How many large (4′×8′) skylights are in the roof?

16. What type of finish is used on the door to the VP Sales office?

17. What is the maximum distance between handrail supports on the stairs?

18. What color glass is used in the doors at the main (east) entrance to the office?

19. The walls around the vault in the office are composed of three layers. What are these layers?

20. In the copy area on the mezzanine level, how many wall-mounted cabinets are needed?

21. What is the minimum allowable width and depth for the elevator car?

22. What types of treads are used in the stairs?

23. What is used for the interior finish of skylight shafts in the office?

24. On the mezzanine level, what is the distance between the finished floor and the finished ceiling?

25. What type of roofing system is used?

Activity C-3

Structural Drawings

Name ______________________________

1. What size columns are used around the perimeter of the office?

2. How are joists supported by exterior precast walls on the north side of the building?

3. What type of member is used for the majority of the framing in the canopy above the employees' entrance?

4. What is the loading dock slab specification?

5. How thick are the exterior precast wall panels?

6. At what elevation is the top of the grate for the trench drain?

7. How are the roof beams supported at the east and west exterior walls?

8. What is placed above the roof joists in the warehouse?

9. What size is the wall footing under the office building?

10. What type of joists are used to support the mezzanine floor above the lunchroom?

11. What is the change in elevation from the west edge of the loading dock slab to the trench drain?

12. How is the 1 1/2″ metal roof deck attached to the roof joists?

13. What size columns are used in the warehouse?

14. What is the size of the footing under the column located at column lines B and 4?

15. What reinforcement is included in the slab at the bottom of the elevator pit?

Activity C-4

Plumbing Drawings

Name ______________________________

1. How many connections to the storm sewers are needed?

____

2. What size pipe is connected to the drinking fountains?

____

3. How many roof drains (including overflow drains) are needed?

____

4. How many hose bibbs are needed?

____

5. How many water closets are needed?

____

6. Where is the water meter located?

____

7. Where is the water heater located?

____

8. Where is the sump located?

____

9. How many cleanouts are provided in the waste piping?

____

10. What size pipe is to be connected to the sanitary sewer?

____

Activity C-5

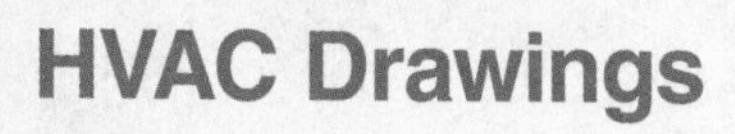

HVAC Drawings

Name ______________________________

1. How many thermostats are in the warehouse?

2. How many exhaust fans are in the warehouse?

3. What type of liner is used in supply ducts?

4. Which roof-mounted HVAC units supply air to the first floor?

5. Where is the thermostat controlling RTU-3 located?

6. What size stack is needed for the ceiling-mounted heating units (UH-1)?

7. What is the maximum amount of air that can be supplied to the lunchroom?

8. What is the minimum allowable distance between outdoor intake openings and exhaust openings?

9. What rooms are serviced by the two EF-2 exhaust fans?

10. Which units are fueled by natural gas?

Activity C-6

Estimating

Name ____________________

Estimate the total construction cost using the approximate "square footage" estimate. Then, select one aspect of the construction, such as foundation, framing, interior finishes, roofing and siding, etc. Make a detailed estimate of the cost for that portion of the project. Calculate material and labor costs separately. Use the estimate sheet below.

Estimate Takeoff and Cost Sheet

Project ____________________

Item no.	Item identification	Location on job	Cost			Total cost
			Labor	Material	Equipment	

Advanced Project D

Parking and Office Building

This project is based on drawings for the parking and office building shown in **Figure D-1**. *This 9-story building has a parking garage and 145,000 square feet of office space. Refer to Prints CMB-1 through CMB-35 from the Large Prints supplement to answer the following questions. An index is provided below. Note: The prints provided do not represent a complete set of drawings for the building.*

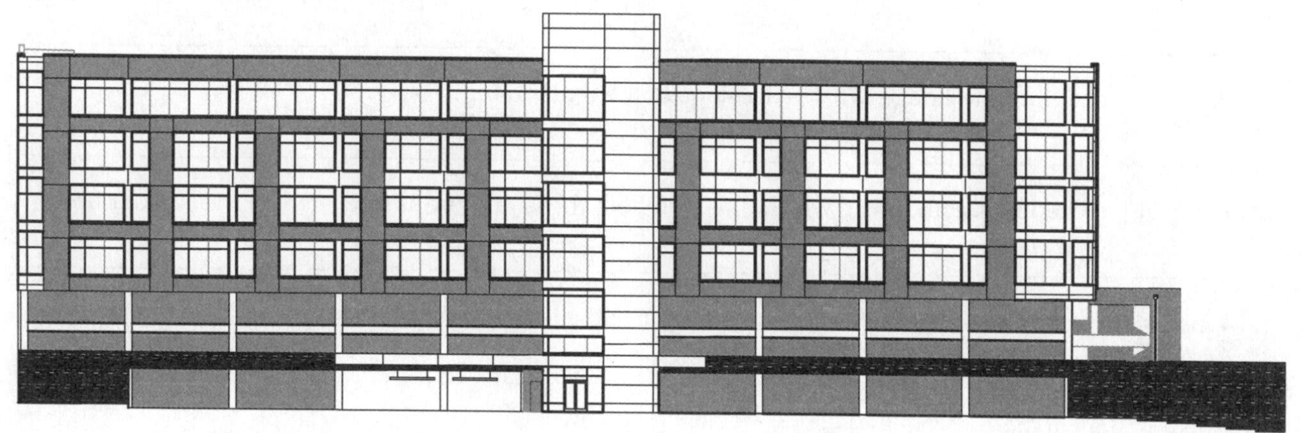

Neyer Architects, Inc.

Figure D-1. Elevation of the 9-story parking and office building.

PRINT INDEX • CINCINNATI MANOR PARKING AND OFFICE BUILDING

PRINT LABEL	SHEET NUMBER	SHEET TITLE
CMB-1	TS001	TITLE SHEET
CMB-2	C1.0	EXISTING CONDITIONS AND DEMOLITION PLAN
CMB-3	C1.1	SITE DIMENSION PLAN
CMB-4	C1.2	MASS EXCAVATION PLAN
CMB-5	C1.3	SITE GRADING PLAN
CMB-6	C1.4	SITE UTILITY PLAN
CMB-7	A101	PARKING LEVEL P1 FLOOR PLAN
CMB-8	A103	PARKING LEVEL P3 FLOOR PLAN
CMB-9	A106	GARAGE LEVEL P6/OFFICE LEVEL 1 FLOOR PLAN
CMB-10	A107	OFFICE LEVEL 2 FLOOR PLAN
CMB-11	A110	ROOF PLAN
CMB-12	A201	ENLARGED PLANS
CMB-13	A203	ENLARGED PLANS
CMB-14	A212	PLAN DETAILS
CMB-15	A303	GARAGE LEVEL P3 CEILING PLAN
CMB-16	A401	EXTERIOR ELEVATIONS
CMB-17	A402	EXTERIOR ELEVATIONS
CMB-18	A601	WALL SECTIONS
CMB-19	A611	SECTION DETAILS
CMB-20	A615	SECTION DETAILS
CMB-21	S001	STRUCTURAL GENERAL NOTES
CMB-22	S101	LEVEL P1 FOUNDATION PLAN
CMB-23	S103	LEVEL P3 FRAMING PLAN
CMB-24	S106	LEVEL P6 AND OFFICE LEVEL 1 FRAMING PLAN
CMB-25	S107	OFFICE LEVEL 2 FRAMING PLAN
CMB-26	S110	OFFICE LEVEL 5 (ROOF) FRAMING PLAN
CMB-27	S201	FOUNDATION DETAILS
CMB-28	S202	FOUNDATION DETAILS
CMB-29	S301	GARAGE FRAMING DETAILS
CMB-30	S302	GARAGE FRAMING DETAILS
CMB-31	S321	OFFICE FRAMING DETAILS
CMB-32	S401	GARAGE COLUMN SCHEDULE AND DETAILS
CMB-33	S403	OFFICE COLUMN SCHEDULE AND DETAILS
CMB-34	S504	TYPICAL PT BEAM DIAGRAMS
CMB-35	S505	NEW GARAGE PT BEAM SCHEDULE

Activity D-1

Civil Drawings

Name ________________________

1. Which sheet provides information to do the demolition work? What is the scale of the plan?

2. What is designated to remain on the demolition plan?

3. What is the bearing and length of the property line on the north side of the property?

4. What is the lowest elevation level of excavation and how many total cubic yards are to be excavated?

5. At the entrance into the garage, what is the slope of the center ramp?

6. Referring to Grading Detail B-B, in which direction does the land slope as indicated by the contour lines?

7. What is the entry elevation into the garage and what is the lowest level finish floor elevation of the garage?

8. What do the following symbols represent?

A. ———— G ———— ________________

B. ———— W ———— ________________

C. ———— SAN ———— ________________

D. ——— — — ——— ________________

E. ________________

F. ________________

9. What service should be called before any excavation starts?

____

10. Give the distances the building is set back from the following locations.

A. Vernon Place _______________________________

B. East University Avenue _______________________

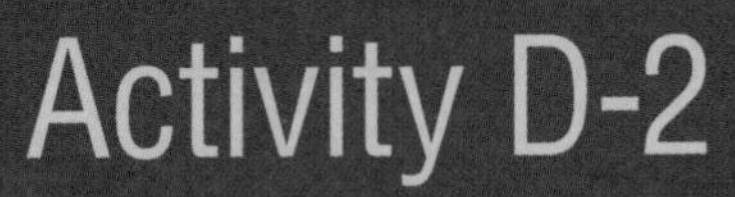

Activity D-2

Architectural Drawings

Name ______________________________

1. Which way is plan north indicated on the architectural drawings?

____

2. What is the typical center-to-center spacing of columns along the west side of the building?

____

3. What elevation is set to the reference elevation 100′-0″? What floor of the building is at this elevation?

____

4. What is the elevation of the first-floor lobby area at the tower level?

____

5. What is the T/Metal Panel elevation at the Stair A lobby tower?

____

6. What is the main exterior finish material of the office building?

____

7. On which sheets would you find the following?

   A. Wall sections______________________________

   B. Stair sections______________________________

   C. Schedules______________________________

8. What are the main two wall sections cut for the Stair A lobby tower?

____

9. Referring to the sections referenced in Question 8, answer the following questions.

   A. What is the difference between the two section cuts? ______________________________

   ____

   B. What detail is referenced at the top of the section through the lobby? ______________________________

   C. What detail shows the lobby entrance door? ______________________________

10. On which sheet is the roof plan shown?

____

11. How many roof drains are there on the roof?

____

12. What is the typical slope of the roof?

____

13. Which stair enlarged plan is for Stair B Room P302?

____

14. What is the handrail size and type for this stair?

____

15. Which note is for the pipe bollard at the door entry at Door P302B?

____

16. Referring to the plan details on Sheet A212, answer the following questions.

    A. Which notes refer to the insulation? _______________

    B. Which note refers to the structural steel columns? _______________

    C. Which note refers to the brick veneer? _______________

    D. What is the typical metal stud size and spacing? _______________

17. What is the ceiling finish in Lobby Room P306?

____

18. What is the ceiling finish in Fire Comm Center Room P307?

____

19. What are the finish materials noted for the exterior of the parking garage?

____

____

____

20. Referring to Section Detail C on Sheet A615, answer the following questions.

    A. What sheet does the detail refer back to? _______________

    B. What is the actual elevation of Parking Level P6? _______________

    C. Referring to the metal edge angle specified in Note 55, what is the edge of slab dimension from the centerline of the beam?

    ____

Activity D-3

Structural Drawings

Name ________________________________

1. What is the psi rating for the following items?

   A. Garage elevated slabs, beams, and girders ________________

   B. Stair pan fills________________

   C. Drilled piers ________________

2. How much extra concrete should be assumed for slabs on metal deck?

   ____

3. Identify the sheet showing the following.

   A. Foundation plan ________________

   B. Level P5 framing________________

   C. Post-tensioned beam schedule________________

4. What enlarged partial plan detail is indicated at the northwest corner of the structure on the foundation level?

   ____

5. What is the minimum clearance between reinforcing steel and concrete for the following members of the structure?

   A. Columns________________

   B. Foundation walls ________________

   C. Footings ________________

   D. North barrier wall Level P3________________

6. Referring to Section 20 on Sheet S302, what is the horizontal reinforcement specified for the foundation wall and where is this information found?

   ____

7. Answer the following questions regarding the drilled piers.

   A. What size pier is specified for Column D-9?________________

   B. Is there anything special to note about this pier? ________________

   C. What is the reinforcing required? ________________

8. Referring to the foundation plan, what is spanning from Column C-9 to Column D-9?

   ____

9. What is the typical slab-on-grade thickness and reinforcement?

10. What is the typical shear wall thickness and reinforcement?

11. What is the typical foundation wall detail referenced along the east wall, and how thick is the footing?

12. Referring to Section 3A/S201, what is the specified thickness of the foundation wall?

13. What is the difference in the steel beams that carry the grating along the east side of the building and span east to west?

14. Where is the P.T. Slab Schedule located?

15. Referring to Garage Level P3, give the mark and size of the post-tensioned beams at the following locations.

A. Grid line 1, A to B

B. Grid line 5, D to E

C. Grid line 12, B to C

16. Referring to the post-tensioned beam schedules, what do the abbreviations T_a, T_b, and T_c refer to? What is the information given underneath the abbreviations in the schedules?

17. What is the typical slab thickness for Garage Level P3?

18. Referring to Garage Level P3, what is indicated by the dashed lines running east to west between grid lines 8 and 9?

19. What is the slab thickness and reinforcement for Garage Level P6/Office Level 1?

20. What is the typical size for steel beams running east and west?

21. What is the typical U.N.O. section around the building perimeter for Garage Level P6/Office Level 1?

____

22. What is the largest steel beam running north and south? Where is it located?

____

23. Referring to the concrete garage column and office steel column schedules, answer the following questions regarding Column D-10.

    A. What is the schedule group designation for each type of column?

    ____

    B. What is the size and reinforcement of the concrete column from Level P2 to Level P3?

    ____

    C. What is the size of the steel column from Office Level 2 to Office Level 3?

    ____

24. Answer the following questions regarding Column D.9-11.

    A. What is the schedule group designation for office steel columns?

    ____

    B. What is the size of the steel column from Office Level 2 to Office Level 3?

    ____

    C. What kind of connection is required between beams and columns?

    ____

25. Refer to the roof framing plan to answer the following questions.

    A. What is the steel joist size specified between grid lines 8 and 9 from D to E?

    ____

    B. What are the typical sections indicated around the building perimeter?

    ____

    C. What type of steel angle is specified in the U.N.O. sections to provide continuous support at the perimeter and how is it connected?

    ____

Activity D-4

Estimating

Name ______________________________

Estimate the total construction cost using the approximate "square footage" estimate. Then, select one aspect of the construction, such as foundation, framing, interior finishes, roofing and siding, etc. Make a detailed estimate of the cost for that portion of the project. Calculate material and labor costs separately. Use the estimate sheet below.

Estimate Takeoff and Cost Sheet

Project ______________________________

Item no.	Item identification	Location on job	Cost			Total cost
			Labor	Material	Equipment	

SECTION 7
Reference Section

OrchidFlower/Shutterstock.com

Abbreviations, Symbols, and Tables

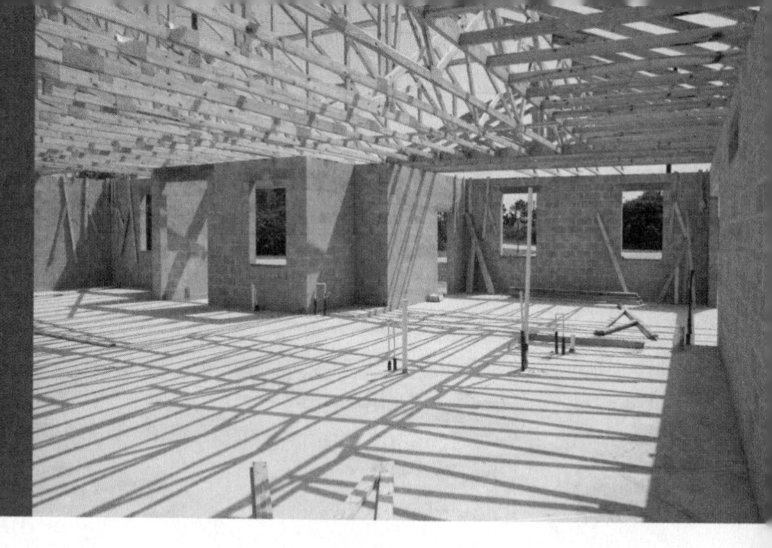

A

AB	Anchor Bolt
ABC	Aggregate Base Course
ABS	Acrylonitrile Butadiene Styrene
AC	Alternating Current
ACI	American Concrete Institute
ACOUS	Acoustical
ACP	Asbestos Cement Pipe
ACT	Actual
ADD'L	Additional
AFF	Above Finished Floor
AGGR	Aggregate
AH	Air Handling Unit
AIA	American Institute of Architects
AIR COND	Air Conditioning
AISC	American Institute of Steel Construction
AISI	American Iron and Steel Institute
ALT	Alternate
AL	Aluminum
AMP	Ampere
AMT	Amount
ANSI	American National Standards Institute
AP	Access Panel
APPD	Approved
APPROX	Approximate
ARCH	Architectural
ASB	Asbestos
ASPH	Asphalt
ASTM	American Society for Testing and Materials
AUTO	Automatic
AUX	Auxiliary
AWG	American Wire Gage
AV	Air Vent
@	At

B

B	Bathroom, Bottom
B/	Bottom of
BASMT	Basement
BBL	Barrel(s)
BBR	Base Board Radiation
BC	Bolt Circle
BD	Board
BD FT	Board Feet
BEV	Beveled
BF	Bottom of Footing
BLDG	Building
BLK	Block
BLKG	Blocking
BLR	Boiler
BM	Beam, Bench Mark
BOT	Bottom
BP	Base Plate
BR	Brass, Bedroom
BRG	Bearing
BRK	Brick
BRKT	Bracket
BRZ	Bronze
BTU	British Thermal Unit
BUS	Busway

C

C	Celsius
C or COND	Conduit
C/C, C to C	Center to Center
CAB	Cabinet
CB	Catch Basin
CD	Ceiling Diffuser
CEM	Cement
CER	Ceramic
CFM	Cubic Feet per Minute

CHAM	Chamfer
CHR	Chilled Water Return
CHS	Chilled Water Supply
CI	Cast Iron
CIP	Cast-in-Place Concrete
CIR	Circuit
CIR BKR	Circuit Breaker
CIRC	Circumference
CJ	Construction Joint, Control Joint
CKT	Circuit
CL	Center Line, Closet
CLG	Ceiling
CLK	Caulk
CLR	Clear
CMU	Concrete Masonry Unit
CO	Cleanout
COL	Column
COM	Common
COMP	Composition
CONC	Concrete, Concentric
CONN	Connection
CONT	Continuous
CONTR	Contractor
CP	Candle Power, Concrete Pipe
CR	Ceiling Register
CSK	Countersink
CT	Ceramic Tile
CU	Copper, Cubic
CW	Cold Water
CWS	Cold Water Supply

D

DC	Direct Current
DEC	Decorative
DEG	Degree
DET	Detail
DF	Drinking Fountain
DH	Double-Hung
DIA or ∅	Diameters
DIAG	Diagram, Diagonal
DIM	Dimension, Dimmer
DISC	Disconnect
DIV	Division
DMPR	Damper
DN	Down
DO	Ditto
DP	Duplicate
DR	Dining Room, Drain, Door
DS	Downspout
DW	Dishwasher, Dry Wall
DWG	Drawing
DWL	Dowel

E

EA	Each
EB	Expansion Bolt
EF	Exhaust Fan
EL, ELEV	Elevation
ELEC	Electric
ELVR	Elevator
EMER	Emergency
EMT	Electrical Metallic Tubing
ENAM	Enamel
ENC	Enclosure
ENGR	Engineer
ENT	Entrance
EQ	Equal
EQUIP	Equipment
EQUIV	Equivalent
EST	Estimate
EW	Each Way
EWC	Electric Water Cooler
EXC	Excavate
EXCL	Exclude
EXH	Exhaust
EXIST	Existing
EXP	Expansion
EXP JT	Expansion Joint
EXT	Exterior
EXTN	Extension

F

F	Fahrenheit
F BRK	Fire Brick
F EXT	Fire Extinguisher
FAB	Fabricate
FAM RM	Family Room
FAO	Finish All Over
FBO	Finish by Owner, Installed by General Contractor
FD	Floor Drain
FDN	Foundation
FDR	Feeder
FH	Fire Hydrant
FIG	Figure
FIN	Finish
FIN FL	Finish Floor
FIX	Fixture

FL	Flashing, Floor
FLEX	Flexible
FLG	Flange, Flooring
FLR	Floor
FLUOR	Fluorescent
FP	Fireplace
FR	Frame
FRG	Furring
FS	Floor Sink
FT	Feet
FTG	Fitting, Footing
FURN	Furnish
FWH	Frostproof Wall Hydrant
FX WDW	Fixed Window

G

GA	Gage
GALV	Galvanized
GAR	Garage
GC	General Contractor
GFCI	Ground Fault Circuit Interrupter
GI	Galvanized Iron
GL	Glass, Glazed
GND	Ground
GR	Grade
GRAN	Granular
GRTG	Grating
GYP	Gypsum
GYP BD	Gypsum Board

H

H	Hall
HB	Hose Bibb
HC	Heating Coil
HDW	Hardware
HDWD	Hardwood
HEX	Hexagonal
HGT	Height
HM	Hollow Metal
HOR	Horizontal
HP	Horsepower, High Point
HR	Hour
HTR	Heater
HU	Humidifier
HVAC	Heating, Ventilating, and Air Conditioning
HW	Hot Water
HWR	Heating Water Return
HYD	Hydraulic
HZ	Hertz

I

ID	Inside Diameter
IF	Inside Face
INCAND	Incandescent
INCL	Include
INCR	Increaser
INSUL	Insulation
INT	Interior
INV	Invert
ISO	International Standards Organization

J

J	Junction
JAN	Janitor
JB	Junction Box
JST	Joist
JT	Joint

K

K	Kip (1000 pounds), Kitchen
KP	Kick Plate
KS	Kitchen Sink
KVA	Kilovolt Amperes
KW	Kilowatt
KWY	Keyway

L

L	Left
L&PP	Light and Power Panel
LAB	Laboratory
LAD	Ladder
LAM	Laminated
LAQ	Lacquer
LAT	Lateral
LAU	Laundry
LAV	Lavatory
LB	Pound
LBR	Lumber
LDG	Landing
LEV	Level
LG	Long
LH	Left-Handed
LIN	Linear
LLH	Long Leg Horizontal
LLV	Long Leg Vertical
LP	Low Point
LPG	Liquefied Petroleum Gas
LR	Living Room

LT	Light
LTD	Limited
LTH	Lath
LTR	Letter
LV	Low Voltage
LVR	Louver

M

MAGN	Magnesium
MAINT	Maintenance
MAN	Manual
MAS	Masonry
MATL	Material
MAV	Manual Air Vent
MAX	Maximum
MB	Mixing Boxes (or Units)
MCM	Thousand Circular Mils
MDP	Main Distribution Panel
MECH	Mechanical
MED	Medium
MET	Metal
MEZZ	Mezzanine
MFR	Manufacturer
MH	Manhole
MID	Middle
MIN	Minimum
MISC	Miscellaneous
MK	Mark
ML	Metal Lath
MLDG	Molding
MN	Main
MO	Masonry Opening, Motor-Operated
MOD	Modular, Modification
MOR	Mortar
MS	Manual Starter
MT	Metal Threshold
MTD	Mounted
MTG	Mounting
MTL	Metal
MULL	Mullion

N

N	North
NA	Not Available
NEC	National Electrical Code
NEMA	National Electrical Manufacturers Association
NF	Near Face
NO	Number
NOM	Nominal
NOR	Normal
NTS	Not to Scale

O

OC	On Center
OD	Outside Diameter
OF	Outside Face
OFF	Office
OPNG	Opening
OPP	Opposite
OUT	Outlet
OVHD	Overhead

P

P	Pump
PAN	Pantry
PAR	Parallel
PART	Partition
PASS	Passage
PAV	Paving
PC	Pull Chain
PCS	Pieces
PERIM	Perimeter
PERP	Perpendicular
PH	Phase
PL	Pilot Light, Plate, Property Line
PL GL	Plate Glass
PLAS	Plaster
PLAT	Platform
PLBG	Plumbing
PLWD	Plywood
PNEU	Pneumatic
PNL	Panel
PORC	Porcelain
PR	Pair
PREFAB	Prefabricated
PRESS	Pressure
PRI	Primary
PROP	Property, Proposed
PRV	Pressure Reducing (Regulating) Valve
PSF	Pounds per Square Foot
PSI	Pounds per Square Inch
PSIG	Pounds per Square Inch Gage
PTD	Painted
PTN	Partition
PVC	Polyvinyl Chloride
PWR	Power

Q

QUAL	Quality
QT	Quarry Tile
QTY	Quantity

R

R	Radius, Risers
RA	Return Air
RAD	Radiator
RCP	Reinforced Concrete Pipe
RD	Roof Drain, Round
REC	Recessed
RECP	Receptacle
REF	Reference, Refrigerator
REG	Register
REINF	Reinforcing
REQ'D	Required
RES	Resilient Edge Strip
RET	Return
RETG	Retaining
REV	Revision
RFG	Roofing
RGH	Rough
RGH OPNG	Rough Opening
RH	Right-Handed
RIO	Rough-In Opening
RM	Room
RV	Relief Valve
RWD	Redwood

S

S	Stretcher
S or SW	Switch
SA	Supply Air
SAD	Supply Air Diffuser
SAN	Sanitary
SCH	Schedule
SCR	Screen, Screw
SCUP	Scupper
SDG	Siding
SDL	Saddle
SEC	Secondary
SECT	Section
SEL	Select
SER	Service
SEW	Sewer
SG	Supply Air Grille
SH	Sheet, Shower
SHLP	Shiplap
SHT	Sheet
SHTHG	Sheathing
SI	Storm Water Inlet
SIM	Similar
SJ	Steel Joist
SK	Sink
SL	Slate, Sleeve
SM	Sheet Metal
SOV	Shutoff Valve
SP	Soil Pipe, Sump Pump
SPAN	Spandrel
SPEC	Specification
SPKR	Speaker
SPR	Sprinkler
SQ	Square
SR	Supply Air Register
SST	Stainless Steel
ST	Storm
STD	Standard
STIFF	Stiffener
STIR	Stirrup
STL	Steel
STM	Steam
STN	Stone
STOR	Storage
STR	Straight, Strainer
STRUCT	Structural
SUR	Surface
SUSP	Suspended
SV	Safety Valve
SWG	Standard Wire Gage
SYM	Symbol, Symmetric

T

T	Thermostat, Tread
T/	Top of
T&B	Top and Bottom
T&G	Tongue and Groove
TAN	Tangent
TB	Top of Beam
TC	Terra Cotta, Top of Concrete, Top of Curb
TEL	Telephone
TEMP	Temperature, Temporary
TEMPL	Template
TERM	Terminal
TERR	Terrazzo
TF	Top of Footing

THK	Thick or Thickness
THR	Threaded
THRU	Through
TJ	Top of Joist
TM	Top of Masonry
TMB	Telephone Mounting Board
TOIL	Toilet
TP	Top of Pier, Top of Pavement
TRANS	Transformer
TS	Time Switch, Top of Slab
TV	Television
TW	Top of Wall
TYP	Typical

U

UG	Underground
UH	Unit Heater
UL	Underwriters Laboratories
UNFIN	Unfinished
UNO	Unless Noted Otherwise
UR	Urinal

V

V	Vent, Volts
VAC	Vacuum
VAN	Vanity
VAR	Varies
VC	Vitrified Clay
VCJ	Vertical Control Joint
VERT	Vertical
VEST	Vestibule
VF	Vinyl Wall Fabric
VOL	Volume
VP	Vitreous Pipe
VTR	Vent Through Roof

W

W	Washer (laundry), Waste, Water, Watts
W or WTH	Width
W/	With
W/O	Without
WC	Water Closet
WD	Wood
WDW	Window
WF	Water Fountain, Wide Flange
W GL	Wire Glass
WH	Water Heater, Weep Hole
WHR	Watt Hour
WI	Wrought Iron
WM	Water Meter
WP	Waterproofing, Work Point
WR	Washroom
WR BD	Weather Resistant Board
WS	Weatherstrip
WT	Weight
WV	Water Valve
WWF	Welded Wire Fabric

Y

Y	Wye

Z

Z	Zinc

Building Material Symbols

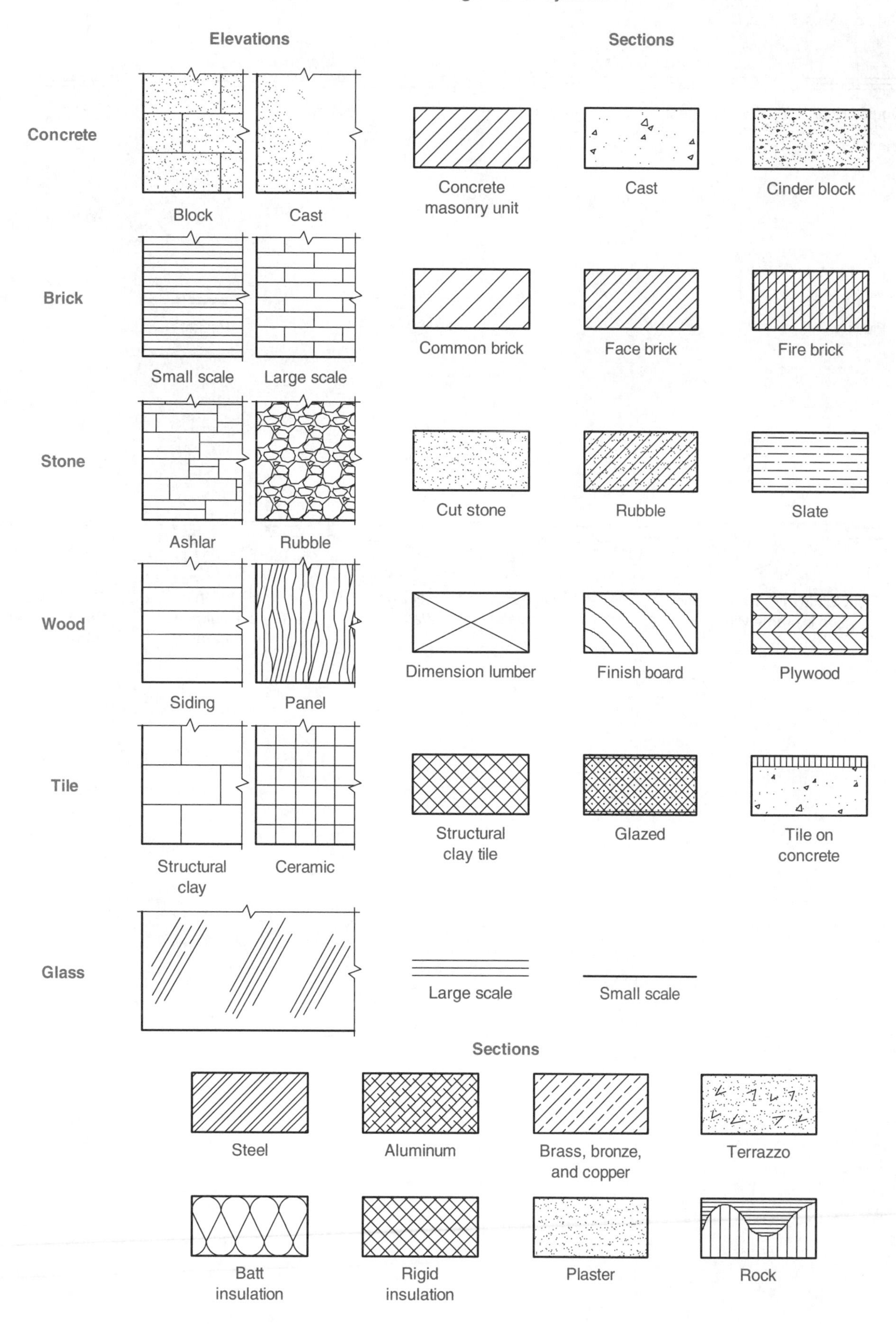

Window and Door Symbols

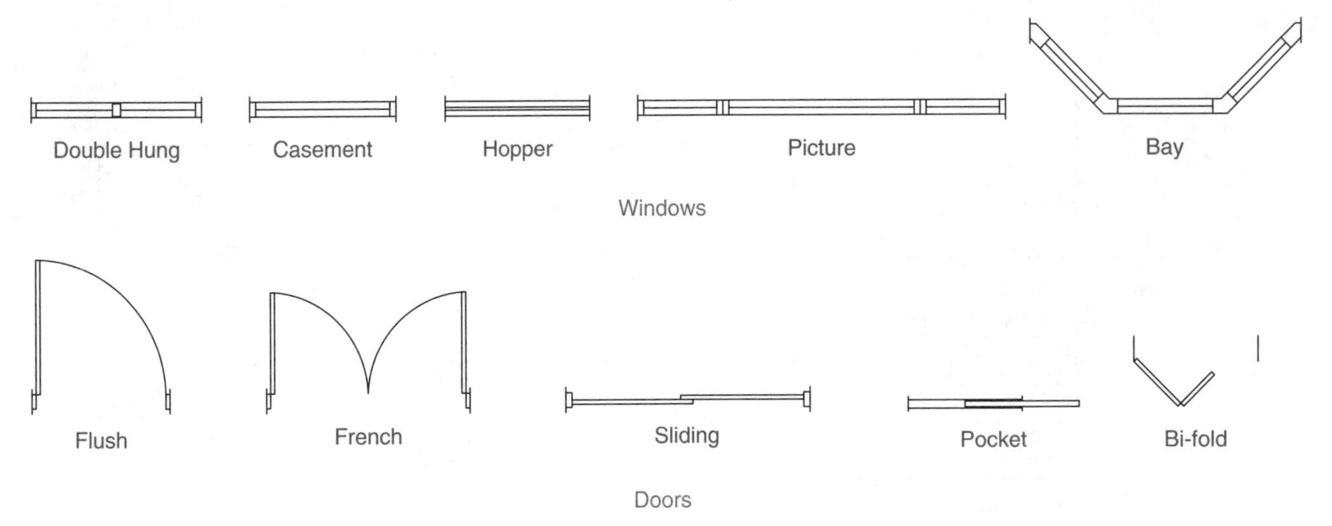

Exterior Door Symbols

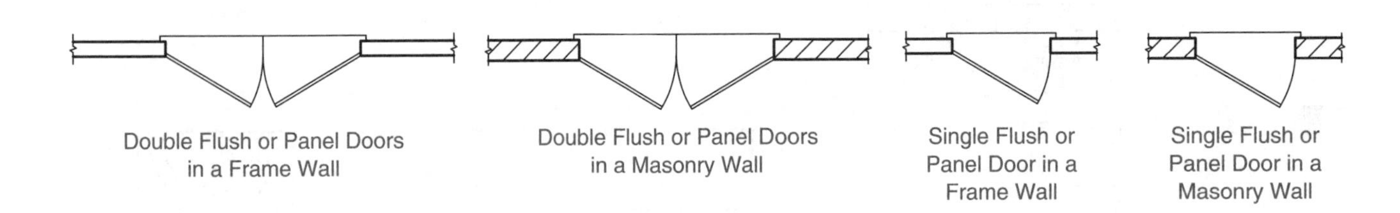

Kitchen Plan Symbols

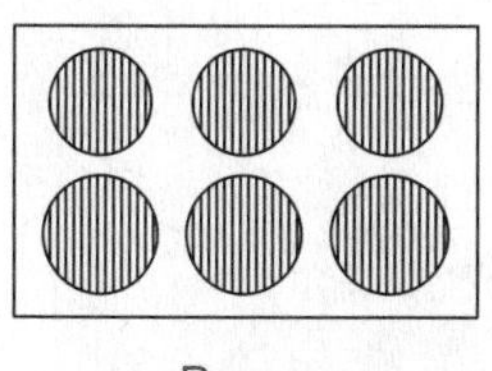
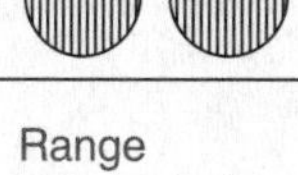
Range

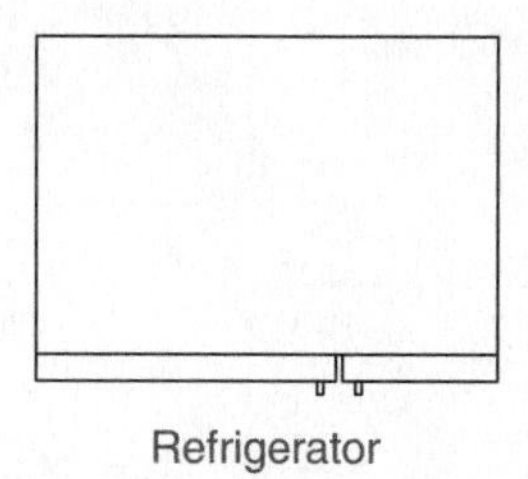
Refrigerator

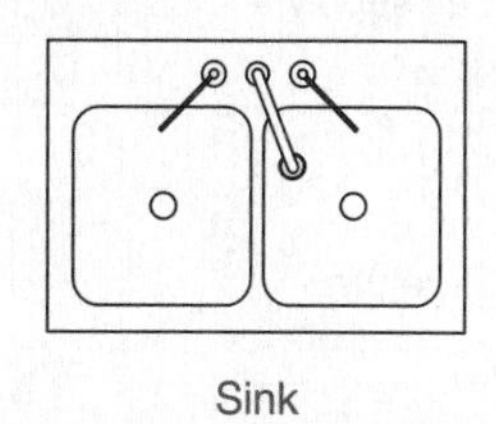
Sink

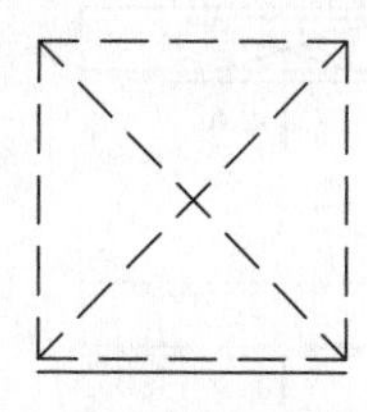
Dishwasher

Bathroom Plan Symbols

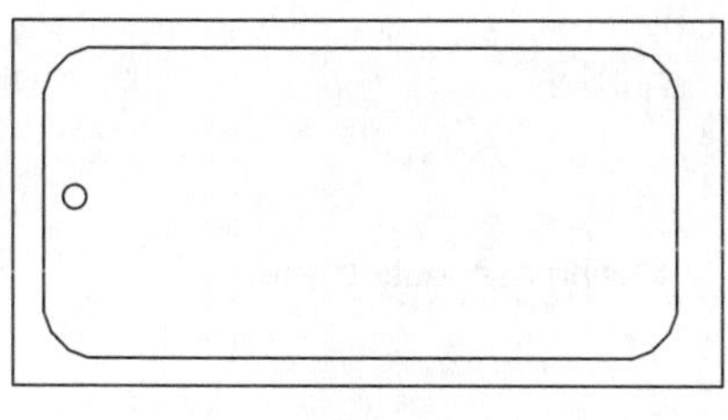

Bathtub

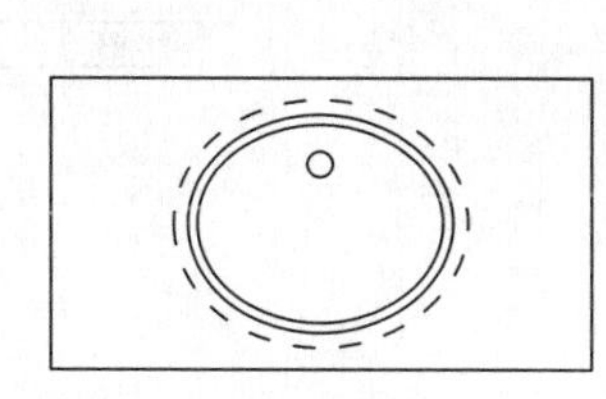
Vanity

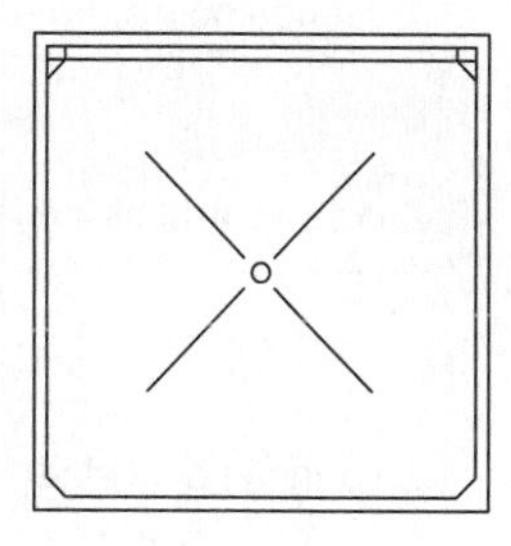
Shower

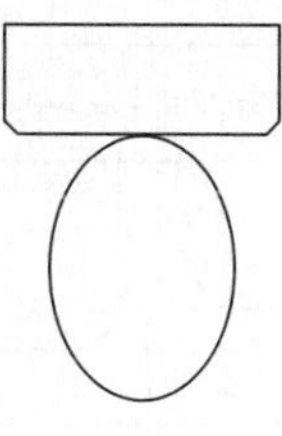
Toilet

Climate Control Symbols

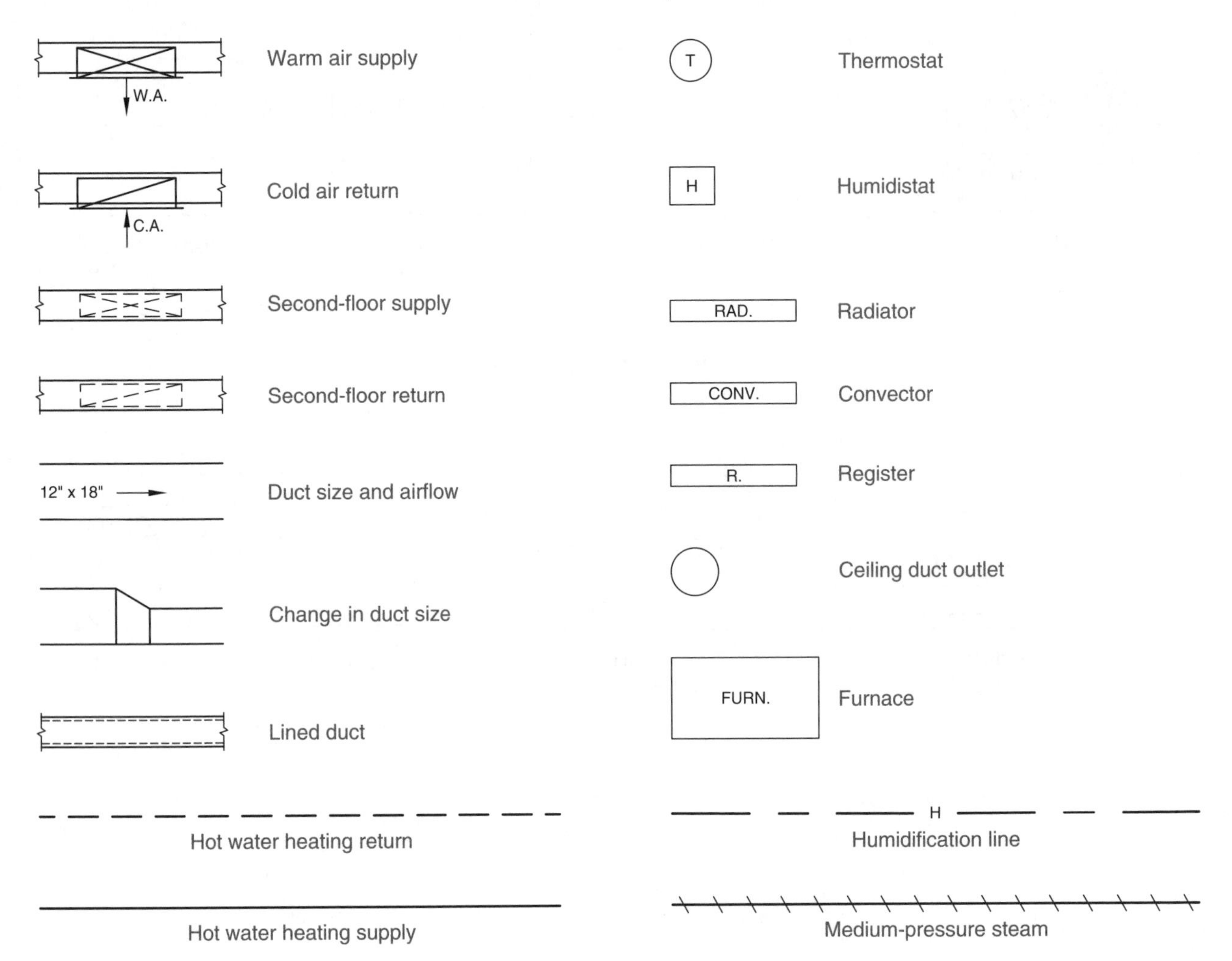

Plumbing Symbols

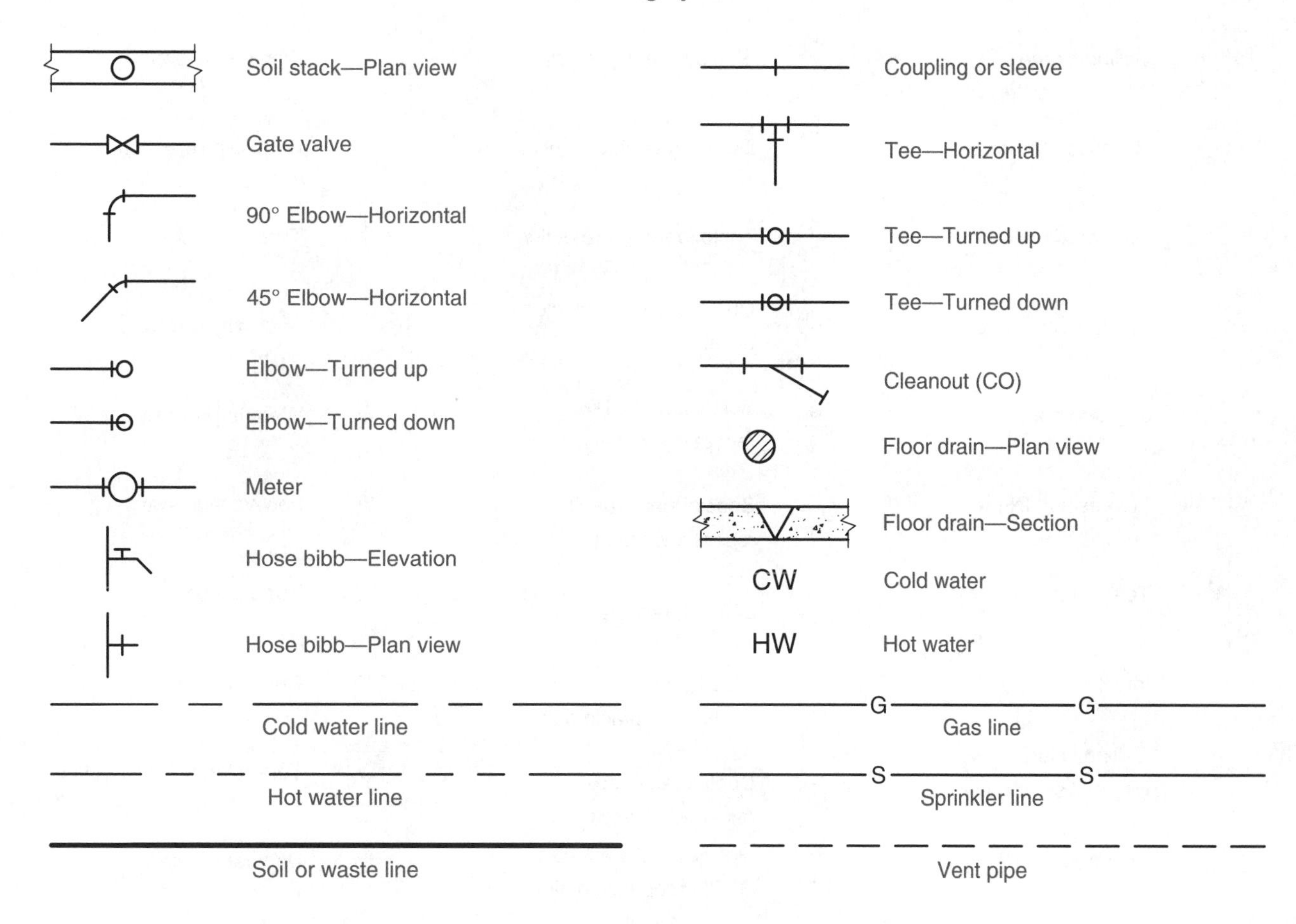
Soil stack—Plan view
Gate valve
90° Elbow—Horizontal
45° Elbow—Horizontal
Elbow—Turned up
Elbow—Turned down
Meter
Hose bibb—Elevation
Hose bibb—Plan view
Cold water line
Hot water line
Soil or waste line
Coupling or sleeve
Tee—Horizontal
Tee—Turned up
Tee—Turned down
Cleanout (CO)
Floor drain—Plan view
Floor drain—Section
CW
Cold water
HW
Hot water
G
G
Gas line
S
S
Sprinkler line
Vent pipe

Electrical Symbols

- Ceiling fixture
- Recessed fixture
- Drop cord fixture
- Fan hanger fixture
- Junction box
- Fluorescent fixture
- Telephone
- Intercom
- Ceiling fixture with pull switch
- Thermostat
- Special fixture (A, B, C, Etc.)
- Flush-mounted panel box

- Single receptacle outlet
- Duplex receptacle outlet
- Triplex receptacle outlet
- Quadruplex receptacle outlet
- Split-wired duplex receptacle outlet
- Special-purpose single receptacle outlet
- 240-volt receptacle outlet
- Weatherproof duplex receptacle outlet (WP)
- Duplex receptacle outlet with switch (S)
- GFCI receptacle outlet (GFCI)
- Special duplex receptacle outlet (A, B, C, Etc.)

- Single-pole switch ($)
- Double-pole switch ($_2)
- Three-way switch ($_3)
- Four-way switch ($_4)
- Weatherproof switch ($_WP)
- Low-voltage switch ($_L)
- Push button
- Chimes (CH)
- Television antenna outlet (TV)
- Dimmer switch ($_D)
- Special switch ($_A, B, C, Etc.)

Topographical Symbols

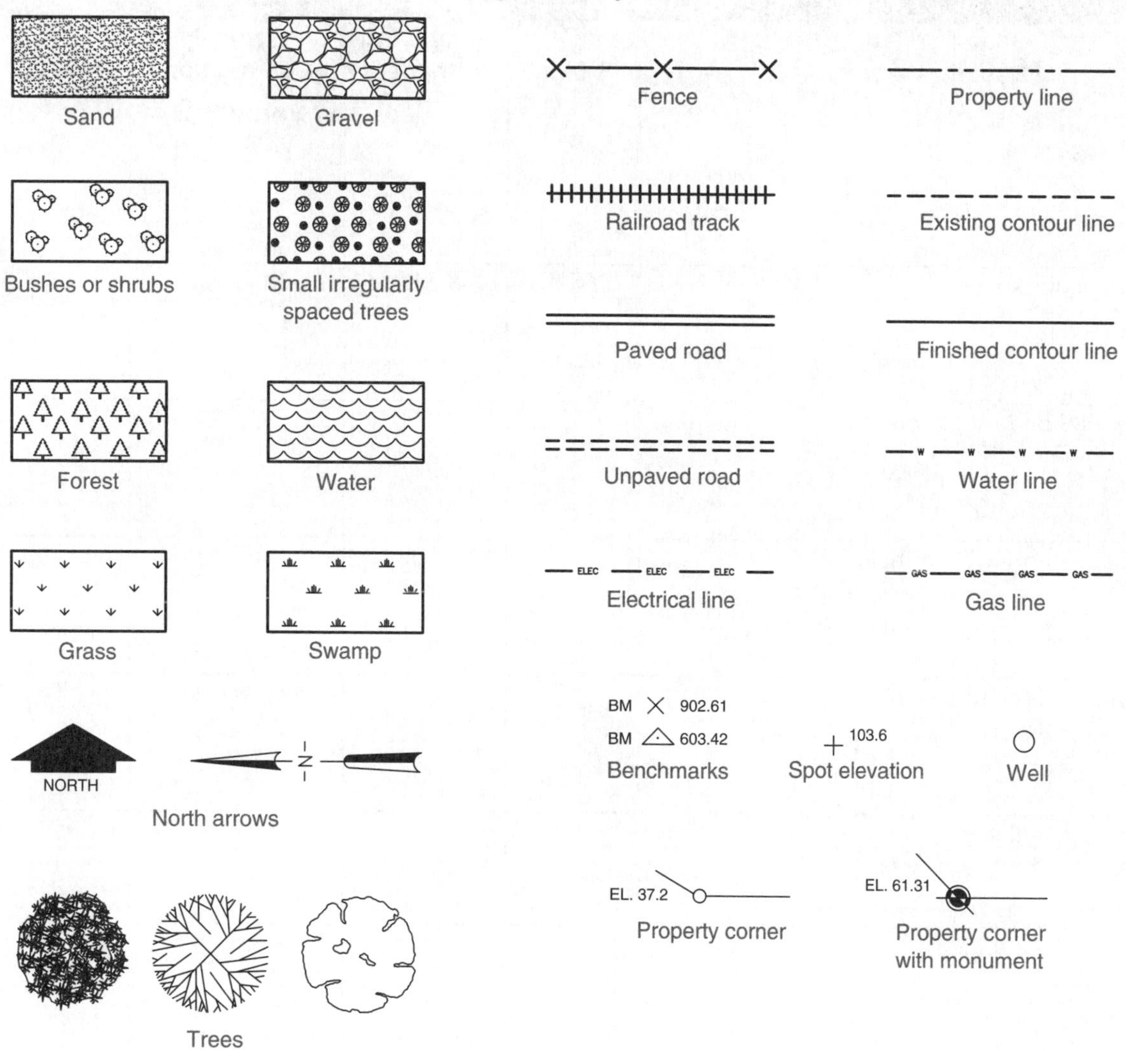

Structural Steel Shapes

Descriptive Name	Shape	Identifying Shape	Typical Designation (Depth Wt/Ft in Lb.)	Nominal Size (Depth Width)
Wide flange shapes		W	W21 × 147	21 × 13
Miscellaneous shapes		M	M8 × 6.5	8 × 2¼
American Standard beams		S	S8 × 23	8 × 4
American Standard channels		C	C6 × 13	6 × 2
Miscellaneous channels		MC	MC8 × 20	8 × 3
Angles — equal legs		L	L6 × 6 × ½*	6 × 6
Angles — unequal legs		L	L8 × 6 × ½*	8 × 6
Structural tees (cut from wide flange)		WT	WT12 × 60	12
Structural tees (cut from miscellaneous shapes)		MT	MT5 × 4.5	5
Structural tees (cut from Am. Std. beams)		ST	ST9 × 35	9
Zees		Z	Z6 × 15.7	6 × 3½

*Leg thickness

ASTM Standard Reinforcing Bars			
Bar Size Designation	**Area Sq. Inches**	**Weight Pounds per Ft.**	**Diameter Inches**
3	.11	.376	.375
4	.20	.668	.500
5	.31	1.043	.625
6	.44	1.502	.750
7	.60	2.044	.875
8	.79	2.670	1.000
9	1.00	3.400	1.128
10	1.27	4.303	1.270
11	1.56	5.313	1.410
14	2.25	7.650	1.693
18	4.00	13.600	2.257

Wire Bar Support Types and Sizes			
Symbol	**Bar support illustration**	**Type of support**	**Sizes**
SB		Slab bolster	3/4″ to 3″ heights in 5′ lengths
SBR		Slab bolster upper	3/4″ to 3″ heights in 5′ lengths
BB		Beam bolster	1″ to 5″ heights in 5′ lengths
UBB		Beam bolster upper	1″ to 5″ heights in 5′ lengths
BC		Individual bar chair	3/4″ to 2″ heights in increments of 1/4″
JC		Joist chair	3/4″ to 2″ heights in increments of 1/4″
HC		Individual high chair	2″ to 40″ heights in increments of 1/4″
HCMD	Type A Type B	High chair for metal deck	2″ to 15″ heights in increments of 1/4″
CHC		Continuous high chair	2″ to 20″ heights in 5′ lengths
UCHC		Continuous high chair upper	2″ to 20″ heights in 5′ lengths
CHCM	Type A Type B	Continuous high chair for metal deck	1″ to 5″ heights in 5′ lengths
UJC		Joist chair upper	14″ Span Heights 1″ through 3 1/2″ vary

Illustrations courtesy of MeadowBurke

Metric Modules for Construction

The International Standards Organization (ISO) established 100 millimeters as the basic module for the construction industry. This is about the same as the 4 inch module used in the customary system of measurement. The illustration below shows the metric modules and the customary measurement equivalent.

Metric - Feet Conversion*

Feet	1	2	3	4	5	6	7	8	9	10
Metric Module	300 mm 30 cm	600 mm 60 cm 0.6 m	900 mm 90 cm 0.9 m	1200 mm 120 cm 1.2 m	1500 mm 150 cm 1.5 m	1800 mm 180 cm 1.8 m	2100 mm 210 cm 2.1 m	2400 mm 240 cm 2.4 m	2700 mm 270 cm 2.7 m	3000 mm 300 cm 3.0 m
Feet	**20**	**30**	**40**	**50**	**60**	**70**	**80**	**90**	**100**	**200**
Metric Module	6000 mm 600 cm 6 m	9000 mm 900 cm 9 m	12 000 mm 1200 cm 12 m	15 000 mm 1500 cm 15 m	18 000 mm 1800 cm 18 m	21 000 mm 2100 cm 21 m	24 000 mm 2400 cm 24 m	27 000 mm 2700 cm 27 m	30 000 mm 3000 cm 30 m	60 000 mm 6000 cm 60 m

* Recommended modular conversions

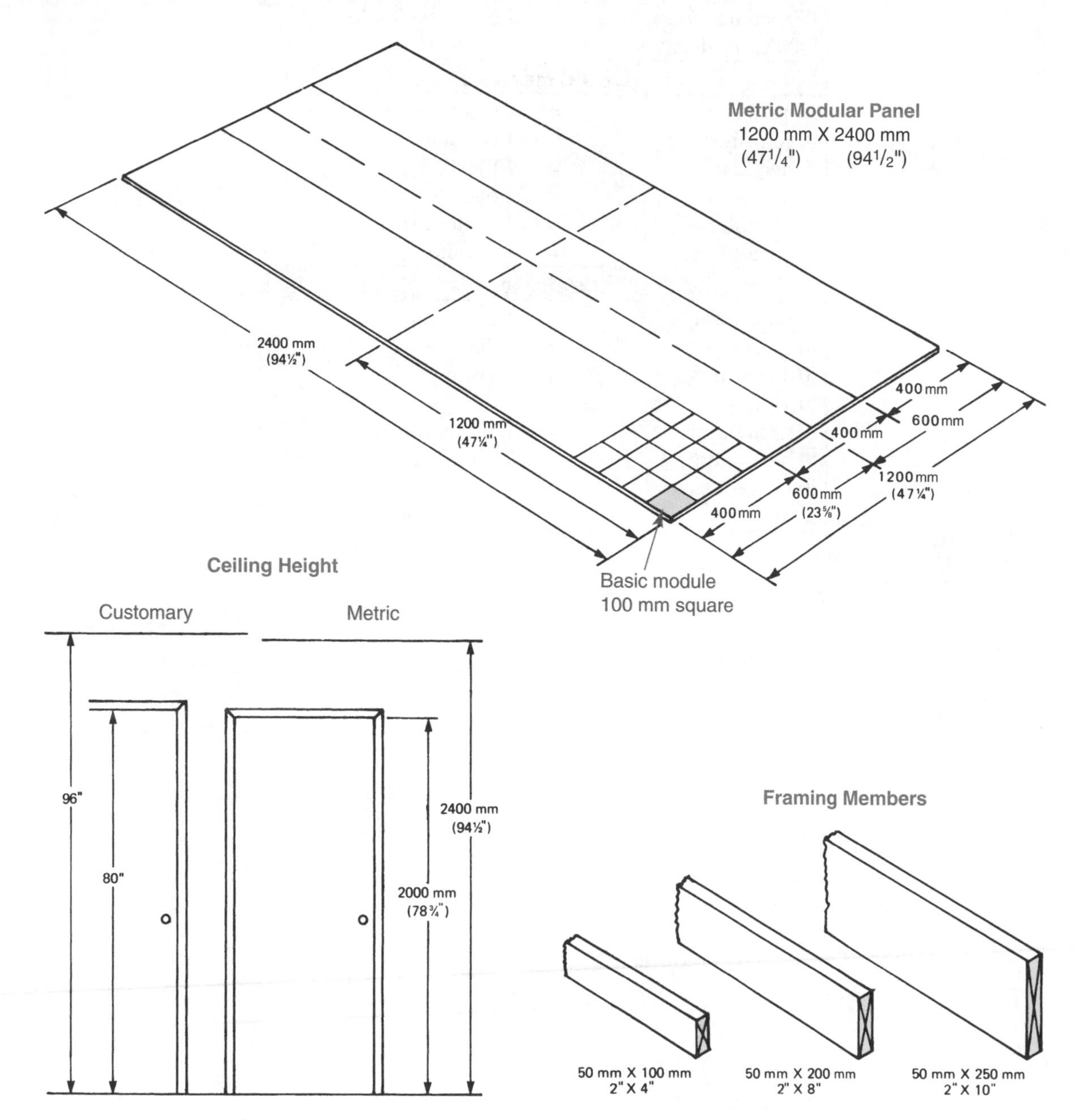

The Metric System		
Linear Measure		
10 millimeters	=	1 centimeter
10 centimeters	=	1 decimeter
10 decimeters	=	1 meter
10 meters	=	1 decameter
10 decameters	=	1 hectometer
10 hectometers	=	1 kilometer
Square Measure		
100 square millimeters	=	1 square centimeter
100 square centimeters	=	1 square decimeter
100 square decimeters	=	1 square meter
100 square meters	=	1 square decameter
100 square decameters	=	1 square hectometer
100 square hectometers	=	1 square kilometer
Cubic Measure		
1000 cubic millimeters	=	1 cubic centimeter
1000 cubic centimeters	=	1 cubic decimeter
1000 cubic decimeters	=	1 cubic meter
Liquid Measure		
10 milliliters	=	1 centiliter
10 centiliters	=	1 deciliter
10 deciliters	=	1 liter
10 liters	=	1 decaliter
10 decaliters	=	1 hectoliter
10 hectoliters	=	1 kiloliter
Weights		
10 milligrams	=	1 centigram
10 centigrams	=	1 decigram
10 decigrams	=	1 gram
10 grams	=	1 decagram
10 decagrams	=	1 hectogram
10 hectograms	=	1 kilogram
100 kilograms	=	1 quintal
10 quintals	=	1 metric ton

Weights and Measures Conversion Table

Linear Measure				
1 inch	=		=	2.54 centimeters
1 foot	=	12 inches	=	0.3048 meters
1 yard	=	3 feet	=	0.9144 meters
1 rod	=	5-1/2 yards	=	5.029 meters
1 rod	=	16-1/2 feet	=	5.029 meters
1 furlong	=	40 rods	=	201.17 meters
1 mile (statute)	=	5280 feet	=	1609.3 meters
1 mile (statute)	=	1760 yards	=	1609.3 meters
1 league (land)	=	3 miles	=	4.83 kilometers
Square Measure				
1 square inch	=		=	6.452 square centimeters
1 square foot	=	144 square inches	=	929 square centimeters
1 square yard	=	9 square feet	=	0.8361 square meters
1 square rod	=	30-1/4 square yards	=	25.29 square meters
1 acre	=	43,560 square feet	=	0.4047 hectare
1 acre	=	160 square yards	=	0.4047 hectare
1 square mile	=	640 acres	=	259 hectares
1 square mile	=	640 acres	=	2.59 square kilometers
Cubic Measure				
1 cubic inch	=		=	16.387 cubic centimeters
1 cubic foot	=	1728 cubic inches	=	0.0283 cubic meters
1 cubic yard	=	27 cubic feet	=	0.7646 cubic meters
Chain Linear Measure (For Surveyor's Chain)				
1 link	=	7.92 inches	=	20.12 centimeters
1 chain	=	100 links	=	20.12 meters
1 chain	=	66 feet	=	20.12 meters
1 furlong	=	10 chains	=	201.17 meters
1 mile	=	80 chains	=	1609.3 meters
Chain Square Measure				
1 square pole	=	625 square links	=	25.29 square meters
1 square chain	=	16 square poles	=	404.7 square meters
1 acre	=	10 square chains	=	0.4047 hectare
1 square mile	=	640 acres	=	259 hectares
1 section	=	640 acres	=	259 hectares
1 township	=	36 square miles	=	9324.0 hectares
Angular and Circular Measure				
1 minute	=	60 seconds		
1 degree	=	60 minutes		
1 right angle	=	90 degrees		
1 straight angle	=	180 degrees		
1 circle	=	360 degrees		

Metric - Inch Equivalents

Inches		Millimeters	Inches		Millimeters
Fractions	Decimals		Fractions	Decimals	
	.00394	.1	15/32	.46875	11.9063
	.00787	.2		.47244	12.00
	.01181	.3	31/64	.484375	12.3031
1/64	.015625	.3969	1/2	.5000	12.70
	.01575	.4		.51181	13.00
	.01969	.5	33/64	.515625	13.0969
	.02362	.6	17/32	.53125	13.4938
	.02756	.7	35/64	.546875	13.8907
1/32	.03125	.7938		.55118	14.00
	.0315	.8	9/16	.5625	14.2875
	.03543	.9	37/64	.578125	14.6844
	.03937	1.00		.59055	15.00
3/64	.046875	1.1906	19/32	.59375	15.0813
1/16	.0625	1.5875	39/64	.609375	15.4782
5/64	.078125	1.9844	5/8	.625	15.875
	.07874	2.00		.62992	16.00
3/32	.09375	2.3813	41/64	.640625	16.2719
7/64	.109375	2.7781	21/32	.65625	16.6688
	.11811	3.00		.66929	17.00
1/8	.125	3.175	43/64	.671875	17.0657
9/64	.140625	3.5719	11/16	.6875	17.4625
5/32	.15625	3.9688	45/64	.703125	17.8594
	.15748	4.00		.70866	18.00
11/64	.171875	4.3656	23/32	.71875	18.2563
3/16	.1875	4.7625	47/64	.734375	18.6532
	.19685	5.00		.74803	19.00
13/64	.203125	5.1594	3/4	.7500	19.05
7/32	.21875	5.5563	49/64	.765625	19.4469
15/64	.234375	5.9531	25/32	.78125	19.8438
	.23622	6.00		.7874	20.00
1/4	.2500	6.35	51/64	.796875	20.2407
17/64	.265625	6.7469	13/16	.8125	20.6375
	.27559	7.00		.82677	21.00
9/32	.28125	7.1438	53/64	.828125	21.0344
19/64	.296875	7.5406	27/32	.84375	21.4313
5/16	.3125	7.9375	55/64	.859375	21.8282
	.31496	8.00		.86614	22.00
21/64	.328125	8.3344	7/8	.875	22.225
11/32	.34375	8.7313	57/64	.890625	22.6219
	.35433	9.00		.90551	23.00
23/64	.359375	9.1281	29/32	.90625	23.0188
3/8	.375	9.525	59/64	.921875	23.4157
25/64	.390625	9.9219	15/16	.9375	23.8125
	.3937	10.00		.94488	24.00
13/32	.40625	10.3188	61/64	.953125	24.2094
27/64	.421875	10.7156	31/32	.96875	24.6063
	.43307	11.00		.98425	25.00
7/16	.4375	11.1125	63/64	.984375	25.0032
29/64	.453125	11.5094	1	1.0000	25.40

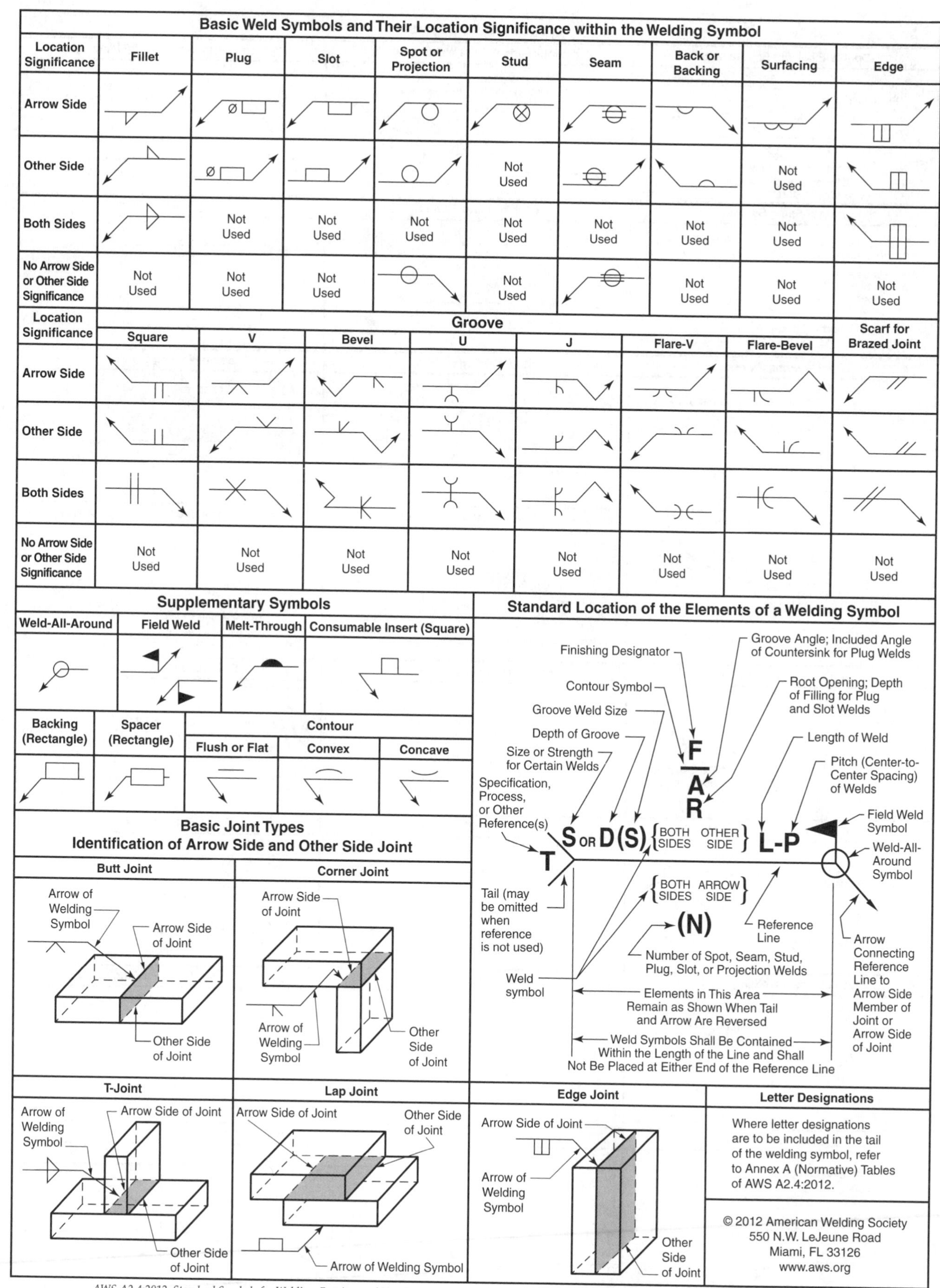

AWS A2.4:2012, Standard Symbols for Welding, Brazing, and Nondestructive Examination, reproduced with permission from the American Welding Society (AWS), Miami, FL USA

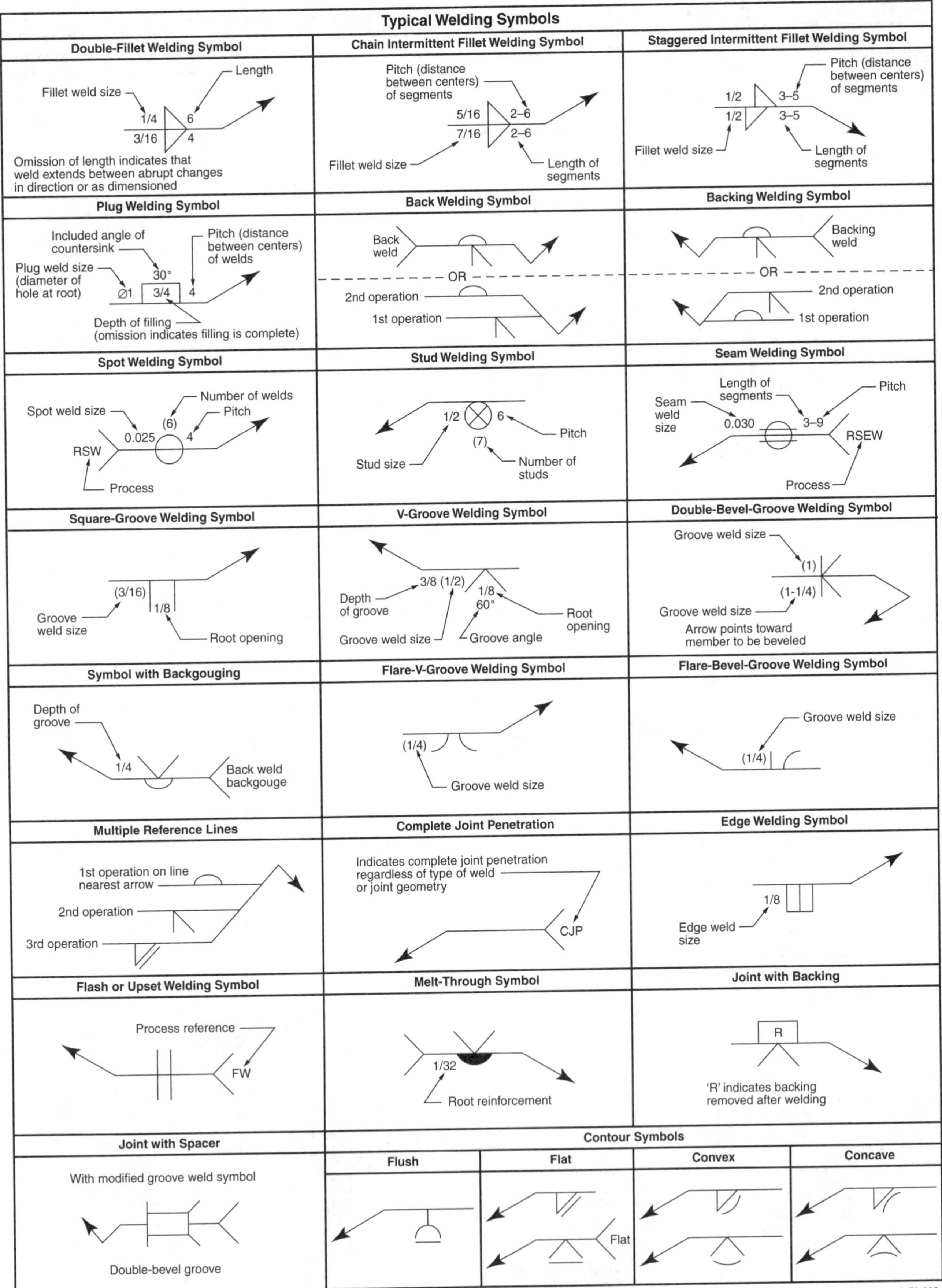

Typical Welding Symbols
Double-Fillet Welding Symbol
Fillet weld size
Length
1/4
6
3/16
4
Omission of length indicates that weld extends between abrupt changes in direction or as dimensioned
Chain Intermittent Fillet Welding Symbol
Pitch (distance between centers) of segments
5/16
2–6
7/16
2–6
Fillet weld size
Length of segments
Staggered Intermittent Fillet Welding Symbol
Pitch (distance between centers) of segments
1/2
3–5
1/2
3–5
Fillet weld size
Length of segments
Plug Welding Symbol
Included angle of countersink
Pitch (distance between centers) of welds
Plug weld size (diameter of hole at root)
30°
Ø1
3/4
4
Depth of filling (omission indicates filling is complete)
Back Welding Symbol
Back weld
OR
2nd operation
1st operation
Backing Welding Symbol
Backing weld
OR
2nd operation
1st operation
Spot Welding Symbol
Spot weld size
Number of welds
Pitch
(6)
0.025
4
RSW
Process
Stud Welding Symbol
1/2
6
Pitch
(7)
Stud size
Number of studs
Seam Welding Symbol
Seam weld size
Length of segments
Pitch
0.030
3–9
RSEW
Process
Square-Groove Welding Symbol
(3/16)
1/8
Groove weld size
Root opening
V-Groove Welding Symbol
Depth of groove
3/8 (1/2)
1/8
60°
Root opening
Groove weld size
Groove angle
Double-Bevel-Groove Welding Symbol
Groove weld size
(1)
(1-1/4)
Groove weld size
Arrow points toward member to be beveled
Symbol with Backgouging
Depth of groove
1/4
Back weld backgouge
Flare-V-Groove Welding Symbol
(1/4)
Groove weld size
Flare-Bevel-Groove Welding Symbol
Groove weld size
(1/4)
Multiple Reference Lines
1st operation on line nearest arrow
2nd operation
3rd operation
Complete Joint Penetration
Indicates complete joint penetration regardless of type of weld or joint geometry
CJP
Edge Welding Symbol
1/8
Edge weld size
Flash or Upset Welding Symbol
Process reference
FW
Melt-Through Symbol
1/32
Root reinforcement
Joint with Backing
R
'R' indicates backing removed after welding
Joint with Spacer
With modified groove weld symbol
Double-bevel groove
Contour Symbols
Flush
Flat
Flat
Convex
Concave

AWS A2.4:2012, Standard Symbols for Welding, Brazing, and Nondestructive Examination, reproduced with permission from the American Welding Society (AWS), Miami, FL USA

Designation of Welding and Allied Processes by Letters

Welding and Allied Processes	Letter Designation
adhesive bonding	AB
arc welding	AW
arc stud welding	SW
atomic hydrogen welding	AHW
bare metal arc welding	BMAW
carbon arc welding	CAW
gas	CAW-G
shielded	CAW-S
twin	CAW-T
electrogas welding	EGW
flux cored arc welding	FCAW
gas metal arc welding	GMAW
pulsed	GMAW-P
short circuit	GMAW-S
gas tungsten arc welding	GTAW
pulsed	GTAW-P
plasma arc welding	PAW
shielded metal arc welding	SMAW
submerged arc welding	SAW
series	SAW-S
brazing	B
block brazing	BB
carbon arc brazing	CAB
diffusion brazing	DFB
dip brazing	DB
furnace brazing	FB
induction brazing	IB
infrared brazing	IRB
resistance brazing	RB
torch brazing	TB
braze welding	BW
arc braze welding	ABW
flow brazing	FLB
flow welding	FLOW
electroslag welding	ESW
high energy beam welding	HEBW
electron beam welding	EBW
high vacuum	EBW-HV
medium vacuum	EBW-MV
nonvacuum	EBW-NV
laser beam welding	LBW
induction welding	IW
oxyfuel gas welding	OFW
air acetylene welding	AAW
oxyacetylene welding	OAW
oxyhydrogen welding	OHW
pressure gas welding	PGW
percussion welding	PEW
resistance welding	RW
flash welding	FW
projection welding	PW
resistance seam welding	RSEW
high frequency	RSEW-HF
induction	RSEW-I
resistance spot welding	RSW
upset welding	UW
high frequency	UW-HF
induction	UW-I
soldering	S
dip soldering	DS
furnace soldering	FS
induction soldering	IS
infrared soldering	IRS
iron soldering	INS
resistance soldering	RS
torch soldering	TS
wave soldering	WS
solid-state welding	SSW
coextrusion welding	CEW
cold welding	CW
diffusion welding	DFW
explosion welding	EXW
forge welding	FOW
friction welding	FRW
hot pressure welding	HPW
roll welding	ROW
ultrasonic welding	USW
thermal cutting	TC
arc cutting	AC
carbon arc cutting	CAC
air carbon arc cutting	CAC-A
gas metal arc cutting	GMAC
gas tungsten arc cutting	GTAC
plasma arc cutting	PAC
shielded metal arc cutting	SMAC
high energy beam cutting	HEBC
electron beam cutting	EBC
laser beam cutting	LBC
air	LBC-A
evaporative	LBC-EV
inert gas	LBC-IG
oxygen	LBC-O
oxygen cutting	OC
flux cutting	OC-F
metal powder cutting	OC-P
oxyfuel gas cutting	OFC
oxyacetylene cutting	OFC-A
oxyhydrogen cutting	OFC-H
oxynatural gas cutting	OFC-N
oxypropane cutting	OFC-P
oxygen arc cutting	OAC
oxygen lance cutting	OLC
thermal spraying	THSP
arc spraying	ASP
flame spraying	FLSP
plasma spraying	PSP

CSI MasterFormat Standard

The MasterFormat system, developed by the Construction Specifications Institute (CSI), is a standard for writing specifications and organizing data in construction projects. The major categories in the MasterFormat system are called divisions. The divisions are organized under groups and subgroups and are further divided into specific numbers and subject titles. A full list of divisions in the MasterFormat system is given on the following page. A full list of the numbers and titles in the MasterFormat system is available at www.masterformat.com.

The MasterFormat system uses a numbering system to organize information into a hierarchy of levels. The divisions are classified as the top level in the system (Level 1). The titles organized under the divisions are classified as Levels 2, 3, and 4. Numbers used in the system are arranged in pairs. In general, three pairs of numbers are used to define three levels of detail and establish a 6-digit numbering system. Level 1 numbers identifying divisions are represented by the first two digits. Level 2 numbers are represented by the second pair of digits and Level 3 numbers are represented by the third pair of digits. When used, Level 4 numbers are represented by a fourth pair of digits separated from the third pair with a decimal point. Level 4 numbers are used when more specific information is required. If needed, Level 5 numbers are added to allow for user-assigned information.

An example of numbers and titles in the MasterFormat system with levels indicated is given below. This example is a portion taken from the Maintenance of Concrete title in Division 03—Concrete. Note that four levels of detail are represented.

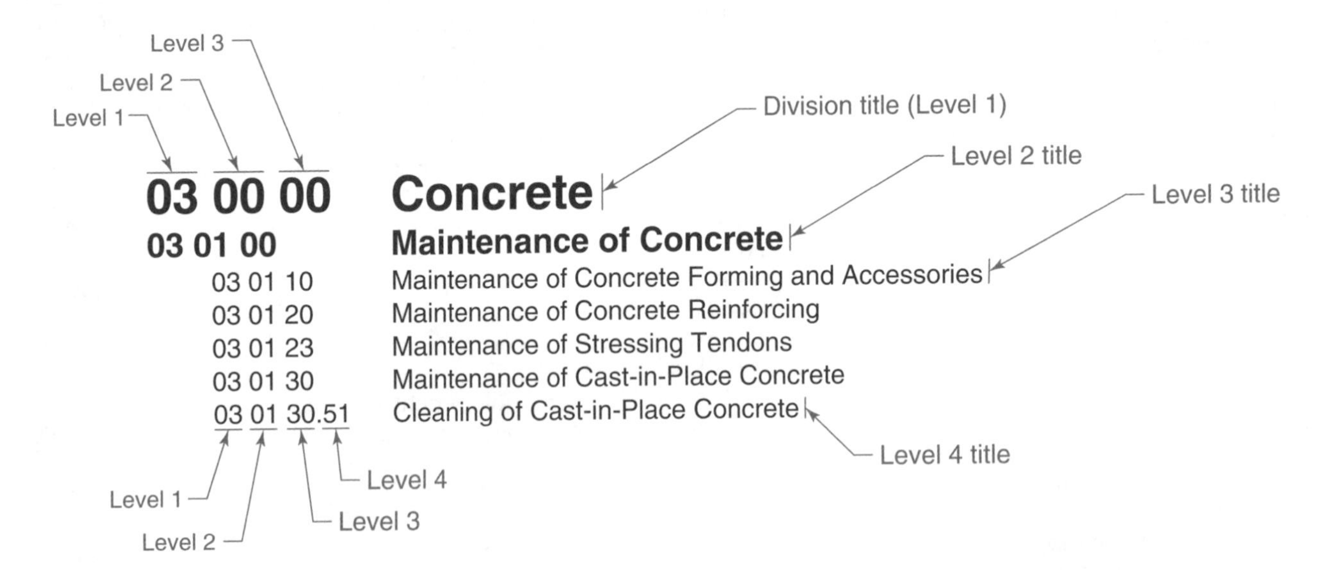

MasterFormat Division Numbers and Titles

Procurement and Contracting Requirements Group

Division 00 Procurement and Contracting Requirements

Specifications Group

General Requirements Subgroup

Division 01 General Requirements

Facility Construction Subgroup

Division 02 Existing Conditions
Division 03 Concrete
Division 04 Masonry
Division 05 Metals
Division 06 Wood, Plastics, and Composites
Division 07 Thermal and Moisture Protection
Division 08 Openings
Division 09 Finishes
Division 10 Specialties
Division 11 Equipment
Division 12 Furnishings
Division 13 Special Construction
Division 14 Conveying Equipment
Division 15 Reserved for future expansion
Division 16 Reserved for future expansion
Division 17 Reserved for future expansion
Division 18 Reserved for future expansion
Division 19 Reserved for future expansion

Facility Services Subgroup

Division 20 Reserved for future expansion
Division 21 Fire Suppression
Division 22 Plumbing
Division 23 Heating, Ventilating, and Air Conditioning (HVAC)
Division 24 Reserved for future expansion
Division 25 Integrated Automation
Division 26 Electrical
Division 27 Communications
Division 28 Electronic Safety and Security
Division 29 Reserved for future expansion

Site and Infrastructure Subgroup

Division 30 Reserved for future expansion
Division 31 Earthwork
Division 32 Exterior Improvements
Division 33 Utilities
Division 34 Transportation
Division 35 Waterway and Marine Construction
Division 36 Reserved for future expansion
Division 37 Reserved for future expansion
Division 38 Reserved for future expansion
Division 39 Reserved for future expansion

Process Equipment Subgroup

Division 40 Process Interconnections
Division 41 Material Processing and Handling Equipment
Division 42 Process Heating, Cooling, and Drying Equipment
Division 43 Process Gas and Liquid Handling, Purification, and Storage Equipment
Division 44 Pollution and Waste Control Equipment
Division 45 Industry-Specific Manufacturing Equipment
Division 46 Water and Wastewater Equipment
Division 47 Reserved for future expansion
Division 48 Electrical Power Generation
Division 49 Reserved for future expansion

Glossary

A

addendum: A change to the contract documents made by the architect or design team prior to the construction bid. (17)

adhesive: A cement, glue, or other material used to hold two or more items together.

admixture: Material added to concrete or mortar to alter it in some way. (7)

adobe brick: Building units made from natural sun-dried clays or earth and a binder. (7)

aggregate: Sand, gravel, rock, or other material used along with cement and water to make concrete. (7)

air infiltration barrier: A material such as a sheet of plastic placed in floors and walls to prevent the passage of air.

Alphabet of Lines: The accepted drafting practice of using specific line thickness and style to represent various drawing features. (4)

American bond: Masonry pattern in which every sixth or seventh course of stretchers is interrupted by a header course. Also called "common bond."

American Welding Society (AWS): Organization responsible for welding standards. (16)

ampere: The unit of measurement of electrical current. (15)

anchor: A device, generally made of metal, used to fasten plates, joists, trusses, and other building parts to concrete or masonry.

anchor bolt: A metal bolt with one threaded end and one L-shaped end. It is embedded in concrete and is used to hold structural members in place.

angle: With regard to structural steel, an L-shaped member with two perpendicular legs. Common leg lengths are 2″–7″, with common thicknesses of 1/8″–5/8″. (7)

anodize: An electrolytic means of coating aluminum or magnesium by oxidizing.

approximate method: An estimating method in which the size of the building is multiplied by a unit cost. Used to determine a rough estimate. (17)

apron: A piece of trim below the window stool and wall support used to conceal the edge of the wall material. Also: a concrete ramp in front of the garage door.

arc: A portion of a circle. (2)

arc fault circuit interrupter (AFCI): An electrical device that detects arcing conditions and opens the circuit if an arc fault occurs. (15)

architect's scale: Device that can be used to measure distances on a drawing by matching the drawing scale to the appropriate scale listed on the instrument. (3)

architectural drawings: Drawings showing the materials and construction processes that define the structure, typically created as plan, elevation, section, and detail drawings. (9)

area: A measurement made by multiplying length times width. The value is given in square units (square feet or square meters, for example). (2)

areaway: The open space around foundation walls, doorways, or windows to permit light and air to reach the below-ground-level floors. (2)

arrow side: On a welded joint, the side nearest the welding symbol arrow on the drawing. (16)

Note: The number in parentheses following each definition indicates the unit in which the term can be found.

arrowheads: Marking devices placed at the ends of a dimension line to define the distance between points (extension lines) on the drawing. (4)

ashlar: A stone cut by sawing to a rectangular shape. (7)

asphalt: A mineral pitch used for waterproofing roofs and foundation walls. It also is used with crushed rock to pave drives and parking areas.

assumed benchmark: An arbitrarily chosen benchmark. (8)

ASTM International: Organization responsible for setting criteria for the quality of structural materials. Formerly known as the American Society for Testing and Materials (ASTM). (6)

auger cast pile: A type of foundation system that is drilled with a 12″–14″ steel auger. As the auger is being raised out of the completed hole, concrete is pumped down the shaft to fill the cavity. (10)

B

backfill: To replace ground that has been excavated during construction. Also, the material used to replace excavated material.

backing wall: In masonry, a wall hidden behind a veneer wall.

balloon framing: A wall framing method in which wall studs run the entire height of the building, from foundation to roof plate. (12)

balusters: Vertical stair members used to support a hand rail.

balustrade: A row of balusters supporting a common rail.

basement plan: A drawing showing the top view of the basement. (10)

basic weld symbols: Representations of various types of welds, used on plans to guide the welder.

batten: A narrow strip of wood placed across the joint between two boards, such as siding.

batter boards: A temporary framework of stakes and horizontal members used in laying out a foundation.

beam: A horizontal structural member used between posts, columns, or walls. (11)

bearing: A measure of direction expressed as degrees east or west of north or south. (8)

bearing partition: An interior wall that transmits a load from above to a wall, columns, or footings below.

benchmark: A point of known elevation, such as a mark cut on a permanent stone or plate set in concrete, from which measurements are taken. (8)

bent glass: Glass building material produced by heating annealed glass to the point where it softens so it can be pressed over a form. (7)

bevel: A cut on the edge of a board at an angle other than 90°.

bevel siding: A siding material that is tapered from a thick edge to a thin edge.

bird's mouth: A notch formed by placing two cuts in a rafter, allowing it to sit on the top plate. (12)

blocking: Bridging members placed between wall studs to provide structural support and to prevent fire from spreading from floor-to-floor. (12)

blueprint: A print consisting of white lines against a blue background. The term is used interchangeably with *print*. (1)

bolster: A bent wire device used in holding reinforcing bars in place during the pouring of concrete. (11)

bond: The holding or gripping force between reinforcing steel and concrete. Also, the pattern in which masonry units are laid. (7)

bond beam: A reinforced concrete beam running around a masonry wall to provide added strength. Vertical bond beams are formed by inserting reinforcing bar in a cell after the wall is laid and filling with grout. (7)

border lines: Lines located near the edge of the sheet of drawing paper, serving as the drawing border. (4)

branch circuit: A path that distributes electricity from the source to outlets in the building. (15)

break lines: Lines used to indicate that only a portion of an object is shown. (4)

brick masonry: Construction method that uses units (*bricks*) that are manufactured, rather than removed from quarries. (7)

brick veneer: A brick wall of single brick, usually covering a frame structure.

bridging: The bracing of joists by members that connect between the sides of adjacent joists. (12)

budget parameter estimating: Estimating method used by owners, architects, or contractors to determine a rough estimate of cost. See *approximate method.* (17)

building brick: The most commonly used type of brick, specified for applications where appearance is not important. Normally called "common brick." (7)

building codes: Laws and standards specifying requirements for building construction. (6)

building information modeling (BIM): A process in which a 3D model provides a virtual representation of a building and is used in design, construction, and operation of the building. (1)

C

CADD: Acronym for computer-aided design and drafting. (1)

caisson: A large-diameter shaft (usually from 18″–72″ in diameter) that is drilled, then filled with steel reinforcing material and concrete. (10)

camber: A slight vertical curve (arch) formed in a beam or girder to counteract deflection due to loading.

cant strip: A wooden strip used to raise the first course of roofing shingles. Also: an angular board placed at the junction of the roof deck and wall to relieve the sharp angle when the roofing material is installed.

cantilever: A projecting structural member or slab supported at one end only.

cast-in-place concrete: Concrete that is cast at the construction site in its permanent location. (11)

caulk: A substance used to seal cracks and joints.

cement: The material used in concrete to bind the aggregate together. (7)

centerlines: Thin lines composed of alternating long and short dashes, used to indicate centers of objects. (4)

centimeter (cm): Unit of length measurement used in the metric system. One inch is the same length as 2.54 centimeters. (3)

ceramic tiles: Floor or wall covering units made from nonmetallic minerals, fired at a very high temperature. (7)

chair: A bent wire or plastic device used to hold reinforcing bars in place during concrete pouring. (11)

chamfer: A beveled outside corner or edge on a beam or column.

chords: The top and bottom members of open-web joists, and the principal members of trusses, as opposed to the diagonals.

circuit: The electrical path from the source through the components and back to the source. (15)

circuit breaker: A protective device for opening and closing an electrical circuit. It opens automatically in case of an overload on the circuit. (15)

circumference: The perimeter around a circle. (2)

cleanout: An opening in the waste pipe of a plumbing system for rodding out the drain. Also: an opening at the lower part of a fireplace for removing ashes. (13)

clerestory: A windowed area rising above the lower story that admits light and ventilation.

cold joint: Construction joint in concrete occurring at a place where continuous pouring has been interrupted.

collar beam: A tie between two opposite rafters, well above the wall plate. (12)

column: A vertical structural member. (11)

column schedule: A list of the details and specifications for all of the columns in a structure. (11)

common bond: Masonry pattern in which every sixth or seventh course of stretchers is interrupted by a header course. Also called "American bond."

common rafter: A rafter running between the wall plate and the ridge. (12)

composite wall: A masonry wall with a veneer wall in front of a backing wall of less expensive brick.

computer-aided design and drafting (CADD): Preparation of drawings using computer equipment and software. (1)

concrete: Structural material comprising water, gravel, sand, and cement. (7)

concrete brick: Solid or cored units molded from a mixture of Portland cement and aggregates and hardened chemically. (7)

concrete masonry unit (CMU): A masonry unit made of concrete measuring 8″ × 8″ × 16″ long (including mortar joint). (7, 10)

conductor: A material, usually wire, used to carry electrical current. (15)

conduit: Enclosed tubing used to carry electrical conductors. (15)

construction drawings: Drawings used to communicate the architectural and engineering design of a construction project. (1)

construction joint: Separation between two placements of concrete; a means for keying two sections together.

Construction Specifications Institute (CSI): Organization of professionals responsible for the development of standards for specification writing and data identification in the construction industry. (6)

contour lines: Lines drawn on the site plan to indicate the changing elevation of the land. All of the points along a single contour line are at the same elevation. (8)

contour symbol: Indicator that shows whether a weld is to be flat-, convex-, or concave-faced. (16)

control valve: Valve in a water distribution system used to stop the flow. (13)

coping: The top course or cap on a masonry wall protecting the masonry below from water penetration.

copper piping: A type of pipe used extensively for water distribution. (13)

corbel: A stone, masonry, or wood bracket projecting out from a wall.

cornice: The part of the roof extending horizontally out from the wall.

course: A horizontal layer of masonry unit.

cripple jack: A rafter running between a valley rafter and a hip rafter. (12)

cripple stud: A stud that does not run the full height of the wall due to the presence of a header or a rough sill. (12)

crosshatch lines: Thin lines, usually drawn at a 45° angle, that are used in a sectional view to show material that has been "cut" by the cutting-plane line.

curtain wall: A nonbearing wall between columns.

cut stones: Stones that are cut to size and finished at a mill prior to being used for construction. (7)

D

dead load: The load on a structure resulting from the weight of its own materials and any other fixed loads, such as a roof-mounted air conditioner.

decimal fraction: A fraction with a denominator that is 10 or a multiple of 10. (2)

delta: The central angle formed by the radii of a curve meeting the curve at the points of tangency. (8)

denominator: The bottom number in a fraction. (2)

detail: A type of drawing showing a specific detail of the construction. Details are normally drawn at a larger scale than other drawings. (1, 5, 9)

detailed method: Estimating method in which every aspect of construction (including materials, equipment, and labor) is listed and assigned a cost. (17)

diameter: The length of a line running between two points on a circle, through the center of the circle. (2)

diffuser: A grille or register over an air duct opening into a room that controls and directs the flow of air.

dimension lines: Thin lines used to indicate dimensions, extending the length of a distance being measured. (4)

distribution panel: An insulated panel or box, sometimes called a *breaker panel,* that receives the current from the source and distributes it through branch circuits to various points throughout a building. (15)

distribution pipes: Pipes that deliver supply water to plumbing fixtures. (13)

diverter: A piece, usually metal, used to direct moisture to a desired path or location.

divisions: Major construction categories identified within the MasterFormat® specification writing standard. (6)

dormer: A projection built out from a sloping roof, including one or more vertical windows.

double header: A header consisting of two headers, used when extra strength is desired.

double trimmer: A framing detail in which the side of an opening is supported by two members connected together.

dowels: Straight metal bars used to connect or position two sections of concrete or masonry.

drain/waste/vent (DWV) system: A piping system used for sewage disposal. (13)

drywall: A type of interior wall covering, such as gypsum board.

duct: A round or rectangular pipe, usually metal, used for transferring conditioned air in a heating and cooling system. (14)

E

easement: An area of land marked for use and maintenance of utilities. (8)

eaves: The portion of the roof that overhangs the wall.

electric radiant heating system: A heating system in which electrical resistance is used to generate heat. (14)

electrical metallic tubing (EMT): A type of electrical conduit, also called *thinwall.*

electrical plan: A plan view drawing showing the layout of the electrical wiring and fixtures. Also called *electrical drawing.* (1, 5)

elevation: In surveying, the height of a survey marker above sea level; a measurement on a plot or foundation referenced to a known point. In architectural drafting, a drawing showing the front view, rear view, or one of the side views of a structure. (1, 5, 9, 10)

engineer's scale: Measuring device typically used on civil drawings such as site plans and highway projects. Engineer's scales are referred to in whole numbers and are related to so many feet per inch. (3)

engineering copier: A copy machine used to make reproductions of drawings and designed to handle larger sizes of paper. (1)

equipment schedule: Listing used with an electrical plan. It lists all electrical equipment with detailed wiring and power information. (15)

evaporative cooling system: A cooling system used in areas of low humidity in which air is blown through a damp medium. (14)

excavation: The recess or pit formed by removing the ground in preparation for footings or other foundations.

expansion joint: A joint formed in concrete or masonry units to allow for expansion and contraction in materials caused by temperature changes and shrinkage. (10)

extension lines: Thin lines used in dimensioning, drawn perpendicular to the dimension line to specify the features between which the dimension applies. (4)

F

face brick: A select brick made of clays and chemicals to produce a desired color and effect for use in the face of a wall. (7)

fascia: A finish board nailed to the ends of rafters or lookouts.

ferrous metals: Metals that contain iron as a principal element and typically have magnetic properties. (7)

field weld symbol: Indicator showing that the weld is to be made at the construction site, rather than in the assembly shop. (16)

finish plumbing: The final stage of plumbing installation, in which fixtures are connected to the rough plumbing. (13)

finish symbol: Letter that indicates the finishing method to be used on a weld (C = chipping, G = grinding, M = machining, R = rolling, H = hammering).

fire stop: A block placed between studs of a wall to prevent a draft and the spread of fire. (12)

firebrick: A refractory ceramic brick made to resist high temperatures. (7)

fired-clay tile: A tile used primarily for floor coverings. It is produced from clays and is fired in a kiln to harden the surface. (7)

flanges: The parallel faces of a structural beam, connected by the web.

flashing: Sheet metal or other thin material used to prevent moisture from entering a structure, such as around a chimney.

flexible insulation: Insulating material made of mineral or vegetable fiber and available in blanket and batt form. (7)

float glass: The most common type of glass, produced by floating a continuous ribbon of molten glass on a bath of molten tin. (7)

floating slab construction: A slab-on-grade construction method in which the footings, foundation walls, and slab are cast at the same time. (10)

floor plan: A plan view showing room sizes and locations and many construction details. For simple construction, the floor plan may contain all of the needed information. (1, 5, 9)

flue: The passageway in a chimney that provides for the escape of smoke, gases, and fumes.

flush door: A smooth surface door, without panels or molding.

footing: The base of a foundation wall or column. The footing is wider to provide a larger bearing surface. (10)

forced-air system: A heating system that uses a motor-driven fan to distribute the warm air through ducts. (14)

foundation plan: A plan view showing the dimensions and details of a building's foundation system, including footings, walls, and piers. (5, 10)

fractional rule: A ruler that is divided into feet, inches, and fractional parts of an inch. (3)

framing: The wood or metal structure of a building, which gives it shape and strength.

framing plan: A plan view showing the layout of the structural members supporting a floor or roof. (5)

friction pile: A pile that works on the principle of frictional resistance on the sides of the pile from the soil into which it has been driven. (10)

frost line: The depth below the ground surface subject to freezing. (10)

full-divided scale: A scale in which the main units along the entire length of the scale are subdivided. (3)

furring: Narrow wood strips fastened to a wall or ceiling for use in nailing finished material.

G

gable: The portion of a wall above the eave line and between the slopes of a gable roof.

gage system: Classification method used for metal materials less than 1/4″ in thickness (often called "sheet metal"). (7)

galvanized steel pipe: A type of pipe with great strength and dimensional stability, used for residential and light commercial water distribution systems. (13)

girder: A principal beam supporting other beams.

glass: Translucent ceramic material used in building construction. (7)

glass block: Hollow building units made by fusing two sections of glass together. (7)

glazed brick: Brick finished with a hard, smooth coating, and used for decorative and special service applications. (7)

glazing: Installation of glass in window sashes and doors.

grade: The level of the ground around a building.

grade beam: A low foundation wall or a beam, usually at ground level, used to provide support for the walls of a building. (10)

graph paper: Paper displaying a grid of light lines, used to help keep lines straight and make proportioning easier. (5)

gravel: Rock aggregate material ranging in diameter from 1/4″ to several inches. (7)

gravel stop: The metal strip at the edge of a built-up roof.

green building: A building technology utilizing design strategies and construction methods to build structures that make efficient use of natural resources and energy. Also known as *sustainable design*. (7)

groove angle: The angle of the groove or bevel cut onto the edges of adjoining pieces of metal that are to be welded. (16)

groove weld: A type of weld made in a groove joint between adjoining pieces of metal. (16)

ground: A wire connecting an electrical circuit or device to the earth to minimize injuries from shock and possible damage from lightning. (15)

ground fault circuit interrupter (GFCI): An electrical receptacle that breaks the circuit if an overload occurs. Used when the receptacle is located near a water source, such as in a kitchen or bathroom. (15)

grout: A cementitious mixture of high water content, prepared to pour easily into spaces in a masonry wall. Made from Portland cement, lime, and aggregate, it is used to secure anchor bolts and vertical reinforcing rods in masonry walls. (7)

gusset: The piece of metal or plywood used to reinforce a joint of a truss.

H

haunch: Portion of a beam that increases in depth towards the support.

header: The horizontal structural member over a door or window opening. Also: a structural member nailed to the top of the sill plate at its exterior edge, onto which the floor joists are attached; also called a *rim joist*. (12)

header course: A course of brick laid flat so their long dimension is across the thickness of the wall, and the heads of the course of bricks show on the face of the wall. (7)

heating, ventilating, and air-conditioning (HVAC) system: System that produces the movement of heated or cooled air within a building. (14)

hidden lines: Lines composed of a series of short dashes evenly spaced and consistent in length, used to define edges and surfaces that are not visible in a particular view. (4)

hip jack: A rafter running between the plate and a hip rafter. (12)

hip rafter: A rafter running from an outside corner of the building to the ridge board. (12)

hose bibb: A threaded water faucet suitable for fastening a garden hose.

HVAC: Abbreviation for heating, ventilating, and air-conditioning. (1)

hydration: A chemical reaction between cement and water that results in the hardening of concrete. (7)

hydronic heating system: A heating system that uses hot water to warm the air. (14)

I

I-beam: A structural steel shape with narrow top and bottom flanges forming a cross-sectional shape like the letter "I." (11)

improper fraction: A fraction in which the numerator is larger than the denominator. (2)

inclined lines: Straight lines that are neither horizontal nor vertical. (5)

insulating glass: A window or door glass consisting of two sheets of glass separated by a sealed air space to reduce heat transfer. (7)

interior elevations: Elevation views of a building's interior. (1)

interpretation: The ability to understand lines, symbols, dimensions, notes, and other information on working drawings. (1)

isometric drawing: A pictorial drawing positioned so that its principal axes make equal 30° angles with the plane of projection. Used for some detail and schematic layouts on construction drawings. (13)

J

jamb: The top and sides of a door or window frame.

joist: One of a series of wood or metal framing members used to support a floor or ceiling. (12)

K

keyway: Slotted joint in concrete used to connect portions cast at different times, such as a groove in the footing where the foundation wall is to be poured. (10)

kiln-burned brick: Building units made from natural clays or shales (sometimes with other materials added, such as coloring) and molded to shape, dried, and fired for hardness. (7)

kiln-dried lumber: Lumber that has been dried to a desired moisture content by controlling heat and humidity.

kilo: The metric system prefix meaning one thousand. A kilometer, for example, is 1000 meters; a kilogram is 1000 grams.

knee wall: A short wall. (12)

L

labor cost: Amount required to cover employee salaries and wages and related expenses on a construction project. (17)

lally column: A vertical steel pipe, usually filled with concrete, used to support beams and girders.

laminating: A method of constructing by bonding layers of material with an adhesive, used to form products such as plastic laminates and glue-laminated wood beams. (7)

landing: A platform in a flight of stairs to change direction or break a long run.

leader: A thin line that has an arrow at one end and a text note or symbol at the other end. The arrow points to the feature or detail to which the note refers. (4)

ledger: A horizontal strip of wood attached to the side of a beam to support joists. (12)

legend: A list of symbols and their corresponding meanings, used in a set of prints. (4, 8)

lighting schedule: A listing of the brand, quantity, and electrical details of the lamp fixtures to be used in a building. (15)

lintel: Support member for a masonry opening, usually made of precast concrete or steel. (7)

live load: All movable objects, including people and equipment, in a building.

loadbearing partitions: Interior walls that carry the ceiling or floor load from above. (12)

longitudinal section: Section taken through the long dimension of a building. (1)

lookout: The structural member running from the outside wall to a rafter end, which carries the soffit.

loose-fill insulation: Insulation made from materials such as rock wool, glass wool, and cellulose. It is poured, blown, or packed in place by hand. (7)

lumber: Wood that has been cut to specific dimensions for structural use. (7)

M

main: The primary supply line from the municipal water meter (or other source of supply) to the building. (13)

MasterFormat®: A standard identification system for organizing content documenting requirements, products, and activities used in the construction industry. (6)

mechanical plan: A plan view drawing showing the location of mechanical equipment in a building, such as the HVAC system and plumbing system. Also called *mechanical drawing.* (1, 5)

melt-through symbol: Indicator showing that 100% joint or member penetration is required in a weld made from one side. (16)

meter (m): Metric unit of measurement, equal to 39.37″. (3)

metric rule: A ruler that is divided into meters, centimeters, and millimeters. (3)

metric scale: Scale graduated in millimeters and centimeters. (3)

mil: Unit of thickness measurement, equal to 1/1000 of an inch (.001″).

millimeter (mm): A measuring unit in the metric system equal to 1/1000 meter. (3)

minute: A unit of angular measure equal to 1/60th of a degree. (8)

mixed number: A number that consists of a whole number and a proper fraction. (2)

modular measurement: The design of a structure to use standard size building materials.

monolithic slab: A concrete slab system in which the slab and footings are cast as one continuous unit. (10)

mortar: Cementitious substance used as a binding agent for masonry units. (7)

mosaic tiles: Small tiles, typically 1″ square, that can be laid to form a design or pattern. (7)

N

National Electrical Code (NEC): Handbook of standards and guidelines for electrical installation. (15)

nominal size: A general classification term used to designate the size of commercial products but not identical to the actual size of the item. For example, a 2×4 (nominal) board is actually 1 1/2″ × 3 1/2″.

nonferrous metals: Metals containing little or no iron. (7)

north arrow: An arrow shown on a plan view drawing indicating the orientation of the structure. (8)

numerator: The top number in a fraction. (2)

O

object lines: Lines that represent the main outline of the features of the object. Object lines are heavy, continuous lines showing all edges and surfaces. (4)

on center (O.C.) Spacing measurement referring to the distance between the centers of adjoining studs, joists, or other building components. (12)

one-line diagram: An electrical schematic drawing. (15)

one-way slab system: A concrete floor system constructed with the main reinforcing steel spanning in one direction. (11)

open-divided scale: A scale with whole number units along the length of the scale and subdivisions at the end representing fractional parts of the whole number unit. (3)

open-web steel joist: A truss type joist with top and bottom chords and a web formed of diagonal members. Some manufacturers make a joist with chords of wood and a steel web and refer to it as a truss joist. (7, 11)

orthographic projection: The projection of an object as viewed from six different perpendicular directions. The basis of architectural plan and elevation drawings. Also called *orthographic drawing.* (1, 5)

other side: On a welded joint, the side opposite the welding symbol arrow on the drawing. (16)

overhead: The cost of doing business for a company. (17)

P

panel door: A door of solid frame members with inset panels.

panel schedule: A list of the electrical details of a distribution panel (breaker panel), including the voltage entering the box, the numbers and sizes of the breakers, and the circuits controlled by the breakers. (15)

parapet wall: A wall extending above the roof line.

parge: To coat with mortar.

partition: An interior wall that divides an area.

patterned glass: Flat glass with a pattern rolled into one or both sides to diffuse the light and provide privacy. (7)

paving brick: Hard-surfaced brick used in driveways or areas where abrasion is a concern. (7)

perpendicular projectors: In orthographic projection, lines between views that connect common points. (5)

pervious concrete: Highly porous concrete that allows water to pass through it. (7)

pi: The ratio of circumference of a circle to its diameter, equal to 3.1416. Used when determining the area of a circle. (2)

pictorial drawing: A drawing in which an object is shown as it actually appears. (1)

pier: A heavy column of masonry used to support other structural members.

pilaster: A column projecting on the outside or inside of a masonry wall to add strength or decorative effect.

pitch: Roof slope expressed as the ratio of the rise to the span.

plan drawing: A drawing that shows one floor of a building from directly above. Also called *plan view.* (1)

Plan North: An assumed north direction (differing from True North) that is aligned with the building orientation. (8)

plan view: A drawing that shows one floor of a building from directly above. Also called *plan drawing.* (5)

plans: Another term for prints. The term *plan* also refers to a view that shows the features of a building from directly above, such as the floor plan. (1)

plaster: A material consisting of gypsum and water (or Portland cement, sand, and water) used for interior wall surfaces.

plastic pipe: A type of pipe used extensively for both water distribution and sewage disposal applications. (13)

plat: A drawing of a parcel of land indicating lot number, location, boundaries, and dimensions. Contains information about easements and restrictions.

platform framing: A wall framing method in which wall studs run the length of a single story. Also called *western framing.* (12)

plenum: A chamber in a climate control system that receives air under pressure before distribution to ducts.

plot plan: A drawing showing the location of a building on its plot of land and various details of the land. Also called *site plan.* (8)

plumb: Perpendicular or vertical. Also: to make something vertical.

plumbing plan: A plan drawing showing the layout of the water distribution and drainage systems. (1, 5)

plywood: A composite lumber material consisting of layers of wood glued together with their grains oriented perpendicular. (7)

pocket door: A door that slides into a partition or wall.

polyester film: A translucent plastic material used for making original drawings. (1)

Portland cement: Cement normally combined with water and aggregates to make concrete.

post-and-beam framing: A framing method that makes use of heavy timber material for vertical posts in wall sections and horizontal beams supporting floor and roof sections. The floor and roof sections are typically made from plank material 2″ thick. (12)

posttensioned concrete: Prestressed concrete in which the pretensioning steel is stressed after the concrete is cast. (11)

precast concrete: Concrete that is cast into members at a precast concrete plant and then transported to the construction site. (11)

prestressed concrete: Concrete in which steel is tensioned (stretched) and anchored to compress the concrete. (11)

pretensioned concrete: Prestressed concrete in which the pretensioning steel is stressed before the concrete is cast. (11)

print: A copy of a drawing. (1)

print reading: The gathering of information from a print or drawing. It involves two principal elements: visualization and interpretation. (1)

production rates: Trade guidelines that establish how much material can be installed per hour or day on a construction site. (17)

profit: The money a company intends to make above and beyond construction costs and overhead costs. (17)

proper fraction: A fraction that has a numerator smaller than its denominator. (2)

property line: A line shown on a site plan representing the limits of a plot of land. (4, 8)

proportion: The size of a portion of an object relative to the entire object or another portion. (5)

purlin: A horizontal roof member used to support long rafters or trusses. (12)

Q

quarry tile: A floor covering material produced from clays that provide a wear-resistant surface. (7)

R

raceway: Work channel used to receive electrical wiring. (15)

radius: The distance from the center of a circle to the edge of the circle. (2)

rafter: One of a series of structural members of a roof. The rafters of a flat roof are sometimes called roof joists. (12)

receptacle: An electrical outlet into which appliances and other electrical devices can be connected by means of a plug. (15)

reference line: The horizontal line portion of a welding symbol. It has an arrow at one end and a tail at the other. (16)

reflected ceiling plan: A drawing that illustrates ceiling-mounted light fixtures.

reflective insulation: Insulation designed to reflect radiant heat, made of a material surfaced with metal foil (usually aluminum foil). (7)

register: A grille used to cover an air duct opening into a room.

reinforced concrete: Concrete that is cast with steel reinforcement bars, which provide additional strength. (11)

reinforcing bars: Steel bars placed within forms and surrounded by concrete. The surface of reinforcing bars is normally deformed to improve the bond between the bar and the concrete. (7)

relay: An electrically operated switch. (15)

remote cooling system: Cooling system that has the condensing unit in a remote space away from the area to be cooled. (14)

rendering: A pictorial drawing showing what a structure will look like when the project is finished. (1)

ribbon: In balloon framing, the member attached to the studs and used to support joists. (12)

ridge board: The board at the peak of a roof. (12)

rigid insulation: A lightweight, low-density product with good heat and acoustical insulating qualities. It is available in large sheets. (7)

rise: In roof construction, the vertical distance from the top plate to the ridge of a roof. In stair construction, the vertical distance between floors. (5, 12)

riser: The vertical member of a stair. (12)

rough sill: A horizontal framing member running between studs below a window opening. (12)

rough-in plumbing: The second stage of plumbing installation, in which water supply pipes and sewage drain pipes are installed. (13)

rough-sawn lumber: Wood that has been cut to size but not dressed or surfaced. (7)

rubble: A type of stone masonry unit taken directly from the quarry, without smoothing or finishing. (7)

run: In roof construction, the horizontal distance from the wall supporting the bottom of the rafter to the ridge board. In stair construction, the total horizontal length of the stairway. (5, 12)

running bond: The most common brick pattern, consisting of offset courses of stretchers.

R-value: A numeric classification for insulating value (resistance to heat transfer). (7)

S

saddle: A small, gable roof constructed between a vertical surface (such as the back of a chimney) and a sloped roof to prevent water from standing.

safety glass: Glass that has been treated to resist breakage, or to break into less hazardous pieces. (7)

sample term. A term used as a sample. Also called *model term.* (1)

sand-lime brick: Building units made from a mixture of sand and lime, molded into shape and hardened under steam pressure and heat. (7)

scale: A measuring device with graduations for laying off distances. Also: the ratio of size that a structure is drawn, such as: 1/4″ = 1′-0″. (3)

schedule: A list of required building components, such as doors, windows, or beams; provides information related to each listed item. (1, 9)

screed: A tool (such as a board) used to guide finishers in leveling off the top of fresh concrete.

second: In angular measurement, a unit equal to 1/60th of a minute and 1/3600th of a degree. (8)

section: View showing the building as if it were cut apart. Sections show walls, stairs, and other details not clearly shown in other drawings. Sections are usually drawn in larger scale than the elevations and plan drawings. (1, 5, 9)

section cutting lines: Lines identifying sectional views. A section cutting line marks the part of the drawing being "cut" to create a sectional view. Arrows on the end of a section cutting line indicate the direction from which the section is being viewed. (4)

section lines: Thin lines, usually drawn at a 45° angle, used in a sectional view to represent a building material. (4)

septic tank: A concrete or metal tank used to reduce raw sewage by bacterial action. Used where no municipal sewage system exists.

series-loop system: Type of hydronic system used in residences to carry the heated water to the convectors. (14)

service entrance: The service equipment including the conductors from the utility pole, the service head, and the mast that bring the electrical current to the distribution panel. (15)

setback: The distance from the property boundaries to the building location, required by zoning. (8)

sheathing: The boards or panels that cover the studs and rafters of a building to which finishing materials are applied. (5)

sill plate: Wood board attached to the top of a foundation wall, onto which the wall frame is attached. (12)

single-wythe wall: A masonry wall composed of one row of bricks. (7)

site plan: A drawing showing the location of a building on its plot of land and various details of the land. Also called *plot plan.* (8)

slab: A flat area of concrete such as a floor or drive.

slab-on-grade: A concrete slab supported by ground. (10)

sleepers: Wood strips laid over or embedded in a concrete floor for attaching a finished floor.

slope: In roof construction, the relationship between the rise and run. (12)

soffit: In framing, the underside of a staircase or roof cornice. In masonry, the underside of a beam, lintel, or arch.

sole plate: A horizontal framing member that serves as a base for the studs in a wall frame, normally the same size as the studs. (12)

span: The distance between supports for joists, beams, girders, and trusses. (12)

spandrel beam: A beam in an exterior wall of a structure.

spandrel wall: The portion of a wall above the head of a window and below the sill of the window above.

specifications: The written directions issued by architects or engineers to establish general conditions, standards, and detailed instructions to be used on a project. (6)

splicing: Method of connecting two reinforcing bars at their ends to produce a single, long bar. (11)

spot weld: Small, round weld, formed by heat and pressure, that is used to join thin metals. (16)

square: A unit of measure referring to 100 square feet. Roofing materials and some siding materials are sold on this basis.

stained glass: Colored glass produced by adding metallic oxides in the molten state. This glass can be used for windows and decorative pieces. Sometimes called *art* or *cathedral glass.* (7)

steel pile: A long H-shaped (H-pile) or round (pipe pile) steel member that is hammer-driven into the earth. (10)

stirrup: In reinforced concrete beams, a thin reinforcing bar wrapped around the main reinforcing bars to hold them in position. Stirrups also provide additional reinforcement against shear. (11)

stone masonry: Construction method using granite, limestone, marble, sandstone, or slate as a building material. Today, stone masonry is usually decorative, rather than structural. (7)

story: The space between two floors of a building or between a floor and the ceiling above.

stretcher: A brick laid in a flat position, lengthwise with the wall. (7)

stringer: Angled member that supports stairs. (12)

structural clay tile: Building units made of materials similar to those used in brick, but in larger sizes. (7)

structural framing plans: Required for more complex structures, these plans show the framing of the roof, floors, and various wall sections. (1)

structural steel: General name for several types of mild steel normally used in construction. (7, 11)

stucco: A plaster material consisting of Portland cement, lime, sand, and water. Used for exterior wall surfaces.

stud: A vertical wall framing member.

subfloor: Surface of plywood or other board material fastened to joists and located below the finished floor material.

surfaced lumber: Wood that has been dressed or finished to size by running it through a planer. (7)

sustainability: The ability of a structure to maintain operational efficiency and have minimal impact on the environment throughout its lifetime. (7)

sustainable design: A building technology utilizing design strategies and construction methods to build structures that make efficient use of natural resources and energy. Also known as *green building.* (7)

sweated fittings: Copper piping joints that are soldered using lead-free solder. (13)

T

tail: The vee-shaped end of the welding symbol. Notes are placed within the tail to designate the welding specification, process, or other reference. (16)

takeoff: Computation of a quantity for a cost estimate. (17)

termite shield: Sheet metal shield placed in or on a foundation wall to prevent termites from entering the structure.

terra cotta: A type of clay tile used for its attractiveness. Literally—burnt earth. (7)

thermal insulation: Material designed to reduce heat transmission through walls, ceilings, and floors. (7)

thermostat: An automatic device controlling the operation of a climate control system.

ties: In reinforced concrete columns, thin reinforcing bars wrapped horizontally around the primary, vertical bars. The ties keep the larger bars from moving as the concrete is cast. (11)

title sheet: A drawing sheet included with a set of working drawings, typically used to identify individual drawings and other information about a building project. (1)

top plate: The horizontal framing member running above wall studs. (12)

topography: The locations and details of land features. (8)

transverse section: Section taken through the narrow width of a building. (1)

tread: The horizontal member of a stair. (12)

triangle-square method: A commonly used method of sketching arcs and circles. (5)

True North: Actual North, as opposed to the arbitrary Plan North. (8)

truss: A structural unit having members such as beams, bars, and ties, usually arranged in triangular form. Used for spanning wide spaces.

two-way slab system: A concrete floor system in which the main reinforcing steel spans in two directions. (11)

typical (TYP): This term, when associated with any dimension or feature, means the dimension or feature applies to the locations that appear to be identical in size and shape.

U

Uniform Drawing System (UDS): A standardization of drawing guidelines that consists of eight interrelated modules for organizing and presenting drawing information used in planning, designing, and constructing facilities. (1)

unit cooling system: Small, individual unit—installed in a window or through an exterior wall—that is used to cool a room. (14)

unit method: A technique useful in estimating proportions. It involves identifying a unit that is common to both the width and height of the object. (5)

V

valley: The internal angle formed by two roof slopes.

valley jack: A rafter running between a valley rafter and the ridge. (12)

valley rafter: A rafter running from an inside corner of the building to the ridge board. (12)

vellum: A translucent paper material used for making original drawings. (1)

veneer wall: A single thickness (one wythe) masonry unit wall tied to a backing wall.

vent: A pipe, usually extending through the roof, providing a flow of air to and from the drainage system.

visualization: The ability to create a mental image of a building or project from a set of working drawings. (1)

vitrified clay tile: A ceramic tile fired at a high temperature to make it very hard and waterproof.

voltage: The electromotive force that causes electrical current to flow through a conductor. (15)

W

waffle slab: A two-way slab system supported by concrete members running in two directions. (11)

waste stack: A large diameter pipe that carries wastewater (sewage) to the building drain. (13)

waterproofing: The process of protecting foundation walls from underground water. (10)

waterstop: A sealing device used at concrete construction joints to serve as a moisture barrier. (10)

watt: The unit of measurement of electrical power. (15)

weep holes: Small holes in a wall to permit water to exit from behind.

weld all-around symbol: Indicator showing that the weld extends completely around a joint. (16)

weld dimensions: Record of required weld dimensions, drawn on the same side of the reference line as the weld symbol. (16)

weld symbol: A standard symbol that identifies a specific type of weld to be made. (16)

welded wire fabric (WWF): Wire mesh fabricated by means of welding the crossing joints of steel wires. Normally used to reinforce concrete slabs. (7)

welding symbol: A symbol, made up of several elements, that provides the information needed to completely specify a required weld. (16)

wide-flange beam: A structural steel shape with top and bottom flanges wider than the flanges of an I-beam. (11)

winders: Triangular (pie-shaped) treads used to change direction of a staircase. (12)

wiring diagram: A separate drawing used when wiring details cannot be shown clearly on the electrical plan. Wiring diagrams correspond to a specific piece of equipment. (15)

working drawings: A set of drawings providing the necessary details and dimensions to construct the object. May also include the specifications. (1)

wythe: A continuous vertical section of masonry, one unit in thickness; sometimes called *withe* or *tier.*

Z

zoning: A system of classifying the use and development of land. (8)

Index

A

B

C

T

U

V

W

Z